Protein Folding

Protein Folding

Thomas E. Creighton, Editor

**European Molecular Biology Laboratory
Heidelberg**

W. H. Freeman and Company

New York

Library of Congress Cataloging-in-Publication Data

Protein folding / Thomas E. Creighton, editor.
    p. cm.
   Includes bibliographical references and index.
   ISBN 0-7167-7027-X
   1. Protein folding. I. Creighton, Thomas E., 1940–
QP551.P69582 1992
547.7'5—dc20

                                               92–10475
                                                 CIP

Cover image provided by the editor.

Printed in the United States of America

1 2 3 4 5 6 7 8 9 0 VB 9 9 8 7 6 5 4 3 2

Contributors

Thomas E. Creighton
European Molecular Biology Laboratory, Heidelberg

Robert B. Freedman
University of Kent, Canterbury

Jean-Renaud Garel
Laboratoire d'Enzymologie du CNRS, Gif-sur-Yvette

David P. Goldenberg
University of Utah, Salt Lake City

M. Karplus
Harvard University, Cambridge, Massachusetts

Peter L. Privalov
Institute of Protein Research, Pushchino

Oleg B. Ptitsyn
Institute of Protein Research, Pushchino

F. M. Richards
Yale University, New Haven

Franz X. Schmid
Universität Bayreuth, Bayreuth

E. Shakhnovich
Harvard University, Cambridge, Massachusetts

Janet M. Thornton
University College, London

Contents

Preface

Proteins are virtually unique in being linear macromolecules with a non-repetitive, specific covalent structure and in being able to adopt a relatively fixed three-dimensional structure, or conformation. The covalent structure is determined by the structures of the 20 different amino acids and by the order in which they are linked together, using the genetic information, into a polypeptide chain. The conformation of a protein is specified by the rotations about all the single bonds of its covalent structure. Different conformations of the same protein have the same covalent structure, and they are interconverted solely by bond rotations. The only exception to this is the disulfide bond, which can be formed reversibly between two Cys residues, depending upon the conformation of the remainder of the protein and the thiol/disulfide status of the environment (Chapter 7). Polypeptide chains differing only in their disulfide bonds can also be said to differ only in their conformations.

The "protein folding problem" refers to the questions how and why a protein adopts a specific native conformation. Any molecule can adopt different conformations, but it is a special problem with proteins because of the vast number of conformations possible. With an average of m equally probable conformations per amino acid residue and n residues in the polypeptide chain, the total number of possible conformations of the polypeptide chain will be m^n. Not all such conformations will be possible, of course, because atoms would overlap in space in some of them, but the numbers are so large that even if only a small fraction of these conformations are permitted, there will still be a very large number. For example, an average protein contains about 300 residues, and each residue on average might be capable of existing in eight distinctly different conformations, which would suggest that 8^{300} (or 10^{270}) conformations are possible. Even with a relatively small protein of only 100 residues, and a very conservative estimate of only two conformations per residue, the number is 10^{30} total possible conformations.

What determines which of these many conformations is adopted by a protein as its "native" conformation? The ultimate determinant of the native conformation is the amino acid sequence, but each amino acid sequence (and there are many more amino acid sequences possible than there are conformations, 20^n for n residues) does not produce a different conformation. What are the rules governing the relationship between amino acid sequence and three-dimensional structure? Given the amino acid sequence, can the native three-dimensional structure be predicted? When the answer to the latter question is generally yes, the protein folding problem will be considered to be solved.

At present (early 1992), a more tractable problem is to determine the process or pathway by which proteins adopt their native conformations. This is not a greatly simpler problem, however, again because of the vast number of conformations possible. A single conformation of a protein cannot be encountered by solely random fluctuations, for it would take much too long to sample even "only" 10^{30} conformations. Proteins are frequently observed to fold on the second to minute time scales. There are very many indications that proteins do not fold by random processes, but pass through a limited number of intermediate conformational states. Nevertheless, it has been difficult to prove the kinetic importance of such pathways. If a particular pathway is crucial, it might be feasible to block that pathway so that a protein could not refold once unfolded but also would not unfold readily under similar conditions. (See p. 331.) Apparent examples of such proteins are known, but their refolding is not blocked and is simply in competition with irreversible aggregation of the unfolded protein.

Two excellent examples of the importance of kinetic pathways of folding have recently been found. One concerns the α-lytic protease, which is synthesized and folds as a pro-form, with an amino-terminal extension of 166 residues. The pro-form unfolds and refolds, but the mature protein does not. The inability of the mature protein to refold might be imagined to be due to its aggregation or to proteolysis by a few remaining folded molecules in the case of a protease, for the pro-segment is an inhibitor of the proteolytic activity of the mature protein. It has recently been shown, however, that the unfolded mature α-lytic protease does not aggregate or get proteolyzed; instead, it remains soluble in a molten globule-like state that retains its competence for refolding when the pro-segment is added (Baker et al., in press). The pro-segment appears to interact with, and stabilize by at least 9 kcal/mol, the transition state for folding.

The second example is the plasminogen activator inhibitor-1, a protease inhibitor of the "serpin" class. This protein folds initially to a form that is active as an inhibitor, but relatively unstable to unfolding. It then changes slowly, with a half-time of 1 h at 37°C, to a much more stable form that is not active as an inhibitor. The inactive form can be converted to the active form again by unfolding it and then refolding (Hekman and Loskutoff, 1985). Clearly, there are kinetic barriers preventing the protein refolding directly to its most stable form. The molecular basis of this phenomenon is clear from the three-dimensional structures of several serpins (Carrell et al., 1991). The more stable forms have a sixth β-strand inserted within the middle of a β-sheet that contains only five strands in the less stable form. Insertion of the sixth strand occurs in most serpins only after proteolytic cleavage of the polypeptide chain, but it is believed to occur spontaneously in the plasminogen activator inhibitor-1, although slowly, in the transition to the inactive form. The crystal structure of the inactive form shows the strand inserted into the β-sheet (Mottonen et al., in press). The five-stranded form of the protein is clearly that favored kinetically by folding, even though the six-stranded form is the more stable.

Besides the time taken for a polypeptide chain to fold, another consideration is that all possible conformations will be approximately equally probable in a fully unfolded protein. A plausible sample of unfolded protein, say of between 1 mg and 1 kg of a protein consisting of 300 residues, will contain only 10^{16} to 10^{22} molecules. Therefore, each molecule is likely to have a unique conformation at any instant of time. Yet such conformationally heterogeneous populations of protein molecules are usually observed to fold to the same native conformation with a single rate constant, indicating that all the unfolded molecules have the same probability of folding. The most frequent exception is when the unfolded molecules differ in some conformational aspect that intrinsically is interconverted only slowly, such as *cis,trans* isomerization of peptide bonds, particularly those preceding Pro residues (Chapter 5). Even in this case, the unfolded molecules with the same incorrect *cis,trans* isomer usually refold with the same rate constant.

It is undoubtedly the case that a fully unfolded, random-coil polypeptide chain is not possible in practice, because of the presence of 20 different side chains, and proteins unfolded in different ways frequently have different average physical properties; such different populations probably represent different subsets of the truly random-coil state. Nevertheless, such different unfolded proteins usually refold at the same rate under the same final folding conditions. The nature of the initial unfolded protein is not crucial in these cases, but the final folding conditions are. It seems clear from these general observations that the initially heterogeneous unfolded protein molecules must rapidly equilibrate through a limited set of nonrandom conformations, so that they may all complete refolding in a short period of time and with the same rate constant. How proteins manage to do this, and the actual pathways they follow, have been the subject of much experimental and theoretical study since the early 1970's.

This book provides a comprehensive review of the entire subject of protein folding and is written by some of the leaders in the field. Fred Richards presents a wide-ranging overview of the entire problem in the first chapter, setting the stage for the remainder of the book. The native conformations of proteins would be expected to have clues as to how they are formed by the folding process, and Janet Thornton describes these aspects in Chapter 2. The physical basis of the stabilities of the three-dimensional folded conformations is fundamental to the folding process, and a complete understanding of folded conformation stability would probably make it possible to understand the folding process also. Peter Privalov summarizes the thermodynamic aspects of protein stability in Chapter 3, dealing especially with the frequently confusing hydrophobic effect. The many theoretical aspects of protein folding are described by Martin Karplus and E. Shakhnovich in Chapter 4.

The cooperativity of protein folding causes partially folded intermediates to be unstable and not populated at equilibrium, where the fully folded and fully unfolded states usually predominate. Consequently, it is necessary to study the folding process kinetically, as a function of time. The observed kinetics of folding of small proteins and individual domains are described in Chapter 5 by Franz Schmid, with emphasis on the slow phenomenon of *cis,trans* isomerization of peptide bonds preceding Pro residues that is the most prominent aspect of such studies. The molten globule state is now recognized to be an important exception to only the fully folded and fully unfolded states being stable, in that it is a third conformational state that can be stable at equilibrium with certain proteins under certain conditions. It is also believed to be an important kinetic intermediate state in folding, and it may be the natural form of unfolded proteins under refolding conditions. Nevertheless, the conformational properties of this state are not yet understood. Oleg Ptitsyn reviews the experimental and theoretical evidence in Chapter 6. Actual partially folded intermediates in protein folding are, at best, transient kinetic intermediates that are difficult to detect and to characterize. Such inherently unstable intermediates can be trapped in a stable form, however, if the folding is coupled to disulfide bond formation. In this case, the folding pathway can be determined by trapping and characterizing the disulfide-bonded intermediates that accumulate kinetically during folding and disulfide formation, as described in Chapter 7. The trapped intermediates can be characterized in detail to determine the conformational basis of the folding pathway.

With the recent revolution in protein engineering technology, site-directed mutagenesis has become a major tool in studying protein stability and folding. In favorable circumstances, random mutations that produce a desired effect can be selected and identified. Both such approaches are being used to characterize intermediates and transition states in protein folding, as described by David Goldenberg in Chapter 8. Large proteins comprising multiple domains or subunits are believed to fold by the individual domains and subunits folding to native-like conformations, then assembling. Such a process has some interesting and practical aspects that are described by Jean-Renaud Garel in Chapter 9.

Protein folding normally occurs within the cell, during or shortly after biosynthesis of the polypeptide chain on the ribosome. The special features of biosynthetic folding, and the many cellular factors discovered recently to participate in it, are described in Chapter 10 by Robert Freedman and interpreted in the light of what is known about protein folding *in vitro*.

The reader will not find the solution to the protein folding problem in these pages, nor is a single viewpoint presented, for the various authors present different conclusions on a number of points, such as the randomness of unfolded proteins (Chapter 1 and 3) and the topology of the molten globule state (Chapters 6 and 7). Nevertheless, this book presents the most authoritative and comprehensive account of the subject currently available. I hope that the information presented here will be used to guide the subject further, so that the remaining questions can be answered, a current consensus can evolve, and the ultimate solution to the protein folding problem can at least be envisaged.

Thomas E. Creighton

REFERENCES

Baker, D., Sohl, J. L., and Agard, D. A. (1992) *Nature 356*, 263–265.

Carrell, R. W., Evans, D. L. I., and Stein, P. E. (1991) *Nature 353*, 576–578.

Hekman, C. M., and Loskutoff, D. J. (1985) *J. Biol. Chem. 260*, 11581–11587.

Mottonen, J., Strand, A., Symersky, J., Sweet, R. M., Danley, D. E., Geoghegan, K. F., Gerard, R. D., and Goldsmith, E. J. (1992) *Nature 355*, 270–273.

Abbreviations

General

2D	two-dimensional
3D	three-dimensional
AK-HDH	aspartokinase-homoserine dehydrogenase
ANS	1-anilinonaphthalene-8-sulfonate
ATP	adenosine triphosphate
ATPase	enzyme activity that catalyzes hydrolysis of ATP
Bip	heavy chain binding protein (or Hsp70)
BPTI	bovine pancreatic trypsin inhibitor
CD	circular dichroism
CMCys	carboxymethylcysteine residue
C_p	heat capacity at constant pressure
CSA	compact, self-avoiding
CyP	cyclophilin
Da	Dalton
DHFR	dihydrofolate reductase
DTT_{SH}^{SH}, DTT_S^S, DTT	dithiothreitol in the thiol form, the disulfide form, or without specifying the state of the thiols, respectively
ER	endoplasmic reticulum
FKBP	FK506 binding protein
FUD	intermediate state that is folded but with unpaired domains
GdmCl	guanidinium chloride (guanidine hydrochloride)
GSBP	glycosylation-site binding protein
GSH, GSSG	reduced and oxidized forms, respectively, of glutathione
H	enthalpy

HA	hemaglutinin
HPLC	high-pressure liquid chromatography
Ig	immunoglobulin
k_B	Boltzmann's constant
k_{cat}	rate constant for catalysis
K_d	dissociation constant
LDH	lactate dehydrogenase
M_w	molecular weight
MBP	maltose binding protein
NEM	N-ethyl maleimide
NMR	nuclear magnetic resonance
NOE	nuclear Overhauser effect
ODH	octopine dehydrogenase
PDI	protein disulfide isomerase
PFK	phosphofructokinase
PGK	phosphoglycerate kinase
PPI	prolyl peptide isomerase
PRAI	phosphoribosyl anthranilate isomerase
RBP	ribose binding protein
REM	random energy model
RNase	ribonuclease
Rubisco	ribulose bisphosphate carboxylase
S	entropy
SDS-PAGE	polyacrylamide gel electrophoresis in the presence of sodium dodecyl sulfate
SRP	signal recognition particle
T	temperature (absolute)
$t_{1/2}$	half time
TS	tryptophan synthase
α-TS, β-TS	α and β_2 subunits of tryptophan synthase
τ	relaxation time for a reaction, which is the reciprocal of the apparent rate constant
UV	ultraviolet
VLDL	very low density lipoprotein
VSV	vesicular stomatitis virus
Xaa, Yaa	unspecified amino acid residue

Amino Acid Residues of Proteins

Ala	alanine
Arg	arginine
Asn	asparagine
Asp	aspartic acid

Cys	cysteine
Gln	glutamine
Glu	glutamic acid
Gly	glycine
His	histidine
Ile	isoleucine
Leu	leucine
Lys	lysine
Met	methionine
Phe	phenylalanine
Pro	proline
Ser	serine
Thr	threonine
Trp	tryptophan
Tyr	tyrosine
Val	valine

1

Folded and Unfolded Proteins: An Introduction

F. M. RICHARDS

1 INTRODUCTION

It has long been known that dramatic changes occur in the properties of proteins upon heating or acidification. Although the end result was frequently visible coagulation, an essential first step in the process was recognized in the first decade of the twentieth century and was called *denaturation.* That some of these early changes can, on occasion, be reversed with recovery of biological function was shown in the early 1930's, the process of *renaturation.* The probable nature of these reactions as conformational changes involving expanding or contracting polymer chains was also pointed out at that time. However, the list of protein amino acids was not yet complete; there was not uniform agreement on the underlying covalent structure of proteins; and the intricate process of protein biosynthesis and its genetic control was not understood at all. This early history of protein chemistry and the major players of the time are elegantly reviewed by Fruton (1972).

In the late 1940's and 1950's, the revolution in modern molecular biology began. The direct connection of "one gene—one enzyme" was established. The basic structure of double-stranded DNA was proposed. Some inspired guessing provided the rationale for the Central Dogma: DNA→RNA→protein. The dissection of the complex machinery involved in the biosynthesis of polypeptide chains was underway. Our current understanding of the processes whose investigation was begun in that era is covered in many modern texts such as Watson et al. (1987), Alberts et al. (1989), and Singer and Berg (1991). The history and excitement of the whole field are marvelously conveyed in *The Eighth Day of Creation* (Judson, 1979).

As a part of these developments, the fundamental significance of the *denaturation/renaturation* reactions became clear through the pioneering studies of Anfinsen, starting in the late 1950's. (To remove the aura of laboratory artefact and to insert the impression of increased understanding, the reactions are most commonly now referred to as *unfolding and folding*.) The first experiments were carried out only a short time after Sanger (1952) had determined the sequence of insulin and thereby proven beyond doubt that the basic covalent structure of a protein is a linear peptide chain and that the amino acid sequence of the chain is unique for a given protein. By showing that an unfolded (denatured) polypeptide chain could spontaneously refold to form a native protein with full biological activity, Anfinsen concluded that this sequence, by itself, contains all of the information necessary to define the three-dimensional structure of the protein and, thus, its biological function (Anfinsen, 1973).

Study of the protein folding problem can really be said to have begun in earnest with Anfinsen's experiments. The work started slowly with a widespread feeling that the problem was intrinsically too difficult to be understood in molecular detail. Although still easy to find, such scepticism is apparently decreasing in the face of the current intense interest and activity in many laboratories around the world. The problem presented by the folding reaction is perhaps the clearest and most general example of the complex interface between chemistry and biology. At least for a large number of polypeptides, folding to the native state is clearly a problem in chemistry. Only the composition and properties of the solvent have to be properly adjusted for the reaction to occur and for biological function to appear in the folded product. There is no requirement for mysterious biological "factors."

Notwithstanding the previous sentence, such factors appeared in the mid-1980s. Some enzymes, catalyzing reactions such as disulfide interchange and proline *cis,trans* isomerization, have been recognized for some time. Others, called chaperonins, have more mysterious functions that are not yet understood, but their presence appears to be necessary to successfully complete some folding reactions. Occasionally ATP cleavage is required. It may be that a major function of some of the chaperonins is to prevent folding initially so as to assist in targeting processes within the cell, thus ensuring that the final folded protein appears in the correct cellular compartment. Such systems are discussed by Freedman in Chapter 10 of this volume.

The presence of these complications does not change the fact that in many folding reactions these extra factors are not required, and in all cases the total necessary information is indeed contained in the polypeptide sequence(s) of the units that fold independently. Such intrinsic folding processes are the principal topic of this book. The theoretical elucidation of these reactions must provide the basic ground rules for the understanding of yet more involved conformational changes and macromolecular associations.

The remainder of this chapter contains a brief review of some general aspects of the structure and folding of globular proteins (Fig. 1–1). Since the early 1980's a number of volumes have appeared addressing these same problems, and they serve as a record of the rapid progress in this field: Jaenicke (1980), Ghelis and Yon (1982), Creighton (1983), Wetlaufer (1984), Fasman (1989), Gierasch and King (1990), and Chadwick and Widdows (1991).

2 THE FOLDING REACTION

The elementary form of the folding reaction is deceptively simple:

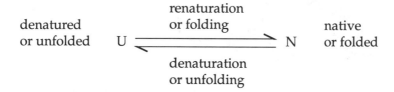

denatured   renaturation or folding   native
or unfolded U ⇌ N or folded
      denaturation or unfolding

FIGURE 1–1. *Changing impressions of protein structure. (A) State of the art in the mid-1940's. The models were the best interpretation of the available physicochemical data. No high-resolution data were even on the horizon at that time. One might note how similar these models are to those of today for systems without high-resolution data. (Reprinted with permission from Oncley, 1949)*

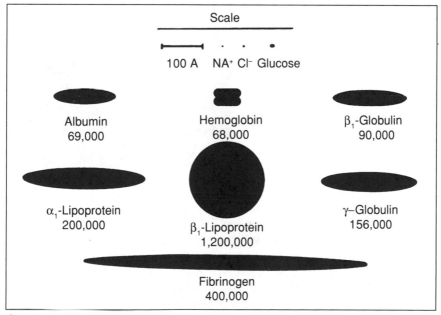

A

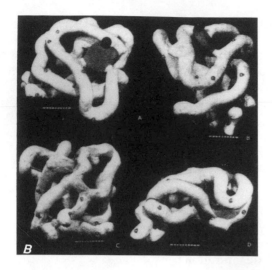

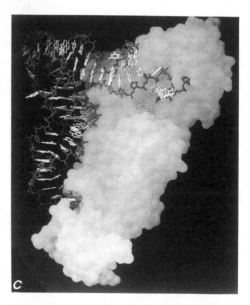

FIGURE 1–1. *Changing impressions of protein structure. (B) The first 6 Å model of myoglobin. Photographs from different sides of a physical model, with no symmetry, and no simplicity. (Reprinted with permission from Kendrew et al., 1958) (C) A modern high-resolution structure. The complex of tRNA*[Gln]*, shown as a stick model of all nonhydrogen atoms, with glutaminyl-tRNA synthetase, pictured as the molecular surface, and ATP, not visible in this reproduction. This figure, as is usual now, is produced entirely by computer graphics. Physical models have become rare. (Reprinted with permission from Rould et al., 1989)*

Read as a chemical process, the equation says: a molecule with structure U is isomerized to one with structure N; the reactions to the right and to the left are both single first-order steps; the ratio of the rate constants will give the equilibrium constant for the reaction; there are no other components in the reaction. This is the description of a two-state process. Given the actual complexity of the reactants, it is quite surprising that not infrequently such an equation is adequate to explain a large body of experimental data.

Given the inherent structural complexity, a logical next assumption, taken directly from polymer chemistry, would be to assume that the reactant and product are not single chemical species, but are collections of related structures that may or may not be easily interconvertible. Thus

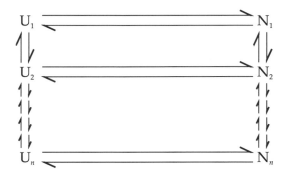

If the various different structural forms of U are in rapid equilibrium with each other and the same is true for those of N, and if these rates of interchange are fast on the time scale of any of the possible U $\rightleftharpoons$ N reactions, then most experiments will appear to follow a simple two-state process even though there are, in fact, a very much larger number of actual molecular states.

If the interchange reactions are slow relative to the folding/ unfolding process, however, each U structure might easily form its own N structure, or collapse to a common N structure if there is only one, in a fast first-order reaction. Studies at equilibrium would still appear two-state, but the kinetics of the folding process, for example, would then be the sum of individual exponential decays and would not represent a first-order reaction. If the U state even approximates a random coil, then the U molecules will indeed represent an ensemble of conformational forms.

In the general case there will be intermediate structures clearly identifiable as neither the U nor the N class. One then has a matrix of possible reactions, all presumed, at least in principle, to be reversible.

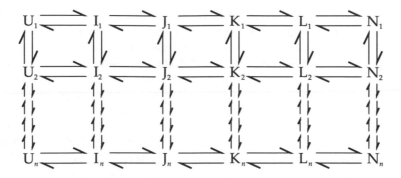

Cross-reactions between components differing in both row and column might also be possible. A matrix such as the preceding was set up for the perennial problem of the association of oxygen with hemoglobin by Koshland et al. (1966). That system was analyzed in detail with special emphasis on the various limiting patterns in the concentrations of intermediates that were produced by specific assumptions on relative rates. There, as here, it was assumed that the structures varied such that the I's are more like the U's than the N's while the K's look more like N's than U's. Thus, for the folding reaction, as one proceeds from left to right the hydrodynamic volume of the components are presumed to be decreasing, with compactness of the structure increasing.

Such a reaction scheme is much too involved to be of any practical use, but most detailed models for the folding reaction can be derived by eliminating certain of the preceding components and/or by putting restrictions on the relative values of the rate constants of the various steps.

2.1 A Sampling of Models for the Folding Reaction

A great many models have been proposed for the folding process in general, or for the specific folding of a particular protein. The terminology in this field is very confusing, and the differences between various models frequently appear to be semantic. A conference on nomenclature may be needed to try to come up with a set of terms whose meaning is unequivocal and that would be accepted by those in the field. The models described next are discussed by Kim and Baldwin (1982).

If the rate constants are such that only the first line of components ever exist, then one has the *sequential folding model:*

$$U_1 \rightleftharpoons I_1 \rightleftharpoons J_1 \rightleftharpoons K_1 \rightleftharpoons L_1 \rightleftharpoons N_1$$

In this extreme case the order of the individual species from left to right is completely specified and there is only one structure at each identifiable step.

If all the rate constants from U to I are slow and all the constants beyond I are very fast, one has the *nucleation/growth model* with the I's representing one or more kinetic nuclei. The initial step is rate limiting. No intermediates beyond the nucleus ever accumulate in sufficient quantity to detect. This limiting case is ruled out for the increasing number of systems where intermediates can be demonstrated.

The problem of the varying use of the words *nucleus* and *nucleation* has been spelled out clearly both by Baldwin (1986b, 1989) and by Wetlaufer (1990). It can have a kinetic definition, as here, or can refer to a small unit of structure with an implied stability, albeit perhaps transient, as implied in some of the following mechanisms. One should have a clear nomenclatural distinction between these two very different meanings. Wetlaufer (1990) has suggested that a qualifier be added to remove any doubt as in "kinetic nuclei" or "structural nuclei."

In either the *diffusion-collision-adhesion model* or the *framework model*, the rate constants are such that various intermediates can occur in amounts depending upon the detailed ratios of the rates. In the first model, the process of diffusion is emphasized, and this could be imagined as applying to a particular step, say $I_i \rightarrow J_i$, where two preformed but unassociated secondary units simply diffuse together in an effective collision. $J_i \rightarrow K_i$ would then be the stabilization of such a complex through local rearrangements, i.e., adhesion. In the framework model, the intermediates are assumed to represent collections of the units of secondary structure that have some intrinsic stability and that occur in the final native conformation. The docking problem of the preformed units, rather than diffusion, is then considered to be the rate-limiting process. Otherwise, the formalism would appear to be the same. Multiple pathways are possible in the folding process in either model. In the extreme limit, both of these turn into the *sequential model*.

The rates may be such that only structures of the type K occur. An example of this is the *hydrophobic collapse model* where the hydrophobic effect is presumed to be the overwhelming driving force. The water is squeezed out as far as possible without any particular regard to the formation of secondary structure. The latter is formed in later steps in a series of local reorganizations, $K_i \rightarrow L_i \rightarrow N$. The structures, K_i, might be the molten globule state(s) for which there is increasing evidence, i.e., close to, but not quite as small as, the final native form(s) N. (See Chapter 6.)

The *jigsaw puzzle model* (Harrison and Durbin, 1985) uses this standard pastime as an indication of how the folding process might occur. If the analogy is to be close, the pieces would have to be tied together with string to represent the chain and sequence, and obviously there can only be a single "native state." (This latter assumption is usually made in all of the other models as well.) For the purposes of this analogy, these are not serious restrictions on either the game or the model. If the pieces have a constant uniform color, then there will be no preferential starting point, each folding attempt will follow a different path, and the full matrix representation would be required. As one knows from personal experience, if the color pattern is loud and simple, almost every folding attempt will follow the same path (i.e. the sequential model). If the color is less definitive but nonetheless present, then a limited selection of paths will be followed with nearly the same "intermediates" developed on each "try" (i.e. the framework or diffusion-collision-adhesion models). The jigsaw model clearly rules out kinetic nucleation, as any puzzle solver knows, but it also does not fit easily into "hydrophobic collapse," which in the case of a real puzzle clearly represents a catastrophe.

The brief descriptions just given are for general folding models. Specific models are usually developed for individual proteins with a given proposal being expanded from the two-state process only as far as necessary to accommodate the data being reported. The box mechanism suggested for thioredoxin by Kelley and Stellwagen (1984) is an example. Here U_1, U_2, J_1, and N_2 are the only components assumed to exist. With the addition of I_1 and I_2 and appropriate cross-reactions, a more complex "cubic" model was later proposed to encompass additional data (Shalongo et al., 1987).

The letters U, I, J, K, L, and N are common symbols given to components in models proposed for the interpretation of kinetic or thermo-dynamic data. While certain physical properties may be specified, the suggested structure of any of these components is frequently vague or not commented on at all. The following discussion is devoted to a consideration of some of the known or surmised details of the structure of these components.

For background and a more expanded discussion of the material in the following brief summary, the reader is urged to consult standard textbooks in biochemistry, such as Stryer (1988) or Zubay (1988), or the more focused single-volume monographs such as Schulz and Schirmer (1979), Creighton (1983), and Brändén and Tooze (1991).

3 RESTRICTIONS ON LOCAL CONFORMATIONS

The vast amount of structural data on small organic molecules, including the amino acids themselves and small peptides, can be taken over directly to proteins. Thus, in a crystallographic study of a protein, differences from the canonical small molecule values in any bond lengths or angles of more than about 0.02 Å or 3° will be due to experimental error in almost all cases. It would be an unusual situation where the structure could support the necessary strain energy to maintain more substantial distortions. Metal binding sites and the associated strong ligand fields may be an exception to that statement, as may occasional products in some current genetic engineering experiments.

The best-known conformational restriction in proteins is provided by the Ramachandran map, a plot of the allowed values of the torsional angles, φ (the N–Cα bond) and ψ (the Cα–C bond), of the main chain of the polypeptide (Ramachandran and Sasisekharan, 1968) (Fig. 1–2.) The usual version of the map assumes that the torsional angle of the peptide bond itself, ω, is 180° or the *trans* form. This angle varies in actual structures by 10° to 15° to either side, but the map is not affected in any major way. The original Ramachandran map was based upon the hard sphere approximation for atoms and provided in two different forms with different assumed atomic radii. More recent versions are based on quantum mechanical calculations and provide more detail in the energy contours, but the overall shape of the "allowed" area is closely equivalent to the original. The map applies to all amino acid residues except Gly and Pro. The absence of Cβ in the former greatly expands the allowed conformational space and removes the asymmetry of a chiral carbon, while for the latter the bond from the side chain to the imino nitrogen markedly reduces the available range of φ. Thus, the value of the maximum conformational entropy that could be contributed by each position of the chain will increase or decrease at the Gly and Pro positions. The fit of the data from carefully refined structures to the Ramachandran map is impressive. The restrictions that it represents are fundamental to the structures of all proteins in all states.

The barriers to rotation around single bonds are relatively low. One might expect a broad distribution of torsional angles in the side chains. Starting with the survey by Janin et al. (1978), however, and the update by Ponder and Richards (1987), the actual distributions in native proteins are found to be quite narrow and to be centered on the positions of the rotational minima expected from both experimental and theoretical studies on isolated small molecules (Vasquez et al., 1983) (Fig. 1–3.) Thus, the rotamer approximation for the allowed conformations of amino acid side chains gives a remarkably good fit to the most highly refined protein structures in almost all cases. The validity of the mean values for the rotamer positions is established by the fact that, starting with the chain stripped back to the α carbon atoms, proteins can be "rebuilt" using the library members, normally without serious steric clashes. The number of rotamers actually found in native proteins is considerably less than the total possible number. Since additional rotamers are appearing in the Protein Data Bank (Bernstein et al., 1977) as new structures are

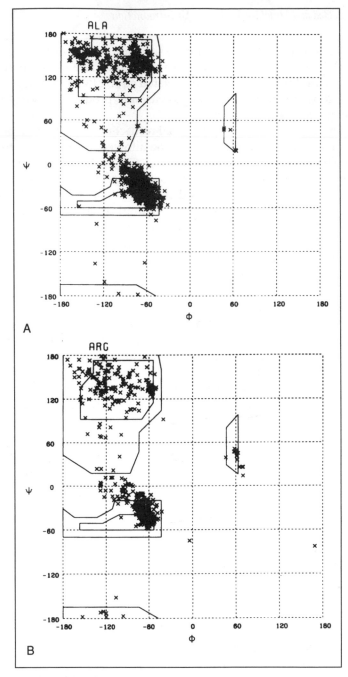

FIGURE 1–2. *Ramachandran φ-ψ plots of main chain conformations for four residue types as found in high-resolution X-ray structures of proteins. The allowed areas based on hard sphere atoms with two different sets of assumed radii are shown by the contours. The fits for all other residues not shown are*

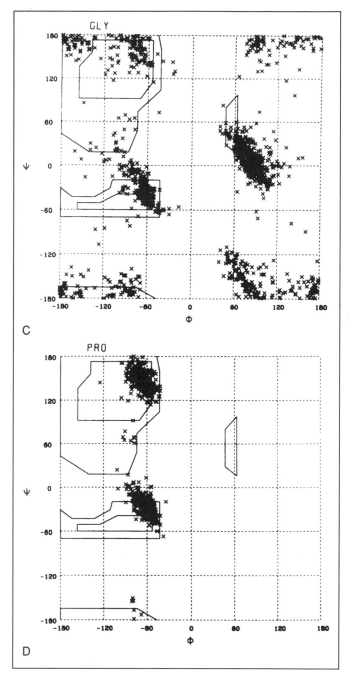

similar to those for the alanine and arginine distributions. The very restricted area for proline, φ close to –60°, and the highly expanded area around glycine, symmetrical about φ = 0, ψ = 0, are shown in the bottom two panels. (These patterns were provided by M. Rould using the program FRAGLE; Finzel et al., 1990.)

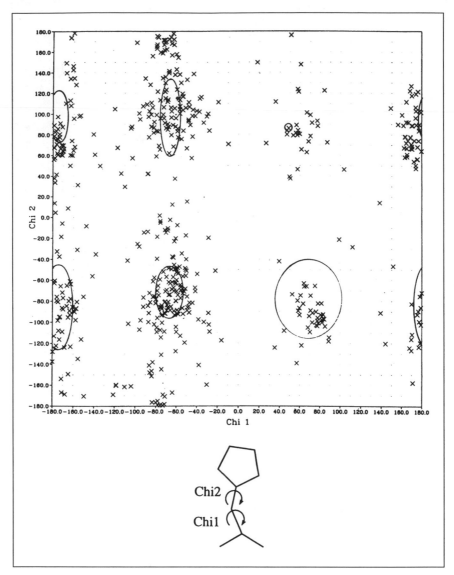

FIGURE 1–3. *Observed side chain χ angles for histidine residues. The ellipses show one standard deviation from the mean position on each axis as found in the statistical survey by Ponder and Richards (1987) of 19 proteins refined at high resolution. The crosses are the individual values from a larger segment of the Protein Data Bank kindly provided by H. Hellinga. The greater spread of values in the χ 2 direction is a reflection of the shallow sixfold torsional potential associated with the trigonal gamma carbon atom. The small circle at (60°,90°) is the position of the single rotamer found in the earlier survey.*

added, it would probably be as well at this time to use the full list for predictive purposes.

If the rotamer approximation is valid for the condensed form of proteins represented by their native states, it is almost certainly appropriate for the unfolded state, where there would be even less tendency to maintain energetically unfavorable conformations. Such restrictions will apply to folding intermediates as well. It would seem that the entire folding process, perhaps omitting actual transition states, is likely to be carried out with a closely defined set of bond lengths, bond angles, and torsion angles centered on the minimum energy positions.

The conformational restrictions previously referred to are all local in nature and will apply to the main chain and side chains under all circumstances. Long-range and excluded volume effects will be considered later in the chapter.

4 THE UNFOLDED (DENATURED) STATE

The term *denaturation* has been used to refer to any change from the biologically active, native state, including poorly understood irreversible alterations such as covalent modification and aggregation. These latter reactions are excluded from this discussion. We will consider only conformational changes that are, at least in principle, reversible. The terms *folding* and *unfolding* have this implication. Dill and Shortle (1991) have suggested that at our present level of understanding it is useful to distinguish between "highly open and solvent exposed states . . . obtained under strongly denaturing conditions," classed as *unfolded,* and a second category of *"compact denatured* states obtained under weaker denaturing conditions." The latter are frequently referred to as *folding intermediates* and have a separate section in this chapter. The principal focus in this section is the unfolded state as just defined. The following issue is covered in much more detail, and with extensive references, in the particularly lucid review by Dill and Shortle (1991).

Looked at simply as an organic chemical, a biologically produced polypeptide is a member of a particular class of linear polymers. It is unusual because, while very heterogeneous with 20 different monomer units, it has both a completely defined sequence and a specific length. Certain members of this class are immediately distinguished from the common industrial polymers, because all of the molecules fold up into essentially identical compact units that can form crystals with a very high degree of molecular order. These unique structures are simultaneously the source both of the magic of biological function and of the extreme difficulty in using current chemical theory to predict structure from sequence.

4.1 The View from Polymer Chemistry

The simplest model for a polymer is known as the freely jointed chain, a series of segments of constant length that are connected end-to-end with a completely "universal joint" at each junction. Since these units are considered to be line segments with no thickness, excluded volume plays no role in the properties of such a chain. Consecutive segments follow a random walk pattern. The mean end-to-end distance, as a measure of the size of such a hypothetical chain, is proportional to the product of the segment length and $n^{1/2}$, where n is the total number of segments. A collection of such chains, an ensemble, will represent a vast array of different conformations, all with about the same energy.

A real polymer molecule, of course, does not have freely jointed connections. The restrictions imposed by proper bond angles and finite atom size introduce a "stiffness" into the chain because of this fixed geometry and the excluded volume effect that is described by the Ramachandran map (Fig. 1–2). The excluded volume is a reflection of the fact that two atoms cannot occupy the same space at the same time. For a given length of the monomer unit, these factors will make the overall size of a real chain larger than that of its freely jointed equivalent. The volume of space occupied by such a chain increases more rapidly with chain length than that of the freely jointed chain, as $n^{2/3}$ rather than $n^{1/2}$. For comparison, the volume of a close packed solid with minimum voids will increase as $n^{1/3}$.

The apparent size of a chain of given length will depend markedly upon the chemical nature of both the polymer and the solvent in which it is dissolved. In a "good" solvent, a polymer chain is highly expanded because the interactions between the solvent and the units of the chain are preferred over the interactions between the chain units themselves. In a "poor" solvent these relations are reversed and the chain contracts in an attempt to exclude contact with the solvent as far as possible. These contracted chains will usually aggregate and precipitate in further attempts to avoid solvent contact. These relations are clear both conceptually and experimentally in such homopolymers as polyethylene or polystyrene.

For each homopolymer there will be one or more solvents, called theta solvents, that have a special property. They will be relatively poor solvents that cause the polymer to contract somewhat from its dimensions in a neutral solvent, where the interactions between two monomer units or between a monomer unit and the solvent are essentially the same. In a theta solvent this contraction will match exactly the expanding effect of excluded volume. In such a solvent the polymer shows the $n^{1/2}$ size proportionality of a freely jointed chain.

In the more complex case of a heteropolymer with varying properties along the chain, one can expect differing stiffnesses in different parts of the chain. The effective exponent will then become an appropriate average over the properties of these differing regions. Note that there will be a very large number of actual segment lengths and stiffnesses that will provide the same average length and measured molecular dimension for the whole chain. The detailed character of the solvent now begins to play a much more selective role in the behavior of the solutions. A given solvent may appear to be "good" for one part of the polymer and "poor" for another. Thus, the swelling or shrinking of a polymer in a particular polymer/solvent pair will reflect the subtle compensation of strongly opposing forces. The apparent stiffness of the various regions of the polymer will vary with the solvent conditions.

From such a point of view, water is actually a poor solvent for the polypeptide chain under conditions where the native folded state is stable. Very little water is normally found within the interiors of globular proteins. They are about as compact in that sense as they can be. These dense molecules are, however, frequently very soluble in water without any evidence of aggregation or precipitation. Such behavior would normally be taken as evidence of a good solvent. In polypeptides this curious behavior is related to the differing chemical properties of the individual amino acid residues. This ambivalent relationship between polymer and solvent for the polypeptide/water pair appears to be at the root of the unusual behavior of this system and of all of the biological functions that follow from it.

The qualitative description just given is presented much more rigorously, and in a form useful for predictions, in the heteropolymer theory of Dill (1985), Dill et al. (1989), and Chan and Dill (1991).

4.2 The Polypeptide Chain

For all polymers in aqueous solution, but especially for polypeptides, the situation is even more complicated than just indicated, because water changes its character as a function of temperature. Many proteins are compact and stable in the range 0° to 40°C. At higher temperatures denaturation occurs, in many cases reversibly. Water changes from being a poor solvent to a good one above the range of thermal stability. For certain proteins, unfolding also occurs at accessible low temperatures, 0° to 10°C, the phenomenon of cold denaturation. Again, the solvent character of water changes substantially. The solutes change in their properties also as one goes from one protein to another. The range of stability varies on the temperature scale from those thermosensitive proteins that are stable only at low temperatures to the proteins from the extreme thermophiles, which may undergo cold denaturation at 60°C. These folding and unfolding effects are related to the temperature dependence of the hydrophobic effect and the underlying fundamental thermodynamic function, the heat capacity. The energetics of these conformational changes are addressed in Chapter 3 by Privalov. Discussion of the various forces from different points of view have been given by Baldwin (1986a), Privalov and Gill (1988, 1989), Dill (1990), Livingstone et al. (1991), and Lee (1991), among others.

The completely unfolded state of a protein (i.e., fully denatured) has usually been thought to be properly described as a random coil within the statistical meaning of that phrase. Indeed, in good solvents, such as strong aqueous solutions of urea or guanidinium chloride (GdmCl), the radius of gyration and/or the Stokes radius as determined by any one of a variety of physical techniques usually indicates a highly expanded polymer chain. (See Tanford, 1968, 1970.) The values agree quite well with calculations of expected chain dimensions that include proper attention to the excluded volume effect. It had been assumed that all of the structure of the native form of a protein is lost under these strenuous denaturing conditions.

Questions remain, however, as to whether *all* structure has been lost and to what extent "random" is really random. As an extreme example, a single conformation of an expanded chain can easily be modeled that would have the correct radius of gyration in all of the usual physical measurements. Such a molecule would have a unique structure and not be the ensemble envisaged by the term *random coil*. The entropic consequences of such a situation would be dramatic. No such material, of course, exists. The ensemble of actual structures, however, could be very different from random without altering the radius of gyration. In view of the different chemical characteristics of the various amino acids, no solvent is likely to be "good" for all parts of a polypeptide chain. Thus, one can imagine a collection of structures in which some parts of the chain are maximally extended, while others are contracted into compact forms in positions that would depend upon the sequence. The average over all of these different regions might give a molecule that behaved hydrodynamically as a random coil of appropriate stiffness, but which, in fact, was far from randome in terms of the distribution and uniformity of conformations throughout the chain.

The disulfide bond presents an extra factor in these considerations. This covalent cross-link may have dramatic effects on the apparent size of an otherwise truly random coil. The effect will always be to reduce the apparent size from that of an unrestricted chain, but the extent of the effect will be variable, depending upon the positions of the two Cys residues in the sequence and, thus, the size of the loop produced by the disulfide bond. Many of the proteins that were initially used for denaturation studies contained several such bonds, which severely constrained the conformational space of the chains. If these bonds are cleaved, the chains behave as would be expected for linear polymers.

Much attention is being given today to the use of disulfide bonds in altering, usually increasing, the thermal stability of a protein. Such bonds can be inserted or removed by appropriate genetic manipulation of the sequence (Chapter 8). From the thermodynamic point of view, the effect on stability should be principally in the entropic effect of the altered conformational space available to the unfolded forms. Assuming insertion of a new disulfide bond without strain, any direct effect on the native structure should be limited to subtle changes in the dynamics of the molecule. In practice, however, disulfide bonds inserted where they do not normally exist frequently do not fit quite correctly. Distortions in the folded structure are produced, and these frequently propagate some distance from the sites of mutation. In these circumstances, there will be a change in the energy of the native structure, and the energy changes on unfolding will be more difficult to interpret (Katz and Kossiakoff, 1990).

Polypeptide chains containing Cys residues are produced biosynthetically in the reduced form. From folding experiments it would appear that formation of a disulfide bond is a late step in producing a compact structure in the neighborhood of the potential disulfide group. In Chapter 7 Creighton discusses the formation of disulfide bonds and their use in the study of the folding process, while Freedman discusses their formation *in vivo* in Chapter 10. In the general discussion in this chapter, we are assuming that no covalent cross-links are present and that the chains are properly considered linear polymers.

In an analysis of the thermodynamics of denaturation and a discussion of the hydrophobic effect, B. Lee (1991) has come to the conclusion that the exposure of a protein to solvent in the unfolded state is only about two thirds of the value that it would have as a fully extended chain. This estimate is consistent with earlier experimental data pointing to the same general conclusion: Brandts (1964), Tanford (1970), Ahmed and Bigelow (1986). Even under extreme denaturing conditions it would appear that contact with the solvent is well short of what might be thought of as full hydration. The free energy change associated with the folding/unfolding reaction is well correlated with changes in solvent accessible areas. The constants of proportionality are usually derived from small molecule transfer data where the areas correspond to the maximum values possible for the given covalent structure. It would seem that the full value of such free energy increments can never be realized with a long polymer chain. While a given residue may be fully buried in the native folded state, it can never have full accessibility in any attainable unfolded form. Presumably the entropic penalty of opening the chain to the extent necessary for full accessibility is simply too large. This fact is normally overlooked when quantitative estimates of the total free energy change on folding are made from area calculations.

It should be noted that polypeptides are almost always studied in solvents containing more than one component; water and sodium chloride, water and alcohols, water and urea, water and GdmCl, water and sugars, and so on. Only rarely does one use pure solvents containing a single molecular species. As a result, one must also consider the problem of a nonuniform distribution of the components of the solvent. It is perfectly possible, indeed likely, that the "solvent" in contact with different parts of an expanded chain will have compositions that differ from the mean solvent composition of the whole solution. Such effects have been measured extensively by Arakawa and Timasheff (1982) and Arakawa et al. (1990), who have analyzed the problem in terms of selective binding of the different solvent components. Thus, in both folded and unfolded forms there will be varying contact between protein and solvent, and the solvent itself will be nonuniform in composition adjacent to different regions of the protein. Such nonuniformity is obvious in the distribution of ions around anionic, cationic, and neutral polar groups on the protein surface.

If one wishes to interpret, or particularly to calculate, the thermodynamics of the folding reaction, then the properties of the reactants on the two sides of the equation must be known. The presence of persistent structure not describable by the random-coil formalism would certainly alter one's view of the reaction. If the stoichiometric amount of such structure is large, then the nature of this structure becomes as important as that of the native state. If the stoichiometric amount is not large, then its effect on the equilibrium thermodynamics of the reaction might only be marginal. Even if such structures varied qualitatively under different denaturing conditions, they would not markedly alter the estimates, by extrapolation, of the equilibrium parameters for the folding reaction in water. As pointed out in an earlier section, the fit of data to the two-state assumption cannot be used *per se* to rule out the presence of such structure(s). The effect of even small amounts of these structures on the kinetics of folding could, of course, be dramatic.

In their review, Dill and Shortle (1991) define the state D_0, the denatured state that is in equilibrium with the native form under physiological conditions. At equilibrium under such conditions, no direct observations of the structure of this state can be made because it represents only 1 in 10^4 to 1 in 10^{15} of the molecules in the sample, depending upon the free energy of unfolding. All of the experiments or calculations referred to in this section are concerned with the open forms of the chain obtained under strongly denaturing conditions. Such forms are very unlikely to occur except transiently, perhaps, at the moment of synthesis. D_0 is almost certain to be a "compact denatured" state and will be referred to briefly in the section on intermediates.

Since water may be considered a poor solvent overall for a polypeptide chain, there will be a strong tendency for the latter to contract and to reduce solvent contact after biosynthesis. Whether such contraction occurs during the process of extrusion from the ribosome is not immediately obvious. As previously suggested, different parts of a particular sequence may have very different tendencies to form fixed structures, and mutual stabilization of structured segments may also be both important and variable. As a result, no general conclusions on the timing of the folding process in biosynthesis appear obvious at this time. This topic is discussed in more detail by Freedman in Chapter 10.

In kinetic measurements and in establishing folding pathways, the existence of structure in an unfolded chain becomes a matter of great importance. Regardless of the stoichiometric amounts, any such structures could easily serve as principal starting points in the folding process and thus control the pathways accessible to a given chain. The details of the kinetics of folding will depend upon the extent and interconvertibility of such initial structures. Since the structures may be expected to reflect the special characteristics of particular solvent/peptide pairs, one might expect that different protocols for unfolding a peptide chain will lead to a different set of structures quantitatively and/or qualitatively. Such structures may or may not bear any close relation to the final structure of that segment in the folded state.

4.3 Experimental Evidence for Structure in the Unfolded State

The studies on denaturation before 1970 are covered in detail by Tanford (1968,1970). It was clear at that time that denaturation was a complex process and that different conditions led to different end points. Some structure was certainly indicated. In contrast to other denaturing solvents or conditions, 6M GdmCl or 8-M urea appeared to produce an unfolded product that had the characteristics expected for a random coil. Thus, by implication no persistent structure survived in these solvents. This statement was made part of the lore of the field. The evidence came largely from hydrodynamic techniques such as sedimentation, diffusion, and intrinsic viscosity and from optical spectroscopy such as rotary dispersion and UV absorbance. None of these measurements would reliably indicate the presence of small amounts of structure, since they all measure the bulk properties of the solution of solute molecules and radius of gyration is not a particularly sensitive measure of the presence of small nonrandom regions. Thus, while most of the molecules on average must approximate a random coil, small amounts of structure, whether stable or fluctuating, could not be ruled out. Murray-Brelier and Goldberg (1989) have, in fact, identified by CD different denatured states in *E. coli* tryptophan synthetase.

Early studies on the relation of helix formation to polymer length in homopolypeptides (Katchalski and Sela, 1958) were followed by host-guest measurements on the helix propensity of individual amino acids in attempts to establish the Zimm-Bragg parameters for helix initiation and propagation (see, for example, Platzer et al., 1972). On the basis of these data, it was generally assumed that short peptides, 20 residues or less, would not form stable helices in aqueous solution.

Since the early 1980's, the very extensive studies of Baldwin and his colleagues have shown that this is not the case. (For example, see Marqusee et al., 1989.) His work was sparked by the early observation of Brown and Klee (1971) that at low temperatures a 13-residue fragment of the S-peptide component of ribonuclease-S did indeed appear to have some helix content. This observation was apparently overlooked by everyone until picked up by Baldwin. The full range of Baldwin's studies cannot be reviewed here, but the general conclusion is important. In the 13 to 16 residue peptides, which represent the bulk of the observations, the formation of detectable helical structure is the rule, not the exception. The amount of structure, as estimated by CD spectra, varies with sequence and covers the range from not detectable to a high percentage. It is favored at low temperature. Aggregation has been ruled out; the helix formation is a property of the isolated peptide. While electrostatic forces certainly play a role, other effects are equally important. General statements on the origin of the stabilizing effects of specific sequences are hard to come by. Nevertheless, the appearance of at least marginally stable helices in these short peptides in water is beyond question.

NMR may be the procedure of choice in the search for residual structure in short polypeptides. In favorable cases one can detect the structure directly, rather than estimate it as the difference between two measured large values. Dyson et al. (1988) have provided evidence for preferred conformations of certain very short peptides. The data strongly indicate the presence of certain sharp turn conformations. In studies on isotopically labeled samples of the 108-residue protein thioredoxin under a variety of denaturing conditions, D. Wishart (personal communication) has observed persistent resonances whose positions do not correspond to those expected for a fully unfolded chain. The stoichiometric amount of these presumably "structured" regions is certainly small. Without the enhancement or editing provided by the specific ^{13}C or ^{2}H labeling of selected amino acids, they would almost certainly have escaped detection. The spectra under various denaturing conditions are different, implying not only the presence of structure, but the presence of specific structures in the different solvents. At the present state of the investigation, it is not possible to state what the actual structures are that are leading to the observed spectra. That there is residual structure even under conditions leading predominantly to the random-coil state seems certain. It appears equally certain that any such structures are only marginally stable and are undoubtedly rapidly fluctuating between conformational states. The relaxation times for all of these processes is likely to be in the submicrosecond region. Even in these unfolded conformations, however, peptide bond isomerization and disulfide interchange reactions are much slower processes than the torsional conformational shifts.

Many other experimental studies supporting the existence of structure in the unfolded state are discussed by Dill and Shortle (1991).

5 THE FOLDED (NATIVE) STATE

Only some general aspects of the structure of soluble proteins will be discussed in this section. The details of chain conformations and structural classes are covered by Thornton in Chapter 2.

The early physicochemical studies on proteins recorded in the landmark volume by Cohn and Edsall (1943) showed that: globular proteins were well approximated as ellipsoids of low axial ratio; roughly a monolayer of water surrounding the molecule was required to explain the hydrodynamic data (the value of about 0.3 grams of water per gram of protein persists to this day); and the interior of the protein must be essentially anhydrous and close packed, because the observed partial specific volume could be approximated by the sum of the volumes of the individual amino acid residues. All of these conclusions correspond well to our current view of these molecules. The results of these noncrystallographic procedures were compatible with a unique structure, but did not prove it. The distributions in either molecular weight or conformation, however, would have to be quite narrow to fit the data. Of course, all of this work preceded any knowledge of amino acid sequences and even of the procedures for accurate amino acid analyses.

With the first X-ray diffraction picture of a "wet" crystal of pepsin obtained by Bernal and Crowfoot (now Hodgkin) (1934), it was clear that all the molecules of pepsin must be remarkably similar to each other. The maximum angle of the diffracted beams indicated a significant degree of order at distances not greater than about 1/50th of the maximum dimension of the molecule or on the order of an Angstrom unit. For most, but not all, crystalline proteins subsequently investigated, the "resolution" of the X-ray pattern indicates a similarity between the molecules in the crystal, expressed as a single distance, in the range of 0.2 to 2 Å, depending upon the particular crystal. Until recently this has been taken as evidence that the native state of the protein is indeed a unique structure, and that defining it more precisely would simply require obtaining and analyzing "better" crystals. This feeling was certainly reinforced when two structures for trypsin appeared from entirely separate studies in two different laboratories on different samples of the enzyme. The root mean square difference in position of the atoms in the two structures was 0.16 Å, i.e., they were essentially identical within experimental error (Chambers and Stroud, 1979).

5.1 *Conformational Substates*

From an entirely different type of experiment (spectroscopic studies of the rebinding of oxygen to myoglobin in samples flash photodissociated at low temperature), Frauenfelder and his colleagues came to the conclusion that the native form of a protein is not a single state but a collection of states that are structurally very similar but are separated by measurable energy barriers. They named these "conformational substates" (Frauenfelder, 1985; Ansari et al., 1985) (Fig. 1–4). The multidimensional energy landscape for a protein is certainly very complicated, with a huge number of minima. At low temperatures these substates make their presence clear by the range of kinetic constants required to match the data for what would appear to be a simple bimolecular recombination.

To a first approximation the energy surface does not depend upon temperature. Are these energy barriers high enough to affect the behavior of the protein near room temperature? Elber and Karplus (1987) have carried out quenching and very "gentle" energy minimizations of different time slices from a molecular dynamics run carried out at room temperature. The minimization procedure carefully avoided searching for a global minimum. They frequently ended up with slightly different structures. It was clear that the potential surface underlying the molecular dynamics trajectory was indeed strewn with many closely spaced minima. A very detailed study by Noguti and Gō (1989) using Monte Carlo procedures on bovine pancreatic trypsin inhibitor (BPTI) suggested a hierarchical collection of multiple substates, that is, clustering on the energy surface with significant barriers between clusters. Energy minimization operates at 0° Kelvin, however, and does not tell us directly what effect the fine structure of the energy surface has on the room temperature behavior. This is a bit like asking whether a vessel steaming east from New York is aware of crossing the mid-Atlantic ridge, which is much closer to the surface than is the abyssal plane. Since the ridge still represents deep water by navigational standards, the answer in that case is no. Frauenfelder (1985) has suggested that there is a whole range of minima as one examines the energy surface in more detail. For certain proteins some of the maxima, within the broad well of the native minimum, may indeed be high enough to provide a number of closely related "native" states that might be detected with appropriate techniques. Whether these states need to be taken into account or not will depend upon the problem at hand. For many purposes it will be adequate to assume that the native protein represents a unique state, but the possibility of problems arising from the assumption of uniqueness should always be kept in mind.

In recent, carefully refined, X-ray structures it is clear that the side chains, and even occasionally the main chain, may occupy more than one position in different unit cells in the same crystal (Smith et al., 1986). (See Figure 1–5.)

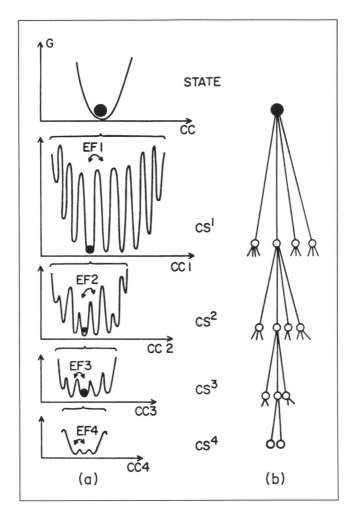

FIGURE 1–4. *Schematic diagram of conformational substates, CS, and equilibrium fluctuations, EF. The ordinate is the Gibbs free energy, the abscissae the conformational coordinates, CC 1–4. At room temperature the state CC appears to be located in a smooth well. As the temperature is lowered, the substructure of the energy landscape begins to appear with many minima representing closely similar forms in thermal equilibrium. As the temperature decreases, molecules are frozen into the individual minima, each of which develops additional structure. Within each tier of substates, the equilibrium fluctuations distribute the molecules among these minima. The final tier, CS^4, would be identifiable only at a temperature near absolute zero. This general description was set up to rationalize the data collected on the kinetics of the photodissociation and reassociation of the myoglobin-carbon monoxide complex as a function of temperature. (Reprinted with permission from Ansari et al., 1985)*

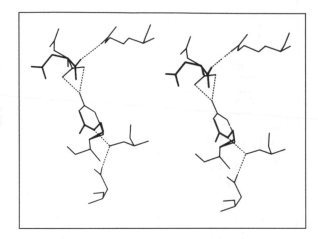

FIGURE 1–5. *Alternate side chain conformations in the same crystal. A stereodiagram of a part of the structure of erabotoxin B, showing two residues that exist in two rotamer conformations each. The main chain and one pair of rotamers are shown in light lines, the alternate pair in heavy lines. The hydrogen bonding schemes are different in these two alternate structures. (Reprinted with permission, Smith et al., 1986.)*

These residues are usually, but not always, on the surface of the protein, and alternate positions can often be accommodated without significant change in the rest of the protein. With roughly equal occupancy, even three separate positions for a residue could be confidently refined if data of sufficient resolution are available. If there were more conformations, the electron density would tend to get smeared out to the point where individual structures would be impossible to identify. The latter effect has long been recognized for the side chains of many lysine residues. The energy barriers between these states cannot be estimated from the X-ray data because there is no time base for the structures derived in the normal diffraction procedures. Such multiple structures probably fit the definition of conformational substates, but they are not what Frauenfelder calls FIMS or *functionally imp*ortant *s*tates (Ansari et al., 1985).

Independent of the structural details, Karplus et al. (1987) have pointed out that the residual configurational entropy of the native state of a protein is actually very high, contrary to the common assumption that it is zero. It is based on internal fluctuations and is about tenfold greater than the incremental configurational entropy of the denatured state. It turns out that the vibrational entropy of a single conformer of the denatured state is very similar to that of the native state. Thus, the entropy change observed on denaturation does, by chance, correspond to that expected for the configurational term representing the multiple unfolded forms, but it is actually the difference between two much larger numbers. Small changes in the internal dynamics of the two states could have large effects on the measured entropy of unfolding and be misassigned to the usual chain entropy term.

5.2 Conformational Flexibility

It is necessary to distinguish between the terms *flexibility* and *dynamics*. The latter implies a time scale and continuing motion, the thermally activated vibrations and rotations of individual atoms and groups, the slower motions of the whole molecule that are at least partially describable as normal modes and so on. Flexibility, on the other hand, has no particular time reference, but implies multiple structures that are of comparable energy and can be interconverted with a series of small changes, so that they are easily accessible. A flag in a breeze is an example of complex dynamic behavior. A billiard ball and a jump rope are extreme examples of flexibility, with an Oriental fan being an intermediate case.

In addition to multiple side chain conformations, X-ray crystallographic studies have brought the problem of the uniqueness of the native state into sharp focus. Part of the apparent uniqueness may be forced by the crystal lattice. In a high-resolution structure there is no doubt that the asymmetric unit of the crystal is repeated accurately by the symmetry operations within the unit cell and by translation from one unit cell to others in the same mosaic block. The asymmetric unit, however, may contain more than one protein molecule, and a given protein may exist in different space groups where there are different intermolecular contacts. When the structures of these molecules are compared in detail, very obvious differences have been found in several instances. The differences are well outside the experimental error of the structure determinations.

Some years ago very large changes were found in hexokinase between the free enzyme and the glucose liganded form by Bennett and Steitz (1978). More recently Wang and Steitz (unpublished) have found a whole series of structural forms of hexokinase with differing relations between the two domains. Faber and Matthews (1990) have found similar large differences in T4 lysozyme between different molecules within the asymmetric unit and between different crystal forms (Fig. 1–6). In both of these cases, two separate domains of the protein each appear to maintain their own structures while the domains themselves change their position with respect to each other by a hinge-like motion. While the maximum shifts in position for groups of selected residues can be large, most of the residues within a single domain experience extremely small changes in their environment during these conformational shifts. These examples show that the chain of a protein may exhibit considerable structural flexibility with very little equilibrium energy penalty. (In some cases two crystal forms can occur together in the

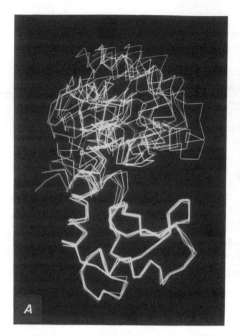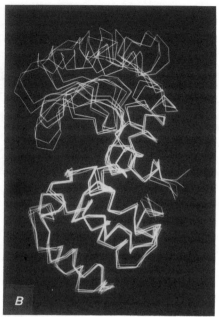

FIGURE 1–6. *An example of protein flexibility in T4 lysozyme, mutant M6I. Five different structures of the same molecule in different parts of the asymmetric unit in one crystal and also in other crystal forms have been overlaid. In (A) the N-terminal domains (residues 15 to 60) have been superimposed. In (B) the C-terminal domains (residues 80 to 160) have been superimposed. The substantial relative movements of the two domains are clear, while the structures of the individual domains show very much smaller changes. (Reprinted with permission from Faber and Matthews, 1990)*

same crystallization vial.) The dimensional shifts between different conformations in a flexible set are well beyond the size of the fluctuations caused by thermal vibrations about the mean atomic positions in one particular conformation. Other examples of flexibility, especially in the serine proteases and the immunoglobulins, are discussed by Huber (1979).

The evidence for chain flexibility in the folded state presented thus far is dependent upon the examples of a few crystal space groups and multiply occupied asymmetric units. There is no experimental procedure that defines the full range of such deformations or the energy barriers, if any, between the forms. Ligand binding in some cases comes close to the rigid lock and key model—for example, glutathione reductase (Schulz, 1991)—while in others substantial structural alteration (induced fit) occurs in one or both of the partners in the complex—for example, the adenylate kinases (Schulz, 1991) and glutamine t-RNA synthetase (Rould et al., 1989). Whether these structural variations represent distinct states separated by significant energy barriers, and thus with some intrinsic stability, or are simply forms that are locked in by the crystal lattice on an otherwise relatively flat energy surface, is not known at this time.

The problem of flexibility has been the subject of theoretical analysis. The report by McCammon et al. (1976) on the hinge-bending mode in lysozyme is an early example. In this and other cases, the probable location of the hinge was established by looking at a model of the protein. Gibrat and Gō (1990) have proposed a procedure based on normal mode analysis where the hinge can be identified without any *a priori* assumptions. In a recent molecular dynamics simulation of the HIV protease (Harte et al., 1991), the analysis of correlation coefficients over a 40-psec interval showed a hinge movement of the "flap" in this acid protease, although the amplitude was small. The maximum length of current computer simulations is still so short that one would not expect to see the large rigid body shifts observed in the T4 lysozyme or hexokinase structures.

5.3 Surface Area

If one is asked for an estimate of the surface area of the trunk of an oak tree, the problems are very much the same as asking for the surface area of a protein molecule. The answer will depend upon definitions and upon the tools available for measurement. Do you simply measure the circumference and length of the trunk with a string or do you allow for all levels of invagination of the bark? Richards (1977) has discussed this problem with regard to proteins and suggested definitions of several "surfaces" that are indicated in Figure 1–7. The smallest common "tool" available at the molecular level is a water molecule, which for this purpose is customarily taken as a sphere of radius 1.4 Å. The simplest surface to calculate is the *accessible surface* which is formed by the locus of the center of the water probe as it contacts all available parts of the protein (Lee and Richards, 1971). The *molecular surface* represents the sum of the surfaces of the protein atoms in contact with the water probe (the *contact surface*) and the surfaces of the probe sphere that are between its contacts with two or three protein atoms (the so-called *re-entrant surface*). Originally the accessible surface was calculated by a procedure of numerical approximation (Lee and Richards, 1971; Shrake and Rupley, 1973; Richards, 1985). Analytical solutions to this problem that are mathematically accurate have been provided subsequently by both Connolly (1983) and Richmond (1984).

To calculate the surface area by any procedure, one must have the atomic coordinates of the structure and assigned radii for the hard spheres used to represent each atom. The accessible area of any atom will depend

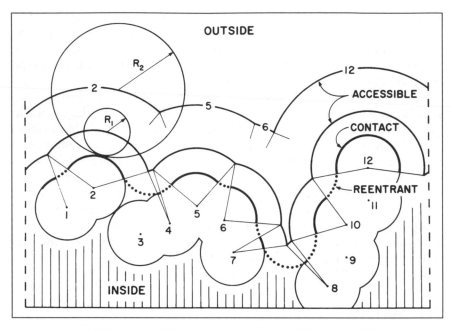

FIGURE 1–7. *Definitions of different surfaces and areas. The* accessible surface *is the locus of the center of a probe sphere that is rolled around the protein making maximal contact without overlap. The* contact surface *is that portion of the protein atom(s) in direct contact with the probe. The* re-entrant surface *is the probe surface facing the protein between two or three contact points with neighboring atoms. The latter two, which are disconnected patches, are summed to make up the continuous* molecular surface. *The areas of these surfaces can be approximated by numerical integration or calculated exactly through differential geometry. The numerical values are very sensitive to the choice of van der Waals radii for the protein atoms and to the size chosen for the probe sphere. (Reprinted from Richards, 1977)*

sensitively not only upon the structure but also upon the list of assigned radii. There are several such lists. Since it is not clear what the radii ought to be, there is no "best" list. It is inappropriate to compare areas of two structures that have been calculated with different sets of atomic radii.

The area calculated for a peptide chain will depend markedly upon its conformation. Even for a compact native protein, with no changes allowed in any of the main chain atom positions, the total surface area will depend significantly upon the particular conformation chosen for the surface side chains. The latter control the effective "roughness" of the surface. Thus, one must be very cautious in both the calculation and the use of these surface area values. A recent example is provided by the work of Northrup et al. (1990). The change in the surface properties of residues during a molecular dynamics simulation was shown to have substantial effects on the estimates of electrostatic fields and forces made on the successive instantaneous structures.

Apart from its use simply as a geometric parameter for a molecule, the accessible area, and particularly changes in accessible area, have taken on special importance in relation to the hydrophobic effect. Chothia (1974) showed that there was a linear relationship between the surface areas of amino acid residues and the free energy changes associated with the transfer of the amino acids from water to an organic solvent (Nozaki and Tanford, 1971). The slope of this line corresponded to a free energy change of about 25 cal/Å^2/mole for nonpolar residues. The appropriate value for use in protein folding problems is the subject of some debate. A higher value of 46 to 47 cal/mole/Å^2 is suggested by Sharp et al. (1991b). Whatever the proper value, the linear nature of the relation is now widely accepted and used.

The area calculations are actually done atom by atom. The values can be combined in any way one wishes. The monograph by Hansch and Leo (1979) describes the division of molecules into fragments and the parameterizing of correlations with a variety of chemical properties, but with special emphasis on partition coefficients. The prediction of partition behavior for molecules not in the basis set is generally good. Following this basic idea Eisenberg and McLachlan (1986) divided the side chain atoms into nonpolar, polar, and charged classes and derived the appropriate free energy increments per unit area for each class. Abraham and Leo (1987) have provided a more detailed fragment set for the side chains. Thus, the extent of burial of a given residue in the folding reaction is listed atom by atom and summed according to type, rather than using a single area change value for the whole residue. Thus, two residues of the same type occurring in different parts of the surface may contribute quite differently to the free energy of folding, even if the total area changes are the same for both locations.

The accessible area of a fully extended polypeptide chain is reduced by a factor of about 3 in folding into the native structure. Since the actual areas for these polymers are large, the area changes for the nonpoplar atoms can easily represent energies of hundreds of kcal/mol. This represents an explicit statement of, and a calculation procedure for, the hydrophobic effect. Its use does not require a knowledge of the underlying origin of the effect, which is the subject of much current debate (see Chapter 3) and it does not divide up the free energy between the enthalpic and entropic terms. In principle, the parameter lists only apply at a specified temperature. Separate lists should be used at other temperatures, although these have not yet been provided.

In folded structures, the amino acid residues Phe, Leu, Ile, and Met tend to be fully buried. The charged groups of Arg, Lys, His, Glu, and Asp residues tend to be exposed on the surface, although buried carboxyl groups are not uncommon. All other residues appear everywhere in the structures of folded proteins, and their extent of burial cannot be usefully predicted. An extensive survey of the distribution of accessible areas for each residue type has been provided by Rose et al. (1985) (Fig. 1–8).

FIGURE 1–8. *Normalized distribution functions by residue type of the fraction of the accessible area buried in the native structures of 23 proteins. The percent buried area is binned in 5% increments and shown as histograms. The cumulative numbers are shown by the solid lines. A very broad, almost featureless distribution is characteristic of most residue types. (Kindly provided by G. Rose)*

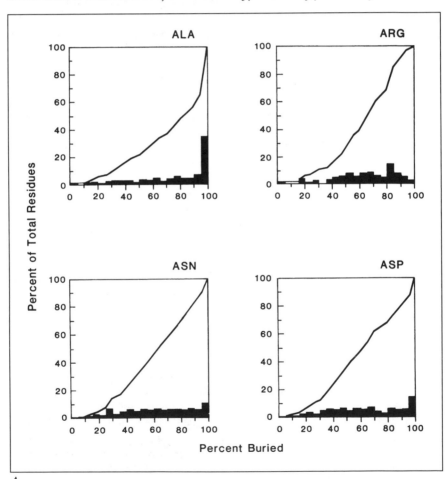

A

In the original paper of Lee and Richards (1971), it was pointed out that roughly half of the accessible area of several folded proteins was provided by nonpolar atoms and half by polar ones. By the nature of the amino acids themselves, it is clear that one cannot bury all of the nonpolar atoms, no matter what the folded structure may be. However, one can, in principle, adjust the distribution of the polar and nonpolar areas on the molecular surface. For monomeric species the distribution of area type is apparently random. The underlying secondary structure is not obviously reflected in this distribution, as can be seen in Figure 1–9. Hydrophobic "patches" may represent sites of oligomerization in the formation of

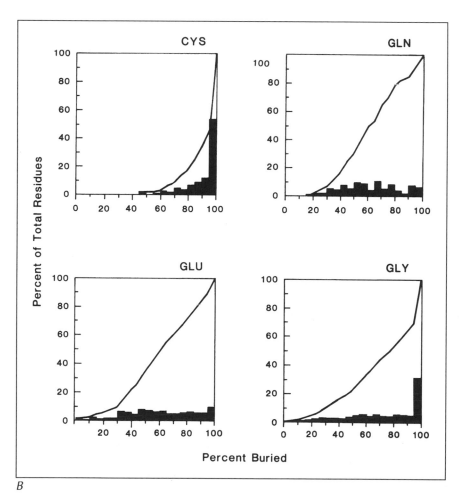

B

(continued on next page)

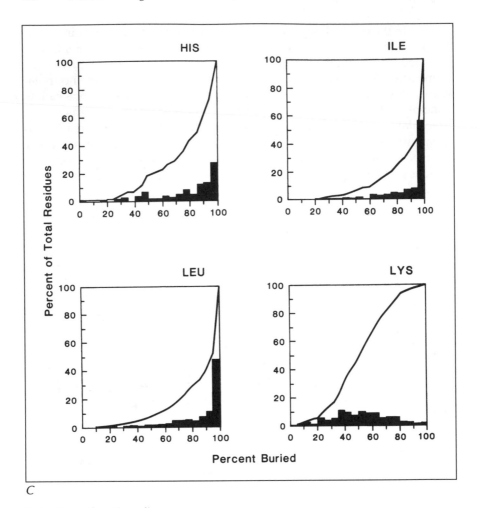

C

FIGURE 1–8. (*continued*)

quaternary structure, but the frequent involvement of specific polar interactions in such associations may make difficult the visualization of such regions in surface projections.

In spite of all of these complications and uncertainties, area changes are frequently used today both qualitatively and quantitatively in working with folding processes and in describing ligand binding reactions with both small and macromolecular ligands. The reliability of the energy changes calculated from the areas remains to be established.

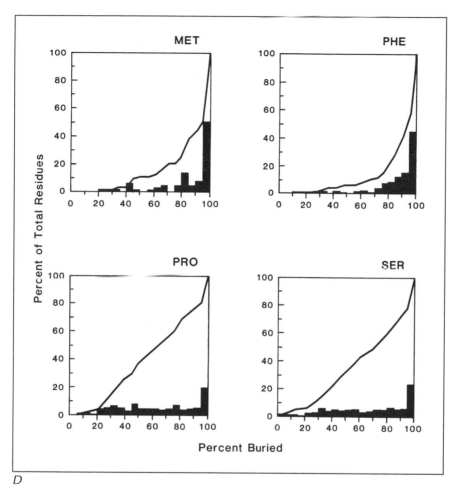

D

(continued on next page)

5.4 Volume, Packing, and Cavities

In the inside of most folded proteins, there are very few water molecules with no connection to the surrounding solvent. The interior is almost anhydrous. As mentioned in the start of this section, it has been known for a long time that there cannot be a great deal of empty space in the protein interior; otherwise, the volume changes upon denaturation would be much larger than they are known to be.

The packing of the interior residues was quantitatively evaluated by Richards (1974) and Finney (1975). *Packing density* is the term used to describe the effectiveness of the molecular arrangement in filling space.

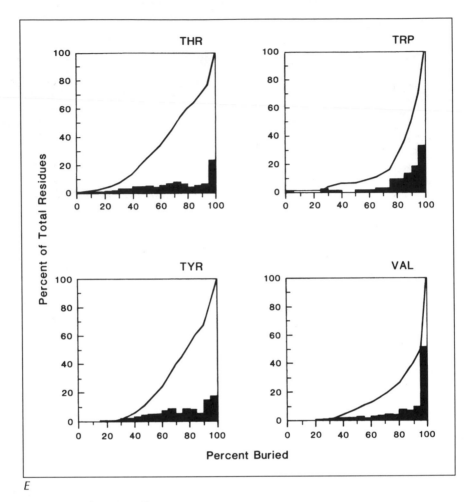

E

FIGURE 1–8. (continued)

FIGURE 1–9. (opposite) Stereographic projections of the nonpolar accessible surface area of two proteins. The intersection with a surrounding sphere of vectors from the center of mass through the nonpolar patches are shown as marks at the appropriate latitude and longitude. The size of the patches is not shown in this presentation. The + signs mark the ±90° and ±180° meridians. On the left is myohemerythrin, a 4-helix bundle protein. On the right is plastocyanin, a β-sheet sandwich. The distributions are essentially random and indistinguishable.

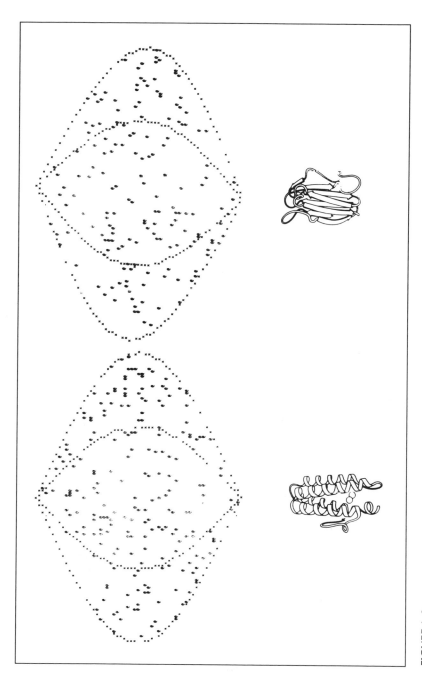

FIGURE 1–9.

This dimensionless number is the ratio of the volume of the van der Waals envelope of the molecule to the volume of space that it actually occupies. For hexagonally close packed spheres, for example, this number is 0.7405. In spite of the variety of molecular shapes, in crystals of small organic molecules this number is usually in the range 0.70 to 0.78. The value for folded proteins is in the same range. Thus, proteins have been said to be as closely packed as they can be. This statement may be somewhat misleading. The main chain of the protein has a very high packing density when considered by itself, due to the many covalent bonds. If the average packing density for the whole protein is about the same as that of close packed spheres, the side chains must be more loosely packed to provide that average.

The importance of packing in crystals of small molecules has been extensively studied by Kitaigorodsky (1961). He came to the conclusion that geometric packing restrictions based solely on the van der Waals terms is by far the most important controlling factor in the observed structures. More recently Hsu and Chandler (1978) and Chandler et al. (1983) examined both the liquid and solid states and also found that the van der Waals terms (i.e., packing) overwhelmed any effects due to factors such as the dipole moment.

For the *origin of the native structure,* one looks for the general forces that operate on the amino acid sequence of a chain to produce a single compact conformation, and packing obviously is a serious contender. The other major noncovalent forces are the hydrophobic effect and electrostatic interactions. The large forces produced by the former lead to general compaction of the chain but are hard to picture as the guide to a unique conformation. Thus, the micelles of even a single detergent species represent an ensemble of structures rather than a single highly ordered assembly. The net charge on many proteins can be varied over wide limits without any significant change in structure. The stability of the protein may be altered by changes in charge, but the conformation appears fixed until major unfolding occurs at the transition point. These points have been buttressed by recent mutations in T4 lysozyme by Dao-pin et al. (1990a,b). The dramatic effects of charge changes on biological activity are usually accompanied by very small changes in overall conformation. (For example, see Johnson and Barford, 1990, on phosphorylase.) Certain toxins and viral coat proteins may be exceptions to this statement. There is much evidence for substantial changes in these cases, although the detailed structural alterations are not known at this point. Thus, for the general folding problem, one is left with packing as the only current candidate for selection of the particular native fold.

In the context used here, "packing" is concerned only with filling the available space with rigid objects that cannot occupy the same space at the same time, i.e., steric repulsion. The attractive part of the van der Waals term is not explicitly considered, except indirectly by the requirement to fill the space with a specified efficiency. The unfilled space is empty and is referred to as void space or cavities. The extent to which packing can exert this control depends upon the *packing density*. This is shown schematically in two dimensions in Figure 1–10. If the objects fill all of the space, the packing density is 1.00. If, in addition, the objects are not uniform in shape, then the "structure" is completely defined. (If some of them are identical in size and shape, some structures will be degenerate.) As the packing density is lowered, cavities appear and relative movements between the pieces become possible. With further decreases, shifts in relative positions become possible, and finally all restriction on movement and position disappears. This latter state occurs at quite a high packing density in the example shown.

The packing density at which structure starts to be imposed upon the system depends upon the asymmetry of the objects. Thus, the alignment of highly asymmetric objects like tobacco mosaic virus particles will occur

FIGURE 1–10. *Schematic diagram of packing in two dimensions. The objects 1, 2, 3, and 4 are of different sizes and shapes. They completely fill the upper left panel with a packing density of unity. In each of the subsequent panels the objects have a constant size and shape, but the area of the bounding square is increased, so that the packing densities are 0.76, 0.60, and 0.50. It can be seen that packing restrictions on the relative positions of the objects are very sensitive to packing density, and, for objects of compact shape, the restrictions essentially disappear at fairly high densities. In three dimensions proteins appear to be represented by a situation somewhere between the upper right and lower left panels.*

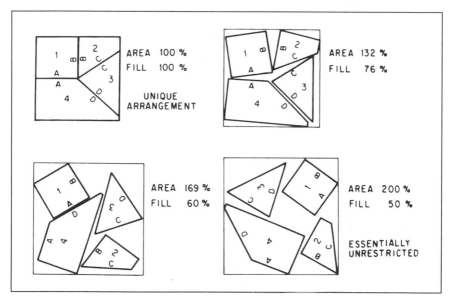

at much lower densities than it would for spherical molecules, where the maximum packing density has to be approached before the cubic or hexagonal packing order is fully enforced. The amino acid side chains vary in their asymmetry, but all of them should be considered rather "chunky." Thus, their steric effects on structure will be maximally important only at fairly high packing densities. Unfortunately, as with all other aspects of the protein structure problem, the actual packing densities of the side chains are just in the region where their significance is difficult to evaluate. It may be that packing is very important in certain regions of a molecule and much less so in others, and that the difference in mean packing densities between different proteins may be significant even when these differences are quite small.

Using the lattice approach for describing polymer conformation, Chan and Dill (1990) have examined the effect of packing density on the allowed conformations. Using an exhaustive search on a 2D lattice, they concluded that at high densities there was a strong preference for the 2D equivalent of α-helices and β-sheets produced solely by the excluded volume effect at high packing density, and without any other attribute assigned to the lattice positions. Gregoret and Cohen (1991) have examined this problem with a different algorithm not involving the use of lattices. They drew the same general conclusion, except that the tendency for secondary structure formation was not as strong. The secondary structure motifs common to all proteins do seem to be favored by packing considerations. This is not the whole answer obviously, since helices and sheets do form in systems where no "packing pressure" exists.

One might note that if stable protein structures required the maximum possible packing density, evolution could not have occurred as it is currently envisaged. Each mutation of an interior residue would have required one or more simultaneous and compensating mutations to maintain such a high density. Thus, for any alteration in sequence, paired or higher-order mutations would be necessary at each step. The probability of success would be reduced to the vanishing point. The contrary appears to be closer to the truth. Proteins adapt remarkably well to single site mutations. Extensive studies by Matthews on T4 lysozyme, by Sauer on λ repressor, by Shortle on staphylococcal nuclease, and by many many others in other systems have provided a huge amount of data on this point. (For example, see Bowie et al., 1990.) Although the sensitivity varies markedly with position in the polypeptide chain, interior residues are by no means excluded from such changes. Should one conclude that packing is or is not an important factor? The answer is not yet proven, but as of late 1991, the odds are that it is. The operational range in packing density means that its influence on structure is going to be subtle. The effect of packing is seen as a property to be averaged over the whole protein, and it is unlikely to represent the simple addition of observable single site effects.

For a detailed investigation of packing, the *inverse folding problem* has been useful. Rather than asking "What structure is specified by a particular sequence?" (the usual statement of the folding problem), one asks "How many sequences are compatible with a particular structure?" (Drexler, 1981). Ponder and Richards (1987) developed an algorithm to address this latter question. The main chain of a particular structure is fixed. A collection of closely associated interior side chains, defined as a *packing unit,* is "removed." The program then provides a list of all possible collections of residues that can be placed back in these positions without steric overlap and that fill a specified fraction of the available space. The linear sequence positions of the residues of such a packing unit usually have no simple relationship to each other.

In order to make these calculations computationally accessible, use of the rotamer approximation for the conformation of side chains was essential. (See earlier Section 3.) The general results obtained are as follows: (1) The volume of space represented by each packing unit can be filled by a number of different sequences. This number, however, is a tiny fraction of the combinatorial possibilities. (2) For a given allowed sequence, a number of rotamer combinations are frequently possible. This latter observation does not correspond with the facts. Interior residues are almost always in single well-defined positions. The results indicate that the present packing assumptions are not the full story. For example, the rotamer library simply specifies angles. No account is taken of the differing energies of the various rotamer wells, and the attractive van der Waals terms have not been optimized. (3) In all of the tests made so far, a few sequences are found that fill the space better than the wild-type sequence. This indicates that the native structure has not attained the maximum possible packing density. The reason for this is unknown, but it is unlikely to be accidental. It might be related to function or other cellular requirements. The dynamics of the molecule will almost certainly be altered by variation in the packing density. The possible relation to evolution has been mentioned previously.

The packing restrictions on the surface residues will obviously be much less than on the interior ones, but there will still be some. Overall, one would guess that a full packing calculation for an entire tertiary structure, not feasible as of early 1992, would yield a very large number of acceptable sequences even when such factors as location of charged groups and polar group hydrogen bonding were properly taken care of. This full range of sequences would probably show little or no sequence conservation, as is found for the naturally occurring members in several protein classes. The globins are a well-known example.

A major fault of the present version of the packing program is that it assumes an absolutely rigid main chain. This is too restrictive. Between the members of a protein class, the secondary units can vary considerably in their relations to each other, all within tertiary structures that would be classed as "the same" or at least very similar. If one had the ability to account for such variations in developing the sequence list, it would only make it longer. The obvious question now is, if even the restricted list is long, is the packing restriction sufficient to be a useful structure predictor at all? At this time one can only guess.

There are 20^{100}, or 10^{130}, different peptide sequences of length 100. For a given tertiary structure, assume that each position in the chain is restricted by packing or other factors to a group of 10 amino acids. The members of the group of 10 vary with position in the chain, and the occupants selected for each position are uncorrelated. The total number of sequences in this mildly restricted situation would then be 10^{100}, a huge number but still only a tiny fraction, 10^{-30}, of the possible total. This number would be much smaller still if fewer than 10 residues were allowed at each position. Another tertiary structure also allowing 10 residues at each position but, with a different grouping of residue types, would yield the same fraction of the total.

The question now is, would these two restricted but different sequence sets be mutually exclusive? If the allowed amino acid groupings differed by only two residues at each position, the probability of picking the same sequence from each set would be only 2×10^{-10}. Any greater difference between the allowed groups and the probability becomes vanishingly small. Thus, for practical purposes only a very modest discrimination at each chain position would be required to produce the different tertiary classes. This elementary example, of course, is unrealistic in assigning equal weight to the importance of each position for the final structure and assuming the substitutions are uncorrelated, but the "flavor" may be correct, and it suggests that sequence analysis based on coarse assignment of residue character to each position may be adequate for class prediction. The work of Bashford et al. (1987) clearly is pointing in that direction, as is the "profile analysis" procedure of Eisenberg (Bowie et al., 1991.)

The results of the packing algorithm are now being checked by experiment. Sauer and his colleagues have provided a vast amount of data from genetic experiments on the λ repressor. This system has the tremendous advantage of a very sensitive selection procedure for identifying folded, biologically active, mutants, and the crystal structure of the N-terminal fragment has been determined by Pabo and Lewis (1982) and Jordan and Pabo (1988). Reidhaar-Olson and Sauer (1988) carried out random muta- genesis at a large number of individual positions along the chain (Fig. 1–11). There was great variability in the number of allowed substitutions at the different positions. The numbers tended to correlate strongly with the surface accessibility of the positions, the interior positions being much more restricted. Lim and Sauer (1989) then picked an interior packing unit and carried out random mutations simultaneously at several of the positions

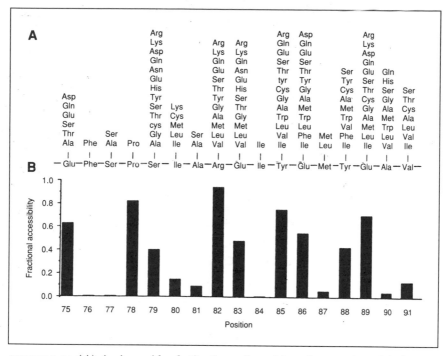

FIGURE 1–11. (A) Amino acid substitutions allowed in a short region of the λ repressor. The wild-type sequence is shown along the bottom line. The allowed substitutions are shown above each position. (B) The fractional solvent accessibility of the wild-type side chain in the protein dimer relative to the same atoms in an Ala-X-Ala model tripeptide. The strong correlation with the number of allowed substitutions in A is clear. (Reprinted with permission from Bowie et al., 1990)

within this unit. The number of active mutants corresponded roughly with what would be expected from the packing algorithm, about 0.5% for changes at three or four sites. A more detailed analysis involved simultaneous specific mutations at three sites where only five nonpolar residues were used as replacements (Lim and Sauer, 1991). The maximum number of mutants was thus 125. Of the 78 thus far synthesized, the agreement between experiment and prediction for the most active mutants is satisfying, but this agreement degrades for some of the less active forms (Fig. 1–12). Most interesting, perhaps, are three sequences predicted to be acceptable (J. Ponder, personal communication), but which have no detectable activity. Since the packing program should be too restrictive rather than too inclusive, this is a curious result. Further investigation will require the crystal structures of a number of these mutants to understand fully what changes have been produced.

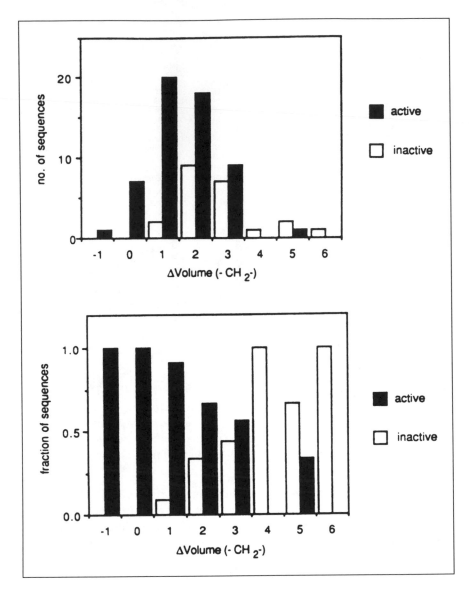

FIGURE 1–12. *Permitted under- and overpacking. Five nonpolar amino acids were each inserted at three sites of an interior packing unit in the N-terminal domain of the λ repressor, with 78 of the possible 125 sequences tested. The distribution of total volumes of these three residues is shown in the histograms in units of methylene groups (about 22 Å³). (Top) Paired bars in each bin. Filled bar is for mutants with any detectable activity. Open bar is for inactive mutants. (Bottom) Same data expressed as fraction of sequences in each bin. (Reprinted with permission from Lim and Sauer, 1991)*

Interior packing has been examined in T4 lysozyme by Karpusas et al. (1989). In an attempt to fill some cavities in the wild-type structure, Leu to Phe and Ala to Val replacements were prepared. The overall structure changed only slightly to accommodate the larger residues, but the stability was reduced slightly in each case. This was apparently due to the strain energy introduced in the χ angles of the side chains of the substituted residues required to minimize alterations in the cavity cages surrounding them. The extent and significance of the structural alterations clearly depends upon the details of the location of the mutation. Not surprisingly, required movements of the main chain are much more pervasive in their effects than are movements restricted to side chain rotations.

Varadarajan and Richards have investigated the association reaction between S-peptide and S-protein to yield the catalytically active ribonucleases. The C-terminal five residues of the 20-residue S-peptide have been known for some time to contribute little to the binding constant. A peptide comprising residues 1 through 15 with a C-terminal amide, designated S-15, has been used as the reference peptide. The 104-residue S-protein component has been used without change. The thermodynamic parameters for the binding of S-15 and of seven mutants altered at position 13 are listed in Table 1–1. Additional data collected over the temperature range of 5° to 25°C provide the heat capacity changes (Varadarajan et al., 1992). X-ray crystallographic data to 1.7 Å on the isomorphous crystals of these complexes has been collected and refined. Sections of certain of the derived models are shown in Figure 1–13.

TABLE 1–1. *Difference Thermodynamic Parameters for the Binding of S-Peptide Analogues to S-Protein Relative to the S-Peptide (1-15): S-Protein (21-124) Interaction*[a]

Peptide	$\Delta\Delta G^a$ Kcal mol^{-1}	$\Delta\Delta H^a$ Kcal mol^{-1}	$T\Delta\Delta S^b$ Kcal mol^{-1}	$\Delta\Delta Cp^b$ Kcal mol^{-1}°K^{-1}
M13A	4.0	4.7	0.7	−0.08
M13G	5.0	(−1.3)	(−6.3)	c
M13ANB	1.7	7.2	5.7	0.16
M13V	0.3	3.2	2.9	0.21
M13I	0.2	4.8	4.6	0.24
M13L	0.6	5.2	4.6	0.16
M13F	2.7	3.5	0.8	−0.03

[a] The measured values for "native" peptide S-15, which has methionine in position 13, have been subtracted from those obtained with derivatives of S-15 to get the listed differences that show the effects of these mutations.
[b] Average error estimates: ±0.2 for $\Delta\Delta G°$, ±0.8 for $\Delta\Delta H$, ±0.9 for $T\Delta\Delta S$, and ±0.1 for $\Delta\Delta Cp^b$. All measurements were made at pH 6.0 and 25°C.
[c] Not measured.
Source: Connelly et al. (1990, 1992).

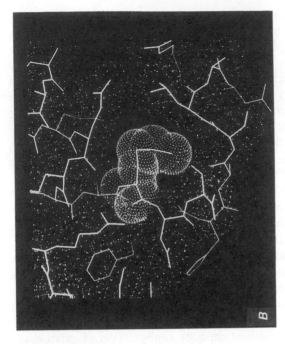

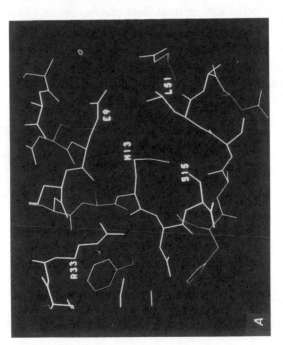

FIGURE 1–13. Sections through the structures of ribonuclease-S and some mutants in the neighborhood of residue 13. The complexes shown are of S-protein (residues 21 to 124), and of S-15 (residues 1 to 15), and of two mutants of S-15: M13V and M13ANB, where residue Met 13 has been replaced by valine and amino-n-butyric acid, respectively. (Residues 16 to 20 of normal S-peptide were deleted.) (A) shows a stick diagram of the region and identifies several residues. In subsequent panels the van der Waals surfaces of the atoms have been dotted at low density for the residues surrounding residue 13 and at high density for the atoms of residue 13 itself. The blank regions are voids or packing defects. The wild-type residue, methionine (B), does not completely fill the cavity but represents normal packing.

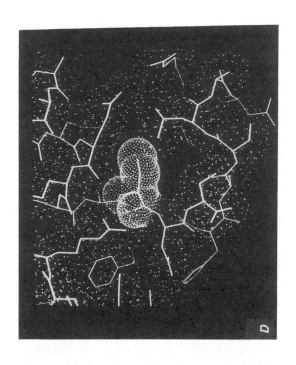

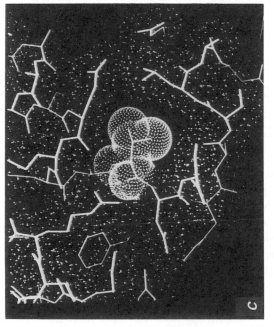

The equivalent of one methyl group is lost in going to valine (M13V) (C) and another in forming amino-n-butyric acid (M13ANB) (D). The void sizes and distributions change, and Leu 51 shifts to a different rotamer in an attempt to fill the space of the second lost methyl group in the M13ANB structure. (Reprinted with permission from Varadarajan et al., 1990)

Surprisingly large enthalpy changes can be seen in some cases where the change in free energy is very small. The heat capacity changes are quite large and cannot be explained within an order of magnitude by the differential changes in accessible area among the different mutants. It is known that upon association of S-peptide and S-protein, the hydrogen exchange rates of a large number of protons in S-protein are decreased by a factor of the order of 1,000 (Rosa and Richards, 1981). If similar changes occur with all of the mutant complexes, this alteration in the low-frequency dynamic behavior may be the source of much of the heat capacity change. It is clear that, while the hydrophobic effect may provide a driving force in this association, it is by no means the only player. No aspect of these single site changes offers easy generalizations.

The X-ray crystallographic results show that the atomic positions in a large part of all of these structures are the same within ±0.2 Å. Positional shifts as large as 1.5 Å occur in a loop region that is remote from position 13 and the surrounding residues. The origin of this movement is unclear. As found in the comparable T4 lysozyme study, the cavity in which residue 13 is located attempts to compensate for the size changes by expanding or contracting its volume. The individual atomic shifts are never greater than 0.7 Å and are much smaller for most. The expansion required to accommodate a Phe residue actually leaves larger voids than are found with the native Met. The contraction around Gly is not extensive and a water molecule appears, but there is not enough water to fill all of the empty space.

As in the Lim and Sauer (1989,1991) studies on λ repressor, it is seen that considerable changes in cavity volume in a small region can be tolerated. It is likely that such variations could not be simultaneously produced all over the protein and still have a stable folded structure. Thus, the packing quality that is characteristic of stable structures is likely to be a global one for the whole molecule, and not easily assessed in studies of single small regions.

6 INTERMEDIATE STATES

An understanding of protein folding pathways requires the identification and structural analysis of the intermediate steps in this process (Kim and Baldwin, 1982). This is perhaps the most intriguing and most difficult part of the entire field. The complexities of the starting, or unfolded, state have already been discussed. Added to those problems are the transient nature and low stoichiometric amounts to be expected for intermediate structures under most conditions. The presence of intermediate states is usually detected only in kinetic studies. Some properties of these intermediates can be inferred from the kinetic data, but little structural information can be directly derived and is usually obtained by inference. On occasion, however, conditions have been found where concentrations of presumed intermediates occur at equilibrium in amounts sufficient for direct structural study.

Certain aspects of the covalent geometry of the polypeptide chain affect and are affected by the folding reaction. The isomerization of the *cis* and *trans* conformations of peptide bonds, and especially those involving Pro residues, has a relatively high activation barrier that permits its isolation and study. The oxidation/reduction of the covalent disulfide bond falls into this category also. The biology of these reactions and their analytical use are covered in Chapters 5, 7, and 10.

A general question for all intermediates is whether they contain types of structure that are not seen in the native state and that do not fall under the definition of a true random coil? The answer is almost certainly yes, even though we cannot specify in any detail what these structures are at this time.

The analysis of helix formation is ordinarily done on the basis of the two-state hypothesis, that is, the helix is either present in its canonical form or the chain is indeed a random coil. This is also the presumption of the Zimm-Bragg theory for the transition. For long helices, the theory also predicts breaks in the helix with the appropriate statistical distribution, but the segments that are helical are presumed to be "perfect helices" and the interhelical regions "random." Sundaralingham and Sekharudu (1989) have pointed out the occurrence of water molecules attached to, or inserted into, the main chain helical hydrogen bonds in several known folded structures. They propose that these reflect possible intermediates in the normal formation or unraveling of a helix (Fig. 1–14). Can one imagine these to be present much more often in the rapidly fluctuating structures that are postulated for the earliest folding steps? CD ellipticity at about 220nm is commonly used as a general monitor of helical content. Would changes in the φ-ψ angles in a specifically hydrated structure of this type alter the CD signal? Could characteristic NOE peaks be defined for such structures as they are for the standard α-helix? Similar questions could be asked about the β-sheet conformation and indeed about all the loops as well. In each case solvent molecules would be presumed to play a central role as an actual component in such structures, if they exist. It would seem prudent at this time to keep an open mind on the possible structures of intermediates and not to be forced automatically into assuming the canonical forms of secondary structure.

The most compelling case for the preceding cautionary notes is the existence of the "molten globule" state. Although the precise definition is the subject of much debate, the existence of this state is no longer questioned. In systems where it is found at all, it is commonly, but not solely, observed in acidic solutions. In all measurements sensitive to radius of gyration or Stokes radius, the molten globule state appears to be somewhat

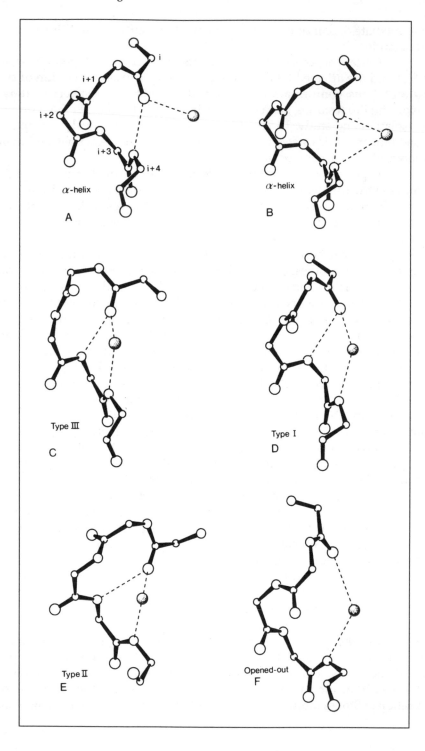

larger than the native one. The interconversion is rapid and reversible on adjustment of pH. This is quite distinct from the irreversible changes induced by acidification in the toxin and viral coat proteins. The secondary structure as derived from CD or NMR spectra is largely intact. The environment of those side chains that can be observed spectroscopically appears to be different. Insertion of substantial amounts of water into the structure seems certain. The most curious aspect of this state is that all of the enthalpy change associated with the N to U transition appears to have occurred in the N to molten globule step. Further alteration of the solvent environment will produce the molten globule to U transition, but there is apparently no heat change involved. It behaves as though the peptide were fully hydrated in the molten globule state. This is hard to picture if the signals for secondary structure are unaffected, because almost half of all of the loss of accessible surface area in the U to N transition occurs in formation of the isolated, but standard, secondary structural units. The detailed nature of MG state, or collection of states, is mysterious indeed. It is discussed in much more detail by Ptitsyn in Chapter 6.

6.1 The Hydrogen Exchange Technique

Application of the amide hydrogen exchange reaction to the detection of peptide conformation was started by Linderstrøm-Lang in the early 1950's. In recent decades it has been developed and extended into a fine art by S. W. Englander and J. J. Englander (see Englander and Kallenbach, 1984). The most recent development by Roder et al. (1988) and by Udgaonkar and Baldwin (1988) has been to combine hydrogen–deuterium exchange during folding with specific proton identification using high-resolution NMR as the analytical tool. Because of the specific proton identification, this procedure has now taken center stage in the study of folding intermediates. It is clear that, when combined with stopped-flow sample preparation procedures, information can now be obtained on conformations that only appear transiently in rate-limiting steps.

FIGURE 1–14. (opposite) Representative hydrated segments observed in protein crystal structures. (A) Water molecule bound externally only to helix carbonyl 0 atom. (B) Water molecule bound to both the carbonyl and amide groups, where the amide proton is involved in a three-center hydrogen bond. (C) Insertion of a water molecule disrupting the helix hydrogen bond and the formation of a type III turn. (D) Same as (C) but forming a type I turn. (E) Same as (C) and (D) but forming a type II turn. (F) Insertion of a water molecule into the helix resulting in the disruption of the helix and turn hydrogen bonds and forming an open turn. These hydrated segments are suggested as possible intermediates in folding. (Kindly supplied by M. Sundaralingam)

The actual hydrogen exchange measurement provides a kinetic rate constant for the exchange process. A particular proton will have an intrinsic rate for a given set of solvent conditions when that proton is maximally exposed to the solvent. The ratio of this rate constant to the measured value in the sample is referred to as the *protection factor* and is the number most directly related to structure. The underlying assumption in its use is that a hydrogen atom involved in a hydrogen bond with another protein atom cannot exchange, and that disruption of this bond and formation of a new one with a solvent molecule is essential for exchange to occur. Since a protein molecule is undergoing dynamic fluctuations all the time, even a deeply buried hydrogen atom will occasionally be available for exchange. The experimental technique is such that protection factors at least as high as 10^8 can be measured with confidence. Thus, very rare events are monitored.

Unfortunately, a given value of the protection factor can arise from a variety of different structures and so, by itself, is not a diagnostic for one particular structure. The usual assumptions are that any structure formed early during folding will resemble that of the same chain segments in the native state, and that a protection factor can be interpreted as the extent to which such a structure is formed. The protection factor for protons in an isolated α-helix could be used as a measure of helix formation. When this helix is incorporated into a higher level of structure, its dynamic fluctuations will be further dampened and its protons will show higher protection factors. If these two events are sufficiently separated on the time scale, they can be separately evaluated, as in any kinetic study of sequential reactions.

If the assumption about the native-like structure content of the intermediate is incorrect, the protection data are not affected, but the structural interpretation becomes much less certain. The hydrogen exchange behavior of some of the possible hydrated folding intermediates is not known, but would be expected to be much faster than that of the less hydrated samples because the solvent would already be incorporated as part of the structure. A detailed three-dimensional understanding of the mechanism of the base-catalyzed hydrogen exchange process in water would be most extraordinarily helpful at this time.

7 THE SOLVENT

While proteins are soluble in a number of polar organic solvents, water is the only one with direct importance to biology. After many thousands of man years of study, water continues to be a mysterious liquid with very unusual properties. This substance is described in the extensive multivolume series edited by Franks (1972 and later), and more recent reviews directed particularly at proteins have been provided by Edsall and MacKenzie (1978, 1983), by Finney (1977) and Finney et al. (1980), and by Rupley and Careri (1991).

7.1 The Interface

It is exceedingly difficult to prepare an anhydrous protein. Some water is always bound with such avidity that removing it without irreversible destruction of the protein may not be possible. A much larger amount of water is associated with the protein with weaker binding as reflected in the adsorption isotherms. All of this water has one or more properties that clearly distinguish it from the pure solvent. The estimates of the amount vary somewhat, but the rule of thumb of 0.3g H_2O/g protein is close. This corresponds roughly to the first monolayer in contact with the protein surface.

Israelachvilli (1985) and Parsegian and Rau (1984) have been documenting "nonbulk" water in certain systems that extends many molecular diameters away from the surfaces. The most dramatic example is a flat mica surface where 8 to 10 layers of ordered water can be demonstrated. Less structured, but comparable, effects can be found in dispersions of lipid bilayers and in solutions of DNA and of collagen. The effects so far have been found adjacent to extensive planar surfaces or long rigid molecules. There is no comparable evidence as of 1991 for such ordering around globular proteins. While keeping a sharp eye on further developments, it is probably more useful, in terms of the proteins discussed in this volume, to concentrate on the first monolayer of solvent.

The monolayer in globular proteins shows no thermal transition even many tens of degrees below the freezing point of bulk water. The NMR relaxation time of the water protons are different from those of the bulk. The self-diffusion coefficients are smaller. Other properties are little affected, however, at least to the extent that they can be measured; for example, the compressibility (Kundrot and Richards, 1988) and the mass density (Low and Richards, 1954) cannot be distinguished from the bulk values. (However, see Levitt and Sharon (1988), who disagree about the mass density. Part of the difficulty is in definitions.)

7.2 Solvent Structure

As the crystallographic work on proteins proceeds to higher and higher resolution, more and more attention is paid to proper inclusion of the water molecules that are an essential component of the crystal and whose removal results in a degradation of the crystalline order. Once the protein atoms have been placed with due allowance for multiple side chain conformations, there are always a plethora of residual electron density peaks in what is termed the "solvent region." Normally, repetitive cycles of refinement are undertaken where putative water molecules are placed in the positions of highest electron density not already occupied. Success is usually measured by both the value of the crystallographic R factor and the difference electron density map for the structure at that stage. When neither estimate improves over a few cycles, the refinement is considered to have converged for that data set. At this point some, but by no means all, of the water molecules known to be present in the crystal will have assigned positions. The only exception to that statement is the small protein crambin which has provided the highest resolution data for any protein crystal so far, the reflections going out to 0.8 Å. In this case it is thought that all of the water molecules, have, in fact, been located (Teeter, 1984).

For most proteins of axial ratio less than about 5 to 1, the amount of water in the crystal is between 30% and 50% by weight. This solvent is not like the water of hydration in small molecule crystalline hydrates, such as $CuSO_4.5H_2O$ where the water is just as "solid" as the salt ions. In protein crystals most, if not all, of the water is much more like a liquid than a solid. The diffusion of small molecules into and out of the crystals proceeds at a rate which, after accounting for pore restriction, is comparable to that expected for diffusion in bulk water, not to that for ice or vitreous water (Wyckoff et al., 1967; Bishop and Richards, 1968). It is thus not surprising that some of the water in the crystal cannot be tied to a specific position. On the other hand, given the liquid behavior of most of the solvent, how can one account for the apparent ability to localize any of the molecules at all? The probable answer has been provided in an analogy used by Felix Franks. The following is an abstract from his course notes:

> The structure in a solid I relate to a dinner party at which the guests sit at small, or not so small tables and they are served by waiters and have no need to get up during the dinner. Therefore, if pictures are taken at intervals all the chairs will be occupied and as the evening goes on they will be occupied by the same people. Structure in a liquid I describe in terms of a buffet supper where the chairs are also occupied, but every once in a while the guests have to get up, walk to the buffet table and replenish their plates and they might then go back and sit with someone else. Therefore, if snapshots are taken during the evening, once again all the chairs will be occupied, but not necessarily by the same people. This of course presupposes that exchange is rapid compared to the time between snapshots. One can therefore identify water molecule locations with specific coordinates but that is not to say that these positions are permanently occupied by the same water molecules.

Such movements of the "located" waters could easily permit the diffusion of larger solutes. It also permits the other bulk properties referred to earlier to resemble liquid rather than solid water.

The refinement of the assignable water structure is still a bit of an art form, with a great deal of investigator choice built in. The individual waters are assigned positions and individual isotropic B factors. These parameters are highly correlated, however, making separate refinement problematical except for very high resolution data, although this point is the subject of disagreement within the X-ray crystallography fraternity.

Except for the relatively infrequent multiple-site problem, the occupancy for protein atoms is fixed at unity. This is not true for the waters. In fact, it is common practice to set up separate nets of water molecules with the expected tetrahedral bonding fulfilled in each net. Occupancies are adjusted so that the total in any region does not exceed the mean density of water. These nets are treated as though they exist simultaneously in view of the time and space averaging represented by the crystallographic analysis. In principle there could be many such nets, but the data do not permit the use of more than two before the uncertainties are overwhelming. For the individual positions, if the occupancy is less than 0.3, one is getting very close to the noise level of all but the very best maps. In the areas where individual atoms are not assigned, the electron density in the model is given the mean value for liquid water, or one's best guess if the mother liquor contains strong concentrations of salt ions.

The result of this effort has been the placement of hundreds of water molecules in many different structures. As expected, the best-located positions are next to charged protein atoms where the electrostriction forces are very strong. Polar atoms frequently have hydrogen-bonded waters in reasonable positions. The nonpolar areas, as one might expect, are the least well ordered. The hydrogen bonding of the water molecules to each other in the nets is usually reasonable. The subjective nature of much of the fitting, however, means that one should be skeptical of any particular detail of a proposed structure, even though the overall impression is probably correct.

The location of specific water molecules by NMR is only just starting. The report by Otting and Wüthrich (1989) shows that in a favorable case NOE cross-peaks can be observed and used to confirm the presence of at least highly localized water. The bulk of the surface water may be moving and exchanging too rapidly to permit detailed structural analysis with present techniques.

7.3 Ligand Binding

In any ligand binding reaction, whether the ligand is a small molecule or another macromolecule, the formation of the complex will require the displacement of first-layer solvent molecules from both the protein and the ligand and, perhaps, the reorganization of some solvent not directly in the binding site. The amount and thermodynamic parameters of such solvent must be known in any serious attempt to calculate binding constants, even where the structures of the two components and the complex are known in detail. In favorable situations the solvent structure will be known in the binding site area for all three structures. The associated energies are much harder to determine. Experimental energies of hydration are known for some simple small molecule systems, but almost never for the protein partners. Although calorimetric procedures provide the overall free energy and enthalpy of binding, the division of these numbers into the various contributing structural factors is not possible. Perturbation-free energy calculations are, however, providing some very interesting and surprisingly accurate estimates.

If the macroscopic concept of surface tension can be used at the molecular level, one can get some qualitative ideas about the solvent–protein interface. The water–hydrocarbon interface has roughly the right magnitude to predict transfer-free energies from accessible surface areas. When such surfaces are curved with radii appropriate to the molecular scale, the chemical potential of the water is significantly affected. Thus, there will be a greater tendency of the water to leave a convex water surface than a concave one. Consequently, the displacement of water from the usual invaginated active site of an enzyme should be easier than from an equivalent site on a flat surface. This will make up in part for the difficulty of removing the convex layer around the ligand partner. (See the discussion by Sharp et al., 1991b.)

The folding problem is not different in principle from ligand binding. It just implies that the "ligand" and the "binding site" are covalently connected. The hydration layer problem will be very similar. A detailed understanding of the water–protein interface would seem to provide a major research focus during the coming years.

The thermodynamics of aqueous solutions of proteins and certain small molecules and the definition and expression of the "hydrophobic effect" are under the most intense discussion at this time. This is a most confusing area where the facts, the analysis, the definitions, the mixing of macro and micro points of view, and semantics all weigh in heavily. There is universal agreement only on the importance of the effects and on the desirability of having both clear definitions and an easily understood explanation. This subject is discussed by Privalov in Chapter 3.

REFERENCES

Abraham, D. J., and Leo, A. J. (1987) *Proteins: Struct. Funct. Genet. 2*, 130–152.

Ahmed, F., and Bigelow, C. C. (1986) *Biopolymers 25*, 1623–1633.

Alberts, B., Bray, D., Lewis, J., Raff, M., Roberts, K., and Watson, J. D. (1989) *Molecular Biology of the Cell*, 2nd ed., Garland Publishing, New York.

Anfinsen, C. B. (1973) *Science 181*, 223–230.

Ansari, A., Berendzen, J., Browne, S. F., Frauenfelder, H., Iben, I. E. T., Sauke, T. B., Shyamsunder, E., and Young, R. D. (1985) *Proc. Natl. Acad. Sci. USA 82*, 5000–5004.

Arakawa, T., Bhat, R., and Timasheff, S. N. (1990) *Biochemistry 29*, 1924–1931.

Arakawa, T., and Timasheff, S. N. (1982) *Biochemistry 21*, 6545–6552.

Baldwin, R. L. (1986a) *Proc. Natl. Acad. Sci. USA 83*, 8069–8072.

Baldwin, R. L. (1986b) *Trends Biochem. Sci. 11*, 6–9.

Baldwin, R. L. (1989) *Trends Biochem. Sci. 14*, 291–294.

Bashford, D., Chothia, C., and Lesk, A. M. (1987) *J. Mol. Biol. 196*, 199–216.

Bennett, W. S., and Steitz, T. A. (1978) *Proc. Natl. Acad. Sci. USA 75*, 4848–4852.

Bernal, J. D., and Crowfoot, D. (1934) *Nature 133*, 794–795.

Bernstein, F. C., Koetzle, T. F., Williams, G. J. B., Meyer, E. F., Brice, M. D., Rodgers, J. R., Kennard, O., Shimanouchi, T., and Tasumi, M. (1977) *J. Mol. Biol. 112*, 535–542.

Bishop, W. H., and Richards, F. M. (1968) *J. Mol. Biol. 38*, 315–328.

Bowie, J. U., Lüthy, R., and Eisenberg, D. (1991) *Science, 253*, 164–170.

Bowie, J. U., Reidhaar-Olson, J. F., Lim, W. A., and Sauer, R. T. (1990) *Science 247*, 1306–1310.

Bränden, C., and Tooze, J. (1991) *Introduction to Protein Structure,* Garland Publishing, New York.

Brandts, J. F. (1964) *J. Am. Chem. Soc. 86*, 4302–4314.

Brown, J. E., and Klee, W. A. (1971) *Biochemistry 10*, 470–476.

Chadwick, D., and Widdows, K. (eds.) (1991) *CIBA Found. Symp.* no. 161, Wiley and Sons, New York.

Chambers, J. L., and Stroud, R. M. (1979) *Acta Cryst. B35*, 1861–1874.

Chan, H. S., and Dill, K. A. (1990) *Proc. Natl. Acad. Sci. USA 87*, 6388–6392.

Chan, H. S., and Dill, K. A. (1991) *Ann. Rev. Biophys. Chem. 20*, 447–490.

Chandler, D., Weeks, J. D., and Andersen, H. C. (1983) *Science 220*, 787–794.

Chothia, C. (1974) *Nature 248*, 338–339.

Cohn, E. J., and Edsall, J. T. (1943) *Proteins, Amino Acids, and Peptides,* Reinhold Publishing, New York.

Connelly, P. R., Varadarajan, R., Sturtevant, J. M., and Richards, F. M. (1990) *Biochemistry 29*, 6108–6114.

Connelly, P. R., Varadarajan, R., Sturtevant, J. M., and Richards, F. M. (1992) *Biochemistry,* submitted.

Connolly, M. L. (1983) *J. Appl. Cryst. 16*, 548–558.

Creighton, T. E. (1983) *Proteins: Structures and Molecular Properties,* W. H. Freeman, New York.

Crowfoot, D., and Riley, D. (1939) *Nature 144*, 1011–1012.

Dao-pin, S., Sauer, U., Nicholson, H., and Matthews, B. W. (1991a) *Biochemistry* (in press).

Dao-pin, S., Söderlind, E., Baase, W. A., Wozniak, J. A., Sauer, U., and Matthews, B. W. (1991b) *J. Mol. Biol.* (in press).

Dill, K. A. (1985) *Biochemistry 24*, 1501–1509.

Dill, K. A. (1990) *Biochemistry 29*, 7133–7155.

Dill, K. A., Alonso, O. V., and Hutchinson, K. (1989) *Biochemistry 28*, 5439.

Dill, K. A., and Shortle, D. (1991) *Ann. Rev. Biochemistry 60*, 795–826.

Drexler, K. E. (1981) *Proc. Natl. Acad. Sci. USA 78*, 5275–5278.

Dumont, M. E., and Richards, F. M. (1988) *J. Biol. Chem. 263*, 2087–2097.

Dyson, H. J., Rance, M., Houghton, R. A., Lerner, R. A., and Wright, P. E. (1988) *J. Mol. Biol. 201*, 161–200.

Edsall, J. T., and McKenzie, H. A. (1978) *Adv. Biophys. 10*, 137–207.

Edsall, J. T., and McKenzie, H. A. (1983) *Adv. Biophys. 16*, 53–183.

Eisenberg, D., and McLachlan, A. D. (1986) *Nature 319*, 199–203.

Elber, R., and Karplus, M. (1987) *Science 235*, 318–321.

Englander, S. W., and Kallenbach, N. R. (1984) *Quart. Rev. Biophys. 16*, 521–625.

Faber, H. R., and Matthews, B. W. (1990) *Nature 348*, 263–266.

Fasman, G. D. (ed.) (1989) *Prediction of Protein Structure and the Principles of Protein Conformation*, Plenum, New York.

Finney, J. L. (1975) *J. Mol. Biol. 96*, 721–732.

Finney, J. L. (1977) *Phil. Trans. R. Soc. Lond. B278*, 3–32.

Finney, J. L., Gellatly, B. J., Galton, I. C., and Goodfellow, J. (1980) *Biophys. J. 32*, 17–31.

Finzel, B. C., Kimatian, S., Ohlendorf, D. H., Wendoloski, J. J., Levitt, M., and Salemme, R. (1990) in *Crystallographic and Modeling Methods in Molecular Design* (S. Ealick and C. Bugg, eds.), Springer Verlag, New York, pp. 175–189.

Franks, F. (1972–1982) *Water—A Comprehensive Treatise*, seven volumes, Plenum, New York.

Frauenfelder, H. (1985) *Structure and Motion: Membranes, Nucleic Acids and Proteins*, Adenine, Guilderland, New York.

Fruton, J. S. (1972) *Molecules and Life*, John Wiley & Sons, New York.

Ghelis, C., and Yon, J. (1982) *Protein Folding*, Academic Press, New York.

Gibrat, J-F., and Gō, N. (1990) *Proteins: Struct. Funct. Genet. 8*, 258–279.

Gierasch, L. M., and King, J. (eds.) (1990) *Protein Folding: Deciphering the Second Half of the Genetic Code*, American Association for the Advancement of Science, Washington, D.C.

Gregoret, L. M., and Cohen, F. E. (1991) *J. Mol. Biol. 219*, 109–122.

Hansch, C., and Leo, A. J. (1979) *Substituent Constants for Correlation Analysis in Chemistry and Biology*, John Wiley & Sons, New York.

Harrison, S. C., and Durbin, R. (1985) *Proc. Natl. Acad. Sci. USA 82*, 4028–4030.

Harte, W. E. Jr., Swaminathan, S., Mansuri, M. M., Martin, J. C., Rosenberg, I. E., and Beveridge, D. L. (1991) *Proc. Natl. Acad. Sci. USA 87*, 8864–8868.

Hsu, C. S., and Chandler, D. (1978) *Molecular Physics 36*, 215–224.

Huber, R. (1979) *Trends Biochem. Sci. 4*, 271–276.

Israelachvilli, J. N. (1985) *Intramolecular and Surface Forces: with Applications to Colloidal and Biological Systems*, Academic Press, Orlando, Fla.

Jaenicke, R. (ed.) (1980) *Protein Folding*, Elsevier/North-Holland Biomedical, New York.

Janin, J., Wodak, S., Levitt, M., and Maigret, B. (1978) *J. Mol. Biol. 125*, 357–386.

Jordan, S. R., and Pabo, C. O. (1988) *Science 242*, 893–899.

Judson, H. F. (1979) *The Eighth Day of Creation*, Simon & Schuster, New York.

Karplus, M., Ichiye, T., and Pettitt, B. M. (1987) *Biophysical J. 52*, 1083–1085.

Karpusas, M., Baase, W. A., Matsumura, M., and Matthews, B. W. (1989) *Proc. Natl. Acad. Sci. USA 86*, 8237–8241.

Katchalski, E., and Sela, M. (1958) *Adv. Protein Chem. 13*, 243–492.

Katz, B., and Kossiakoff, A. A. (1990) *Proteins: Struct. Funct. Genet. 7*, 343–357.

Kelley, R. F., and Stellwagen, E. (1984) *Biochemistry 23*, 5095–5102.

Kendrew, J. C., Bodo, G., Dintzis, H. M., Parrish, R. G., and Wyckoff, H. (1958) *Nature 181*, 662–666.

Kitaigorodsky, A. I. (1961) *Organic Chemical Crystallography*, Consultant's Bureau, New York. (Original Russian edition 1955.)

Kim, Peter S., and Baldwin, R. L. (1982) *Ann. Rev. Biochem. 51*, 459–489.

Koshland, D. E., Nemethy, G., and Filmer, D. (1966) *Biochemistry 5*, 365–385.

Kundrot, C. E., and Richards, F. M. (1988) *J. Mol. Biol. 200*, 401–410.

Lee, B. (1991) *Proc. Natl. Acad. Sci. USA 88*, 5154–5158.

Lee, B., and Richards, F. M. (1971) *J. Mol. Biol. 55*, 379–400.

Levitt, M., and Sharon, R. (1988) *Proc. Natl. Acad. Sci. USA 85*, 7557–7561.

Lim, W. A., and Sauer, R. T. (1989) *Nature 339*, 31–36.

Lim, W. A., and Sauer, R. T. (1991) *J. Mol. Biol. 219*, 359–376.

Linderstrom-Lang, K. (1955) *The Chemical Society Symposium on Peptide Chemistry*, Special Publication no. 2, London.

Livingstone, J. R., Spolar, R. S., and Record, M. T. (1991) *Biochemistry 30*, 4237–4244.

Low, B. W., and Richards, F. M. (1954) *J. Am. Chem. Soc. 76*, 2511–2518.

McCammon, J. A., Gelin, B. R., Karplus, M., and Wolyner, P. G. (1976) *Nature 262*, 325–326.

Marqusee, S., Robbins, V. H., and Baldwin, R. L. (1989) *Proc. Natl. Acad. Sci. USA 86*, 5286–5290.

Murray-Brelier, A., and Goldberg, M. E. (1989) *Proteins: Struct. Funct. Genet. 6*, 395–404.

Noguti, T., and Gō, N. (1989) *Proteins: Struct. Funct. Genet. 5*, 97–138.

Northrup, S. H., Wensel, T. G., Meares, C. F., Wendoloski, J. J., and Matthew, J. B. (1990) *Proc. Natl. Acad. Sci. USA 87*, 9503–9507.

Nozaki, Y., and Tanford, C. (1971) *J. Biol. Chem. 246*, 2211–2217.

Oncley, J. L. (1949) *Conference on the Preservation of the Cellular and Protein Components of Blood*, American National Red Cross, Washington, D.C.

Osterhout, J. J. Jr., Baldwin, R. L., York, E. J., Stewart, J. M., Dyson, H. J., and Wright, P. E. (1989) *Biochemistry 28*, 7059–7064.

Otting, G., and Wüthrich, K. (1989) *J. Am. Chem. Soc. 111*, 1871–1875.

Pabo, C. O., and Lewis, M. (1982) *Nature 298*, 443–447.

Parsegian, V. A., and Rau, D. C. (1984) *J. Cell Biol. 99*, 1965–2005.

Platzer, K. E. B., Anathanarayanan, V. S., Andreatta, R. H., and Scheraga, H. A. (1972) *Macromolecules 5*, 177–187.

Ponder, J. W., and Richards, F. M. (1987) *J. Mol. Biol. 193*, 775–791.

Privalov, P. L. (1979) *Adv. Protein Chem. 33*, 167–241.

Privalov, P. L., and Gill, S. J. (1988) *Adv. Protein Chem. 39*, 191–234.

Privalov, P. L., and Gill, S. J. (1989) *Pure Appl. Chem. 61*, 1097–1104.

Ramachandran, G. N., and Sasisekharan, V. (1968) *Adv. Protein Chem. 23*, 284–438.

Reidhaar-Olson, J. F., and Sauer, R. T. (1988) *Science 241*, 53–57.

Richards, F. M. (1974) *J. Mol. Biol. 82*, 1–14.

Richards, F. M. (1977) *Ann. Rev. Biophys. Bioeng. 6*, 151–176.

Richards, F. M. (1985) *Methods Enzymol. 115*, 440–464.

Richmond, T. J. (1984) *J. Mol. Biol. 178*, 63–89.

Roder, H., Elove, G. A., and Englander, S. W. (1988) *Nature 335*, 700–704.

Rosa, J. J., and Richards, F. M. (1981) *J. Mol. Biol. 145*, 835–851.

Rose, G. D., Geselowitz, A. R., Lesser, G. J., Lee, R. H., and Zehfus, M. H. (1985) *Science 229*, 834–838.

Rould, M. A., Perona, J. J., Söll, D., and Steitz, T. A. (1989) *Science 246*, 1089–1212.

Rupley, J. A., and Careri, G. (1991) *Adv. Protein Chem. 41*, 37–172.

Sanger, F. (1952) *Adv. Protein Chem. 7*, 1–67.

Schulz, G. E. (1991) *CIBA Foundation Symposium on Protein Conformation* No. 161, Wiley and Sons.

Schulz, G. E., and Schirmer, R. H. (1979) *Principles of Protein Structure*, Springer-Verlag, New York.

Shalongo, W., Ledger, R., Jagannadham, M. V., and Stellwagen, E. (1987) *Biochemistry 26*, 3135–3141.

Sharp, K. A., Nicholls, A., Friedman, R., and Honig, B. (1991a) *Biochemistry* (in press).

Sharp, K. A., Nicholls, A., Friedman, R., and Honig, B. (1991b) *Science 252*, 106–109.

Shrake, A., and Rupley, J. A. (1973) *J. Mol. Biol. 79*, 351–371.

Singer, M., and Berg, P. (1991) *Genes & Genomes*, University Science Books, Mill Valley, Calif.

Smith, J., Hendrickson, W. A., Honzatko, R. B., and Sheriff, S. (1986) *Biochemistry 25*, 5018–5027.

Sprang, S. R., Acharya, K. R., Goldsmith, E. J., Stuart, D. I., Varvill, K., Fletterick, R. J., Madsen, N. B., and Johnson, L. N. (1988) *Nature 336*, 215–221.

Stegmann, T., Doms, R. W., and Helenius, A. (1989) *Ann. Rev. Biophys. Biophys. Chem. 18*, 187–211.

Stryer, L. (1988) *Biochemistry*, 3rd ed., W. H. Freeman, New York.

Sundaralingam, M., and Sekharudu, Y. C. (1989) *Science 244*, 1333–1337.

Tanford, C. (1968) *Adv. Protein Chem. 23*, 121–282.

Tanford, C. (1970) *Adv. Protein Chem. 24*, 1–95.

Teeter, M. M. (1984) *Proc. Natl. Acad. Sci. USA 81*, 6014–6018.

Udgaonkar, J. B., and Baldwin, R. L. (1988) *Nature 335*, 694–699.

Varadarajan, R., Richards, F. M., and Connelly, P. R. (1990) *Current Science 59*, 819–824.

Varadarajan, R., Connelly, P. R., Sturtevant, J. M., and Richards, F. M. (1992) *Biochemistry 31*, 1421–1426.

Vasquez, M., Nemethy, G., and Scheraga, H. A. (1983) *Macromolecules 16*, 1043–1049.

Wang, J., and Steitz, T. A. (1991) Personal communication.

Watson, J. D., Hopkins, N. H., Roberts, J. W., Steitz, J. A., and Weiner, A. M. (1987) *Molecular Biology of the Gene*, 4th ed., W. A. Benjamin, Menlo Park, CA.

Wetlaufer, D. (1984) *The Protein Folding Problem in AAAS Selected Symposium 89*, Westview Press for American Academy of Arts and Sciences, Boulder, CO.

Wetlaufer, D. (1990) *Trends Biochem. Sci. 15*, 414–415.

Wyckoff, H. W., Doscher, M., Tsernoglou, D., Inagami, T., Johnson, L. N., Hardman, K. D., Allewell, N. M., Kelly, D. M., and Richards, F. M. (1967) *J. Mol. Biol. 27*, 563–578.

Zubay, G. (1988) *Biochemistry*, Macmillan, New York.

2

Protein Structures: The End Point of the Folding Pathway

JANET M. THORNTON

1 INTRODUCTION

The data base of protein structures has grown rapidly in recent years so that there are now over 400 sets of coordinates in the Brookhaven Databank (Bernstein et al., 1977). Many of these structures have been solved to resolutions of better than 2 Å and have been highly refined, allowing accurate definition of the conformation and of atom·atom interactions. From this wealth of data it seems reasonable to ask if we can extract any clues that will throw light on protein folding pathways. This problem has some similarities to the task that astronomers face of attempting to discover the origins of the universe from present-day observations. We are more fortunate, however, in that we can replay protein folding at will, as described elsewhere in this volume. In this chapter, a summary will be given of the salient points derived from observing the end point of the folding pathway-the native three-dimensional structure. The problem will be addressed by proceeding from local conformational parameters and interactions through to tertiary structure and topology. Such observations can only suggest important factors in folding. The elucidation of the pathways must await the experimental data now being obtained using modern techniques.

2 DIHEDRAL ANGLES

Most of the conformational flexibility of a protein during folding derives from rotations about the main chain φ,ψ dihedral angles and the side chain χ angles. It has long been known that φ,ψ rotations are hindered by interactions of the adjacent residues (Ramachandran et al., 1963). This restriction in rotation is compounded for all residues except Gly by steric clashes with the side chain C_β atoms. What is the extent of this steric hindrance? Many of the early protein structures showed dispersed φ,ψ plots, but with increasing resolution and improvement in refinement techniques, the φ,ψ distributions have become much more tightly clustered (See Fig. 2–1A). Using the observed φ,ψ distribution (Morris et al., 1992), three CORE regions can be defined in φ,ψ space, corresponding to the α,β and α_L regions (Fig. 2–1B). Together these comprise only 11% of the φ,ψ space. For high-resolution, well-refined structures, more than 90% of the residues on average lie in these CORE regions.

FIGURE 2–1. Ramachandran maps of known protein structures. (A) φ,ψ distribution for all residues, except Gly and Pro, taken from 310 proteins in the Brookhaven Protein Databank.

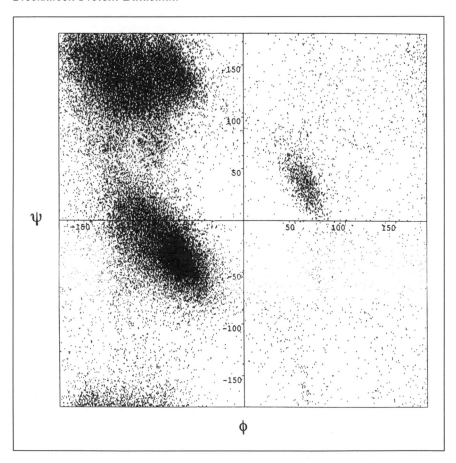

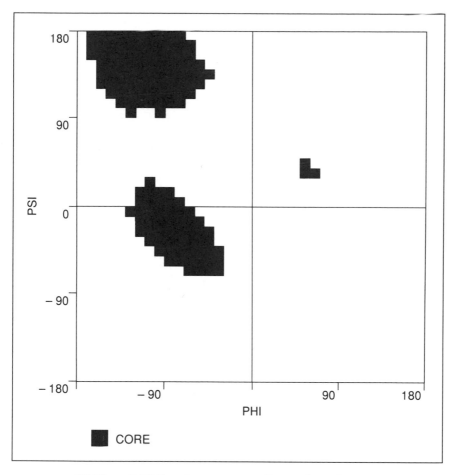

FIGURE 2–1. (B) *Three CORE regions corresponding to the* α, β, *and* α_L *regions in* φ,ψ *space. These were defined from the observed distributions shown in (A).*

Similarly, the side chain χ_1 angles adopt one of three preferred conformers: g^- (+60°), t (+180°), or g^+ (−60°) (Janin et al., 1978). As crystallographic refinement proceeds, the χ_1 distribution generally becomes more tightly clustered into these three idealized energy wells. To calculate a measure of clustering for a whole protein, the standard deviation of χ_1 values from their average can be calculated for each of the three conformers and then pooled. For high-resolution, well-resolved proteins, the standard deviation of pooled χ_1 values was only 15.7° (Morris et al., 1992). Figure 2–2 illustrates the correlation of this value with the resolution of the structure.

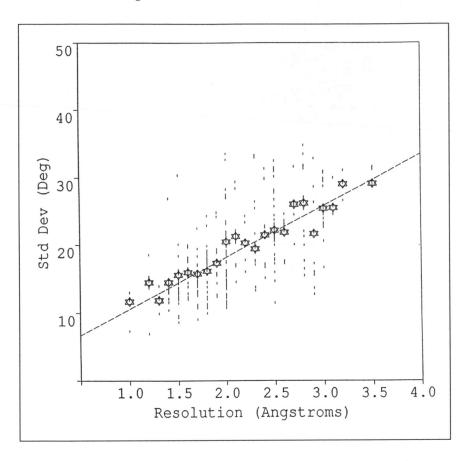

FIGURE 2–2. *Pooled standard deviations of χ_1 values from the classic g^-, g^+, t rotamers plotted against the resolution of the protein structure. The points are for individual proteins; the six-pointed stars are the average for a particular resolution.*

These results are very striking. They suggest that despite the abundance of inter-residue contacts in the protein core, including both van der Waals and hydrogen-bonding interactions, the local conformational parameters still succeed in achieving low-energy states. Very few residues adopt conformations that are "internally" unfavorable. These data suggest that during folding individual residues will be in favorable φ, ψ, and χ_1 states for most of the time. Perhaps unfavorable local conformers are only stabilized in the later stages of folding, when specific side chain tertiary interactions begin to form.

3 GLYCINE RESIDUES

Glycine residues provide a special role in structure and folding for two reasons:

- They can adopt a much wider range of φ,ψ angles than can the other amino acids, especially positive values of φ.
- The lack of a C^β atom allows them to fit in where other residues would be too bulky.

In homologous structures, some Gly residues are highly conserved because no other amino acid can fulfil these two criteria. The φ,ψ plot for all Gly residues in high-resolution structures is shown in Figure 2–3. Surprisingly, although the Gly residue in principle is free to adopt most φ,ψ values, it has a strong preference for $\varphi = +90°$, $\psi = 0°$ (Nicholson et al., 1989). Gly residues are energetically rather a liability because it is more expensive entropically to fix a Gly residue into a single conformation, relative to any other residue. During folding, however, Gly residues do give increased flexibility that facilitates exploration of 3D space and is probably advantageous for folding.

4 PROLINE RESIDUES

Proline residues affect protein conformation in three ways:

- The relatively high intrinsic probability of having a *cis* peptide bond preceding Pro residues (Brandts et al., 1975).
- The lack of a free NH group preventing main chain hydrogen bonding.
- The effect of the conformation of the preceding residue, making it favor strongly the β conformation (Schimmel and Flory, 1968).

These effects are very apparent in the final folded structures of proteins. In the databank, 5.7% of Pro residues have *cis* peptide bonds, compared to less than 0.5% of non-Pro residues. Furthermore, the occurrence of the *cis* peptide bond reflects the amino acid identity of the preceding residue (MacArthur and Thornton, 1991); almost 20% of Tyr-Pro sequences adopt the *cis* peptide conformation. Over half of these exhibit a definite aromatic·Pro interaction (Fig. 2–4), which Hetzel and Wüthrich (1979) suggested may be involved in stabilizing the *cis* conformer.

A further effect of Pro on the ω angle is also apparent in protein structures. The cyclic nature of the side chain reduces the electron delocalization in the peptide bond, making it longer than normal (1.36 Å instead of 1.33 Å) (Kartha et al., 1974). Such an effect will be difficult to detect in proteins, where the bond lengths are restrained and, even in well-refined

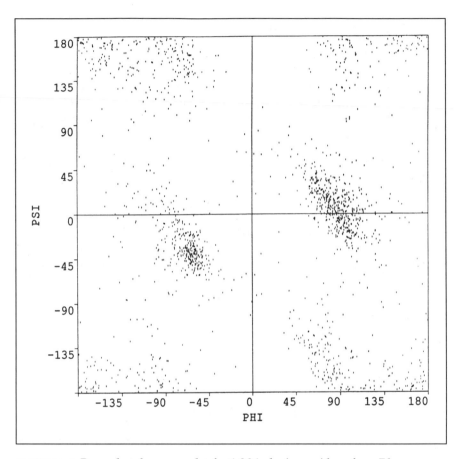

FIGURE 2–3. *Ramachandran* φ,ψ *plot for 1,294 glycine residues from 78 nonhomologous proteins determined to better the 2.5 Å resolution.*

structures, have standard deviations of 0.02 Å. Bond angles are more variable, however, and while all other ω angles have standard deviations averaging ±4.6°, this value rises to ±6.9° in *trans* X-Pro peptide bonds.

In general, the φ,ψ conformation of a residue within a polypeptide chain is independent of adjacent residues. However, Schimmel and Flory (1968) used energy calculations to show that the space available to the residue preceding Pro is severely curtailed by steric conflicts between the $C^{\delta}H_2$ attached to the imide nitrogen and the NH and $C^{\beta}H_2$ atoms of the preceding residue. Inspection of the ψ conformation of residue X in X-Pro (Fig. 2–5B) confirms Schimmel and Flory's prediction, with less than 10% occurring in the α region (MacArthur and Thornton, 1991). In the small X-Pro group in the α region, 85% are in 3_{10} or α-helices where the hydrogen bond network

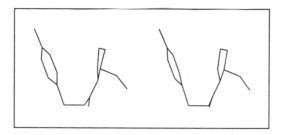

FIGURE 2–4. *Conformation of residues Tyr 94 Pro95 in adenylate kinase, illustrating the stacking interaction between Tyr and Pro that can occur with a* cis *but not a* trans *peptide bond. The figure shows a stereo pair.*

presumably compensates for the unfavorable steric interaction. The consequences of this effect, combined with the lack of a free -NH group, is that Pro residues rarely occur in the center of an α-helix. Multiple Pro residues, as in the sequence -Pro-X-X-Pro-, effectively prevent the formation of an α-helix (MacArthur and Thornton, 1991). These combined effects render Pro a very special residue from the protein folding viewpoint. The slow rate of *cis,trans* isomerization is often the rate-limiting step in folding. (Chapter 5). The other effects, while not so apparent in folding experiments, must also be considered.

5 SIDE CHAIN INTERACTIONS

Because the sequence of a protein determines its fold, the interactions of the side chains with each other and with the main chain are critical. High-resolution protein structures provide a knowledge base from which can be derived the preferred interactions and their geometries (Singh and Thornton, 1990). This is done by extracting all interacting PAIRS of a given type (e.g., Tyr·Pro) from the data base and superposing each PAIR onto a reference amino acid, Tyr in this example. This generates a three-dimensional distribution like that shown in Figure 2–6. Such distributions have been generated for all 400 amino acid PAIR types.

What do these PAIR interactions tell us? It is immediately apparent that the geometric specificity between different types of side chains is very different. Figure 2–6 gives contrasting examples. The arginine·carboxylate interaction (Fig. 2–6A) is very directional, with a strong cluster where two hydrogen bonds can be made simultaneously. Theoretical energy evaluations show that this twin interaction corresponds to the global energy minimum for an arginine·carboxylate interaction (J. Mitchell, private communication). The Tyr·Pro distribution (Fig. 2–6B) is less specific, but still shows a distinct preference for the Pro residue to lie above the face of the aromatic ring and to stack. The exact position is not defined (cf. Arg·COO⁻ of Fig. 2–6A),

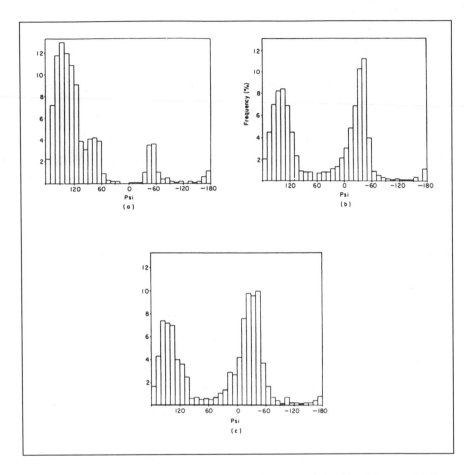

FIGURE 2–5. Distribution of ψ angles. (A) X in X-Pro. (B) All residues excluding Pro and Gly. (C) X in Pro-X, excluding X = Gly.

FIGURE 2–6. (opposite) Side chain PAIR distributions, derived from 62 high-resolution protein structures. For each PAIR, a stereo plot shows all interacting pairs in their observed geometry, but superposed on a REFERENCE side chain. Atom nomenclature is Nitrogen *, Oxygen ○, Carbon •. (A) Arg·Asp carboxylate, superposed upon the reference Arg guanidinium group. (B) Tyr·Pro PAIR distribution, superposed on the reference Tyr and viewed in the plane of the ring. (C) Val·Val PAIR distribution, plotting the positions of the C^β, $C^{\gamma 1}$, and $C^{\gamma 2}$ atoms.

A

B

C

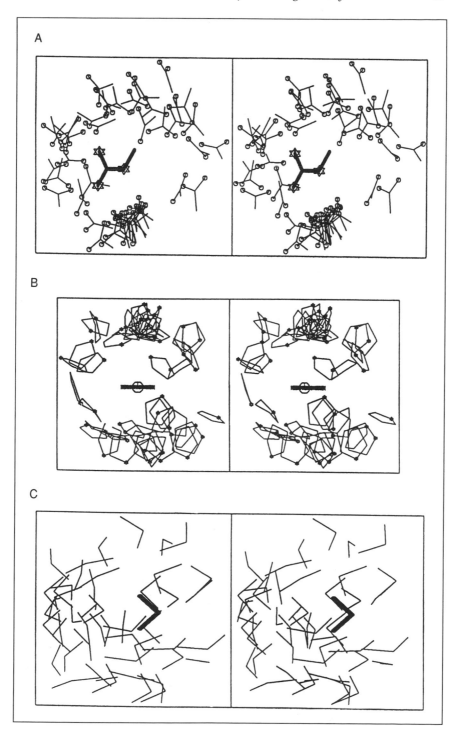

however, and different staggers are observed. At the opposite end of the spectrum lies a typical hydrophobic interaction (Val·Val, Fig. 2–6C) which, although common, shows no preferred interaction geometries.

In simplistic terms, the hydrophobic side chains provide the ubiquitous nonspecific "glue," while the polar groups provide specificity through hydrogen bonding. The aromatics are intermediate. Thus, the "simple" view of packing is upheld by looking at these interactions in detail. During folding, the collapsed "molten globule" (see Chapter 5) may well be stabilized by these nonspecific hydrophobic interactions and local hydrogen bonding, while the specific long-range tertiary hydrogen bonds, observed in the native fold, appear later.

A rather strange observation can be made regarding local side chain···· main chain or side chain····side chain interactions. It has often been suggested that many of these help to stabilize the local structure. For example, side chain···· main chain $i·i+4$ salt bridges have been implicated in α-helix stability (Marqusee and Baldwin, 1987). Similarly, in a β-turn the extra side chain····main chain hydrogen bond between residues i and $i+2$ may provide extra stability (Wilmot and Thornton, 1988). In both of these examples, sequence analyses show a clear preference for the residue types involved (e.g., oppositely charged residues at site i and $i+4$ in a helix). Inspection of the crystal structures, however, shows that only about half of these hydrogen bonds are formed. Often these residues, which are in an ideal position to stabilize the local interaction, are involved in tertiary contacts. Perhaps these residues form "embryonic" local interactions in the folding protein to stabilize the nascent secondary structure, but once this is formed, other tertiary contacts may become more favorable.

6 SECONDARY STRUCTURE

6.1 α-helices

Helices and sheets dominate proteins, with approximately 60% of all residues forming such structures. It has long been suggested that they could be involved in the initial stages of folding, and recent experimental evidence certainly supports the early formation of α-helices (Roder et al., 1988; Miranker et al., 1991). In well-resolved protein structures, the helical φ,ψ angles are very tightly clustered ($\varphi = 65.3 \pm 11.9°$; $\psi = -39.4 \pm 11.3°$). The helix is often distorted, however, being sometimes curved and sometimes kinked (Barlow and Thornton, 1988). The origin of such distortions may reflect the local sequence (e.g., Pro residues), but more often is a result of the influence of asymmetric solvation and tertiary packing effects. The termini of helices are often distorted, especially at the C-terminus, where the α-helix usually tightens to form a 3_{10} helix in the last turn.

6.2 β-strands and β-sheets

β-strands are much more distorted than are α-helices, showing gross twisting and kinking, plus the formation of β-bulges. The β-region of the Ramachandran plot is very broad, allowing large variations in local geometry with little energy cost (Fig. 2–1A). The formation of a β-sheet requires long-range interactions, very different from an α-helix, which involves interactions only between residues that are local in the sequence. The chirality of β-strands and β-sheets results from the chirality of the *L*-amino acids. This gives an asymmetric φ,ψ plot and asymmetric inter-strand contacts between C^{β} atoms (Chou et al., 1982). Consequently, the β-strand always has a right-handed twist when viewed along the strand direction, while adjacent strands in a sheet show a left-handed rotation looking perpendicular to the strand direction (Chothia, 1973).

6.3 β-turns

β-turns are widespread in proteins. They are formed when the polypeptide chain makes a sharp 180° reversal and are best identified using the definition suggested by Lewis et al. (1973), i.e., the distance between C^{α_i} and $C^{\alpha_{i+3}}$ must be less than 7 Å. These turns are usually on the surface of the protein; they involve only four residues and are therefore local interactions. They are classified by the φ,ψ values of the two central residues *i*+1 and *i*+2. The classic turn types (I, II, I′, and II′) were first identified by Venkatachalam (1968). About half of the observed turns are stabilized by a $CO_i \cdots HN_{i+3}$ hydrogen bond. Type I turns may also have an additional side chain $\cdots$ main chain $O_i \cdots HN_{i+2}$ hydrogen bond (Wilmot and Thornton, 1988).

Pro is commonly residue *i*+1, probably for three reasons. First, the φ angle is restricted to −60°, which is perfect for both the classic Type I and II turns; second, it does not have a free −NH group requiring a hydrogen bond acceptor partner; and last, it strongly disfavors α-helix or β-sheet formation, except at their margins. The Type I′ and II′ mirror image β-turns are rare and occur predominantly in β-hairpins (Sibanda et al., 1989). In the Type I′ turn, the twist of the sheet is matched by the twist of this unusual turn. Both of these types of turn require a residue with a positive φ value and so favor the presence of a Gly residue.

Conformationally, turns are not very demanding with respect to sequence. For example, the Type I turn has both central residues in the α conformation, which can be achieved by almost any residue pair. Indeed an α-helix could be considered as a collection of successive Type I turns. The constraints imposed by solvation are probably more important. In a turn there are usually four unsatisfied main chain polar groups (NH_{i+1}, CO_{i+1}, NH_{i+2}, CO_{i+2}). Unless this turn becomes part of a helix, these groups must either be solvated or satisfied by other protein atoms. Although there are a few buried turns where protein atoms provide such an environment (Rose et al., 1983), the vast majority of turns lie on the surface and are surrounded by water. Therefore, it is the hydrophilic amino acids, along with Pro and Gly, that occur most often in β-turns.

The role of turns in protein folding is unknown. Most proteins can be simplistically represented as a collection of secondary structures, which traverse the "core" and pack together, connected by loops on the surface that often contain turns. As such, the turns could be considered as the crucial kink points in the chain that allow the formation of a compact globular core. Since they are "local interactions," they should be kinetically favored and may form early in folding. Recent experimental studies found evidence for turn formation in the early stages of folding (Matouschek et al., 1990).

6.4 Helix and Strand Propensities

What are the factors that determine the preferences of the different amino acids to fold into helix structures, and how are these factors different in peptides/polymers compared to proteins? All amino acid residues, even Pro, are able to adopt an α or β conformation. Their different propensities in proteins must therefore derive from subtle effects, including at least the following:

- Preference for the α or β region of the Ramachandran plot (Fig. 2–1)
- Restriction of side chain conformation by helix formation (entropic effect)
- Local side chain···· main chain interactions that interrupt helix formation
- Side chain····side chain interactions that may stabilize helix or strand formation [adjacent ($i\pm1$) or next turn of the helix ($i\pm3,4$)]
- Effects and N- or C-terminus (N-cap and C-cap residues, respectively)
- Solvation
- Tertiary long-range interactions and packing effects

The requirement to be part of a stable globular structure in proteins will add additional constraints not seen in peptides. What information on these effects can be derived from structural data?

6.4.1 Preferences for α-helix or β-strand
Using simple statistical counting of the known structures, the propensities of the different amino acids for forming α-helix and β-strand conformations have been calculated many times (Chou and Fasman, 1978; Garnier et al., 1978). These values are often compared to similar propensities derived from

solution studies on folding using peptides or host-guest polymers. (For example, see Lyu et al., 1991.)

6.4.2 Side Chain Conformations in the α-Helix

Comparison of χ_1 values of residues in an α-helix with those found in β-strands (McGregor et al., 1987) highlights the restrictions imposed on side chain conformation by helix formation. Two effects are noticeable. First, for most residues (excepting short polar side chains), the gauche− χ_1 conformer becomes effectively forbidden by steric clashes between C_γ of residue i and the carbonyl oxygen of residue i–3 on the helix (Fig. 2–7). Alanine has no C_γ atom, which may be one factor in determining its high helix propensity.

Second, the three g^+, g^-, and t conformers correspond to χ_1 values of −60°, +60°, and 180° for optimal staggering of substituents on the C_α and C_β atoms. In the most common conformer, g^+, adopted by 57% of all residues in α-helices (Janin et al., 1978), the average χ_1 value is notably shifted by about 8° from −60° and from the value observed in β-sheets or coils (Fig. 2–8). This must introduce a strain energy, which will only be absent in Gly and Ala residues. The branched β-carbon amino acids (Thr, Val, and Ile) are special because they cannot relieve clashes with the previous turn by changing χ_1. Consequently, their χ_1 values remain close to −60°, but the evidence from the other amino acids suggests that they must be under considerable strain. Perhaps this also explains why these amino acids are relatively poor helix formers.

6.4.3 Local Side Chain····Main Chain Interactions

Some of the polar amino acid residues, especially Asp and Asn, often form hydrogen bonds between their side chains and local main chain atoms (Baker and Hubbard, 1984). These interactions effectively prohibit helix formation and probably explain why Asp and Asn are poor helix formers. In contrast, Ser and Thr residues often form a back hydrogen bond, OH_i····$O_{i-3,4}$, with the carbonyl oxygen atoms on the previous turn of the helix (Gray and Matthews, 1984). This presumably serves to stabilize the helix and to satisfy hydrogen bonding requirements of both the side chain and the main chain. This appears to be particularly successful in transmembrane helices, where Ser and Thr residues are very common.

6.4.4 Side Chain····Side Chain Interactions

Are there any side chain····side chain interactions that help to stabilize the α-helix? The classic example is the ion-pair between residues i and $i\pm3,4$ that is observed in helical sequences. Solution studies (Marqusee and Baldwin, 1987) suggest that $i,i+4$ salt bridges do stabilize helix formation, but that $i,i+3$ do not. Protein structural data support this conclusion;

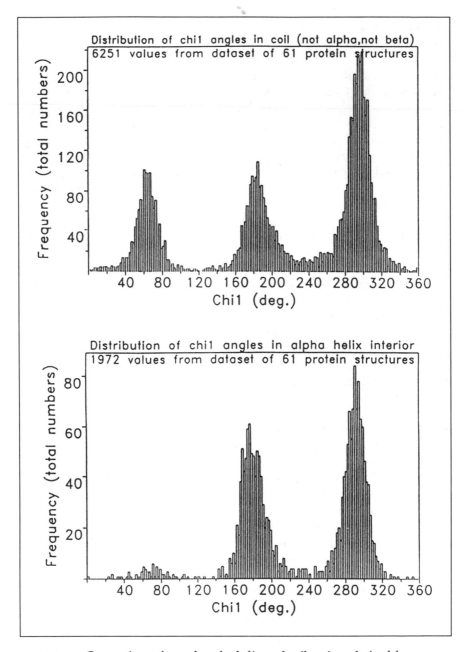

FIGURE 2–7. *Comparison of* χ_1 *values for helix and coil regions derived from a data set of 61 high-resolution protein structures. (A) Values of* χ_1 *for the residues in coil (not helix or strand), using 6,251 values. Half of the residues in the* g^- *region, 60°, are Ser or Thr. (B) Values of* χ_1 *for 1,972 residues in the interior of a helix, defined as more than three residues from either terminus.*

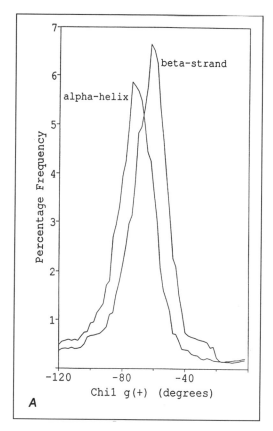

FIGURE 2–8. *Distribution of* χ_1 *gauche$^+$ values. (A) For residues in an α-helix or β-strand.*

preliminary studies reveal that more $i,i+4$ interactions are observed compared with $i,i+3$. These data need further study in the search for such stabilizing interactions.

6.5 Secondary Structure Packing

From analysis of tertiary structures, it is clear that there are preferred modes of packing of secondary structures. (For a review, see Chothia, 1984.) Most of these have been relatively easily rationalized by basic packing constraints, e.g., ridges into grooves, matching sheet twists, and hydrophobic core formation. It is very difficult to imagine such packing, especially for helices, except by requiring formation first of the individual helices, which then pack together and adjust their conformations accordingly. For a β-sandwich

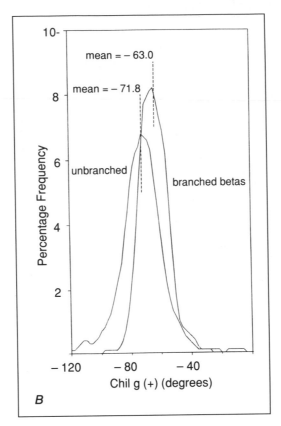

FIGURE 2–8. *Distribution of* χ_1 *gauche*[+] *values. (B) For residues in a helix that have branched (Val, Ile, Thr) and unbranched β-carbons.*

structure, the two sheets on either side of the sandwich may form independently and then pack. Given the convoluted topology often observed (e.g., in a jelly roll), however, these two sheets may alternatively "grow" together, developing a hydrophobic core in the process.

The folding pathway involved for proteins with complex multiple sheets is difficult to envisage, especially when the polypeptide chain wanders between them.

7 *LOOP REGIONS*

The loop regions, connecting the classic secondary structures, are generally short (Fig. 2–9). They are very variable in homologous structures, however, and therefore it has often been implied that they are "passive" during

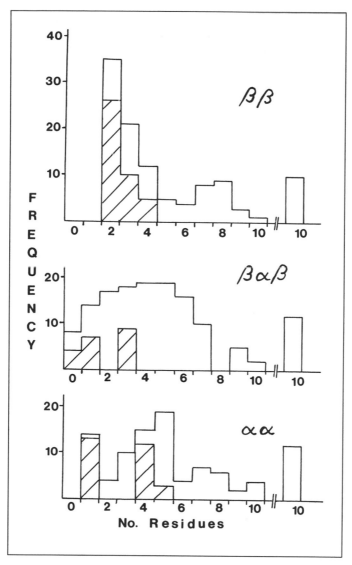

FIGURE 2–9. *Lengths of loop regions in the common supersecondary structures: the β/β hairpin, the β/α/β unit, and the α/α hairpin.*

folding, with only the conserved "core" residues being important in determining the topology. Although some longer loops are flexible, even in the protein crystal structure, loop conformation is often well defined, despite the original description as random coil; also, recurring loop motifs are common (Sibanda et al., 1989; Jones and Thirrup, 1986).

Given a pentapeptide in which each residue may adopt just three states (α, β, other), there are $3^5 = 243$ possible structures. Only 62 of these are observed in real proteins, probably reflecting local steric interactions (M. Swindells, personal communication). In supersecondary motifs (e.g., β-hairpin or $\alpha\alpha$ hairpin), an even more limited set of loop conformers is observed, which often include critical Gly residues that adopt a positive φ value.

For example, β-hairpins are the tightest loops, with only two residues, and form the rare Type I′ and II′ β-turns. Indeed, these turns occur almost exclusively in β-hairpins, where the twist of the strands is matched by the asymmetry of these turns (Sibanda et al., 1989).

The major determinants of loop conformation seem to be the end point separation, location of a few key residues, internal hydrogen bonding, and, occasionally, interactions with framework or core residues. Most loops interact with the solvent and are highly hydrophilic. Perhaps it is the absence of the systematic hydrophobic/hydrophilic patterns seen in secondary structures and the dominance of hydrophilic residues that predisposes a polypeptide segment to be a loop. As with disulfide bridges (Chapter 7) the loops cannot determine the topology, but the final structure must accommodate them comfortably.

Many protein engineering experiments have shown that it is possible to modify loops extensively and to swap loops between structures. This approach has been used to insert various functional loops into virus coat proteins (e.g., poliovirus and baculovirus) (Evans et al., 1989). Only some of these mutants are viable viruses. The extent of permissible loop variation has yet to be fully explored and rationalized, and will depend upon the particular loop. For nonviable viruses, it is not yet known whether the folded protein structure is less stable or if the folding pathway has been disrupted.

8 SOLVATION

One of the strongest "rules" for protein folding derived by inspection of structures is that almost all hydrogen bonding groups must be satisfied by protein or solvent interactions.

The strongest hydrogen bonds involve ion pairs, and it is striking how few exist buried in the core of a protein (Barlow and Thornton, 1983), where their contribution to stability might be expected to be a maximum due to the low effective dielectric constant. The few buried ion pairs observed are usually involved in the function of the protein. This apparent contradiction may well reflect the folding process where, in the unfolded form, the solvent interaction with charged groups is very favorable. To desolvate is energetically expensive and will occur only if the solvent hydrogen bonds are replaced by protein hydrogen bonds. In the melee of the initial molten globular state, where few specific tertiary interactions are made, such pairing will be difficult to achieve and sustain.

To visualize specific solvation, it is simpler to consider the initial stages of protein unfolding. We can suggest possible hypotheses as to how proteins gradually "melt" into the solvent or, alternatively, how solvent molecules invade the native protein globule. It seems reasonable to suggest that the flexible loop regions, with high crystallographic B values, will be the first to melt. Indeed, it is known that such loops must melt before they can be attacked by proteinases (Hubbard et al, 1991). Terminal regions of proteins also have, on average, high B-values (Thornton and Sibanda, 1983) and presumably melt rather easily.

What about the secondary and tertiary interactions? Sundaralingam and Sekhurudu (1989) suggested a model for helix desolvation by considering water molecules bound to helices and turns. They suggest that a helix may form via a water-mediated β-turn, from which the water is subsequently expelled as main chain hydrogen bonds are formed.

In β-sheets, two types of solvation are common (Thanki et al., 1991). Edge β-strands are regularly solvated such that the bound water molecules can almost be considered as forming an extension to the β-sheet. Similarly, at the end of two adjacent hydrogen-bonded β strands, a water molecule often bridges the two strands as they splay apart, forming an extra rung to the β-ladder. This immediately suggests a route for the "invasion" of water molecules as the β-ladder unzips and the protein hydrogen bonds are replaced by solvent molecules. During unfolding, the solvent presumably seeks out the "weak links" in the structure and invades there first. This description of unfolding may be too simplistic to be helpful for detailed analysis, where many alternative states of the protein and solvent are possible.

The distributions of water molecules around individual amino acids show regular patterns of solvation (Thanki et al., 1988). Despite the complex convoluted surface of a protein, the local hydrogen bonding between amino acid residues and water molecules is usually quite strong, and the distributions reflect the hydrogen-bonding capacity of the side chain.

9 TOPOLOGY

It is very easy to construct a hierarchy of protein structure (Fig. 2–10A), with the strands and helices as the basic building blocks that can be assembled in various ways. Sequential strands and helices can form supersecondary structures, which are often observed in unrelated proteins and represent stable folds (e.g., the β-hairpin, the β/α/β unit, and the α/α hairpin). These in turn form the basis of the larger domain structures, such as the four-helix bundle or the eightfold β/α barrel. Levitt and Chothia (1976) and Richardson (1981) have defined essentially five different families of domain

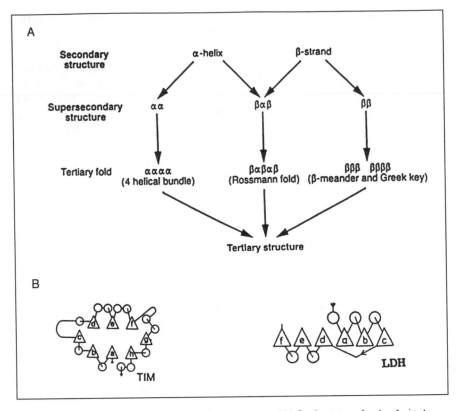

FIGURE 2–10. *(A) Hierarchy in protein structure. (B) β-sheet topologies for triose phosphate isomerase (TIM) and lactate dehydrogenase (LDH), showing the consecutive strand topology in TIM compared to the chain reversal in the Rossmann fold.*

structure: the all α; all β; alternating α/β; α + β (nonalternating); and small irregular proteins whose structures usually include multiple disulfide bridges or ligands.

What, if anything, do these topologies tell us about protein folding? As expected from kinetic considerations, sequential secondary structures are more likely to interact in three dimensions than those that are distant in the sequence. For example, the preference for sequenual strands to be adjacent in a β-sheet of a folded protein is overwhelming. In classic motifs, sequential interactions dominate (e.g., β/α barrel). (See Figure 2–10B.) There is clearly, however, also the requirement to form a hydrophobic core. If the sheet propagates sequentially, this can only be achieved by forming a barrel. The alternative is to reverse direction (as in the Rossmann fold). (See Fig. 2–10B.) It is interesting to consider arrangements of two, three, and four sequential strands. For two sequential strands, hairpins dominate alternative arrangements in one sheet by more than 7:1. For three strands the meander is about three times as common as a ψ loop. In contrast for four strands, the more convoluted Greek key arrangement (Richardson, 1981) is much more common than the straight up-down arrangement, which does not allow formation of a hydrophobic core.

The order of condensation of secondary structures into tertiary folds cannot, however, be defined by inspection of the folded form. It is, of course, tempting to suggest that, because local interactions dominate in the final structure, sequential segments will coalesce first and then join together to form the complete domain. The experimental evidence on cytochrome *c* (Roder et al., 1988) in which the N- and C- terminal helices in the sequence are thought to be the first to interact, is completely contrary to this hypothesis. Clearly other effects, such as secondary structure propensity or interactions with a heme group, will play a role. The effect of the chirality of the *L*-amino acid residue is propagated throughout all levels of protein structure. The secondary structures are chiral (right-handed α-helices, and β-sheets that twist in one preferred direction); the supersecondary structures are also nonsymmetric (e.g., the right-handed βαβ unit; Richardson, 1981; Sternberg and Thornton, 1976); and this asymmetry is also seen in chiral domain structures (e.g., the right-handed β/α barrel and the four-helix bundle). The properties of the basic units are "amplified" into large topological preferences.

10 COMPARISON OF SIMILAR STRUCTURES

With the increasing number of protein structures, it has become apparent that sequences that show no significant sequence homology often adopt the same structure. We now have many examples of four-helix bundles, β/α barrels, and up-down β-barrels that show very little sequence homology. For example, the structures of interleukin-1β (Preistle et al., 1988) and a small trypsin inhibitor ETI (Onesti et al., 1991) have only eight residues identical in over 170, yet their core structures are very similar. Whether these structures represent convergent or divergent evolution remains an open question, as does whether or not their folding pathways are also similar.

Therefore it seems increasingly likely that there will be a limited number of protein topologies that allow β-strands and/or α-helices to pack together to form a tight globular core. The same packing problem can apparently be solved by a combination of many different sets of amino acid residues. For example, the residues in the core of four-helix bundles show few similarities, other than a tendency to be hydrophobic. As information on folding pathways becomes increasingly available, it will be interesting to find out if the folding pathway is conserved as well as the topology or if, as with the individual amino acid residues, many different alternative routes converge to the same solution.

Acknowledgments

I would like to thank Malcolm MacArthur, Jus Singh, Gail Hutchinson, and Mark Swindells for useful discussions and for their contributions to this work.

REFERENCES

Baker, E. N., and Hubbard, R. E. (1984) *Prog. Biophys. Mol. Biol. 44*, 97–179.

Barlow, D. J., and Thornton, J. M. (1983) *J. Mol. Biol. 168*, 867–885.

Barlow, D. J., and Thornton, J. M. (1988) *J. Mol. Biol. 201*, 601–619.

Bernstein, F. C., Koetzle, T. F., William, G. J. B., Meyer, E. F., Jr., Brice, M. D., Rodgers, J. R., Kennard, O., Shimanouchi, T., and Tasumi, M. (1977) *J. Mol. Biol. 122*, 532–542.

Brandts, J. F., Halvorson, H. R., and Brennan, M. (1975) *Biochemistry 14*, 4953–4963.

Chothia, C. (1973) *J. Mol. Biol. 75*, 295–302.

Chothia, C. (1984) *Ann. Rev. Biochem. 53*, 537–572.

Chou, P. Y., and Fasman, G. D. (1978) *Adv. Enzymol. 47*, 45–148.

Chou, K. -C., Pottie, M., Nemethy, G., Ueda, Y., and Scheraga, H. A. (1982) *J. Mol. Biol. 162*, 89–112.

Evans, D. J., McKeating, J., Meredith, J., Burke, K., Katrak, K., John, A., Ferguson, M., Minor, P., Weiss, R., and Almond, J. (1989) *Nature 339*, 385–388.

Garnier, J., Osguthorpe, D. J., and Robson, B. (1978) *J. Mol. Biol. 120*, 97–120.

Gray, T. M., and Matthews, B. W. (1984) *J. Mol. Biol. 175*, 75–81.

Hetzel, R., and Wüthrich, K. (1979) *Biopolymers 18*, 2589–2606.

Hubbard, S. J., Campbell, S. F., and Thornton, J. M. (1991) *J. Mol. Biol. 220*, 507–530.

Janin, J., Wodak, S., Levitt, M., and Maigret, B. (1978) *J. Mol. Biol. 125*, 357–386.

Jones, T. A., and Thirrup, S. (1986) *EMBO J. 5*, 819–822.

Kartha, G., Ashida, T., and Kakudo, M. (1974) *Acta Cryst. B30*, 1861–1866.

Lewis, P. N., Momany, F. A., and Scheraga, H. A. (1973) *Biochim. Biophys. Acta 303*, 211–229.

Levitt, M., and Chothia, C. (1976) *Nature 261*, 552–558.

Lyu, P. C., Sherman, J. C., Chen, A., and Kallenbach, N. R. (1991) *Proc. Natl. Acad. Sci. 88*, 5317–5320.

MacArthur, M. W., and Thornton, J. M. (1991) *J. Mol. Biol. 218*, 397–412.

Marqusee, S., and Baldwin, R. C. (1987) *Proc. Natl. Acad. Sci. 84*, 8898–8902.

Matouschek, A., Kellis, J., Serano, L., and Fersht, A. R. (1989) *Nature 340*, 122–126.

Matouschek, A., Kellis, J. T., Serano, L., Bycroft, M., and Fersht, A. R. (1990) *Nature 346*, 440–445.

McGregor, M. J., Islam, S. A., and Sternberg, M. J. E. (1987) *J. Mol. Biol. 198*, 295–310.

Miranker, A., Radford, S. E., Karplus, M., and Dobson, C. M. (1991) *Nature 349*, 633–636.

Morris, A. L., MacArthur, M. W., Hutchinson, E. G., and Thornton, J. M. (1992) *Proteins: Struct. Funct. Genet. 12*, 345–364.

Nicholson, H., Soderlind, E., Tronrud, D. E., and Matthews, B. W. (1989) *J. Mol. Biol. 210*, 181–193.

Onesti, S., Brick, P., and Blow, D. (1991) *J. Mol. Biol. 217*, 153–176.

Preistle, J. P., Schar, H. P., and Grutter, M. G. (1988) *EMBO J. 7*, 339–343.

Ramachandran, G. N., Ramakrishnan, C., and Sasisekharan, V. (1963) *J. Mol. Biol. 7*, 95–99.

Richardson, J. (1981) *Adv. Protein Chem. 34*, 167–339.

Roder, H., Elove, G. A., and Englander, S. W. (1988) *Nature 335*, 700–770.

Rose, G. D., Young, W. B., and Geirasch, L. M. (1983) *Nature 304*, 655–657.

Schimmel, P. R., and Flory, P. J. (1968) *J. Mol. Biol. 34*, 105–120.

Sibanda, B. L., Blundell, T. L., and Thornton, J. M. (1989) *J. Mol. Biol. 206*, 759–777.

Singh, J., and Thornton, J. M. (1990) *J. Mol. Biol. 211*, 595–615.

Sternberg, M. J. E., and Thornton, J. M. (1976) *J. Mol. Biol. 110*, 269–283.

Sundaralingam, M., and Sekhurudu, Y. C. (1989) *Science 244*, 1333–1337.

Thanki, N., Thornton, J. M., and Goodfellow, J. M. (1988) *J. Mol. Biol. 202*, 637–657.

Thanki, N., Thornton, J. M., and Goodfellow, J. M. (1991) *J. Mol. Biol. 221*, 669–691.

Thornton, J. M., and Sibanda, B. L. (1983) *J. Mol. Biol. 167*, 443–460.

Venkatachalam, C. M. (1968) *Biopolymers, 6*, 1425–1436.

Wilmot, C. M., and Thornton, J. M. (1988) *J. Mol. Biol. 203*, 221–232.

Wüthrich, K., and Grathwohl, C. (1974) *FEBS Letters 43*, 337–340.

3

Physical Basis of the Stability of the Folded Conformations of Proteins

PETER L. PRIVALOV

1 GENERAL SPECIFICATION OF A PROTEIN MOLECULE AS A PHYSICAL OBJECT

In considering the physical problem of formation of the native protein structure, one should bear in mind that a protein is a very unusual molecular object. Its main properties are the following:

1. A protein is a large polymer, consisting of many thousands of atoms; i.e., from the physical point of view it represents a macroscopic system.
2. In the native state, a protein is an exclusively ordered macroscopic system. Each of its atoms occupies a definite position, as in a crystal, but differing in that the position of each of the atoms is unique relative to its neighboring atoms. Therefore, a protein represents a new class of macroscopic systems with an aperiodic order.
3. A protein is synthesized as a linear heterogeneous polymer of different amino acid residues, linked by peptide bonds in a definite, genetically determined order. This polypeptide chain folds into the unique conformation that is determined entirely by the information included in the sequence of amino acid residues. The folding of the polypeptide chain into the three-dimensional structure is in principle a reversible process, depending upon the environmental conditions. In other words, it is a thermodynamically driven transition from the unfolded

macroscopic state to a highly ordered macroscopic state that is called *the native state of the protein molecule* (Anfinsen, 1973).

Folding of the random-coil polypeptide chain into a unique conformation, which entails a tremendous increase in the order of the macroscopic system, should be apparent thermodynamically as a significant decrease in the entropy of the system. An entropy decrease is thermodynamically unfavorable, so one would expect this entropy effect to be compensated by the energy gained as a result of a redistribution of various intramolecular interactions between the protein groups and the environment. Judging by how easily the folding process can be reversed by the slightest change of the environmental conditions, this process of folding must alter only various noncovalent interactions between the protein groups, not the covalent structure. On the other hand, the uniqueness of the native conformation of a protein, in which the position and the state of every atom or group of atoms are strictly determined by the state and position of the other atoms, implies that this molecule is a highly cooperative macroscopic system, and the process of rearrangement of its structure should be a highly cooperative process. This in turn implies that the noncovalent interactions in a protein should cooperate strongly when the molecule is in its native state.

2 INTRAMOLECULAR FORCES IN PROTEINS

One can expect the following noncovalent interactions in proteins: van der Waals interactions, interactions between charged groups (salt links), between polar groups (hydrogen bonds), and between nonpolar groups, the so-called hydrophobic interaction.

In aqueous solution, in which proteins are usually considered, the van der Waals interaction should occur not only between the protein groups but also between these groups and water, and they are usually supposed to be almost identical. If so, they would compensate each other, and their overall contribution to the stabilization of protein structure would be zero. The same could be said about hydrogen bonding, since protein polar groups can form hydrogen bonds not only between themselves, but also with water molecules. As for the interaction between charged groups, there are not very many in proteins. Therefore, it is usually assumed that they play a minor role in stabilization of the native protein structure (Hollecker and Creighton, 1982; Creighton, 1985), although they might be important in directing protein folding (Schoemaker et al., 1987).

By elimination, therefore, one arrives at the conclusion that stabilization of the native compact protein structure is primarily due to the hydrophobic interactions between the nonpolar groups of protein molecule, especially as there are many such groups in proteins and they are clustered together forming the nonpolar core in globular protein. The popularity of hydrophobic interactions was also enhanced by their unusual character:

The hydrophobic interaction is usually assumed to be a force responsible for the low solubility of nonpolar substances in water, their hydrophobicity. This can be regarded as a tendency for these nonpolar substances to aggregate in aqueous media. The low solubility of molecules means thermodynamically that their transfer into the solvent requires the expenditure of substantial work, i.e., that the Gibbs energy of transfer of these substances from the pure liquid phase (l) into water (w) is large and positive (i.e., $\Delta_l^W G = -RT \ln K >> 0$, when the solubility constant K <<0). On the other hand, it was found that the enthalpy of transfer of these substances from the pure liquid phase into water, is almost zero at room temperature. Since $\Delta G = \Delta H - T \Delta S$, it looks as if the large positive Gibbs energy of transfer of the nonpolar molecule into water results from a large negative entropy of transfer $\Delta_l^W S << 0$. (See Table 3–1.)

There were several explanations for this striking experimental observation. According to Frank and Evans (1945), the presence of a non-polar solute leads to an increase in the order of the water. This explanation is in accord with the observed increase in the heat capacity of water in the presence of nonpolar solutes (Edsall, 1935), which can be interpreted as the extra heat of the gradual melting of ordered water as the temperature

TABLE 3–1. Solubility, Gibbs Energy, Enthalpy, Entropy, and Heat Capacity Increment of Transfer of Some Typical Nonpolar Substances from the Pure Liquid Phase to Water at 25°C

Substance and its surface area in $Å^2$	Solubility in mole fractions $\times 10^4$	ΔG	ΔH	ΔS	ΔC_p
		\($kJ \cdot mole^{-1}$\)		\($J \cdot K^{-1} \cdot mole^{-1}$\)	
Benzene 240	4.01	19.4	2.08	-58.06	225
Toluene 275	1.01	22.8	1.73	-70.7	263
Ethylbenzene 291	0.258	26.2	2.02	-81.0	318
Cyclohexane 273	0.117	28.2	-0.10	-94.8	360
Pentane 272	0.095	28.7	-2.00	-102.8	400
Hexane 282	0.02	32.5	0	-109.1	440

Source: Baldwin (1986) and Privalov and Gill (1989).

increases. It was also supported by the correlation found between the entropy and heat capacity effects and the surface area of the solute molecules transferred to water. According to the scaled-particle theory of solutions, the negative entropy of dissolution of nonpolar solutes in water results simply from mixing solvent and solute molecules of different sizes (Lucas, 1976; Lee, 1985). This explanation has been questioned by Pohorille and Pratt (1990). Whatever the interpretation, it appeared clear that transfer of nonpolar molecules or groups into water results in a decrease in entropy, which is unfavorable thermodynamically, and the system expels these molecules. This expelling action of water on the nonpolar solutes causes them to aggregate and was regarded as a hydrophobic interaction between the nonpolar molecules (Kauzmann, 1959; Tanford, 1980).

3 THEORETICAL AND PRACTICAL PROBLEMS IN STUDYING PROTEIN FOLDING

The main driving force in folding of a polypeptide chain into a native structure, i.e., the force working for decreasing the entropy of the polypeptide chain due to the ordering of its conformation, appears from the preceding analysis also to be entropic. This entropic force results from the removal of the nonpolar groups from water, their dehydration upon being buried in the compact protein interior. These two entropic effects are of the opposite sign, however: while the entropy of ordering of the polypeptide chain is negative, the entropy of dehydration of the nonpolar groups is positive; therefore, they compensate each other. But how exact is this compensation?

If only hydrophobic interactions maintain the compact ordered state of a protein, and the conventional understanding of the mechanism of this interaction is correct, one would expect the positive entropic effect of dehydration of nonpolar groups to exceed in magnitude the negative entropic effect of the conformation ordering. Therefore, the overall entropy of folding of a protein native structure in water media should be positive, and not negative as expected for protein folding in a vacuum. But, then, why do proteins unfold at increased temperatures? With an increase in temperature, the entropic force should also increase ($-T\Delta S$, if $\Delta S < 0$); i.e., the hydrophobic interaction should increase, maintaining the compact state of protein.

This discrepancy with reality raised the question of whether our understanding of the hydrophobic interaction is correct and whether protein structures are indeed stabilized only by the hydrophobic interaction.

The point of view that van der Waals interactions do not contribute to the stabilization of protein structure or to the hydrophobicity of nonpolar substances was based on the assumption that the van der Waals interactions between the protein groups are similar to those between the protein and water. The main argument for that was the small volume change upon protein unfolding, as judged by the small influence of pressure. It is far from clear, however, whether the contacts between the highly packed protein groups within a protein interior are indeed the same as the contacts of these groups with water, in view of the relatively open structure of the liquid, with its larger fractional free volume.

Similar concern occurs about hydrogen bonds, especially as the conventional opinion about the similarity of these bonds between protein polar groups and those with water has changed many times during the past several decades.

The problem of the magnitude of the noncovalent interactions in proteins is closely connected with the problem of the extent of cooperativity of these interactions that is required to overcome the dissipative action of entropy. Does cooperativity involve only neighboring groups or does it extend over longer distances in proteins?

After taking into consideration the surrounding water, we actually have no idea of the magnitude of the interactions within folded proteins, of the real entropy balance, and there is no hope of evaluating them solely by the logical analysis of folding an abstract polypeptide chain into an abstract ordered conformation.

Solution of this problem requires precise knowledge of the thermodynamics of folding of real proteins in real aqueous solutions. This information can be obtained only experimentally by direct measurements of the energetics of protein unfolding and refolding. The use of indirect approaches might be misleading, because these approaches are usually based on assumptions that in themselves present problems. Reliable information upon the energetics of protein unfolding/refolding can be obtained only by direct calorimetric measurements of the corresponding heat effects. These heat effects of the intramacromolecular folding processes in proteins must be studied in dilute solutions to avoid intermolecular effects. As a consequence, the observed heat effects are usually very small, and their measurement required development of special superprecise microcalorimetric techniques. (For reviews see Sturtevant, 1974; Privalov and Potekhin, 1986; Privalov and Plotnikov, 1989.)

The main practical problem that arises in experimental studies of folding of a real protein is the following: in theoretical studies of protein folding, the standard initial state is usually considered to be the completely unfolded polypeptide chain in the random-coil conformation, in which amino acid residues do not interact with each other. This ideal state does not, however, exist in reality. There are too many different types of groups in proteins, and correspondingly too many types of interactions between these groups, for there to be any environmental condition that would completely exclude interactions between all protein groups without adding new interactions. For example, the denaturants urea and guanidinium chloride (GdmCl) are believed to unfold a polypeptide chain completely, but they interact with the polypeptide chain.

In practice, the initial state in folding is the disordered state of the protein under denaturing conditions, where there are minimal interactions between protein groups. The validity of this approach is dependent upon determining the extent to which the actual denatured state of the protein deviates from the ideal model—the completely unfolded polypeptide chain.

A problem that also requires special attention is the reversibility of protein unfolding. As mentioned previously (Section 1), protein unfolding/refolding is a reversible process in principle. In practice, however, it may not be entirely reversible. The main reason is usually the various secondary events that can occur with the unfolded polypeptide under extreme denaturing conditions, such as chemical modifications of the amino acid side chains, hydrolysis of the polypeptide chain, cross-linking by covalent bonds, and aggregation. These secondary, irreversible processes are usually much slower than protein unfolding, so they do not affect unfolding significantly, so long as it is performed quickly and the protein is exposed to the extreme conditions for only a short time.

4 COOPERATIVITY OF DENATURATION

Globular proteins usually are sufficiently stable not to lose their native properties when the external conditions vary slightly from optimal. But if the deviation exceeds some critical value, the protein denatures within a relatively narrow range of variation of temperature, pressure, pH, denaturant concentration, and so on. (For a review, see Tanford, 1968.) Denaturation is apparent from changes in all the properties of the protein molecule that are sensitive to its three-dimensional structure, suggesting that this process involves breakdown of the entire native protein structure. In the case of small globular proteins, all these properties change simultaneously, suggesting that the native structure breaks down in an "all-or-none" manner. Thermodynamically, this means that the denaturation process can be considered as a simple transition between the two macroscopic states: the native (N) and denatured (D).

This view gained wide popularity because it opened the possibility of quantitative studies of the equilibrium of this reaction. In that case the equilibrium constant for the denaturation process can be expressed as:

$$K = \frac{Q_X - Q_N}{Q_D - Q_N}$$

where Q_N and Q_D are the values of any observed parameter characterizing the pure native and the pure denatured states, respectively, and Q_X represents the value of this index under any given conditions. The dependence of this effective equilibrium constant on external variables, such as temperature, pressure, and ligand activity, gives the effective parameters characterizing the denaturation process (Privalov, 1979; Wyman and Gill, 1990). Using the equations of equilibrium thermodynamics for conversion of the native (N) to the denatured (D) state, it is straightforward to estimate changes of the:

Gibbs energy
$$\Delta_N^D G^{eff} = G^D - G^N = -RT \ln K^{eff} \qquad (3\text{--}1)$$

enthalpy
$$\Delta_N^D H^{eff} = H^D - H^N = R \frac{\partial \ln K^{eff}}{\partial 1/T} \qquad (3\text{--}2)$$

volume
$$\Delta_N^D V^{eff} = V^D - V^N = -RT \frac{\partial \ln K^{eff}}{\partial P} \qquad (3\text{--}3)$$

amount of bound ligands
$$\Delta_N^D n^{eff} = n^D - n^N = \frac{\partial \ln K^{eff}}{\partial \ln a}. \qquad (3\text{--}4)$$

This approach is valid only if protein denaturation is a two-state transition. The sharpness of the process and the simultaneity in the changes of the observed properties of protein molecules are necessary, but not sufficient, criteria for the process indeed to be a two-state transition (Lumry et al., 1966). For example, if a protein consists of several similar and independent domains, its denaturation might be sharp, and all the properties will change simultaneously, but the process will not be a two-state transition.

The crucial test of the applicability of the preceding equations to protein denaturation, i.e., whether protein denaturation is a two-state transition, is whether the effective characteristics derived by Equations (3–1) to (3–4) correspond to those measured directly. In practice, this became possible with sufficient accuracy only for temperature-induced denaturation, because the conjugate extensive characteristic of temperature is enthalpy (see Eq. (3–2)) which can be measured with sufficient precision using scanning microcalorimetry techniques (Privalov and Potekhin, 1986).

Numerous calorimetric studies of the temperature-induced denaturation of proteins under various solvent conditions showed that, in the case of small globular proteins with a molecular mass (M) less than 20 kDa, denaturation produces a sharp peak of heat absorption (Fig. 3–1). The effective enthalpy of this process can be calculated from its sharpness by Equation (3–2), which is known as the van't Hoff equation. At the midpoint of transition, corresponding to a temperature T_G where

$$\Delta_N^D G(T_G) = \Delta_N^D H(T_G) - T_G \Delta_N^D S(T_G) = 0 \qquad (3\text{--}5)$$

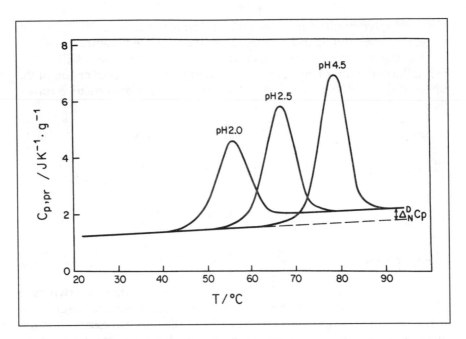

FIGURE 3–1. *Temperature-dependence of the partial heat capacity ($C_{p,pr}$) of a typical globular protein, hen egg-white lysozyme in aqueous solution at different pH values (according to Privalov and Khechinashvili, 1974).*

the van't Hoff equation for a calorimetric curve can be presented as

$$\Delta_N^D H^{vH}(T_G) = 4\,RT_G^2\,\frac{\Delta Cp\,(T_G)}{Q} \tag{3–6}$$

where $\Delta Cp\,(T_G)$ is the height of the heat absorption peak at the midpoint of the transition, which is close to the maximum of the heat absorption peak, and Q is the area of the peak. On the other hand, the parameter $M \times Q$ corresponds to the actual or calorimetric enthalpy of protein denaturation $\Delta_N^D H(T_G)$ (Privalov and Khechinashvili, 1974).

For small compact globular proteins, the effective van't Hoff enthalpy of denaturation was found to be in good agreement with the actual calorimetric enthalpy. The deviation usually does not exceed 5%. This means that the process of temperature-induced denaturation of these proteins can be regarded to a first approximation as a two-state transition (Freire and Biltonen, 1978; Privalov, 1979).

With large globular proteins, there is a large discrepancy between the effective van't Hoff enthalpy and the actual enthalpy of denaturation, and in many cases the denaturation heat absorption has a complex profile.

Therefore, denaturation of these proteins does not represent a single two-state transition, but includes several stages (Privalov, 1982).

When denaturation of a protein is induced by some other variables, such as pressure or ligand concentration, one cannot draw a conclusion about the two-state nature of the process with the same certainty as for the temperature-induced process, because of the difficulty of direct measurement of the corresponding extensive parameters, the volume change or the amount of bound ligand. (See Equations (3–3) and (3–4).) On the other hand, if denaturation induced by temperature at various fixed solvent conditions (solvent composition, pressure) is a two-state transition in all cases, the denaturation induced by varying this other condition at a fixed temperature should also be a two-state transition (Pfeil and Privalov, 1976c). From such an analysis, it was concluded that a small globular protein molecule represents a single cooperative macroscopic system, and its native structure breaks down under extreme conditions in a highly cooperative manner with only very unstable intermediate states (Privalov, 1979).

Detailed analysis of the heat absorption profiles upon denaturation of large proteins and of their proteolytic fragments demonstrated that these very large systems are subdivided into more or less independent cooperative subsystems, i.e., domains, with thermodynamic properties and sizes that are similar to those of small globular proteins. The subdivision of protein structure into domains and the discreteness of their structure are likely to be general principles of protein architecture (Privalov, 1990).

5 DENATURED STATES OF PROTEINS

Proteins can be denatured by changes of various external parameters: increasing the temperature, the pressure, or a denaturant concentration, and decreasing or increasing the pH. Proteins denatured by these various means usually differ in their optical and hydrodynamic properties, which raises the question of whether the denatured state of a protein is a universal macroscopic state or whether there are different macroscopic states in each case: the acid-denatured state, GdmCl-denatured state, and so on. This question can be answered only comparing the basic thermodynamic properties of a protein denatured in various ways, because only these characteristics (the enthalpy, entropy, partial volume, and amount of ligands bound) are essential to specify the macroscopic state (Wyman and Gill, 1990). Viscosity and optical properties, which are usually considered in discussing this problem, are not thermodynamic characteristics and do not specify the macroscopic state; therefore, even a significant difference in these parameters is not very meaningful.

Temperature-induced denaturation of a globular protein proceeds with extensive heat absorption and, what is most remarkable, a significant increase in heat capacity $\Delta_N^D Cp$ (Fig. 3–1). The heat capacity is the temperature derivative of the enthalpy:

$$\Delta_N^D Cp = \frac{\partial \, \Delta_N^D H\,(T)}{\partial T} \tag{3–7}$$

so the increase in heat capacity of a protein upon denaturation means that the enthalpy change is temperature-dependent and should increase with increasing temperature.

This is just what is observed when protein denaturation occurs at different temperatures (Fig. 3–1). With increasing stability of the protein and an increase in the temperature at which the protein denatures, there is an increase in the area of the heat absorption peak, i.e., the enthalpy of protein denaturation increases. This enthalpy increase is not caused by the change in pH, i.e., the ionization of protein groups, because the experiment was carried out in a buffer that has the same heat of ionization as protein groups, and the heats of ionization of buffer and protein groups compensate each other (Privalov and Khechinashvili, 1974; Privalov, 1979). Therefore, Figure 3–1 illustrates the net heat effects of the conformational transitions of a protein induced by increasing the temperature.

Figure 3–2 represents the calorimetric titration of the same protein (lysozyme) at different, but constant, temperatures. At room temperature, lysozyme does not denature at any pH, and the slight heat effect observed calorimetrically is due to ionization of acidic groups on the surface of the native protein. At higher temperatures, the protein denatures with a very significant enthalpy change, which depends upon the temperature and increases as the titration is carried at higher temperatures. The number of groups protonated upon protein denaturation can be determined by potentiometric titration and, because the heats of ionization of all these groups are known, one can easily estimate the enthalpy effect due to ionization upon protein denaturation. This effect is small in comparison with the observed enthalpy change (Pfeil and Privalov, 1976a). Subtracting it from the observed effect gives the net enthalpy of the conformational transition of the protein induced by pH at the given temperature.

Figure 3–3 represents the results of calorimetric titration of the same protein by GdmCl at various constant temperatures and pH values. In contrast to the titration by acid, the enthalpy changes from the very beginning, due to solvation of GdmCl by the native protein. The enthalpy of solvation is large and positive, indicating that the denaturant interacts strongly with the native protein. The heat effect of denaturation has a sign opposite to that of the solvation effect, showing that denaturation proceeds

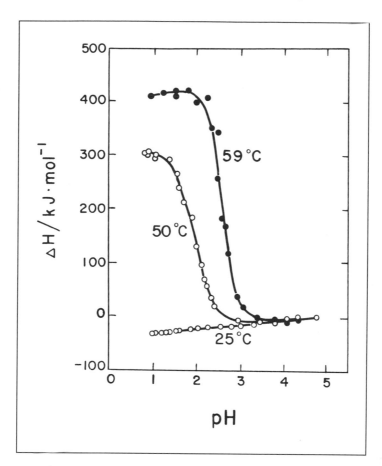

FIGURE 3–2. *The results of calorimetric titration of lysozyme by acid at different temperatures (according to Pfeil and Privalov, 1976a).*

with a positive enthalpy change. Following denaturation, the negative heat effect is due to solvation by GdmCl of the denatured protein. Knowing the heat effects of GdmCl solvation of the native and denatured proteins, one can easily determine the heat effect of preferential binding of the denaturant to the protein upon its unfolding. Subtracting it from the observed overall denaturation heat effect gives the net enthalpy of the conformational transition of protein induced by GdmCl (Pfeil and Privalov, 1976b).

The net enthalpies of the conformational transitions of a protein into the denatured state induced by GdmCl, by pH, and by temperature are plotted as a function of temperature in Figure 3–4. It is clear that the same universal function of temperature is apparent for all three types of transitions.

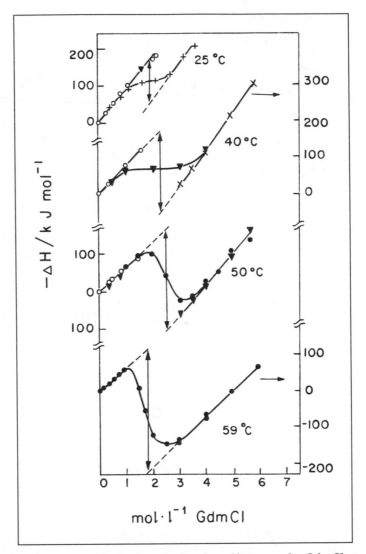

FIGURE 3–3. *The results of calorimetric titration of lysozyme by GdmCl at different temperatures and pH values (according to Pfeil and Privalov, 1976b). +, pH 1.3; ×, pH 2.1; ▼, pH 2.5; ▽, pH 3.0; ○, pH 3.5; •, pH 4.5.*

The slope of the enthalpy change as a function of temperature is just the heat capacity change upon protein denaturation, which is measured by the scanning microcalorimeter in the presence of buffer, compensating the ionization effects; i.e., it is the increase in heat capacity due to the conformational transition of the protein into the denatured state.

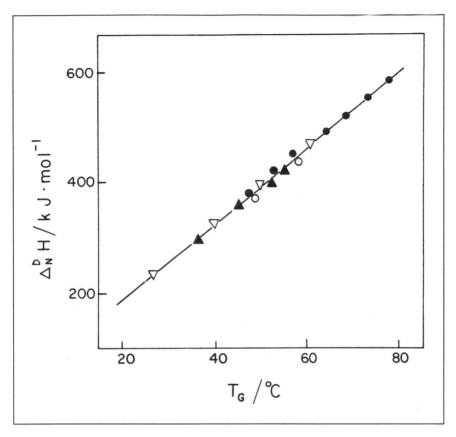

FIGURE 3–4. Temperature-dependence of the enthalpy of the conformational transition of lysozyme that occurs upon its denaturation by temperature, pH, and GdmCl. Circles indicate solutions without guanidinium chloride (GdmCl): open circles, denaturation by temperature at constant pH; filled circles, denaturation by pH at constant temperature. Triangles indicate solutions with GdmCl: filled triangles, denaturation by temperature at constant concentration of GdmCl; open triangles, denaturation by GdmCl at constant temperature (according to Pfeil and Privalov, 1976c).

It was originally thought that the heat capacity increment is different for various types of denaturation, for example, that it is larger for GdmCl-induced denaturation and smaller for temperature-induced denaturation. It was believed that a protein unfolds completely only in the presence of high concentrations of a denaturant (Tanford and Aune, 1970). We now see that this is not the case. The mistake originated from an incorrect account of the effects of preferential binding of the denaturant

(Pfeil and Privalov, 1976b). If the heat capacity effects of preferential binding of denaturant and proton are taken into account correctly, there is no difference in the heat capacity increments upon denaturation induced by heat, acid, GdmCl, or urea. This type of study has been carried on a number of globular proteins in the presence of both GdmCl and urea, and in no case was any difference found in the enthalpies and the heat capacity increments associated with the conformational transitions caused by temperature, pH, or denaturants (Pfeil and Privalov, 1976a,b). All attempts to detect heat effects associated with any cooperative change of the unfolded protein upon transfer between the various denaturing conditions were unsuccessful (Pfeil et al., 1986). If there is any difference between proteins denatured in different ways, this difference is not qualitative from the thermodynamic point of view, i.e., it is not associated with a qualitative difference in the macroscopic states. Therefore, all these forms of protein should be assigned to the same macroscopic state, which we call the *denatured state of the protein* (Privalov, 1979; Privalov et al., 1989).

It should be noted that, in contrast to the rigid native state of a protein with a definite three-dimensional structure, the denatured state of protein has no definite structure, but only an averaged, fluctuating conformation that is very labile. It can change upon the slightest change in the environmental conditions: expanding or squeezing, increasing or decreasing helicity, and so on. But all of these changes are gradual and noncooperative. They are associated with a continuous change in the thermodynamic parameters that specify that state. Therefore, they all should be considered as proceeding within the same macroscopic state, due to a redistribution of the population of its microscopic states.

In contrast to the gradual changes of the denatured protein, the denaturation of a protein is a transition between two macroscopic states that differ qualitatively in their thermodynamic characteristics and correspondingly are described by two different surfaces in phase space (Fig. 3–5).

Transitions between these states occur with a discontinuity in the enthalpy and entropy functions, i.e., there is a discontinuity in the first derivative of the thermodynamic potential. According to the conventional definitions of statistical thermodynamics, this transition should be considered as a first-order phase transition. Correspondingly, the native and denatured states of a protein can be regarded as the two phases of a macroscopic system with different symmetries (Privalov, 1989).

A protein in a concentrated solution of denaturants is usually considered to be in the completely unfolded random-coil state. Denatured proteins under different denaturing conditions, such as high temperatures or concentrated solutions of GdmCl, are likely to be indistinguishable thermodynamically (Privalov et al., 1989). In addition, the partial specific heat capacity of a denatured protein is very close to that expected for the completely

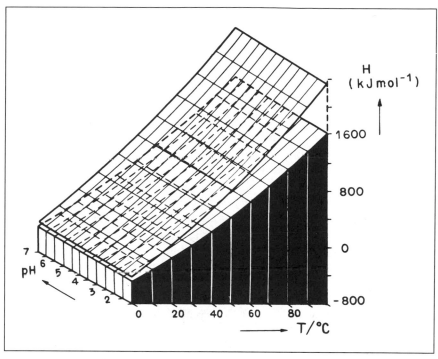

FIGURE 3–5. *The enthalpy of native and denatured lysozyme as functions of pH and temperature (according to Pfeil and Privalov, 1976c). The upper surface is that of the denatured state. The enthalpy is expressed relative to that of the native state at pH7 and 25°C.*

unfolded polypeptide chain, estimated by summing the partial specific heat capacities of all the amino acid residues that make up protein polypeptide chain (Makhatadze and Privalov, 1990; Privalov and Makhatadze, 1990).

Therefore, with present experimental techniques, we cannot distinguish thermodynamically between the denatured state of a protein and its fully unfolded state. Thus, in considering protein folding, we can take the denatured state of protein as its unfolded state, at least to a first approximation. Correspondingly, we can regard denaturation as the process of protein unfolding.

6 ENTHALPY AND ENTROPY DIFFERENCES OF THE NATIVE AND DENATURED STATE OF PROTEINS

Integrating Equation (3–7), one can determine the enthalpy difference between the native and denatured states over a temperature range in which the heat capacity difference of the states is known:

$$\Delta H(T) = \Delta H\,(T_G) + \int_{T_G}^{T} \Delta_N^D\, Cp\,(T)\, dT. \tag{3–8}$$

When $\Delta_N^D\, Cp$ is constant, we get a linear function:

$$\Delta_N^D\, H(T) = \Delta_N^D\, H\,(T_G) + (T - T_G)\, \Delta_N^D\, Cp. \tag{3–9}$$

As was shown in Figure 3–4, the enthalpy change upon denaturation of lysozyme is a linear function of temperature in the temperature range 30° to 80°C. In this temperature region, the denaturation heat capacity increment of lysozyme does not depend detectably upon temperature. This is also observed upon direct calorimetric determination of the protein heat capacity (Fig. 3–1). The same was found for all other globular proteins studied: their enthalpy curves appeared to be linear but their slopes were found to be different for each protein (Fig. 3–6).

Most surprising was the finding that if one plots versus temperature, not the molar, but the specific enthalpy values, i.e., the values calculated per gram of protein or per mole of amino acid residues, these functions for the very different small compact globular proteins converge to a single point at about 110°C (Privalov and Khechinashvili, 1974; Privalov, 1979) (Fig. 3–7A).

A similar situation was found with the entropy functions. The entropy of denaturation can be obtained from the enthalpy change at the transition midpoint temperature, T_G, by Equation (3–5), from which it follows that

$$\Delta_N^D\, S(T_G) = \Delta_N^D\, H\,(T_G)/T_G. \tag{3–10}$$

Bearing in mind that

$$d\Delta S = \frac{\Delta Cp}{T}$$

we have

$$\Delta_N^D\, S\,(T) = \frac{\Delta_N^D\, H\,(T_G)}{T_G} + \int_{T_G}^{T} \Delta_N^D\, Cp\,(T)\, d\ln T. \tag{3–11}$$

When $\Delta_N^D\, Cp$ is constant, we get

$$\Delta_N^D\, S\,(T) = \frac{\Delta_N^D\, H\,(T_G)}{T_G} + \Delta_N^D\, Cp\, \ln\,(T/T_G). \tag{3–12}$$

In contrast to the enthalpy function, the entropy function is nonlinear even when the heat capacity increment does not vary with temperature. Figure 3–7B shows that the specific entropy differences between the native

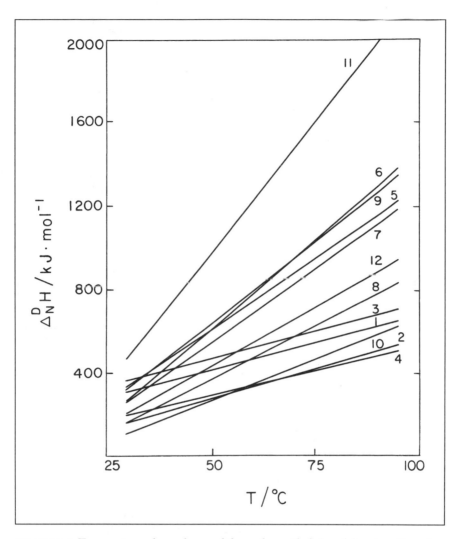

FIGURE 3–6. *Temperature-dependence of the molar enthalpies of denaturation of various globular proteins (according to Privalov, 1979, and Privalov and Gill, 1988). 1, ribonuclease A; 2, parvalbumin; 3, egg-white lysozyme; 4, fragment X4 of plasminogen; 5, trypsin; 6, chymotrypsin; 7, papain; 8, staphylococcal nuclease; 9, carbonic anhydrase; 10, cytochrome c, 11, pepsinogen; 12, myoglobin.*

and denatured states of various globular proteins converge at about the same temperature as the specific enthalpies, 110°C.

The convergence of the specific enthalpy and entropy values of very different globular proteins with increasing temperature shows that

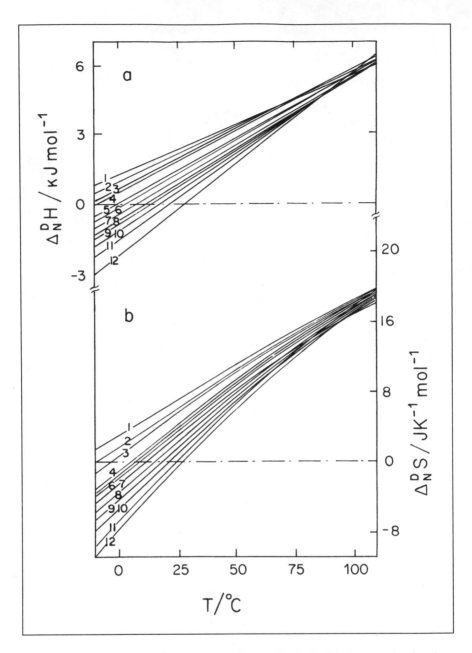

FIGURE 3–7. *Temperature-dependence of the specific (calculated per mole of amino acid residue) enthalpy (A) and entropy (B) of the conformational transitions of various globular proteins from the native to denatured state, obtained with the assumption that the denaturation heat capacity increment is temperature-independent (according to Privalov, 1979, and Privalov and Gill, 1988). The numbers of the proteins are the same as in Figure 3–6.*

there should be some general relationship between the magnitude of these functions at any given temperature and the specific denaturation heat capacity increment, which determines the slopes of these functions. On the other hand, the specific enthalpy and specific entropy are no longer extensive characteristics of the protein, as are the molar enthalpy and entropy, but are intensive characteristics. It is clear, therefore, that the observed regularity reflects some intrinsic property of protein structure, not related to the overall size or shape of the protein molecule. The question is, however, whether the specific enthalpy and entropy convergence at 110°C has any physical meaning, for these functions diverge upon further extrapolation. To obtain the answer, it is first necessary to analyze critically the extrapolation procedure used for the enthalpy and entropy curves, which were determined experimentally over a much smaller temperature range, and to test the assumption that the denaturation heat capacity increment is indeed temperature-independent up to 110°C, i.e., that the partial heat capacity of the native and denatured protein changes in parallel with increasing temperature.

Detailed calorimetric studies of aqueous solutions of proteins in a broader temperature range up to 130°C, which are possible under increased pressure, showed that the partial heat capacities of the native and denatured proteins vary with temperature in slightly different ways. While the partial heat capacity of the native protein is almost a linear function of temperature, that of the denatured protein has a definite curvature (Fig. 3–8).

Therefore, the difference between the heat capacities of the native and denatured states of a protein should decrease at increasing temperature, and one can expect that it should disappear at about 130° to 150°C. In that case, the enthalpy difference between the native and denatured states of protein should clearly be a nonlinear function of temperature: it should approach asymptotically some limit at 130° to 150°C. The same would be expected for the entropy difference. Figure 3–9 shows the enthalpy and entropy functions for only two proteins, myoglobin and ribonuclease A, respectively, with the largest and smallest denaturation heat capacity changes among all the compact globular proteins studied. The functions for all the other proteins lie between these two extreme cases.

It is most intriguing that the asymptotic limit for the specific enthalpy and the specific entropy differences of the native denatured states are very similar for all the compact globular proteins studies; it is about 5.0 kJ per mole of amino acid residues for the enthalpy difference and 17 JK^{-1} per mole of amino acid residues for the entropy difference. In fact, the asymptotic convergence to some universal values of the enthalpy and entropy functions for different globular proteins is not too surprising, because one would not expect the enthalpy and entropy of protein conformational transitions to increase infinitely with increasing temperature; they should

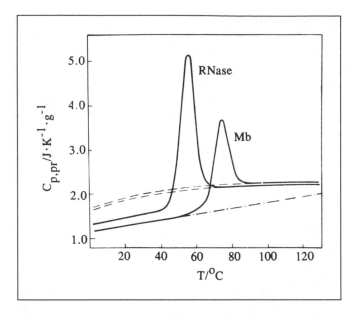

FIGURE 3–8. *Temperature-dependence of the partial heat capacity of pancreatic ribonuclease A (RNaseA) and sperm whale metmyoglobin (Mb) in solution at pH 4.4 (according to Privalov et al., 1989). The dashed line corresponds to the function for a completely unfolded protein: RNaseA without disulphide cross-links and the apo-form of Mb in acidic solution (pH 2.2).*

have some limit. On the other hand, the densities of all the compact globular proteins considered are almost the same. That is, their van der Waals contacts are very similar, and all these proteins are almost equally saturated by internal hydrogen bonds, which amount to about (0.75 ± 0.07) bond per amino acid residue (Privalov, 1979). Therefore, one would expect the specific enthalpies and entropies of their unfolding to be similar. Much more surprising is that this universal level is reached only at temperatures greater than 110°C. At lower temperatures, the specific enthalpy and specific entropy values of different proteins are very different, because these functions have different slopes, determined by the different values of their denaturation heat capacity increments.

7 STABILITY OF THE NATIVE STATE

Stability of any structure is measured by the work required for its disruption. In the case of small single-domain globular proteins, which have only two stable macroscopic states, the work for their transition from the native to the

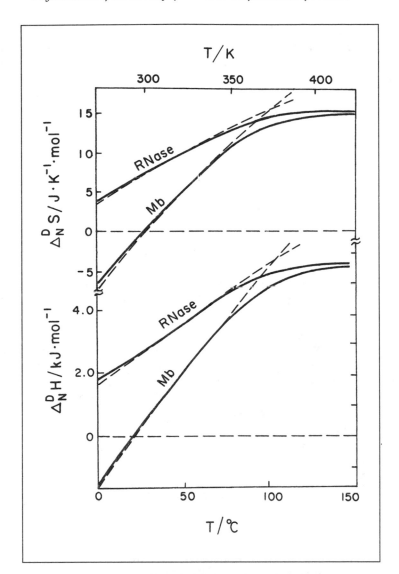

FIGURE 3–9. *Temperature-dependence of the specific enthalpy and entropy differences between the native and denatured states of Mb and RNaseA (per mole of amino acid residues) under solvent conditions providing the maximum stabilities of these proteins (Privalov et al., 1989). Dashed lines correspond to the case where the denaturation heat capacity increment is supposed to be temperature-independent.*

denatured state is simply the Gibbs energy difference of these states, which is presented by a simple combination of the enthalpy and entropy differences of these states. (See Equation (3–2).) Therefore, it is determined by the temperature-dependent balance between two factors, which themselves depend upon temperature. Substituting Equations (3–8) and (3–10) into Equation (3–2), we get for the Gibbs energy difference of the native and denatured states:

$$\Delta_N^D G\,(T) = \Delta_N^D H(T_G)\,{}^{(T_G - T)}/_T + \int_{T_G}^{T} Cp(T)\,\mathrm{d}T - T \int_{T_G}^{T} Cp(T)\,d + 1. \tag{3–13}$$

When $\Delta_N^D Cp$ is constant, this simplifies to

$$\Delta_N^D G(T) = \frac{T_G - T}{T_G} \cdot \Delta_N^D H(T_G) - (T_G - T)\Delta_N^D Cp + T \Delta_N^D Cp \ln{(T/T_G)}. \tag{3–14}$$

Figure 3–10 demonstrates that the value of $\Delta_N^D G(T)$ is not very sensitive to the variation of $\Delta_N^D Cp$ with temperature, at least within the temperature range close to physiological, which is most important.

The most remarkable feature of $\Delta_N^D G$ as a function of temperature is that, in contrast to $\Delta_N^D H$ and $\Delta_N^D S$, which increase monotonously with temperature, the Gibbs energy function is not monotonous, but has a clear maximum. This could be expected, because the enthalpy and entropy functions change their sign at some temperature, which means that their contribution to the Gibbs energy also changes in sign.

In general,

$$\frac{\partial \Delta G}{\partial T} = -\Delta S$$

and at the temperature where ΔG is at a maximum

$$\frac{\partial \Delta G}{\partial T} = 0. \tag{3–15}$$

Therefore, the Gibbs energy difference of the native and denatured states has a maximum at temperature T_S, where $\Delta_N^D S(T) = 0$.

At temperature T_S, the native protein conformation is stabilized only by the enthalpy term. With a further decrease in temperature, the enthalpy decreases to zero and then changes its sign, changing from a factor stabilizing the native state to a destabilizing one. At a sufficiently low temperature, T'_G, the Gibbs energy decreases to zero and the protein should unfold (Brandts, 1969). According to the thermodynamic prediction (Becktel and Schellman, 1987; Privalov, 1990), this should occur at the temperature

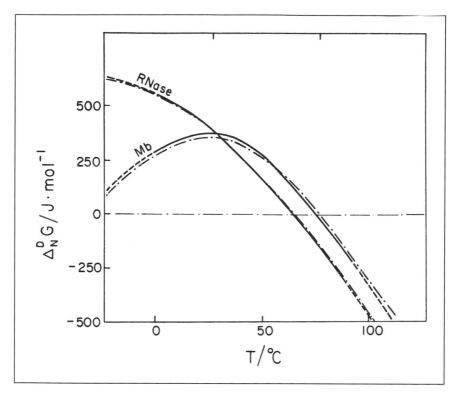

FIGURE 3–10. *The Gibbs free-energy difference between the native and denatured states of myoglobin and ribonuclease, in glycine buffer solutions providing maximum stability, calculated with the assumption that the heat capacity increment does not (dashed line) or does (solid line) depend upon temperature (Privalov and Gill, 1988).*

$$T'_G = \frac{1}{2}T_G^2 \left[\frac{\Delta_N^D H (T_G)}{\Delta_N^D Cp} + T_G \right]. \tag{3–16}$$

In contrast to high-temperature unfolding, which proceeds with heat absorption and therefore with an increase in enthalpy and entropy, low-temperature unfolding should occur with a heat release, because both the enthalpy and entropy have inverted their sign before the temperature where cold unfolding is expected. The predicted decrease in entropy upon protein unfolding contradicts the common sense expectation that protein unfolding involves a disordering of the highly ordered native conformation. Therefore, experimental verification of this prediction was of crucial importance for understanding how proteins fold into a native conformation.

8 COLD DENATURATION OF PROTEINS

The main difficulty in experimentally verifying the predicted cold denaturation was the temperature at which this phenomenon is expected. According to Equation (3–16) for all known proteins, it should take place below $273°K$, i.e., below the freezing point of water. In frozen aqueous solution, one cannot observe any protein conformational transitions.

There were some experimental observations that protein stability indeed decreases upon cooling below room temperature, but most of them were in the presence of denaturants, which complicates their interpretation. For reviews see Jaenicke, 1990; Privalov, 1990.) Attempts to depress the freezing point of water by using antifreeze agents also were not successful, as these substances influence the properties of not only water, but also of the protein. Supercooling of dust-free protein solutions was found to be much better. After removing the dust, which serves as nuclei for ice formation, an aqueous solution can be supercooled to relatively low temperatures (to at least -15°C). On the other hand, as implied by Equation (3–16), the temperature of cold denaturation should be highest for proteins with the largest $\Delta_N^D Cp$ and the smallest $\Delta_N^D H (T_G)$ values. Therefore, the expected temperature of cold denaturation can be raised by an appropriate choice of the protein and solvent conditions. For example, myoglobin has the largest $\Delta_N^D Cp$ among the proteins studied, and in acetate buffer solutions the low-temperature side of its Gibbs energy function should decrease to zero at temperatures above 0°C (Fig. 3–11).

Using various experimental techniques, myoglobin was indeed shown to unfold, with a decrease in helicity, disruption of the tertiary structure, an increase in the intrinsic viscosity, and a significant increase in mobility of all the groups, not only upon heating, but upon cooling as well (Figs. 3–12 and 3–13).

Protein cold denaturation was found to be a highly reversible process.

Using scanning microcalorimetric techniques, it was shown that protein unfolding upon cooling proceeds with a release of heat, while upon subsequent heating the protein folds back in the same temperature range by absorbing heat. Upon further heating, the protein unfolds again, but with heat absorption (Fig. 3–14).

Therefore, cold and heat denaturation occur with heat effects that have opposite signs, as they should according to the thermodynamic prediction. Upon raising the stability of the protein by increasing the pH, heat and cold denaturation shift in opposite directions along the temperature scale, with an increase in magnitude of the corresponding heat effects (Fig. 3–15).

Similar results have been obtained with a number of other globular proteins: apomyoglobin (Figs. 3–11 to 3–15), staphylococcal nuclease (Fig. 3–15), and T4 phage lysozyme. (For a review, see Privalov, 1990.) Cold denaturation is a very general property of globular proteins. If

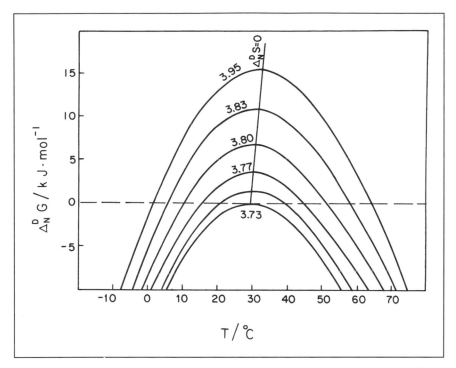

FIGURE 3–11. *The Gibbs free-energy difference between the D and N states ($\Delta_D^N G$) for metmyoglobin in acetate buffer solution at the pH values indicated on the curves (according to Privalov et al., 1986).*

it has not been observed with some of them, it is only because it occurs at too low temperatures. The presence of denaturants facilitates its observation (Griko et al., 1988a; Chen and Schellmann, 1989), because denaturants by preferential binding decrease the overall enthalpy of unfolding and therefore, in accord with Equation (3–16), increase the temperature at which cold denaturation occurs. Studies of cold denaturation in the presence of denaturants confirmed that cold denaturation results from the strong temperature dependence of the enthalpy and entropy of unfolding, which itself results from the denaturation heat capacity increment, i.e., from the difference in the heat capacities of the native and denatured states of a protein.

9 DENATURATION HEAT CAPACITY INCREMENT

Several explanations for the denaturation heat capacity increment can be suggested. The heat capacity might increase because of (1) gradual melting of residual structure in the protein with continued increase in temperature,

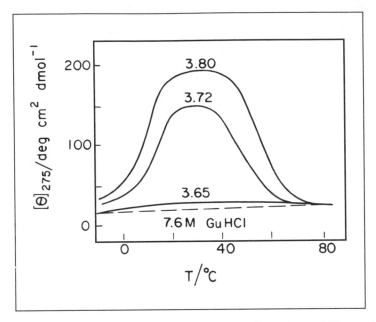

FIGURE 3–12. *Temperature-dependence of the CD ellipticity at 275nm of metmyoglobin, which is a measure of the native conformation, in acetate buffer solutions of different pH values (according to Privalov et al., 1986).*

(2) increase of the configurational freedom of the polypeptide chain upon disruption of the rigid compact native structure of the protein molecule, or (3) hydration of the groups that are exposed to water upon unfolding of the polypeptide chain. The possible contribution of residual structure to the thermodynamic properties of the denatured protein has been discussed in Section 5. It was shown there that the partial heat capacities of a protein denatured by temperature, acid, and denaturants are very similar and close to the partial heat capacity of the ideal unfolded polypeptide with noninteracting groups (Privalov and Makhatadze, 1990). This completely excludes suggestion (1) from consideration. Even without these experimental facts, however, it is clear that this suggestion is incorrect because it leads to the absurd conclusion that at low temperatures, where the enthalpy and entropy of denaturation become negative, the residual structure in the denatured protein is more extensive than the structure of the native protein.

Many attempts have been made to estimate the possible heat capacity effect of the increase in the configurational freedom of the polypeptide chain upon disruption of the rigid native protein structure. According to the indirect estimates of Sturtevant (1977) and Velicelebi and Sturtevant

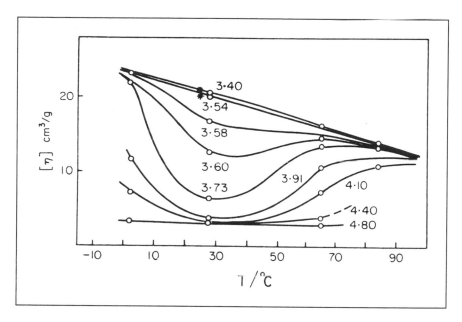

FIGURE 3–13. *Temperature-dependence of the intrinsic viscosity ([η]) of metmyoglobin in acetate buffer solutions of the indicated pH (according to Privalov et al., 1986).*

(1979), the contribution of this effect to the observed denaturation increment of protein heat capacity cannot be large. It was estimated experimentally by comparing the total denaturation heat capacity increment with the heat capacity effect caused by hydration of the protein groups that are exposed to water upon protein unfolding (Privalov and Makhatadze, 1990). The hydration heat capacity effect was found to amount to more than 80% of the heat capacity change of unfolding. Therefore, the heat capacity effect of the configurational freedom gained upon protein unfolding is less than 20% of the total heat capacity increase, and hydration plays the major role in determining the heat capacity change upon unfolding. According to this analysis, the main contributors to the hydration heat capacity effect are the nonpolar groups of proteins that are exposed to water upon unfolding. Their contribution is especially large at low temperatures and decreases with increasing temperature. As for the hydration contribution of the polar groups, it is negative at low temperatures, becomes zero at room temperature, and slowly increases at higher temperatures. The superposition of these two tendencies results in a bell-shaped function of the total hydration heat capacity effect (Fig. 3–16).

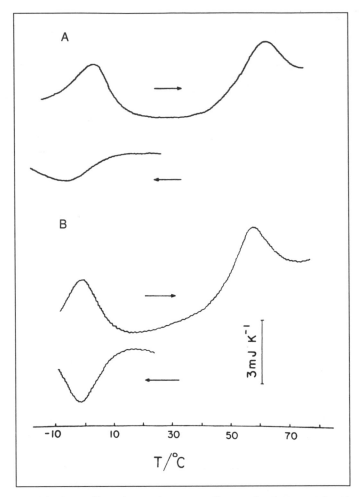

FIGURE 3–14. *The heat effect observed upon cooling and subsequent heating of metmyoglobin (A) and of apomyoglobin (B) by the scanning microcalorimeter (according to Privalov et al., 1986, and Griko et al., 1988b). The shift of the heat effects upon cooling and heating of metmyoglobin at low temperatures is due to the slow kinetics of refolding.*

The specific hydration heat capacity effect is different for different proteins, and it is noteworthy that its magnitude correlates well with the surface area of the nonpolar groups exposed to water upon protein unfolding (Fig. 3–17). Thus, it is clear that the denaturation heat capacity increment is caused primarily by transfer of the internal nonpolar groups to water upon protein unfolding.

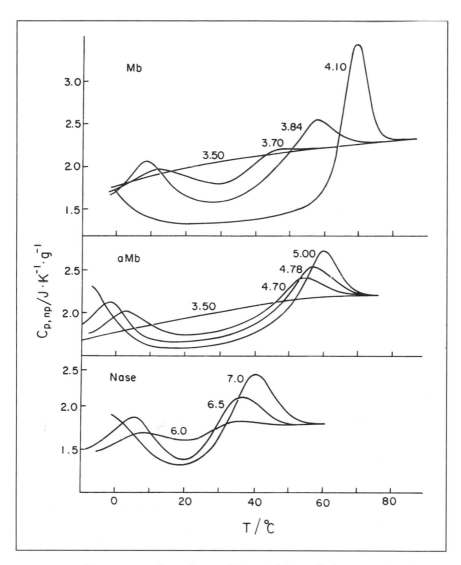

FIGURE 3–15. *Temperature-dependence of the partial specific heat capacity of metmyoglobin (Mb) (according to Privalov et al., 1986), apomyoglobin (aMb) (according to Griko et al., 1988b), and staphylococcal nuclease (Nase) (according to Griko et al., 1988a).*

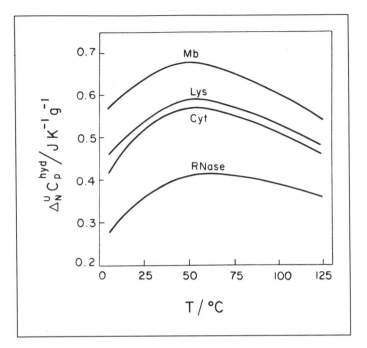

FIGURE 3–16. *Temperature-dependence of the hydration heat capacity increment for four globular proteins: metmyoglobin (Mb), hen egg-white lysozyme (Lys), cytochrome c (Cyt), and pancreatic ribonuclease A (RNase) (according to Privalov and Makhatadze, 1990).*

10 HYDRATION OF NONPOLAR GROUPS

That transfer of nonpolar molecules into water leads to a significant increase in heat capacity has been known for a long time (Edsall, 1935; Gill and Wadsö, 1976). As was mentioned in Section 2, this is often taken to indicate that the nonpolar groups are increasing the order of the surrounding water. The heat capacity increment provides evidence not only for the existence of additional ordering of water in the presence of the nonpolar solute, but also that the extent of this ordering decreases as the temperature increases, i.e., that the solvated water melts gradually upon heating the solution. Correspondingly, one would expect that a temperature increase should result in a decrease in the absolute value of the negative entropy of transfer of the nonpolar solute to water; at some sufficiently high temperature, T_S, it should become zero (Fig. 3–18). It is pertinent that this temperature T_S is universal for all the substances studied. Its value is about 140°C if one takes into account the decrease in the heat capacity increment

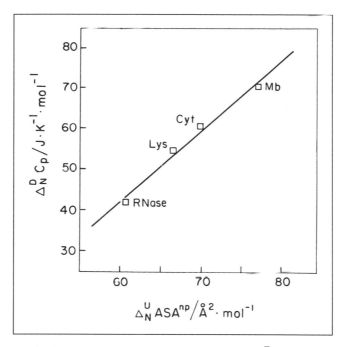

FIGURE 3–17. *The denaturation heat capacity increment* ($\Delta_N^D Cp$) *as a function of the surface area of the nonpolar groups exposed to water upon protein unfolding* ($\Delta_N^u ASA^{np}$) *(according to Privalov and Makhatadze, 1990). Both values are calculated per mole of amino acid residue.*

of these substances with increasing temperature (Baldwin, 1986; Privalov and Gill, 1988, 1989).

The temperature increase should also lead to an enthalpy increase, according to Equation (3–8). If the enthalpy of transfer of a nonpolar substance from the pure liquid phase to water is zero at room temperature (see Section 2), it becomes large and positive at 140°C (Table 3–2).

Unfortunately, this enthalpy of transfer cannot be determined with sufficient accuracy in all cases. Being estimated by extrapolation of the value obtained at room temperature, its accuracy depends upon the precision of the heat capacity determination, which decreases with the decreased solubility of the solute. The precision is especially low for pentane and hexane, due to their very low solubilities. (See Table 3–1.) But even from these data, the close correspondence between the enthalpy of transfer and the enthalpy of vaporization of the nonpolar substance at 140°C is apparent and can hardly be accidental. From the formal thermodynamic point of view, this means that the hydration effect of the nonpolar groups disappears at the temperature T_S.

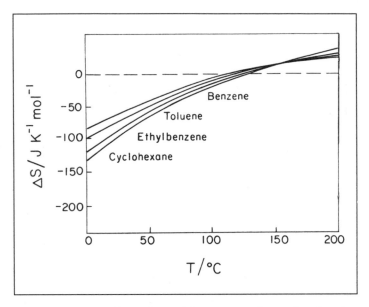

FIGURE 3–18. *Temperature-dependence of the entropy of transfer of various nonpolar substances from the pure liquid phase to water (according to Privalov and Gill, 1988).*

By hydration effect, we mean here all the complex effects associated with the transfer of noninteracting nonpolar molecules into water. This includes the cavity formation in water, the interaction between water and the nonpolar molecules inserted into the cavity, and any rearrangement of the water structure around the cavity. Zero hydration means that the components of this complex effect compensate each other at about 140°C. As a result, transfer of the nonpolar substances into water at 140°C is

TABLE 3–2. *Enthalpy of Transfer from the Pure Liquid Phase to Water and Enthalpy of Vaporization from This Phase for Some Nonpolar Substances at 140°C*

Substance	Benzene	Toluene	Ethylbenzene	Cyclohexane	Pentane	Hexane
ΔH^{trans}/kJ·mole^{-1}	25	28	35	36	39	46
ΔH^{vap}/kJ·mole^{-1}	24	32	36	29	21	26

Source: Privalov and Gill (1989).

determined entirely by the van der Waals interactions between the solute molecules in the liquid phase. It follows then that the van der Waals interactions between the nonpolar molecules and between them and water do not compensate each other as was originally supposed. (See Section 2.) Otherwise, it is not possible to explain the large enthalpy of transfer of nonpolar solutes into water at 140°C.

Let us, however, consider the Gibbs energy of transfer at this temperature, T_S. If we take into account that the entropy is a temperature derivative of the Gibbs energy (see Equation (3–15)) and that zero entropy means that the Gibbs energy has an extremum at the corresponding temperature, we come to the important conclusion that the Gibbs energy of transfer of a nonpolar solute from the pure liquid phase to water, $\Delta_1^W G$, is maximal at T_S, i.e., at about 140°C (Fig. 3–19).

The Gibbs free energy of transfer of a nonpolar solute from the pure liquid phase to water is a measure of the expelling action of water on this solute, i.e., a measure of the hydrophobic interaction. As we see, its maximum value is reached at a temperature where the solvation effect of water by the nonpolar solute is zero and the entire value of $\Delta_1^W G$ is provided by the van der Waals interactions between the nonpolar molecules. The decrease of $\Delta_1^W G$ from its maximum value with decreasing temperature

FIGURE 3–19. *Temperature-dependence of the Gibbs free energy of transfer of various liquid nonpolar substances to water (according to Privalov and Gill, 1988).*

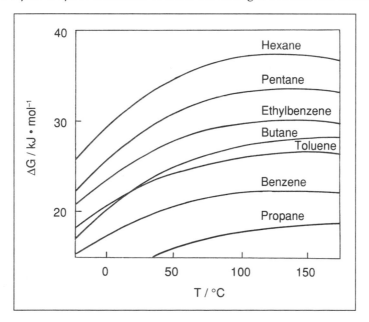

occurs simultaneously with the appearance of signs of water solvation of the nonpolar solute: the negative entropy, the lower enthalpy, and the significant heat capacity increment of transfer. Without hydration, without the heat capacity increment caused by water solvation, the Gibbs energy of transfer would not change as the temperature decreases from T_S. This means that the Gibbs free energy of hydration of nonpolar solutes is negative: it equals zero at T_S and increases in absolute value at decreasing temperatures (Fig. 3–20). The dominant role in determining the hydration Gibbs energy value is clearly played by the enthalpy of hydration, which should also be negative, because the hydration entropy is negative and, therefore, its contribution to the Gibbs energy is positive.

It follows from the preceding that the hydrophobic interaction is caused by the van der Waals interactions between the nonpolar molecules. The hydration of these molecules has only a negative effect, and it increases the solubility of these solutes in water (Shinoda, 1977; Privalov and Gill, 1988, 1989).

As was mentioned previously, the Gibbs free energy of hydration of nonpolar solutes is in itself an integral quantity, consisting, as one can

FIGURE 3–20. Temperature-dependence of the hydration Gibbs free energy calculated per mole of water molecules in the hydration shell for several nonpolar substances (according to Privalov and Gill, 1988).

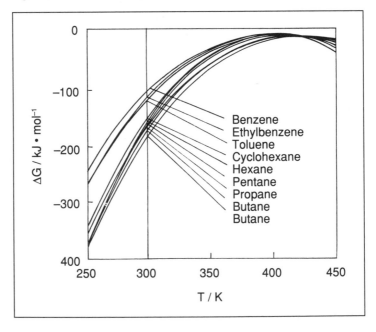

imagine, of many components, e.g., the work associated with cavity formation, the energy of the solute–solvent interactions, and the work involved in the rearrangement of the solvent molecules (water ordering) around the cavity. One can evaluate these components only by using molecular models, which in themselves are far from being flawless (Pohorille and Pratt, 1990). Thus, according to the scaled particle theory for the hard spherical molecules, the Gibbs energy of cavity formation in the solvent is positive, depends upon the relative size of the solvent and solute molecules, and increases with increasing temperature (Lucas, 1976; Lee, 1985). This component is expected to be especially large in water, which has the smallest molecules of any liquid. Nevertheless, as we have seen, in aqueous solution the total Gibbs free energy of hydration of a nonpolar solute is negative. This means that upon the dissolution of a nonpolar solute in water, the other components of the hydration process more than compensate for the work required for cavity formation.

Including within the "hydrophobic interaction" the combined effect of the van der Waals interaction between nonpolar groups and of the hydration of these groups, one can see that this integral effect should increase with increasing temperature. The "hydrophobic interaction" increases, however, due to a decrease in the hydration contribution at increasing temperature (and not to its increase, as originally believed) because the van der Waals and hydration contributions differ in their signs.

An important consequence of the fact that the van der Waals and hydration effects contribute to the hydrophobic effect with opposite signs is the biphasic character of their combined effect. Indeed, the van der Waals interaction is a short-range one and the hydration effect is a long-range one, so it is evident that the "hydrophobic interaction" should be attractive at a short distance and repulsive at a long distance (exceeding the size of a water molecule).

11 HYDRATION OF PROTEIN GROUPS UPON UNFOLDING

There are clear similarities and distinct differences between the enthalpy and entropy functions for protein denaturation and those for transfer of a nonpolar substance to water. In both cases, these functions increase with increasing temperature, in both cases they converge, and, what is most impressive, the temperature of their convergence is the same, about 140°C. As discussed previously, this is the temperature at which the net hydration effects of nonpolar groups are zero. Thus, one can assume that the enthalpy and entropy of protein denaturation at 140°C do not include the effects of hydration, but correspond only to the conformational transition of the polypeptide chain, involving a disruption of the intramolecular bonds that maintain the compact protein structure. The principal difference between protein denaturation and transfer of a nonpolar substance to water is in the entropy values: while the entropy of transfer of a nonpolar solute from the pure liquid phase to water is zero at 140°C, the entropy of protein unfolding is large and positive. This difference in the entropy values at a temperature where hydration effects are absent could indicate that the protein interior is not a liquid-like nonpolar phase, but a crystal-like phase that is specified by a definite positive entropy of melting. The magnitude of this entropy is 17 JK^{-1} per mole of amino acid residues (Privalov, 1979), a value that corresponds to an eightfold increase of the number of possible configurations for each residue and is close to the value expected for the helix-coil transition of polypeptides (Schellmann, 1955).

The main thermodynamic consequence of this aspect of a protein molecule is that the entropy of protein unfolding becomes zero at a much lower temperature than the entropy of transfer of nonpolar substances from the liquid phase to water. For most of the known proteins, this temperature is between 0° and 30°C, and at this temperature the compact native protein structure is most stable (Fig. 3–11).

If protein denaturation were to proceed without a heat capacity increase, the enthalpy and entropy change would not depend upon temperature, but would have constant values $\Delta_N^D H°$ and $\Delta_N^D S°$, respectively. The Gibbs free energy difference between the native and denatured states ($\Delta_N^D G° = \Delta_N^D H° - T \Delta_N^D S°$) would be a linearly decreasing function of temperature (Fig. 3–21), because the dissipative force of the thermal motion ($T \Delta_N^D S°$) is proportional to temperature. When protein denaturation proceeds with an increase in heat capacity, the Gibbs free energy difference between the native and denatured states deviates from linearity. This deviation is always negative and increases in magnitude at lower temperatures, tending to decrease the overall Gibbs energy value. As a result, at some sufficiently low temperature the stability of the native state of a protein can decrease to zero, and the protein will denature upon cooling.

The larger $\Delta_N^D Cp$ is for a given protein, the larger is the destabilizing hydration effect. Therefore, with an increase of $\Delta_N^D Cp$ the overall stability of protein $\Delta_N^D G$ decreases, and its maximum value shifts to a higher temperature. This is illustrated in Figure 3–10 where the Gibbs free energy functions for ribonuclease and metmyoglobin are given. These two proteins differ to a great extent in the $\Delta_N^D Cp$ value, which is much larger for metmyoglobin than for ribonuclease. (See Figure 3–9.) Correspondingly, the Gibbs energy difference for myoglobin is smaller, and its maximum is at a higher temperature. It appears that proteins have an upper limit for a denaturation heat capacity increment as well as an upper limit for the transition temperature, i.e., stability (Fig. 3–22). Therefore, one would suppose that extremely thermostable proteins from thermophilic organisms

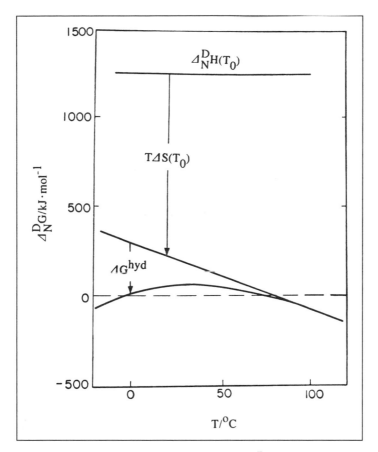

FIGURE 3–21. *Contribution of the dissipative force, $T\Delta_S^D S$, and hydration effect, ΔG^{hyd}, to the stabilization of an abstract globular protein (according to Privalov and Gill, 1988).*

should have some additional special devices for stabilization of their structure, e.g., disulfide cross-links, and salt links.

As shown previously, the denaturation heat capacity increment results mainly from hydration of protein nonpolar groups that are exposed to water upon protein unfolding (Fig. 3–17). Therefore, the decrease in the Gibbs free energy difference between the native and denatured states of a protein is caused mainly by hydration of internal nonpolar groups of the protein that occur upon unfolding. As we have already seen in the previous paragraph, the Gibbs free energy of hydration of nonpolar groups is indeed negative and increases in magnitude with decreasing temperature.

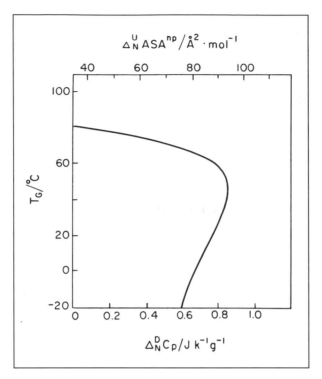

FIGURE 3–22. *Protein denaturation temperature as a function of its denaturation specific heat capacity increment ($\Delta_N^D C_p$) or of the relative surface area of the nonpolar groups ($\Delta_N^u ASA^{np}$) (per amino acid residue) that are exposed to water upon unfolding of the compact native conformation.*

Therefore, the native compact protein structure appears to be stabilized only by the intramolecular bonds, which provide the positive enthalpy of protein denaturation in the absence of hydration effects, i.e., at 140°C. Water solvation of nonpolar groups has only a destabilizing effect, and this sets the upper limit on the content of the nonpolar groups in a globular protein (Fig. 3–22).

The existence of two opposite components in stabilizing the compact native protein structure was illustrated by kinetic studies of protein unfolding. It has been shown by Segawa and Sugihara (1984) on hen egg-white lysozyme and by Chen et al. (1989) on phage T4 lysozyme that the unfolding of these proteins proceeds through an unstable transition state. The enthalpy of the native protein proceeding to this transition state is positive, while that from this state to the unfolded one is accompanied by negative enthalpy and entropy changes; it is at this stage that the heat capacity increase that is specific for protein unfolding occurs. Therefore, it is clear that the second stage is associated with hydration of internal groups of the protein, while the first stage should correspond to disruption of the internal bonds, maintaining a compact native structure of a protein. The transition state can be considered as a "dry molten globule" state of the protein, i.e., a state with diminished noncovalent interactions, but without water inside. The instability of this state is further evidence that the hydration of protein nonpolar groups is a thermodynamically favorable process. All attempts to find experimentally a stable "dry molten globule" state, which was postulated from the opposite point of view of the hydration of nonpolar groups, have failed (Pfeil et al., 1986).

12 MECHANISM OF STABILIZATION OF THE NATIVE PROTEIN STRUCTURE

From the preceding analysis, the compact native protein structure is stabilized by temperature-independent enthalpic forces, which can be revealed by raising the temperature to a point where hydration effects become insignificant. Similar forces are displayed by the process of transfer of nonpolar substances from the pure liquid phase to water. In that case the close correspondence between the enthalpies of transfer of nonpolar substances from the pure liquid phase into water and their enthalpies of vaporization suggested that these forces, which are responsible for the low solubility of nonpolar substances in water, are simply the van der Waals interactions between the nonpolar molecules (Privalov and Gill, 1988,1989). One can then assume that the similar van der Waals interactions within proteins are contributing to the enthalpy of protein denaturation at 140°C. In the case of a protein, however, one cannot consider the van der Waals interactions to be the only interactions responsible for the denaturation enthalpy at this temperature. If the interactions with water do not compensate completely for the van der Waals interactions between the nonpolar atoms, as was originally supposed to have occurred (see Section 2), why should they compensate completely for the contribution of the hydrogen bonding between the protein polar groups? The largely fluctuating hydrogen bond of the protein polar group with relatively mobile water can hardly be on average identical to the bonds between the polar groups of proteins, which are fixed in its rigid native structure (Creighton, 1985). Recent calorimetric studies of dissolution of cyclic dipeptide crystals in water show that the contribution of hydrogen bonds to the formation of a crystal are substantial, and an amino acid crystal clearly is the closest model of a native protein interior (Murphy and Gill, 1990). There is insufficient experimental data as of 1991, however, to estimate the probable contribution of hydrogen bonds and van der Waals interactions to the protein denaturation enthalpy. This is also impractical, with current accuracy of the microcalorimetric technique, by analyzing thermodynamic functions of various proteins, because globular proteins differ little in their hydrogen bond content or in their contacts between nonpolar groups per amino acid residue (Privalov, 1979; Privalov and Gill, 1988).

For estimating the contribution of individual residues to stabilization of protein structure, little information has been obtained yet by using site-directed mutagenesis, because substitution of a residue in the cooperative native structure leads to the deformation of this structure, and it is not easy to take this factor into account (Alber, 1989).

Therefore, at the present time, one can only state that the native compact protein structure is stabilized by intramolecular enthalpic bonds, primarily the van der Waals interaction and hydrogen bonds. As for the effect of water solvation, particularly of nonpolar groups, it has only a destabilizing action. Thus, the protein compact native structure is stable at temperatures where the destabilizing actions of the water solvation and of dissipative forces are relatively small. At higher temperatures this structure breaks down due to an increase in the dissipative force, whereas at lower temperatures it breaks down as a result of an increase of the solvation tendency caused by the hydration of nonpolar groups (Fig. 3–21).

Considering the van der Waals interaction between nonpolar groups and their hydration as a combined effect, the "hydrophobic interaction," one can state that upon increasing the temperature up to 140°C, the "hydrophobic interaction" between protein nonpolar groups increases. On the other hand, one can regard the breakdown of the protein compact structure upon cooling as a result of weakening of the "hydrophobic interaction" of the nonpolar groups in a protein. Although the term *hydrophobic interaction* is widely accepted, its use hardly clarifies the real situation in proteins: indeed, as we have already seen, the hydrophobic interaction not only decreases with decreasing temperature, but it can also change its sign at low temperature and *invert* from a factor stabilizing the compact state to a destabilizing factor. On the other hand, we cannot at present separate the van der Waals and hydrogen bonding contributions to stabilization of protein structure and, hence, cannot estimate quantitatively the net effect of the so-called hydrophobic interaction. All of this raises the question of whether the use of the concept of hydrophobic interaction is indeed justified in considering protein stability (Privalov and Gill, 1988).

13 CONCLUSION

We have seen that a temperature decrease leads to increased hydration of protein nonpolar groups, which is apparent in the negative entropy and negative enthalpy effects. Domination of the hydration enthalpy by the entropy results in a decrease in protein stability upon decreasing the temperature. It is remarkable, however, that at temperatures close to physiological, the entropy of hydration of protein internal groups that are exposed upon protein unfolding almost completely balances the entropy due to the conformational transition of the polypeptide chain into a random-coil conformation. Therefore, the overall entropy change of protein unfolding and refolding is close to zero at physiological temperatures. Correspondingly, the energy expenditure for folding of the native conformation should not be too great. The Gibbs energy value for folding and stabilization of the native protein structure of a single-domain protein does not exceed 50 kJ mole^{-1} (Privalov, 1979; Pfeil, 1981). A cooperative domain usually includes about 100 amino acid residues, so it appears that the contribution of each of the residues to stabilization of the native structure does not exceed 500 J per mole of residue. This value is five times smaller than the energy of thermal motion at room temperature, RT, where R is the gas constant. It follows that a protein has an ordered native structure only because it is a cooperative system, and its components can change their state only simultaneously. In other words, the stability of such a system is determined by the total contribution of all the components of the system.

Thus, the secret of the stability of the native structure of protein is not in the magnitude of intramolecular interactions, which are always too weak to withstand individually the dissipative action of thermal motion, but in the effective cooperation of these interactions.

As of late 1991, we know little about the mechanism of cooperativity of the intramolecular interactions in proteins. Extreme cooperativity, when all the elements of system are integrated into a single unit, seems to be achieved only in molecules with a tight and unique packing of groups. In other words, extreme cooperativity is a peculiarity of the aperiodic structure (Privalov, 1979,1982). It looks as if only such a structure can provide the complex interlacing of all the short- and long-range interactions between groups in the polypeptide chain that is necessary for their cooperation. Also, it is likely that the individual interactions will depend significantly upon their direct surroundings.

For example, one can expect hydrogen bonding in a polar pair to be stronger if this pair is isolated from water by the nonpolar groups. Screening of the polar pair from water by the nonpolar groups requires an expenditure of some work, i.e., proceeds through an activation barrier, which makes the formation of such a system cooperative. Its cooperativity is also enhanced by the biphasic character of the hydrophobic interactions between nonpolar groups outlined previously, which are likely to be attractive at short distances and repulsive at long distances. (See Section 10.) The increase of the number of nonpolar contacts should result in an increase in the magnitude of both components of the hydrophobic interaction:

the attractive one caused by the van der Waals interactions and the repulsive one caused by hydration of the nonpolar groups. In other words, it should increase the activation barrier for assembling and disassembling such a system, i.e., the cooperativity of its formation. The increase in cooperativity seems to be the main impact of nonpolar contacts in the formation of the compact native protein structure (Privalov, 1982).

Extreme cooperativity should be realized only when all the groups of the polypeptide chain are in a certain configuration. All other configurations of these groups must be far less stable and hence must be rapidly replaced, until only the correct configuration is reached, which corresponds to the native state. This is especially clear with the example of the polar groups along the polypeptide chain, which, if not correctly placed in pairs, cannot form hydrogen bonds in a nonpolar surrounding. However small the positive contribution of a hydrogen bond might be in the maintenance of protein structure, the negative contribution of unpaired polar groups in the protein interior should be very large, about -20 kJ mole^{-1}, i.e., one third of the entire energy of stabilization of the cooperative domain structure.

As follows from the preceding, the lower limit on the size of the cooperative domain is determined by the requirement that it should be stable. To be sufficiently stable at physiological temperatures, a domain should include at least 50 amino acid residues. The upper size limit of the cooperative unit is likely to be determined by difficulties in the formation of a completely integrated cooperative aperiodic structure, which increases rapidly with an increase in the number of amino acid residues. Therefore, the cooperative domain usually does not include more than 200 amino acid residues (Privalov, 1982). Thus, aperiodicity and discreteness are closely interrelated properties of a protein: the ordered aperiodic structure must be discrete, while a discrete structure cannot be regular and must be aperiodic (Privalov, 1989). Both of these properties are based upon the ability of weak intramolecular forces to cooperate. Studying the detailed mechanism of their cooperation is, perhaps, the most pressing physical problem in current studies of protein research.

Acknowledgments

This chapter was written in Regensburg during my stay at the Institute of Biophysical Biochemistry as a Humboldt Awardee. I am greatly thankful for Professor Rainer Jaenicke and Professor Hans-Jürgen Hinz for the stimulating atmosphere that I enjoyed while working on it.

REFERENCES

Alber, T. (1989) *Ann. Rev. Biochem. 58*, 765.
Anfinsen, C. B. (1973) *Science 181*, 223.

Baldwin, R. L. (1986) *Proc. Natl. Acad. Sci. USA 83*, 8060.

Becktel, W., and Schellman, S. A. (1987) *Biopolymers 26*, 1859.

Brandts, J. F. (1969) in *Structure and Stability of Biological Macromolecules*, Timasheff, S. N., and Fasman, G. D., eds., Marcel Dekker, New York, p. 213.

Chen, B. -L., and Schellman, J. A. (1989) *Biochemistry 28*, 685.

Chen, B. -L., Baase, W. A., and Schellman, J. A. (1989) *Biochemistry 28*, 691.

Creighton, T. E. (1985) *J. Phys. Chem. 89*, 2452.

Edsall, J. T. (1935) *J. Am. Soc. 57*, 1506.

Frank, H. S., and Evans, M. V. (1945) *J. Chem. Phys. 13*, 507.

Freire, E., and Biltonen, R. L. (1978) I. *Biopolymers. 17*, 463.

Gill, S. J., and Wadsö, I. (1976) *Proc. Natl. Acad. Sci. USA 73*, 2955.

Griko, Yu. V., Privalov, P. L., Sturtevant, J. M., and Venyaminov, S. Yu. (1988a) *Proc. Natl. Acad. Sci. USA 85*, 3343.

Griko, Yu. V., Privalov, P. L., Venyaminov, S. Yu., and Kutyshenko, V. P. (1988b) *J. Mol. Biol. 202*, 127.

Hollecker, M., and Creighton, T. E. (1982) *Biochim. Biophys. Acta 701*, 395.

Jaenicke, R. (1990) *Phil. Trans. Roy. Soc. London B 326*, 535.

Kauzmann, W. (1959) *Adv. Protein Chem. 14*, 1.

Lee, B. (1985) *Biopolymers 24*, 813.

Lucas, M. (1976) *J. Phys. Chem. 80*, 359.

Lumry, R., Biltonen, R. L., and Brandts, J. F. (1966) *Biopolymers 4*, 917.

Makhatadze, G. I., and Privalov, P. L. (1990) *J. Mol. Biol. 213*, 375.

Murphy, K. P., and Gill, S. J. (1990) *J. Chem. Thermodynamics*, in press.

Pfeil, W. (1981) *Mol. Cell. Biochem. 40*, 3.

Pfeil, W., Bychkova, V. E., and Ptitsyn, O.B. (1986) *FEBS Letters 189*, 287.

Pfeil, W., and Privalov, P. L. (1976a) *Biophys. Chem. 4*, 23.

Pfeil, W., and Privalov, P. L. (1976b) *Biophys. Chem. 4*, 33.

Pfeil, W., and Privalov, P. L. (1976c) *Biophys. Chem. 4*, 41.

Pohorille, A., and Pratt, L. R. (1990) *J. Am. Chem. Soc. 112*, 5066.

Privalov, P. L. (1979) *Adv. Protein Chem. 33*, 167.

Privalov, P. L. (1982) *Adv. Protein Chem. 35*, 1.

Privalov, P. L. (1989) *Ann. Rev. Biophys. Biophys. Chem. 18*, 47.

Privalov, P. L. (1990) *CRC Crit. Rev. Biochem. Mol. Biol. 25*, 281.

Privalov, P. L., and Gill, S. J. (1988) *Adv. Protein Chem. 39*, 193.

Privalov, P. L., and Gill, S. J. (1989) *Pure Appl. Chem. 61*, 1097.

Privalov, P. L., Griko, Yu. V., Venyaminov, S. Yu., and Kutyshenko, V. P. (1986) *J. Mol. Biol. 190*, 487.

Privalov, P. L., and Khechinashvili, N. N. (1974) *J. Mol. Biol. 86*, 665.

Privalov, P. L., and Makhatadze, G. I. (1990) *J. Mol. Biol. 213*, 385.

Privalov, P. L., and Plotnikov, V. V. (1989) *Thermochim. Acta 138*, 257.

Privalov, P. L., and Potekhin, S. A. (1986) *Methods Enzymol. 134*, 4.

Privalov, P. L., Tiktopulo, E. I., Venyaminov, S. Yu., Griko, Yu. V., Makhatadze, G. I., and Khechinashvili, N. N. (1989) *J. Mol. Biol. 205*, 737.

Schellman, J. A. (1955) *C. R. Trav. Lab. Carlsberg, ser. chim. 29*, 230.

Segawa, S., and Sugihara, N. (1984) *Biopolymers 23*, 2473.

Shinoda, K. (1977) *J. Phys. Chem. 81*, 1300.

Shoemaker, K. R., Kim, P. S., York, E. Y., Stewart, J. M., and Baldwin, R. L. (1987) *Nature 326*, 563.

Sturtevant, J. M. (1974) *Ann. Rev. Biophys. Biophys. Bioeng. 3*, 35.

Sturtevant, J. M. (1977) *Proc. Natl. Acad. Sci. USA 74*, 2236.

Tanford, C. (1968) *Adv. Protein Chem. 23*, 122.

Tanford, C. (1980) *The Hydrophobic Effect: Formation of Micelles and Biological Membranes*, Wiley Interscience, New York.

Tanford, C., and Aune, K. C. (1970) *Biochemistry 9*, 206.

Velicelebi, G., and Sturtevant, J. M. (1979) *Biochemistry 18*, 1180.

Wyman, J., and Gill, S. J. (1990) *Functional Chemistry of Biological Macromolecules* University Science, Mill Valley, Calif.

4

Protein Folding: Theoretical Studies of Thermodynamics and Dynamics

MARTIN KARPLUS
EUGENE SHAKHNOVICH

1 INTRODUCTION

The properties of proteins, as of all molecules, are governed by their potential energy surfaces. Because a protein subunit is a single molecule, all of the configurations for the polypeptide chain (folded, denatured, and the transition from one to the other) are accessible without bond making or bond breaking steps. In saying this, we neglect the synthesis and degradation of the polypeptide chain. Thus, the conformational space of a given primary sequence is composed of a vast set of configurational isomers, whose relative free energies are determined by the form of the potential surface. (Fig. 4–1A). Among these conformations there is a small subset that is confined to a rather well defined region of low energy that corresponds to the folded or native state. Much has been learned in recent years from experiment and theory about the nature of the potential surface in the neighborhood of the native state (Brooks et al., 1988). It is known that under appropriate conditions (temperature solvent) a given amino acid sequence has a unique average structure (that observed by X-ray crystallography or in solution [NMR]) corresponding to the native state and that significant motions occur relative to that average structure at room temperature. The conformational space sampled by the native protein (see Figure 4–1B) has multiple minima (substates), which have rms coordinate differences in the

---

Note: This chapter is dedicated to Cyrus Levinthal, whose originality and insight foreshadowed many of the recent developments in theoretical approaches to protein folding.

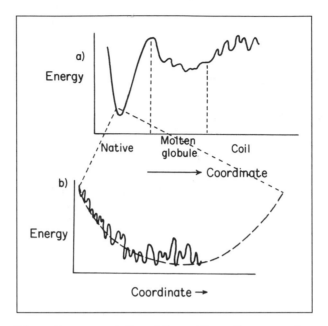

FIGURE 4–1. *Schematic representation of the configuration space of a protein giving the energy as a function of a configurational coordinate: (A) complete space. (B) enlarged view in the vicinity of the native state.*

range of a few tenth of angstroms to a few angstroms (Austin et al., 1975; Elber and Karplus, 1987). These are sampled on a time scale of 0.1 psec to 10 or so nsec at room temperature.

Since the fluctuations in the native state of a protein are relatively small and fast, it is possible to perform molecular dynamic simulations that treat the protein at an atomic level of detail. As a result, a relatively complete description of the thermodynamics and dynamics of the native state is available (Brooks et al., 1988). For cases where detailed comparison between simulations and experiment are possible (e.g., incoherent neutron scattering (Smith et al., 1990), geminate recombination kinetics of myoglobin (Case and Karplus, 1979; Elber and Karplus, 1990; Petrich et al., 1991), the simulation results agree with experiment and are able to provide details to supplement the inferences based on the experimental data. In all cases the multiple minima associated with the native state have essentially the same fold, as indicated by neighbor contact maps, for example. In myoglobin, which has been studied in some detail by molecular dynamics (Elber and Karplus, 1987), it has been shown that the minima correspond to different orientations of the helices, which move as essential rigid bodies, and that these reorientations are associated with rearrangements of the loop regions connecting them and with correlated motions of the side chains that are involved in the interhelix contacts. The different structures found on ligand binding are generally in the range sampled by the liganded or the unliganded protein in its room temperature motions. Also, for all the proteins with a given fold (e.g., the globin fold), the known X-ray structures determined for different homologous sequences are found to be within the same range.

Although these fluctuations are small, they often play an essential role in protein function. In myoglobin, for example, the ligand cannot get in or out of the heme pocket in the rigid protein (Perutz et al., 1965; Case and Karplus, 1979; Elber and Karplus, 1990). In most other proteins, the small rapid fluctuations serve as a "lubricant" to make possible the larger-scale motions (Brooks et al., 1988). Perturbation of the average structure by the binding of ligands, as in the hinge bending mode found in many enzymes (Bruccoleri et al., 1986) and in the quaternary transition in hemo-globin (Gelin et al., 1983; Perutz, 1989), are made possible by the existence of the rapid fluctuations.

The substates associated with the native state represent only a small portion of the conformation space available to the polypeptide chain(s) of which a protein is composed. (See Figure 4–1A.) Although the native state is the one that governs protein function, the remainder of the conformation space plays an essential role as well, since it corresponds to the denatured state and the reaction path or paths between the denatured and the native state that lead to protein folding and unfolding. Much less is known about the details of the potential surface governing this vast region of conformation space, which includes a wide range of structures that may differ by tens of angstroms in rms values. Concomitantly, instead of the time scale of picoseconds to nanoseconds that is required for exploring the native states, the characteristic times corresponding to the motions in the full conformation space are in the range from nsec to hours. The existence of such a hierarchy of time and length scales makes it possible to introduce two very important concepts that yield a simplified but useful description of a protein chain, which serves as a basis for much of the theoretical work on protein folding. The first is the potential of mean force and the second is a discretized description of the polypeptide chain isomers. Both of these concepts are based on the idea of "pre-averaging" of the small-scale motions to obtain a "coarse-grained" model that treats a molecule at "medium resolution," i.e., on the time and length scales at which protein folding occurs.

The concept of a potential of mean force is widely used in statistical mechanics and simulations of complex systems (McQuarrie, 1976; Brooks et al., 1988). A protein molecule has many different types of internal degrees of freedom (e.g., bond lengths, bond angles, main chain and side chain dihedral angles). In addition, the properties of a protein molecule or the polypeptide chain of which it is composed depend upon the solvent environment. For many aspects of protein thermodynamics and dynamics, some of these degrees of freedom need not be treated explicitly, since their relaxation times are much faster and their amplitudes are much smaller than the motions of interest. For protein folding, for example, the conformational properties are mainly governed by the dihedral angles of the polypeptide chain, which tend to have significantly longer relaxation times than the other protein degrees of freedom, as well as those of the solvent. Further simplifications that have been made for polypeptide chains treat only one main chain dihedral angle per amino acid residue (Levitt, 1976; McCammon et al., 1980). Such a model, which uses a pseudodihedral angle based on the α carbon positions that is approximately equal to $\phi + \psi$ (Peticolas and Kurtz, 1980) has been widely employed as the simplest approach that can provide a realistic representation of the polypeptide chain. If the allowed values of the pseudodihedral angles are limited to those that satisfy the constraints corresponding to regular grids in two or three dimensions, lattice models of the type that are commonly used in protein folding studies are obtained (Gō, 1975). Such lattice models sacrifice some of the properties of the polypeptide chain, but have the advantage that the required statistical mechanical averages and dynamics are much simpler to calculate. Lattice representations of polymer conformations have a long history (Flory, 1953; De Gennes, 1979) and many important results in polymer theory were obtained by the use of such lattices.

The lack of a detailed knowledge of the potential-of-mean-force surface and the astronomical cost of a full exploration of the conformation space makes a direct approach to the thermodynamic or dynamic properties of a polypeptide chain prohibitive (and perhaps not that interesting). For the native state, where detailed simulations are useful, the number of accessible conformers is on the order of γ^{100} for a chain of 100 amino acids, where γ is $1+\varepsilon$; for $\varepsilon = 0.2$, there would be 10^8 conformers. This contrasts with the characteristics of the conformation space of the non-native state. Here γ is on the order of 5 or 10, so that one is concerned with 10^{70} conformers; even if excluded volume is taken into account, the number of conformers is still very large (Dill, 1985). Moreover, because the barriers are expected to be in the range of several k_BT, a simple molecular dynamics simulation or even stochastic dynamics simulation is likely to be adequate only for sampling local regions of the conformational space.

It is important to note, however, that for equilibrium properties in a homogeneous system like liquid argon or water, it is often necessary to simulate only a small number of particles (say, 256 argon atoms or water molecules) for a short time (say, 10 psec), so that only a very small part of configuration space is sampled to obtain reliable values for many thermodynamic and dynamic properties. Such ideas are at the basis of molecular dynamics and Monte Carlo simulation methods, and they may well be applicable to the two extreme regions of polypeptide chains (native and completely denatured). A judicious combination of Monte Carlo sampling and molecular dynamics could well provide useful results.

To proceed further in examining the conformation space of a polypeptide chain and elucidating the nature of protein folding, it is useful to complement the detailed simulations by more general approaches. The inherent complexity of the system, including the large size of the configuration space, suggests that theoretical analyses based on statistical mechanics should provide useful results. For the most effective utilization of approaches that introduce essential simplifications (e.g., a single coordinate that describes the entire protein conformational space as illustrated in Figure 4–1A), it is important to have a clear understanding of the questions that are to be addressed. The choice of models will be dictated, at least in part, by the nature of the problems to be addressed. For proteins, the basic questions are:

1. What is the nature of the potential of mean force surface and the phase diagram of a protein?
2. What is required for a polypeptide chain to have a unique free energy minimum or, better, a range of very similar minima that together correspond to the native state?
3. How does the polypeptide chain find this minimum or range of minima in a finite time (the so-called "Levinthal paradox" (Levinthal, 1968, 1969); experimentally the time is in the range of 10^{-2} to 10^2 sec (Jaenicke, 1987).

It is our purpose to discuss these questions from a conceptual point of view and to provide possible answers (or, perhaps more realistically, to provide a framework from which the answers can be determined) that are based on general considerations. As appropriate we will make use of specific, more detailed, models to illustrate some of the general concepts and the importance of the various parameters involved. Because of calculational limitations, most available examples suffer from the fact that they use highly simplified representations of the polypeptide chain or are limited studies of more realistic models. One of the most important aspects of the more detailed simulations is to establish that the general conclusions obtained from simplified models are applicable to proteins. Only by a combination of analytic theory and computer simulations will progress be made in solving the protein folding problems that we have outlined.

The next section (Section 2) provides a brief outline of some of the methods used to simplify the system for the study of the larger-scale phenomena involved in protein folding. We focus on the thermodynamics of polypeptide chains in Section 3. Approaches used for determining the phase diagram of a polypeptide chain and for analyzing the relation between the sequence and the uniqueness of the native structure are described. In Section 4, we consider some of the conceptual models that have been introduced to obtain insights into the nature of the folding mechanism and then discuss the approaches that have been used to attempt to go beyond the conceptual description to actual calculations of the dynamics of folding. A concluding discussion is given in Section 5.

We have tried to mention most types of published work concerned with theoretical studies of protein folding, although we often use our own results to illustrate the possible approaches. In describing the available results, we have tried to place them in perspective; sometimes our view of the significance of a study is not in accord with that of the author(s) and we have so stated since we believe that this is one of the functions of a review. The extensive list of references should aid the reader in obtaining more information about the work we discuss. One subject that we do not discuss is what is often considered to be "the" protein folding problem. This refers to the prediction of the folded conformation from a knowledge of the amino acid sequence. In some ways this problem is a corollary of the answer to two of the questions raised earlier in this chapter. An understanding of how uniqueness is introduced in thermodynamic and evolutionary terms (question 2) and how the unique structure is found in the folding process (question 3) could serve as an approach to the prediction problem. Since the search for predictive algorithms is a very active area in protein research, stimulated in part by the siren song of biotechnology, it has been reviewed extensively in recent years (e.g., Fasman, 1988; Levitt, 1991).

The less theoretically inclined reader might do well to go directly to the concluding section, find the subjects of interest to him or her, and then obtain more details in the text.

2 METHODS

In this section, we briefly outline some of the basic concepts of the methods that are of particular interest for studying the thermodynamics and dynamics of protein folding. More details can be found in texts such as Brooks et al., (1988), McCammon and Harvey (1987), MacQuarrey (1976), and Allen and Tildseley (1987). Additional theory is sketched as needed later in the chapter.

As mentioned in the Introduction, there are two essential simplifications that are widely employed in the treatment of protein folding and polymers more generally. The first is the potential of mean force and the second is a discretized description of the polypeptide chain isomers. Both of these concepts are based on the idea of "pre-averaging" of the small-scale motions to obtain a "coarse-grained" model that treats a molecule at "medium resolution," i.e., on the time and length scales at which protein folding occurs.

As an example of the introduction of a potential of mean force we consider a protein molecule in the presence of solvent. The total configurational partition function of the system, which involves summation over coordinates of the polypeptide chain [R] and the coordinates of the solvent molecules [r], is

$$Z = \sum_{[R]} \sum_{[r]} \exp\left[-E(R_1 \ldots R_N; r_1 \ldots r_s)/k_B T\right] \tag{4-1}$$

where E is the total potential energy function, k_B is the Boltzmann constant, and T is the temperature. In Equation (4–1), as well as much of the following, we are considering continuous variables, but we do not distinguish between summation signs and integrals for simplicity of writing. The configurational partition function is the reduced partition function that can be used for conformational equilibra when the kinetic energy contribution is independent of the conformation (Gō and Scheraga, 1969). For the protein folding problem it is very useful not to treat the "solvent degrees of freedom" explicitly and to write

$$Z = \sum_{[R]} \exp\left[-V(R_1, \ldots R_N)/k_B T\right] \tag{4-2}$$

with

$$V(R_1 \ldots R_N) = -k_B T \ln \sum_{[r]} \exp\left[-E(R_1 \ldots R_N, r_1 \ldots r_s)/k_B T\right] \tag{4-3}$$

where $V(R_1, \ldots R_N)$ is the potential of mean force (McQuarrie, 1976), which is defined as the free energy of a given polypeptide chain configuration (i.e., a fixed set of coordinates) with the solvent coordinates canonically averaged over all possible values at the temperature of interest. It is clear that the potential of mean force is both temperature- and environment-dependent. In many problems it is useful to reduce further the coordinates considered by also doing a canonical average over certain protein degrees of freedom. For a study of the conformations of the polypeptide chain, it is most useful to consider only the soft dihedral angles of the main chain, for which we use the general symbol $(\theta_1 \ldots \theta_M)$. The partition function is written

$$Z = \sum_{[\Theta]} \exp\left[-V(\theta_1 \ldots \theta_M)/k_B T\right] \tag{4-4}$$

where

$$V(\theta_1 \ldots \theta_M) = \sum_{[\underset{\sim}{h}]} \exp[-V(R_1, \ldots R_M)/k_BT] \tag{4-5}$$

where [h] represents the "hard" degrees of freedom (e.g., bond lengths, bond angles, main chain peptide group angles), as well as the side chain dihedral angles in appropriate cases. Given the potential of mean force, all of the thermodynamic and dynamic properties of the protein can be determined, in principle, by well-known procedures based upon statistical mechanics and the theories of reaction rates. At equilibrium, for example, the probability of any given conformer, specified by $(\theta_1 \ldots \theta_M)$, is given by

$$P(\theta_1 \ldots \theta_M) = \exp[-V(\theta_1 \ldots \theta_M)/k_BT]/Z. \tag{4-6}$$

The second concept is based on the fact that for many applications, in which the temperature is sufficiently low that the barriers separating the individual wells are significantly larger than k_BT, it is useful to rewrite Equation (4–4) as a sum over different conformers; that is,

$$Z = \sum_{[\underset{\sim}{\theta^i}]} \exp[-V(\theta_1^i \ldots \theta_M^i)/k_BT] \int_i \exp[-V(\theta_1 \ldots \theta_M)/k_BT] \tag{4-7}$$

where $\theta_1^i \ldots \theta_M^i$ are the values of the dihedral angle in the local minimum of $V(\theta_1 \ldots \theta_M)$ corresponding to conformer i and the integral is over the range of conformations associated with that minimum. The resulting description of the polypeptide chain in terms of discrete conformations is similar to the rotational isomeric model (Flory, 1969), although long-range interactions not present in that model may be included in Equation (4–7). In certain cases the integrals can be approximated by use of a harmonic expansion about $V(\theta_1^i \ldots \theta_M^i)$ (Stillinger and Weber, 1982; Karplus et al., 1987). Further, in the case of polypeptide chains, the averages over the main chain degrees of freedom within a minimum and over the side chain degrees of freedom for all side chains, other than glycine, are approximately constant (Karplus et al., 1987). This considerably simplifies the evaluation of Z as given by Equation (4–7) and the corresponding probabilities $P(\theta_1^i \ldots \theta_M^i)$. The "coarse graining" exemplified by Equations (4–4) and (4–5) has often been used in the study of polymers (Flory, 1969; Helfand et al., 1980) and biopolymers (Gō and Scheraga, 1969).

Extensions of Equations (4–1), (4–4), and (4–7) to include quantum effects can be made, but for most situations of interest here, quantum corrections are expected to be small. The dihedral angle degrees of freedom in peptides and proteins have a characteristic temperature of about 50°K

(Brooks et al., 1988; Zheng et al., 1988; Perahia et al., 1990). Further, the contribution of the bond angle and bond length degrees of freedom (for which quantum effects are more important) to the conformational energy or free energy differences are expected to be small.

In studying such a dynamic process, the validity of the separation of time scales of the solvent and protein motions has to be considered; i.e., if the time scales are not well separated (i.e., the solvent does not relax rapidly relative to the protein conformation change), the details of the solvent dynamics have to be considered, and nonexponential behavior may be observed (Brooks et al., 1988). For the overall dynamics of protein folding/unfolding, it is likely that the solvent does not have to be considered explicitly, so that its average effect on the interactions can be included in terms of a potential of mean force, as for the thermodynamics. The dynamic contributions of the solvent to the motions of the polypeptide chain can then be approximated by including stochastic and dissipative forces in the equations of motion. To treat the effects of solvent in a simple fashion, the Langevin equation (Brooks et al., 1987,1988),

$$M_i \, a_i \, (t) = -\nabla_i \, [V \, (R_{1,...} \, R_N) \,] - \beta_i \, M_i v_i(t) + M_i F_i(t), \qquad (4\text{--}8)$$

is used for each degree of freedom i with mass M_i, velocity v_i, and acceleration a_i. The potential of mean force is $V(R_1...R_N)$ and the dissipative force is given by $\beta_i M_i v_i(t)$, where β_i is the friction coefficient of particle i and the Langevin random force $F_i(t)$ satisfies $<F_i(t)> = 0$ and $<F_i(t)F_i(0) > = k_B T \, \beta_i \delta(t)$. Stochastic dynamics has been found to be particularly useful for introducing simplified descriptions of the internal motions of complex systems. When applied to small molecules in solution (e.g., a peptide or an amino acid side chain), it is possible to do simulations that extend into the microsecond range, where many important phenomena occur. Such an approach does have the limitation that it precludes the detailed study of the dynamics of explicit solute-solvent interactions because the solvent (bath) degrees of freedom have been eliminated. In some cases (e.g., a study of the replacement of protein-protein hydrogen bonds by protein-water hydrogen bonds) an explicit treatment of the solvent may be necessary. Also, if the random force cannot be expressed as simply as in Equation (4–8), a generalized Langevin formulation may have to be introduced (Adelman, 1983; Hynes, 1983; Brooks et al., 1987).

The time scale of many chemical and physical processes occurring in biomolecules is limited by the rate of overcoming an energy barrier. One example occurs in the binding of oxygen to myoglobin, where the ligand must pass several energy barriers of varying size before arriving at the binding site (Austin et al., 1975; Case and Karplus, 1979; Elber and Karplus, 1990). Another is provided by the well-studied case of the $180°$ rotation of the aromatic side chains ("ring flips") in proteins (Northrup et al., 1982).

The phenomenological timescale of such activated events is often as long as a microsecond; i.e., while these processes can be intrinsically fast, they occur only infrequently (with an average frequency of 10^{10} sec^{-1} or even less). Thus, they are not adequately sampled in conventional simulation approaches. Activated dynamics methods provide one alternative that overcomes this sampling problem, although stochastic dynamics has also been applied to barrier crossing phenomena. Activated dynamics is a synthesis of molecular or stochastic dynamics methods and transition state theory. It must be used when the barriers between conformers are large with respect to k_BT, so that a straightforward simulation would not lead to a sufficient number of barrier crossing events to provide converged rate constants. In such calculations, a "reaction coordinate," θ, typically a set of n atomic coordinates that carry the system from a reactant configuration to a product configuration, is identified first, and the potential of mean force associated with motion along the reaction coordinate is determined by free energy simulation methods (e.g., umbrella sampling) (Brooks et al., 1988).

For a one-dimensional problem (e.g., the *trans-gauche* isomerization of butane) the rate constant k for the reaction can be written

$$k = \frac{1}{2} \kappa < |\dot{\theta}| >_{\theta_+} \rho\,(\theta_+) \int_i \rho(\theta)\,d\theta \qquad (4\text{--}9)$$

where θ is the reaction coordinate, and the quantity $< |\dot{\theta}| >_{\theta_+}$ is the average absolute value of the crossing velocity, $\dot{\theta} \equiv d\theta/dt$, evaluated at θ_+, the transition state corresponding to the highest point on the potential of mean force barrier. The constant κ is the transmission coefficient, and the integral is over the reactant well i. The quantity $\rho(\theta_+) / \int_i \rho(\theta)d\theta$ is the probability of "finding" the system at the top of the barrier. If κ is set equal to unity and the equilibrium value is used for $< |\dot{\theta}| >_{\theta_+}$, the rate constant reduces to that obtained from transition state theory. Deviations from the ideal transition state rate often occur, however, so that the reactive flux is, in general, not equal to this simple result, which is determined by the equilibrium properties of the system. It is then necessary to evaluate the transmission coefficient, which accounts for the probability of multiple crossings, and to calculate the (nonequilibrium) velocity distributions at the transition state. Both of these quantities, which are expected to be particularly important in condensed phase reactions, may be computed from trajectories that originate at the transition state (Pechukas, 1976; Chandler, 1978; Brooks et al., 1988).

In determining the dynamics of a polypeptide chain under conditions where the barriers between conformers are large with respect to k_BT, it is possible to separate the librations within a well from the transitions between wells, as is done in descriptions of the thermodynamics. The former can be studied by ordinary molecular dynamics techniques (Brooks et al., 1988) and in some cases the harmonic approximation can be employed (Gō and Scheraga, 1969; Karplus et al., 1987). For the transitions between wells, activated dynamics methodologies can be used in many cases. A major difficulty in applying activated dynamics methods to polypeptide chains is the determination of an optimal reaction coordinate and the transition states. This problem is rooted in the conceptual and computational complexity associated with finding the minimal number of atomic coordinates that adequately specify the transition from a "reactant" configuration to a "product" configuration. A poor choice of reaction coordinate or transition state does not necessarily invalidate the method. The efficiency of the activated dynamics transition state sampling rapidly decreases, however, as the reaction coordinate becomes less than optimal. In polymer studies, it has been found that the reaction coordinate is usually not a single dihedral angle but some combination of dihedral angles (Helfand et al., 1980). The nature of the reaction coordinates involved in the isomerization transitions of the polypeptide chain is one of the essential questions that have yet to be answered.

3 PROTEIN THERMODYNAMICS

In this section we consider primarily aspects of protein thermodynamics that are of fundamental importance for an understanding of protein folding. The first is concerned with the nature of the transition between the folded (native) and unfolded (denatured) states. The second deals with the requirement that proteins have a unique structure and the conditions on the protein sequence that leads to such a unique structure (i.e., that there exists an isolated free energy minimum).

3.1 Protein Denaturation as a First-Order Phase Transition

It was understood very early (see, e.g., Chick and Martin, 1911; Wu, 1929) that protein stability is relatively low so that mild deviations from the physiological environment (heating to 40°–80°C, change of pH from physiological to 2 to 3, and addition of denaturants like urea) can cause the inactivation of a protein. This was attributed to unfolding already in 1929 (Wu, 1929). More quantitative experimental results on protein denaturation became available in the early 1970's when a number of different physio-chemical techniques began to be applied to proteins. Adiabatic scanning calorimetry has played a particularly important role (Privalov and Khechinashvili, 1974), but other methods, such as viscosimetry, far- and near-UV CD, correlation spectroscopy, and NMR, have also contributed significantly. For detailed reviews of experimental data on protein

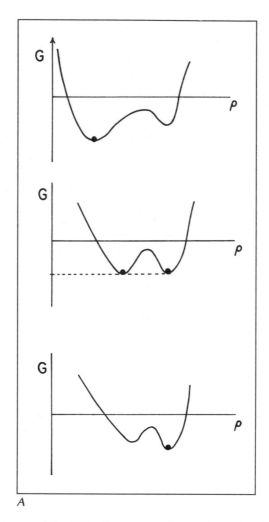

FIGURE 4–2. *Diagram of the Gibbs free energy, G, as a function of density ρ at different temperatures for (A) a first-order and (B) a second-order transition: (top plot) above the transition temperature, (middle plot) at the transition temperature, and (bottom plot) below the transition temperature; the solid dot indicates the equilibrium state.*

denaturation, see the contributions by P. L. Privalov (Chapter 3) and O. B. Ptitsyn (Chapter 6) in this volume.

The most important experimental result that appears to be a universal characteristic of protein denaturation is that the transition in a small one-domain globular protein is of the "all-or-none" type analogous

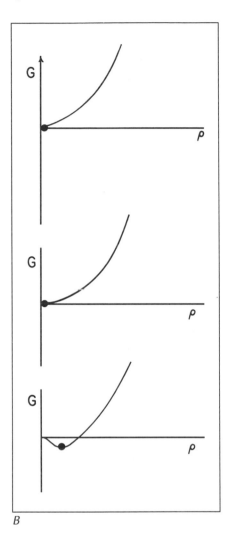

to a first-order phase transition for an infinite system (Privalov, 1979); i.e., "unfolding" can be described at equilibrium in terms of two states that have a barrier between them, with the population of intermediates at equilibrium not exceeding a few percent. (See Figure 4–2.) The "all-or-none" character of protein denaturation is all the more surprising because a protein is a highly inhomogeneous system that structurally corresponds to a disordered solid rather than to a crystal. This "all-or-none" property of the transition perhaps justifies Schroedinger's use of the term *aperiodic crystal* for a biomolecule in the influential book *What is Life?* (Schroedinger, 1945), at least as far as protein denaturation is concerned.

The cooperativity of the denaturation transition has very important biological implications. For example, it is important for the protection of the active site of a protein from thermal fluctuations. Native proteins have very low stability; the free energy difference between the native and denatured protein is only ~10 kcal/mol on a molecular basis or ~0.1 kcal/mol per amino acid residue. Thus, if the amino acids were free to fluctuate independently, each of them would spend about 55% of the time in the "denatured" conformation, according to a Boltzmann distribution; e.g., an active site triad (as in the serine proteases) would be in the "native" conformation with a probability of only 10%. A consequence of the cooperative denaturation is that each of the amino acids "feels" the full 10 kcal/mol of denaturation free energy and so spends only 10^{-7} of the time in a nonnative conformation.

The denaturation transition is accompanied by an increase in the molecular volume and by a significant latent heat (on the order of $\sim k_B T$ per amino acid residue). In terms of a potential of mean force, which gives the free energy of the polypeptide chain as a function of a single coordinate, the molecular volume, there are two minima, corresponding to two different thermodynamic states of a molecule that are separated by a maximum (Fig. 4–2A). There is some evidence that the transition state is compact and that the protein interior is not accessible to solvent (Sugihara and Segawa, 1984; Creighton, 1988), although the universality of this result is not established.

The native state of a protein does not change significantly when external conditions (pH, temperature, and so on) are changed, i.e., so long as the native state is stable, it has a unique structure. In contrast to the native state, the essential features of the denatured state vary strongly from protein to protein and depend upon the conditions under which denaturation is observed. The denatured state may be rather compact (30% larger in volume than the native state) with a significant amount of secondary structure, preserved hydrophobic core (Semisotnov et al., 1987) and some hydrophobic clusters in the "native" positions (Ptitsyn, 1987; Baum et al., 1989; Williams et al., 1991) or it may be a completely unfolded, random coil (Tanford, 1968, 1970) (see Chapter 6). There is no experimental evidence for a well-defined transition between denatured states obtained under different conditions (high or low pH, high or low temperature, added denaturant). Moreover, it has been shown by Privalov (1989) that the thermodynamic functions of a protein denatured under different conditions are the same when extrapolated to the same external conditions. Although it has been argued that there exists a first-order transition in β-lactamase between a compact denatured state and the unfolded state at low pH when the ionic strength is changed (Goto and Fink, 1990), this conclusion may not be correct. It is based only on the observation that the unfolding curve is sigmoidal. Such an unfolding curve is not necessarily connected with a first-order transition. (See Figures 4–2 and 4–3 for a schematic illustration of the difference between a first-order and a second-order transition.) Detailed van't Hoff analysis of the conjugated thermodynamic parameters is required to determine the order of transition; this has not been given by Goto and Fink (1990).

FIGURE 4–3. (A) Gibbs free energy of the native (N) and denatured states (D) as a function of temperature. (B) Density of native and denatured states as a function of temperature. The thick curve denotes equilibrium behavior, and T* is the transition temperature.

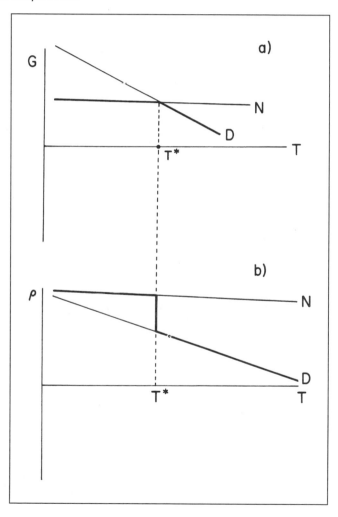

The observed behavior of the denatured state is not surprising. It implies that it is essentially a normal polymer, whose properties (e.g., volume) vary significantly when conditions change from those corresponding to a "poor" solvent to a "good" solvent (De Gennes, 1975; Lifshitz et al., 1978). It is the native state of a protein that is the special case that requires new approaches and ideas for its description. Its unusual thermodynamic features, which appear to be universal for single-domain proteins (multidomain proteins may have separate transitions for each domain [Privalov, 1982]), have led to theoretical attempts to explain them in terms of simple physical models. We describe some of them in the following pages.

3.1.1 "Globule-Coil" Theories of Protein Denaturation

The first theoretical approach to protein denaturation was based on the assumption that a simple globule-to-coil transition is involved. Although such a model does not lead to a first-order transition, it serves as a useful starting point for the analysis of denaturation and folding. A generalization (Kron et al., 1968) of the Flory self-consistent field theory was used, and unfolding was treated as a transition from the regime where attraction between monomers dominates (i.e., below the Θ-point) to the regime where repulsion is dominant. The free energy of a chain in this model is written in the form

$$F = F_{el}(\alpha) + F_{int}(\alpha) \tag{4-10}$$

where $\alpha = R/R_g$ is a parameter that gives the ratio of the mean size (or end-to-end distance) of a molecule; R, to R_g, the size of an unperturbed molecule under Θ-condition; R_g is of the order of $lN^{1/2}$, where l is the length of a Kuhn segment and N is the number of Kuhn segments (Flory, 1969).

The first term in Equation (4-10) describes the change of chain entropy due to the collapse of the chain. This term was estimated by Kron et al. (1968) with the assumption that the collapsed chain still obeys Gaussian statistics and that the end-to-end distance of a Gaussian coil is characteristic for its overall size, R. The second term in Equation (4-10) takes into account interactions between monomers. A virial expansion (Kron et al., 1968), as well as other approaches (Sanchez, 1970), have been used to approximate this contribution to the free energy. Minimization of the free energy with respect to α in Equation (4-10) solves the model (i.e., determines the stable state). The results are:

1. The globule-coil transition may be only a weak (with latent heat $\sim N^{-1/3}$ per monomer) first-order transition for stiff and short chains.
2. As chain length increases, the transition becomes a gradual second-order transition.

3. Even for short and stiff chains, the globule-coil transition as given by this theory is preceded by significant pretransitional swelling of the globule, and only a small part of the density change is due to the globule-coil transition itself.

Corresponding results were obtained by a somewhat different approach by De Gennes (1975). This approach was also used for the analysis of DNA collapse by Post and Zimm (1979) and for a recent analysis of protein collapse (Bryngelson and Wolynes, 1990). In Birstein and Pryamitsyn (1991), the expression for the elastic entropy in Equation (4–11) was evaluated more correctly based on the behavior of a chain immersed in a spherical pore. The main results are qualitatively the same, however.

A more rigorous and general approach to the problem of globule-coil phase transitions in homopolymers was proposed by Lifshitz et al., (1978). They analyzed the globule-coil transitions in homopolymers quantitatively and compared the results with experimental data on simple nonbiological polymers (Birstein et al., 1987). The main qualitative conclusions remain the same as those derived within simple Flory-type approximation; i.e., the globule-coil transition is not a first-order transition and even in case where there exists some discontinuity of density (i.e., for short and stiff chains) the transition occurs between a loose swollen globule and a coil, so the transition state is noncompact and accessible to solvent.

Thus, considerations that treat a protein as a simple homopolymer apparently fail to explain the most important features of protein folding thermodynamics.

3.1.2 Phenomenological Theory of Heteropolymers

A very important feature that was not taken into account in the early theories of denaturation is the heterogeneity of proteins; i.e., the fact that they are composed of 20 different amino acids. In a series of papers (Dill, 1985; Alonso et al., 1989, 1991; Alonso and Dill, 1991) an attempt was made to take into account chain heterogeneity in a simplified manner without considering the fact that each protein has a specific sequence and not only amino acid content. In this approach the native state of a protein is considered to be formed by folding along a special pathway: (1) random collapse of a chain without any structuring and (2) subsequent rearrangement of monomers in the compact state to place hydrophobic residues inside and hydrophilic residues outside. The fact that formation of a hydrophobic core (stage 2) occurs on this pathway after complete compactization (stage 1) implies that heterogeneity plays a role only in the native state; i.e., the denatured state in the theory is structureless and analogous to the collapsed state of a homopolymer (without a hydrophobic core and hydrophilic surface).

Based on this special folding pathway, Alonso et al. (1989,1991) came to the conclusion that denaturation is a first-order transition. The reason for this is that the intermediate state on the chosen pathway (randomly collapsed state) has a free energy that is higher than that of both the native state and the coil conformations. Although a barrier exists on the chosen pathway, a transition is first-order if, and only if, all pathways between native and denatured states involve a barrier. The pathway chosen by Alonso et al. (1989,1991), although convenient for calculations, seems very unlikely to be the lowest energy pathway in their model where the driving force for compactization is hydrophobic. Collapse of a chain without simultaneous separation into hydrophobic core and hydrophilic surface leads to a pure loss of entropy without energetic compensation. This is the reason for the occurrence of a barrier on the pathway.

The free energy of a protein as a function of the density of monomers, ρ, and the parameters, Θ, which is the fraction of surface sites occupied by solvophobic residues and x, which is the fraction of core sites occupied by solvophobic residues (i.e., the free energy of folding in the model of Dill and coauthors) is

$$\Delta G(\rho,\Theta,x) = \Delta G_I (\rho) + \Delta G_{II} (\Theta,x) \tag{4-11}$$

with

$$\Delta G_I(\rho) = -\frac{1-\rho}{\rho} \ln (1-\rho) + \frac{7}{2n}\left[\rho_o{}^{2/3} - \left(\frac{\rho_o}{\rho}\right)^{2/3} \right] - \frac{2}{n} \ln \rho - \chi \, \varphi^2 \, (f_i + \sigma f_e) \, (1-\rho) \tag{4-12}$$

$$\Delta G_{II} (\Theta,x) = -\chi \, [f_i \, (x^2 - \varphi^2) + \sigma f_e \, (\Theta^2 - \varphi^2)]$$

$$+ f_i\left[x \ln \frac{x}{\varphi} + (1-x) \ln \frac{1-x}{1-\varphi} \right] + f_e\left[\theta \ln \frac{\theta}{\varphi} + (1-\theta) \ln \frac{1-\theta}{1-\varphi} \right]$$

$$\tag{4-13}$$

where ρ_o is the density of a coil, n is the number of monomeric units, χ is the parameter characterizing solvophobic interactions, ϕ is the fraction of solvophobic residues in the sequence, and f_i and f_e are the fractions of spherical volume that correspond to interior and exterior sites, respectively. The parameter σ is the fractional exposure of surface residues to contacts with the protein. The function ΔG_I is a free energy of "random condensation" (as for a homopolymer) and so is the same as that derived in earlier treatments of this problem (Kron et al., 1968; Sanchez, 1970) considered previously. The function $\Delta G_{II} (\Theta,x)$ is an estimate of the free energy of formation of a hydrophobic core in the compact state at $\rho = 1$.

The equilibrium density ρ^* corresponds to solutions of the equation of state

$$\frac{\partial \Delta G}{\partial \rho} = 0. \tag{4–14}$$

It was argued (Alonso et al., 1989; Alonso and Dill, 1991) that the theory predicts a first-order transition between native and denatured states, but this conclusion is not correct. To describe a first-order transition, the equation of state, Equation (4–14), must have three solutions: two corresponding to free energy minima and one corresponding to a maximum (see Figure 4–2A). However, Equation (4–14) with ΔG from Equations (4–11), (4–12), and (4–13) applied to a chain of 100 monomers ($n\sim100$) has only one solution ($\rho = \rho^*$) for any set of parameters. This solution corresponds to a denatured protein that may be globular if $\chi < 0$ (so that $\rho^* > 0$). The native state is always thermodynamically unstable in this theory (i.e., $\frac{\partial \Delta G}{\partial \rho} = \infty$ at $\rho = 1$) so that the native state does not exist as a thermodynamically stable state in this model.

Thus, the approach of Dill and coworkers appears to be inconsistent and may not allow one to address the problem of protein stability. However, it should still be a useful starting point for consideration of the gradual transition between a coil and a structureless molten globule. A treatment of the coil–molten globule transition based on the theory of Lifshiftz et al. (1978) has been suggested (Finkelstein and Shakhnovich, 1989).

3.1.3 *Protein Denaturation as Intramolecular "Melting"*

Since the denaturation transition in proteins appears to be very different from the melting transition in ordinary linear polymers, it is appropriate to take into account details of polypeptide chains that differentiate them from such polymers. The most striking difference between proteins and ordinary homopolymers and heteropolymers is that in proteins each monomer (except Gly) possesses a side chain and that a significant fraction of these side chains are tightly packed in a well-defined conformation in the interior of the native protein. As has frequently been pointed out (for a review, see Richards, 1977), the packing of the protein interior is similar to that of a molecular crystal.

A theory of protein denaturation based on the idea that it corresponds to the intramolecular "melting" of such "crystal" was developed by Shakhnovich and Finkelstein (1982,1989). It was shown in this theory that disruption of such a tight packing is always a first-order transition independent of the degree of subsequent rearrangement of the backbone. This is somewhat analogous to the disruption of crystals, which can take place by melting to a liquid (in proteins, to the molten globule) or sublimation to a vapor (in proteins, to the coil).

In considering the transition from the native to the denatured state, it was assumed in their model that the secondary structural segments (which are stable until water penetrates inside) are displaced as units. This implies that the change in free energy depends, to a first approximation, on a single macroscopic parameter, the volume V. This is in accord with simulation results (see Section 1), which show that the motions in the native state of myoglobin, for example, involve small rigid-body diplacements of the helices coupled with loop rearrangements and side chain reorientations (Elber and Karplus, 1987). Since the folding/unfolding transition state is thought to be close to the native state, most of the secondary structural elements and their hydrogen bonds are likely to be present, so the principal forces that take part in denaturation (i.e., the forces that exhibit the most rapid changes as the structure deviates from the native state) are non-bonded interactions (e.g., van der Waals forces) in the core of the globule, which oppose swelling, and entropic forces due to torsional motions of side chains in the core and loops; they facilitate swelling because the entropy increases as the side chain constraints are reduced.

On this basis, the free energy of the protein in the denaturation transition can be approximated as

$$F_{PR}(V) = E_{vdW}(V) + F_{tors}(V) \tag{4–15}$$

where E_{vdW} and F_{tors} are contributions from van der Waals forces and from the entropy of side chain and loop degrees of freedom, respectively; whether neglect of electrostatic contributions is possible, even when secondary structural elements are retained, remains to be verified. These contributions were analyzed by Shakhnovich and Finkelstein (1989) for a simple model system consisting of 50 monomers with a total of 100 degrees of freedom in the side chains. The strength of van der Waals interactions was estimated from experimental data that the free energy of sublimation of a 50-dalton "hydrocarbon molecule" would be about 7 to 8 kcal/mol. The results in Figure 4–4 show that there exists a barrier (free energy maximum) between the native and denatured states so that the transition is predicted to be first-order. The magnitude of the barrier for the chosen set of parameters is about 20 kcal/mol.

The origin of the barrier and a scheme of protein melting is illustrated schematically in Figure 4–5. In the transition state, side chain oscillations within one rotational isomer are allowed, but there is not enough space at this degree of swelling to permit rotational isomerization of side chains in an independent fashion, while a significant part of the van der Waals energy has already been lost. Since the transition barrier state is quite compact (i.e., this state does not have numerous pores of the size of a methyl group (CH_3) to permit independent rotational isomerization), the protein interior is not accessible to water, which is in accord with experimental results on lysozyme (Sugihara and Segawa, 1984).

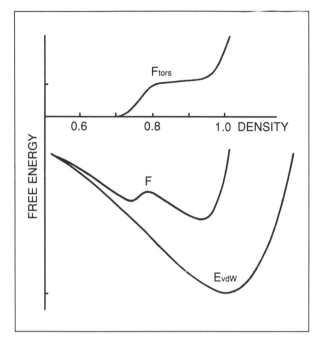

FIGURE 4–4. *Different contributions to the free energy of a chain versus density in the model described by Shakhnovich and Finkelstein (1982, 1989); F, the total free energy, exhibits two minima corresponding to the native and denatured states.*

This analysis, which considered a protein in vacuum, leads to the conclusions that:

1. Denaturation due to "side chain" unfreezing is a first-order phase transition with a compact transition state.
2. The barrier for unfolding is energetic, but for folding is entropic.
3. The protein in its denatured state is compact with a volume about 30% larger than the native state.
4. The protein in vacuum is very stable, with a transition temperature estimated to be 400 to 500K.

Consideration of a protein in a vacuum may be sufficient for a folded protein and when one is interested in swelling from the native to the transition state for denaturation, i.e., for the investigation of the reason for the existence of a barrier to unfolding. After the transition state has been crossed, however, pores into which water molecules can penetrate are

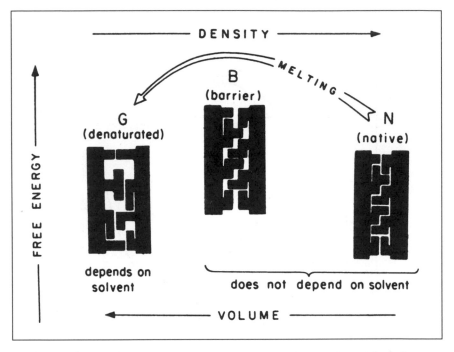

FIGURE 4–5. *Scheme of protein melting suggested by Shakhnovich and Finkelstein (1982, 1989).*

formed inside the protein, and the influence of solvent must be considered. This was done in an extension of the model (Finkelstein and Shakhnovich, 1989), in which the denatured protein was represented as a porous medium that is able to absorb water molecules.

Analysis of this model was done in an approximation that takes account of interactions between segments of the protein, between protein and solvent molecules, and between solvent molecules, as well as the entropy of the placement of solvent molecules inside a protein. The results shown that there could exist two regimes: a "dry" denatured state, when all pores inside a swollen protein are relatively small and empty, and a "wet" denatured state, when the majority of the pores are occupied by the solvent. Denaturation to the dry state is the same as denaturation in vacuum, which has been shown to occur at 500°K (see above). Since proteins denature at 320° to 370°K, the second regime is realized and the denatured state must be wet. This implies that the space not occupied by protein in the denatured state is occupied by water. This is in accord with the experimental finding that the partial specific volume of a protein does not change drastically upon denaturation (Privalov et al., 1988).

The results of the theory are summarized in a schematic phase diagram for a protein molecule in solution (Fig. 4–6), which is plotted as a function of the temperature *versus* solvent quality Ψ. The parameter Ψ, which characterizes the interaction between amino acids and solvent in this theory, is the standard free energy of transfer of amino acids from vacuum to solvent and so depends upon the quality of solvent; it depends on the temperature and the concentration of denaturant or ions. The phase diagram shows that a denatured protein is very susceptible to a change in external conditions. Depending upon such features of a protein as thermostability and hydrophobicity, the protein in the denatured state may be either a

FIGURE 4–6. *Phase diagram of a protein plotted as solvent quality (Ψ) versus temperature. The solid line is the line of first-order phase transitions; the dashed line denotes gradual (second-order) transitions. C is a triple point where native state, coil, and molten globule coexist and that separates different regimes of denaturation. The "experimental window" outlines the limits of the conditions under which a protein can be observed experimentally. (From Finkelstein and Shakhnovich, 1989)*

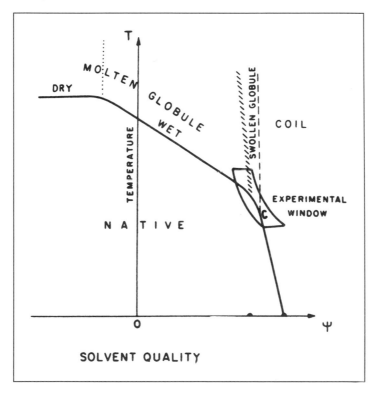

rather dense globule (Ptitsyn, 1987), a loose swollen globule analogous to that described in (Lifshiftz et al., 1978), or an extended coil. This is depicted in Figure 4–7, where the interior of the experimental window (i.e., the range of conditions accessible by experiment; see Figure 4–6) is shown in an expanded view for different kinds of proteins.

Also it is interesting to note that there exists a triple point (labeled C in the phase diagram, Figure 4–6) so that at low temperatures, solvent denaturation is a first-order phase transition from the native state to a coil, while at higher temperatures this transition is split into two: a first-order transition from the native to the molten globule and a second-order transition from the molten globule to coil. Since the position of the triple

FIGURE 4–7. *View through the experimental window (C is concentration of denaturant) (A) corresponds to a "generalized" protein, (B) corresponds to a less hydrophobic protein, and (C) corresponds to a less thermostable protein.*

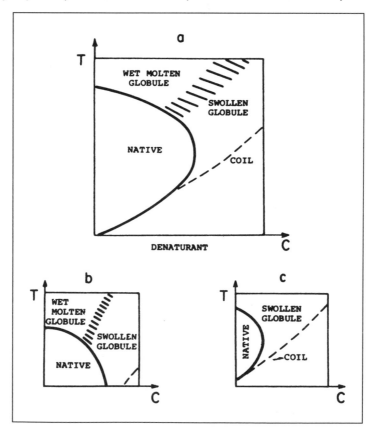

point is expected to differ from protein to protein, solvent denaturation under standard conditions may give qualitatively different results for different proteins.

The theory also gives a simple explanation of the paradox (Baldwin, 1990) that the unfolding transition is always "two-state" while the folding transition usually involves an intermediate (Kim and Baldwin, 1982). Figure 4–8 shows profiles of the free energy (potential of mean force) *versus* the density of a molecule under conditions at which refolding occurs (A) and at which unfolding occurs (B). Comparison of Figures 8A and 8B implies that the molten globule must be an intermediate state on the folding

FIGURE 4–8. Profile of free energy versus *density for (A) a typical refolding experiment and (B) unfolding experiment. The asterisk indicates the state of the protein immediately after the conditions were drastically changed.*

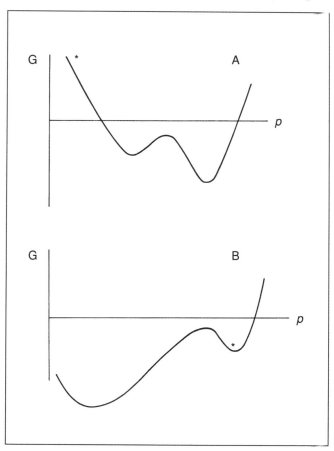

pathway, but not on the unfolding pathway. Experimental arguments in favor of this conclusion are given in Ptitsyn et al. (1990).

Results from the theory lead also to a somewhat surprising conclusion about the role of water in protein stability. Water exerts a pressure that leads to contraction of the native protein but swells the denatured protein. Since the presence of water decreases the free energy of the denatured state, it makes the protein less stable than when it is in vacuum. This is in accord with the simple argument that a denatured protein in a vacuum cannot replace the van der Waals and hydrogen bonding interactions that exist in the folded state by interactions with water and may be one reason for the stability of vacuum molecular dynamics simulations. What is not clear from the model, however, is whether the native structure or a different compact structure would be stable in the absence of solvent. The destabilizing effect of water is less for more hydrophobic proteins than for hydrophilic ones. In this sense hydrophobic interactions are crucial in determining protein stability via their influence on the denatured state. This is in accord with other considerations of the hydrophobic effect (Kauzmann, 1959).

3.1.4 "Minimal Frustration" and Cooperativity of Protein Structure Formation

Alternative attempts to explain the cooperativity of the denaturation transition appear in the work of Gō et al. (Ueda et al., 1975; Gō and Abe, 1981), Bryngelson and Wolynes (Bryngelson and Wolynes, 1987, 1990) and Skolnick, Kolinski, and Yaris (Skolnick et al., 1990). Although two of the studies are based on lattice models (Ueda et al., 1975; Gō and Abe, 1981; Skolnick et al., 1990), while the other (Bryngelson and Wolynes, 1987,1990) is an analytic approach, they are similar in that special properties of the polypeptide chain that bias it toward the native state are an essential part of the folding mechanism.

In Ueda et al. (1975) and Gō and Abe (1981) a lattice model of a protein was considered and it was assumed that in any conformation only monomers, which are lattice neighbors in the known native state, interact attractively. Monte Carlo simulations yield a first-order transition between the native state (where all "native" contacts are present) and the completely disordered denatured state (where less than 10% of the native contacts exist). This result was confirmed in analytical investigations of the same model (Shakhnovich and Gutin, 1989). Thus, the denatured state in this model may be globular but it must correspond to a collapsed homopolymer, without any definite pathway for the chain. In Skolnick et al. (1990), the equilibrium behavior of various chain models on a diamond lattice are described. It was found that if native secondary structural propensities (e.g., β-turns and β-strands corresponding to a β-barrel) were introduced plus appropriately distributed hydrophobic and hydrophilic residues (e.g., alternating for a β-barrel), an all-or-none transition to the native state was obtained in Monte Carlo simulations.

In the analytic theory of Bryngelson and Wolynes (1987,1990) it was assumed that each monomer, independent of other monomers, has an intrinsic energetic preference for the native conformation (Bryngelson and Wolynes, 1987,1990). An energy is associated with the state of each amino acid, so that if a residue is in its native state, this energy is $-\varepsilon$; it is zero otherwise. The tendency of each individual amino acid to be in its native state, independent of interaction with other amino acids in the protein, was attributed in (Bryngelson and Wolynes, 1990) to propensities corresponding to Ramachandran plots. Further, it was assumed that favorable hydrogen bonds are present only between monomers in their "native states"; the energy attributed to bond per monomer was designed as $-J$. Finally, long-range (with respect to distance along the main chain) interactions were also taken to be favorable if two interacting amino acids are in their native states; an energy $-K$ was associated with such interactions. These three terms in the energy function introduce strong biases toward the native state. Moreover, even if one omits the interaction terms (J and K), the model still appears to fold to the native state due to the intrinsic $-\varepsilon$ term. A mean field analysis of the model (Bryngelson and Wolynes, 1987,1990) led the authors to conclude that there exists a first-order transition to the native state, in which all monomers are in the native conformation. As in the other models, the denatured state appears to be disordered without residual native-like structure. This aspect of the analysis requires clarification since the limiting case ($K=0$) in the theory of Bryngelson and Wolynes corresponds to the helix coil-transition (with the "native" state being helical and the "nonnative" state being a coil), and one would expect that the phase transition would disappear as the system becomes truly one-dimensional. Apparently, the theory predicts a first-order transition even at $K=0$ (i.e., when long-range interactions are switched off completely).

Bryngelson and Wolynes (1987,1990) make the point that their model encapsulates the concept that they call "the principle of minimal frustration." This corresponds to the situation where the secondary structural propensities are not in conflict with the tertiary structure of the native state. Such a requirement is in accord with the results of secondary structure prediction schemes. The role of this principle was stressed in the work of Bryngelson and Wolynes (1989), who showed that under some conditions the type of protein described by their model is able to fold rapidly. However, since the biases of the model toward the native conformation are very strong, it is unclear to what extent it describes the folding of proteins. This question is similar to that raised by the model of Skolnick and co-workers (1990).

3.2 Uniqueness of the Native State

It is usually assumed that a given protein sequence has a unique native structure as its free energy minimum; by "unique" is meant a manifold of substates all having the same "fold." (See Section 1.) The thermodynamic requirements for such uniqueness have been investigated (Shakhnovich and Gutin, 1989a,b), and the probability of sequences having a unique structure has been estimated (Shakhnovich and Gutin, 1990). In the theory, a minimal assumption is made in the sense that a heteropolymer with a random sequence of monomers is considered. This corresponds physically to a C_α bead model and so neglects the details of side chain interactions. Only less specific interactions (e.g., hydrophobic, electrostatic, and the attractive part of van der Waals forces) are included in considering a unique structural organization of the backbone. Thus, the model appears more appropriate for a molten globule with a unique backbone conformation than for the native state.

To give a thermodynamic definition of "uniqueness" of a structure, we use the discretized polymer model described in Section 2. A polymer molecule possesses $M = \gamma^N$ conformations, where γ is the number of conformations per monomer. An energy $V_m = V(\Theta_1{}^m....\Theta_N{}^m)$ can be attributed to a conformation m. This "energy" is really a potential of mean force, since it includes all interactions between the monomers and between the monomers and solvent. At temperature T, a Boltzmann probability of finding a molecule in state m is (see Equation (4–6)):

$$p_m = \frac{\exp\left(-\dfrac{V_m}{k_B T}\right)}{\displaystyle\sum_{m=1}^{M} \exp\left(-\dfrac{V_m}{k_B T}\right)}. \tag{4–16}$$

The fact that a molecule possesses a unique structure means that there exists some conformation m_0, for which

$$p_{m_0} \approx 1 \quad \text{or} \quad 1 - p_{m_0} = \varepsilon << 1 \tag{4–17}$$

while for all other conformations $p_{m \neq m_0} << 1$. More generally, there may be a situation where, not one, but $s << M$ conformations have similar Boltzmann probabilities with $p_{m_i} \approx 1/s$ ($i = 1,2...s$) and $p_{m \neq m_i} << 1$.

In addition to determining the conditions for the validity of Equation (4–17), it is interesting to know something about the other conformations that may serve as kinetic "traps" or kinetic intermediates in the folding process. For this purpose it is important to obtain as much information as possible about the configuration space of a heteropolymer of the type considered here and especially to find out whether the configuration space of a "typical" heteropolymer is "smooth" or "rugged." Smoothness of a configuration space means that conformations that have the lowest energies have some structural similarity (e.g., in terms of the number of intermonomer contacts). For rugged landscapes the conformations with the lowest energies are structurally very different. A quantitative parameter associated with the form of the landscape is given in Equation (4–18)

The configuration space of a random heteropolymer was investigated analytically by Shakhnovich and Gutin (1989a,b). A simplified model was used in which the configuration of a polypeptide chain is represented as a set $\{ r_i \}$ corresponding to the coordinates of the C_α atoms. (See Sections 1 and 2.) This model takes into account two key features of proteins: their polymeric structure and the heterogeneity of the primary structures due to amino acid composition. Both local (between neighbors on the chain) and nonlocal interactions are included. The partition function (Equation (4–7)) has the form

$$Z = \sum_{r_i} \prod_i g\,(r_{i+1} - r_i)\exp\left(-\frac{H\{r_i\}}{k_B T}\right) \tag{4–18}$$

where the function $g(r_{i+1} - r_i)$ accounts for local interactions and shows what placements are allowed for monomer $i+1$, provided that monomer i has coordinates r_i (Lifshitz et al., 1978). *Coordinates* may have a meaning not only as 3D coordinates but also as orientations of bonds, and so on. For example, for a freely joined chain with a bond length l, $g(r_{i+1} - r_i) = \delta(\,|\,r_{i+1} - r_i\,| - l)$.

The function $H\{ r_i \}$ gives the energy of the nonlocal interactions. They are represented (Shakhnovich and Gutin, 1989a,b) in the form:

$$H\,\{r_i\} = \frac{1}{2}\sum_{i \neq j} B_{ij}\Delta\,(r_i - r_j) + \frac{1}{6}C\sum_{i \neq j \neq k}\Delta\,(r_i - r_j)\,\Delta\,(r_j - r_k). \tag{4–19}$$

Here Δ is 1 if monomers i and j are lattice neighbors and 0 otherwise; it is introduced to account for the short-range character of the interactions. The constant C is the three-particle interaction constant, which is taken to be positive to account for steric repulsion.

All information about the amino acid sequence is contained in the two-particle term B_{ij}, which reflects the fact that the interaction between monomers i and j depends upon the types of amino acids involved; for simplificity $g(r_{i+1} - r_i)$ is taken to be independent of the amino acid type. Shakhnovich and Gutin (1989a,b) assumed that the interaction energies B_{ij} have random values with a Gaussian distribution:

$$P(B_{ij}) = \frac{1}{(2\pi\,B^2)^{\frac{1}{2}}}\exp\left(-\frac{(B_{ij} - B_0)^2}{2B^2}\right) \tag{4–20}$$

B_0 in Equation (4–20) is the average interaction between monomers; it is taken to be negative to provide for compactization of the polymer; B is the standard variance of interactions and indicates how heterogeneous the chain is (e.g., $B=0$ corresponds to a homopolymer).

In principle, to determine whether a given chain has a unique structure, it is necessary to evaluate its partition function, Equation (4–18), and the Boltzmann probabilities p_m of each conformation. This is impossible, as we discussed in Sections 1 and 2. It is necessary, therefore, to find a nonspecific parameter that characterizes the existence of a unique structure and can be evaluated. Such a parameter was introduced for proteins and discussed by Shakhnovich and Gutin (1989a). The analysis is based on the fact that if there exists a unique low-energy structure, it is expected to have only small fluctuations; i.e., two chains with the same sequence are likely to have the same conformation if they possess a unique structure. Correspondingly, a given chain will have the same structure independent of time. A similarity parameter for the conformations of two copies of the same chain can be defined by the relation:

$$q_{12} = \frac{1}{N} \sum_i^N \delta\,(\mathbf{r}_i^1 - \mathbf{r}_i^2) \tag{4–21}$$

where the two chain conformations are $\mathbf{r}_i^1$ and $\mathbf{r}_i^2$, δ is 1 if the coordinates of monomers coincide up to some microscopic length scale r_0 and is 0 otherwise. For a chain with a fixed unique structure, $q_{12}=1$, independent of time. If the chain explores a large part of its configuration space in the process of thermal motion, it is likely that a comparison of the conformations of two samples will give $q_{12} \ll 1$. It is also possible, as mentioned previously, that a small number of conformations have significant Boltzmann probabilities. In this case one cannot predict the result of an experiment on the comparison of the structures of two chains with the same sequence: they may be in the same conformation or they may not. Thus, it is useful to introduce a probabilistic description of possible results of this "experiment"; i.e., to introduce a probability density $P(Q)$ that two chains picked at some moment will exhibit the same overlap Q. The thermodynamic definition, $P(Q)$, of this function at equilibrium is

$$P(Q) = \sum_{ll'}^M p_l\, p_{l'}\, \delta\,(Q - Q_{ll'}) \tag{4–22}$$

where the summation in Equation (4–22) is taken over all conformations, p_l is the Boltzmann probability for a chain to be in the state l, and $q_{ll'}$ is the structural overlap between conformations l and l' defined by Equation (4–21). The function $P(Q)$ characterizes the configurational space of the molecule. If the molecule has s states with significant Boltzmann probabilities, the function $P(Q)$ will have $s(s-1)/2$ peaks concentrated at $Q = Q_{kk'}$ with the heights of the peaks equal to $p_k p_{k'}$ and a peak at $q=1$ with height $Y = \sum_{k}^{s} p_k^2$. The positions of the peaks denote possible structural overlaps, $Q_{kk'}$, between states (see Figure 4–9), and the effective number of thermodynamically relevant states is $M_{\text{eff}} \approx 1/Y$.

FIGURE 4–9. *An example of (A) a rugged configuration space and (B) the function P(Q) that characterizes this space.*

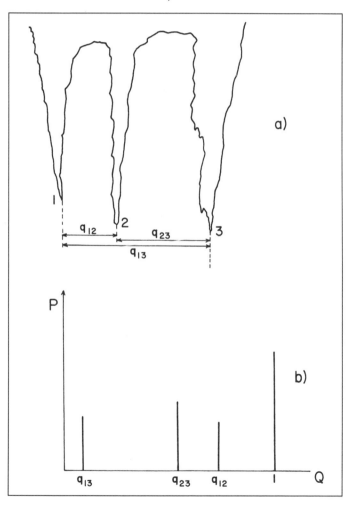

Since the parameters Q and function $P(Q)$ are not specific, any sequence with a unique structure will have $Q \approx 1$, independent of the nature of the structure. This implies that if, in an ensemble of sequences, a significant fraction possess a unique structure, then $<Q> \approx 1$, where $<>$ denotes averaging over all sequences. This makes it possible to consider averages over sequences and to simplify the evaluation of the partition function (Equation (4–18). Such an approach, which makes use of the so-called "replica trick" (Binder and Young, 1986; Parisi et al., 1988) has been employed in a detailed analysis of random heteropolymers, as a model for proteins. Some of the important results are:

1. There exists a critical temperature, T_c, for the system that separates two distinct regimes. At $T > T_c$ the chain behaves like a homopolymer; i.e., it can be either in a coil state or in a globular state (depending upon the "average" interaction energy B_0). The chain does not have a unique structure in this regime, and the number of thermodynamically accessible conformations is of the order of $\sim \exp(vN)$, where v is ~ 0.1. For $T < T_c$, the situation changes drastically. There are on average $M_{eff} = \frac{1}{(1 - T/T_c)} \sim N^0$ conformations per chain that have nonnegligible Boltzmann probabilities and therefore contribute at equilibrium. The effective number of conformations below T_c is of the order of unity and does not depend upon the chain length N.

The transition from one regime to another occurs thermodynamically as a second-order transition, and the free energy of a chain depends upon the temperature as

$$
\begin{cases}
F_{homo} & \text{for } T > T_c \\[2mm]
F_{homo} + N \, k_B \, T_c \ln (\gamma) \left(1 - \dfrac{T}{T_c} \right)^2 & \text{for } T < T_c
\end{cases}
\tag{4–23}
$$

where F_{homo} is the free energy of a homopolymer with "averaged" interaction $B_{av} = B_0 - B^2/2k_B T$. The transition temperature T_c depends upon the chain density, ρ, the number of conformations available for a monomer, γ, and the chain heterogeneity, B (see Equation (4–20) for the definition of B), as

$$
T_c = \frac{B \rho^{\frac{1}{2}}}{2k_B \, (\ln \gamma)^{\frac{1}{2}}} .
\tag{4–24}
$$

2. Conformations with the lowest energies (the only ones relevant at $T < T_c$) have small (vanishing as $N \to \infty$) structural overlaps. This is reflected in the behavior of the function $P(Q)$

$$P(Q) = \frac{T}{T_c} \delta(Q) + \left(1 - \frac{T}{T_c}\right) \delta(1-Q) \qquad (4\text{-}25)$$

which demonstrates that the bottom portion of the conformation space is rugged; i.e., low-energy conformations have very few common contacts.

The transition that occurs at $T=T_c$, which is connected with the formation of a unique structure, is rather unusual: it combines features of first-order and second-order transitions. From the thermodynamic point of view, this transition is second-order since there is no latent heat, as can be seen from Equation (4–23). If, however, we consider the overlap Q with the ground-state conformation as the main order parameter, we see that this parameter is discontinuous at the point of the transition; it jumps from $Q \approx 1$ ($T<T_c$) to $Q \ll 1$ ($T>T_c$) in a narrow transition region. (See Equation (4–25).) Only species with large overlaps with the ground state, or absolutely different from the ground state, are present in the transition region $T \approx T_c$. This fact has important kinetic implications that we consider in Section 4.

3.2.1 Random Energy Model

The ruggedness of the conformation space of such a random heteropolymer means that low-energy conformations have different sets of contacts, and therefore their energies are statistically independent. This result makes the model described by Shakhnovich and Gutin (1989a,b) equivalent to the phenomenological random energy model (REM) suggested by Derrida (1981) for spin glasses. The probability density to find a state with energy E in this model is

$$P(E) = \frac{1}{2\pi\sigma^2} \exp\left(-\frac{E^2}{2\sigma^2}\right) \qquad (4\text{-}26)$$

where σ is equal to $B\rho^{\frac{1}{2}}$.

The calculated "energy spectrum" (i.e., the set of energies of the protein conformations in this model) is shown in Figure 4–10. There is a high-energy quasi-continuous part to which the majority of the total number (γ^N) of conformations belong. They represent compact but disordered conformations without especially favorable contacts. In addition there are a few ($\sim N$) low-energy conformations, which form the discrete part of the spectrum in which many of the stabilizing contacts are present. This type of energy spectrum is a general attribute of REM-type thermodynamic behavior (Parisi et al., 1985).

When $T>T_c$, the energy of a chain corresponds to the quasi-continuous part of the spectrum and it fluctuates from one disordered conformation to another. When the temperature falls below T_c, the chain acquires one of the few conformations that belong to the discrete part of the spectrum, and only these (for some sequences, only one) conformations are thermodynamically

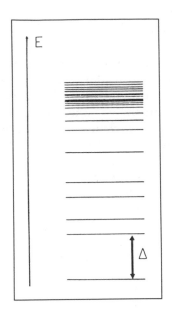

FIGURE 4-10. *A typical energy spectrum of a protein in the random energy model.* Δ *is the difference in energy between the ground state and the first excited state.*

relevant. The transition occurring at $T=T_c$ can be identified as the folding transition, since the backbone may achieve a unique conformation, although, as already mentioned, the model may be more appropriate for the coil to molten globule transition than that to the native state.

Knowing the energy spectrum, it is possible to estimate the probability that a random sequence will have a unique structure; i.e., that *one* conformation m_0 will have overwhelming Boltzmann probability: $p_{m_0} = 1 - \varepsilon$ with $\varepsilon \ll 1$ instead of $M_{eff} > 1$ conformations. The probability of finding such a sequence, P_ε, obtained by using the results of the REM (Shakhnovich and Gutin, 1990a) is

$$P_\varepsilon = \frac{\sin\left(\pi\dfrac{T}{T_c}\right)}{\pi\dfrac{T}{T_c}}\varepsilon^{T/T_c} \tag{4-27}$$

at $T < T_c$ and $P_\varepsilon = 0$ at $T \geq T_c$. Taking $\varepsilon = 0.01$, which corresponds to 99% dominance of the native conformation, and taking $T/T_c = 0.9$, we obtain $P_\varepsilon \sim 10^{-3}$, which shows that one in 1,000 random sequences will have unique backbone conformation at temperature T. This probability is very high on an evolutionary scale. The result suggests that known proteins could have evolved from random heteropolymers in a simple way. It would be of interest to consider the relation of this result to the "exon shuffling" model of Dorit et al., (1990).

It should be noted in evaluating the evolutionary significance of Equation (4–27) that a unique backbone conformation is a necessary, but not a sufficient, condition for an active protein. Much more is needed (e.g., tight packing of side chains in the core and a functioning active site). Moreover, a protein must be able to fold rapidly to such a conformation. (See Section 4.) The probability of finding such sequences relative to those satisfying Equation (4–27) is still to be determined.

The REM can be used to estimate the probability of a neutral mutation (Shakhnovich and Gutin, 1991). The formulation of the problem is the following: Suppose that a random amino acid residue in a protein is substituted by some other amino acid. What is the probability that the native conformation remains unchanged after such a mutation? In other words, what is the probability that the native conformation of a "mutated" sequence will be the same as that of the "wild-type" sequence? On the basis of the "REM-protein" analogy, the probability depends only upon the chain stiffness γ and equals γ^8. Comparison of this type of result with studies of the frequencies of stable mutations (Reidhaar-Olson and Sauer, 1988), as well as experimental (Alber, 1989) and theoretical (Tidor and Karplus, 1991) analyses of mutant free energy changes, would be of interest.

3.2.2 Lattice Models with Full Enumeration

Lattice models are of particular interest for the simulation of protein thermodynamics when it is possible to enumerate exhaustively all conformations. This makes it natural to address the question of the existence and stability of a unique structure and its properties.

An exhaustive search has been made for two types of models, both of which take into account excluded volume. One of these used short chains ($N \le 14$) and a two-dimensional square lattice (Lau and Dill, 1989). The other considered longer chains, but restricted the enumeration to fully compact self-avoiding conformations (Chan and Dill, 1990, 1991; Covell and Jernigan, 1990; Shakhnovich and Gutin, 1990b); since the native structure of a protein is compact, it is presumed to be necessary to search only the set of compact conformations to find the ground state. The restriction to compact self-avoiding (CSA) conformations leads to an essential decrease in the total number of conformations. The total number of unrestricted conformations of a polymer of N monomers positioned on a lattice with coordination number z is $(z - 1)^{N-1}$. The total number of maximally compact conformations is much smaller. Although an exact number is not available, different estimates (Huggins, 1942; Flory, 1949; Orland et al., 1985) give γ^{N-1} with γ

ranging from $^{(z-1)}/e$ (Flory, 1949) to $^{z}/e$ (Orland et al., 1985), where $e = 2.718\ldots$ is the base of natural logarithms. To illustrate the importance of this reduction in the number of conformations, we consider a chain of 27 monomers on a simple cubic lattice (Shakhnovich and Gutin, 1990b). The number of unrestricted conformations of this chain is 5^{26}, or $\sim 10^{18}$. Thus, in spite of the lattice simplification, this number is still far too large for an exhaustive enumeration. For the CSA conformations (i.e., those that fill completely a 3×3×3 fragment of a cubic lattice, see Figure 4–11) the estimated number of conformations is $(5/e)^{26}$, or $\sim 10^{7}$, a more reasonable figure. In fact, the exact number of conformations of such a 27-monomer chain has been determined; it is 4,960,608 without consideration of symmetry (Chan and Dill, 1990) and 103,346 with only those unrelated by symmetry conformations taken into account (Shakhnovich and Gutin, 1990b). The difference between the two numbers is due to the fact that each conformation is 48 times degenerate.

An exhaustive enumeration of all chain conformations (not only compact) was made (Lau and Dill, 1989; Chan and Dill, 1991) for short chains ($N \leq 14$) on a square 2D lattice for all possible sequences based on two types of "amino acids": hydrophobic and hydrophilic residues; the number of sequences for $N = 14$ is $16,384 = 2^{14}$. No preferences were introduced for local structure, and the only interactions were nearest neighbor attractions for a hydrophobic and a hydrophilic side chain, and repulsions otherwise. It was shown that, even for such a simplified model (only two

FIGURE 4–11. An example of a completely compact self-avoiding conformation on 3×3×3 fragment of a simple cubic lattice.

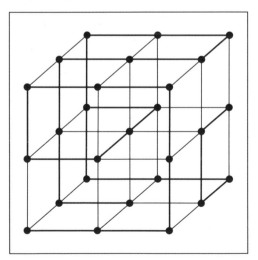

letters in the "protein alphabet"), a significant fraction of all sequences have low-energy conformations that are compact, and some of them have a nondegenerate lowest conformation (i.e., a unique structure). Among all 1,024 sequences of a 10-monomer chain, six were found to have nondegenerate native conformations.

It was stated by Lau and Dill (1989) that the ability of a sequence to form a unique structure increases as the chain length increases. This conclusion was based upon comparison of results with $N=10$ and $N=13$. Since such chains are still short, the observed trend may be due to end effects or specific features of compact structures on 2D lattices. For longer sequences ($N > 10$), it was not possible to enumerate all conformations and all sequences; i.e., only compact conformations and a subset of random sequences (200 sequences [Chan and Dill, 1990; Shakhnovich and Gutin, 1990b]) were generated. Again, a significant number of these sequences possessed a unique structure. It appears (Shakhnovich and Gutin, 1989b) that formation of a unique structure in heteropolymers is very sensitive to space dimensionality (d) and occurs in a different way in $d<2$ and $d>2$, with $d=2$ being a marginal and nonuniversal case that strongly depends upon the type of lattice, type of sequence alphabet, and so on. This is due to the fact that for $d>2$ the majority of contacts are nonlocal, while for $d<2$ the majority of contacts are local. This implies that, although some of the conclusions obtained from the analysis of $d=2$ case may be qualitively correct, care has to be used in their interpretation.

Another result from the study of short sequences on 2D lattices is of considerable interest. Chan and Dill (1990) concluded that compactness induces formation of "secondary structure" (α-helices, parallel and antiparallel β-strands, β-turns) and that neither (ϕ,ψ) propensities nor hydrogen bonds are essential for the widespread occurrence of secondary structure in proteins. Secondary structural elements were defined as patterns of intrachain contacts, e.g., two antiparallel β-strands exist if parts of the chain have contacts $(i,j; i+1,j-1...i+s; j-s)$, while α-helices exist if parts of the chain have contacts of the type $(i,i+4; i+1,i+5...i+s,i+s+4)$. This conclusion appears to contradict Flory's theorem (Flory, 1953; De Gennes, 1979) that the local structure of a polymer in a compact state or in a melt is the same as that of an unrestricted polymer. Since no energetic aspects are involved in the results of Chan and Dill (1990), Flory's theorem should apply. Since Flory's theorem neglects surface effects and is strictly applicable only to polymers of infinite length, further investigation of this point would be helpful (Shakhnovich and Karplus, to be published). In a recent investigation of a "native-like" off-lattice model of a polypeptide chain (Gregoret and Cohen, 1991), it was found that for long enough chains (proteins with 58, 135, and 275 residues were studied), the secondary structure contents of compact chains (with radii of gyration corresponding to those of proteins) and of unrestricted chains differ by only a few percent (e.g., ~18% secondary structure in unrestricted chains and 22% to 23% in compact). A significant increase in secondary structure content was observed for "supercompact" structures (~50% more compact than real proteins), where the surface is very important.

Three-dimensional compact lattices are expected to provide a more realistic representation of protein thermodynamics. All compact self-avoiding conformations of a 27-monomer chain on a 3x3x3 fragment of a cubic lattice were enumerated by Chan and Dill (1990) and Shakhnovich and Gutin (1990b). A model of a heteropolymer with a random sequence of monomers was analyzed by Shakhnovich and Gutin (1990b), and the "numerically exact" results were compared with the analytical theory for a unique structure (Section 3.2). Random "sequences" were characterized by sets of interaction energies B_{ij} between monomers i and j that are nearest neighbors on the lattice. For comparison with the analytic theory, random sets B_{ij} were generated with a Gaussian probability density. The aim of this work was to test the main results of the analytical theory and to determine whether $N = 27$ is sufficiently large, or whether finite-size effects are still important.

Two conclusions of the analytical theory were tested by the lattice simulations. First, is there some temperature, T_c, at which the number of relevant conformations drops drastically so only few (~1) are thermodynamically significant for $T<T_c$? As a measure of the number of relevant conformations, the quantity $X=1-Y$ (see Section 3.2)

$$X = 1 - \sum_{i=1}^{M} p_i^2 \tag{4-28}$$

was used. Here M is the total number of compact conformations (103,346) for a 27-monomer chain on the cubic lattice, and p_i is the Boltzmann probability of a conformation i. Theory predicts

$$<X> = \begin{cases} T/T_c & \text{at } T < T_c \\ 1 & \text{at } T \geq T_c \end{cases} \tag{4-29}$$

where <> denotes an average over all possible sequences. The results of the numerical simulation for a 27-mer are shown in Figure 4–12. They are in good agreement with the theory except in the transition region, which is smooth due to the finiteness of the chain. Thus, both the theory and the lattice simulations indicate that there is a transition below which only a few conformations with low energies become dominant.

Another conclusion for which the analytic theory and lattice simulations are in agreement concerns the landscape corresponding to the configurational space of the model protein. The "ruggedness" of the

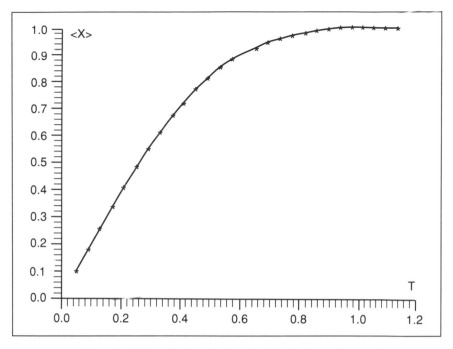

FIGURE 4-12. *Plot of <X(T)> obtained in lattice simulations. (See text.)*

landscape was described in the theory in terms of function P(Q) defined by Equation (4–22). This function was evaluated numerically in the simulations, and the result is shown in Figure 4–13. It is clear that P(Q) is bimodal, with a peak at Q = 1 and another peak at Q « 1. The second peak is concentrated around the value 0.2 to 0.3, which corresponds to overlap between two randomly chosen conformations in the 27-mer compact chain. This result confirms the conclusion that the conformational space of a typical random heteropolymer is rugged and is in accord with the applicability of the REM for heteropolymers in three dimensions (Section 3.2.1).

In a similar study (Covell and Jernigan, 1990) the conformations of several small proteins were enumerated. The polypeptide chain was represented as C_α atoms positioned on a face-centered cubic lattice. Such a lattice, according to Covell and Jernigan (1990), provides the best representation of the stereochemistry of a polypeptide chain in terms of pseudo-bond angles and pseudo-dihedral angles. Several proteins were studied, with lengths from 23 to 46 amino acid residues. The total number of conformations was restricted to a reasonable number by the requirement that all

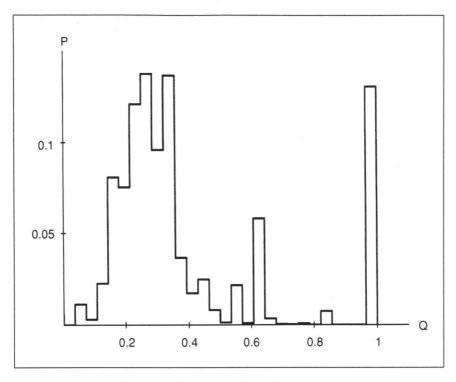

FIGURE 4–13. *P(Q) obtained in lattice simulation for most common sequences. (See text.)*

confomations fit exactly into the lattice fragment corresponding to the native protein; this determines a very special surface for the fragment. The number of conformations obtained varied from 1,000 to 15,000; only for the smaller systems was an exhaustive enumeration made. The energies of the conformations were determined by use of an effective potential function (Myazawa and Jernigan, 1985). It was shown that the native conformation is included in the 2% of conformations with the lowest energies. This suggests that the set of energy parameters used by Myazawa and Jernigan (1985) is appropriate for this type of study, although it may not be sufficient for folding from an extended state (Holley and Karplus, unpublished). It was also shown that the configuration space is "rugged"; i.e., there exist conformations with very large root mean square deviations from the native one (up to 14 Å) with energies close to, or sometimes even lower, than that of the native structure corresponding to the lattice model.

4 DYNAMIC ASPECTS OF PROTEIN FOLDING

Since the seminal analysis of Levinthal (1968,1969), it has been realized that alternatives to a random search must be operative in protein folding. Although the conformation space of the denatured state is vast, there would be no search problem if each of the amino acid residues could find its correct conformation independent of the others, or if only interactions with nearest neighbors were involved. Rapid folding would be expected, as in the helix-coil transition of polymers. Protein folding has been compared to crystallization (Harrison and Durbin, 1985), but it is important to realize that it is significantly different. In the former, once a nucleus is formed, there is no problem; i.e., there are many similar sites at which subsequent molecules can coalesce independently. In protein folding, even if there were a stable nucleus, it is not clear that the protein could continue to fold by the independent condensation of residues. The fundamental distinction between the helix-coil transition or crystallization and protein folding is that long-range interactions play an essential role in determining the native state. In the limit, this implies that the conformational energy of each amino acid residue depends upon all of the others in the polypeptide chain. It is this aspect of the search of the full configuration space, with its vast numbers of conformers, that leads to the Levinthal paradox.

A recent study (Zwanzig et al., 1992) argues that the Levinthal paradox can be resolved simply by introducing local mainchain propensities. They examined a bead model for the polypeptide chain that has only local interactions and found that a small bias of the mainchain propensities (see Section 3.1.4) results in a very large reduction in the apparent first passage time to a folded state. Such an analysis misses the essential point, outlined in the previous paragraph, concerning the difference between the helix-coil transition, which corresponds to the case studied by Zwanzig et al., and the cooperative transition of a protein. Moreover, the parameters used by Zwanzig et al., lead to a polypeptide chain with a vanishingly small probability of being in the ground state.

One can argue naively that the best way to find out how a protein folds is to start with a model for the denatured state of a protein and to simulate the transition to the native state. The simulation itself would solve the Levinthal paradox if the model used were sufficiently detailed and accurate. An approach that should work, in principle, is to use an atom-based model for the potential energy function and to follow a molecular dynamics trajectory from one or more denatured conformers to the native state in the presence of the appropriate solvent. With the available methodology and computing power, such a simulation would require approximately 10^{11} hr (or 10^7 years) for a 100-residue protein where the transition to the folded state takes place in on the order of 1 sec. With the teraflop super-computers that are on the horizon (Deng et al., 1990), such simulations would still take 10^3 years, although it would be possible to examine the faster portions of the folding transition. Clearly when such simulations become possible, they should be done, even though it is likely that a very large fraction of the trajectory would be uninteresting and that a very large amount of human time would be required to interpret the results.

The practical difficulties is doing such brute force simulations has led to several types of theoretical approaches to the dynamics of protein folding. One approach is based on the complete atomic model, but biases the system to speed up the folding transition by many orders of magnitude (as in simulated annealing with NMR distance constraints) or instead studies unfolding under special conditions (e.g., use of a forcing potential, high temperature) that reduce the time involved from microseconds to picoseconds. Related studies deal with small protein fragments (e.g., α-helices, β-turns) for which either the conformation space is sufficiently small so that full searches can be accomplished and/or transitions of interest occur on a manageable time scale. Alternatively, the entire polypeptide is considered, but a simplified description is introduced. One approach is to approximate each amino acid residue by one or two quasi-particles and to use the C_α pseudo-dihedral angles as the only variables (Sections 1 and 2). The problem can be further reduced by restricting the quasi-particles to move on a lattice (e.g., a cubic lattice) and using Monte Carlo procedures to follow the folding process. Such lattice calculations have the advantage over the physically more correct ("off-lattice") atomic models that the total number of conformations is reduced, as already described in Section 3.2.2, and that a larger fraction of all the conformations can be examined in a given time.

In addition to methods that are based on following the dynamics of a polypeptide chain with varying levels of detail, there exist a range of models for protein folding that are phenomenological in character. These models provide a qualitative description of the folding process, often including a conceptual approach to the solution of the Levinthal paradox. Most phenomenological models stop at this point. Some attempts have been made to go further, however, to estimate the time scales of various steps in the folding process or, at least, provide a way of characterizing the parameters involved in determining the time scales. Finally, there are kinetic models that do not attempt to give a full description of protein folding but serve primarily as a means of codifying experimental data. Since experimental studies are reviewed elsewhere in this volume, we mention such results here only in relation to the theoretical analysis.

4.1 Phenomenological Models

Phenomenological models of protein folding generally focus on the Levinthal paradox and introduce ways in which only a very small fraction of the total number of conformers participate in folding from the denatured to the native structure. The search of conformation space by the polypeptide chain under folding conditions must be such that the protein molecule is divided into parts whose information is used independently of the overall folding process. It appears that the soft dihedral angles that determine the chain are biased through relatively local interactions to lead to a reduction of the available configuration space. It is clear from Ramachandran plots of the main chain dihedral angles ϕ and ψ (Schulz and Schirmer, 1979) and the conformational preferences of the side chains (Gelin and Karplus, 1979) that the local conformation space of the amino acid residues is restricted in terms of high probability regions. The individual amino acid biases and their near-neighbor interactions are not sufficient to fold a protein, in accord with the limited success in *a priori* local predictions of secondary structure. Thus, longer-range interactions are required to determine the native structure and to reduce the space that has to be included in a search. A simple solution would be that a protein consisting of 100 amino acid residues is composed of 10 or so approximately equal parts, each of which finds its own unique stable structure, and that these parts then come together to form the complete native protein. For this case, instead of γ^{100} configurations, with γ equal to between 5 and 10, there would be only on the order of 10^{10} γ^{10} configurations, for which a random search can be envisaged. It is evident from studies of peptides in solution, as well as approximate energy calculations, that most sequences short enough to be searched through rapidly do not have a folded structure that is stable by itself. Thus, alternatives to this simple idea must be examined. All of the available phenomenological models use this concept in one way or another, although the details differ significantly. In what follows, we shall discuss a number of the proposed variants and point out their relationships.

One of the earliest discussions was that of Anfinsen (1973), who was concerned with explaining the experimental result that the enzyme staphyloccus nuclease folds spontaneously in solution. He proposed that portions of a protein chain can "flicker" in and out of their native format and serve as nucleation centers that coalesce through noncovalent interactions to give the native protein structure. Ptitsyn and Rashin (1975) considered myoglobin as composed of stable helices and, on the basis of a simple model for hydrophobic interactions between the helices, compared different paths to the most likely folded structure. A formalization of some of these ideas is contained in the diffusion-collision model (Karplus and Weaver 1976,1979), which considers the protein to be divided into a number of parts (microdomains), each short enough for all conformational alternatives to be searched rapidly. As already mentioned, this implies that the microdomain be so small that they are unlikely to be stable. Consequently, several (two or more) of them have to diffuse together and to collide in order to coalesce into a structural entity with the native or near native conformation. This model, whose quantitative aspects are considered shortly, has elements in common with other models that have been proposed more recently. One of these is the so-called framework model (Baldwin, 1989) in which the correct hydrogen-bonded secondary structure is assumed to be formed before the formation of tertiary structure. In a 1990 survey of folding, Kim and Baldwin (1990) updated the framework model to incorporate more recent experimental results. Baldwin (1989) recast the framework model in a form essentially identical to the diffusion-collision model. He proposed that folding begins with the formation of individual, transient secondary structural elements (i.e., elementary microdomains) that are stabilized by packing against each other (i.e., collision and coalescence), and that folding is a hierarchial process in which simple structures are formed first and these interact to give more complex structures. It is implied in the framework model, and demonstrated explicitly in the diffusion-collision model (see Section 4.2), that the number of folding pathways is small.

Harrison and Durbin (1985) proposed the jigsaw puzzle model as a conceptual description of protein folding. They suggested that proteins fold by a number of different, parallel pathways, rather than by a single defined sequence of events, and that this would make folding more robust to mutations that do not adversely affect the native structure. As these authors point out, folding kinetics of this type may be obtained with the diffusion-collision model (Karplus and Weaver, 1976, 1979) if all the elementary microdomains have similar properties so that the multi-microdomain intermediates have similar folding and unfolding rates. Under such conditions, even a small protein with only a few microdomains could have many alternative pathways. Because of the sensitivity of kinetics to differences in free energies, however, most possible pathways in a typical protein will be unimportant and a small number are expected to dominate. As is true for the framework model, the jigsaw puzzle model was not developed for quantitative calculations.

Nucleation models have been suggested (Levinthal, 1966; Tsong et al., 1972; Wetlaufer, 1973) in which a portion of the polypeptide chain, unstable by itself and possibly equivalent to a microdomain, serves as a nucleus for chain propagation to obtain the native structure. The model requires the nucleation unit to be small enough for a random search and, once this unit has the native structure, it assumes that a sequential, essentially independent folding of subsequent amino acid residues is possible. As has already been mentioned, it is not clear how this analog of crystallization would work, since for a nucleus of the required small size the remaining polypeptide chain would still seem to suffer from the Levinthal paradox. In the simplest version of the model, folding is assumed to proceed along a unique pathway in which formation of the nucleus would dominate the time dependence. This approach has been elaborated and modified by Gō and coworkers (1975) into the noninteracting local structure or growth-merge model, and by Kanehisa and Tsong (1979) into the cluster model. A cluster or "embryo" is assumed to grow until it merges with another growing nucleus. By this division into small parts, the Levinthal paradox is overcome. The growth-merge mechanism has many elements in common with the diffusion-collision model. A difference may arise in whether the growth of a single embryo (microdomain) or the coming together (diffusion-collision-coalescence) of two or more of these elements governs the kinetic of the folding process. In both models, nearest neighbor microdomains are more likely to coalesce first, everything else being equal, whether by growth or by collision. This is in accord with the distribution of contacts observed between secondary structural elements in many proteins. It is possible in the diffusion-collision model, however, that more distant microdomains coalesce. In an NMR folding study of cytochrome c, the N and C-terminal helices appear to coalesce first (Roder et al., 1988); this result has been analyzed in terms of the diffusion-collision model (Bashford et al., 1990).

There is experimental evidence that in many proteins under folding conditions a "hydrophobic collapse" takes place before formation of any tertiary structure. (See Chapter 6 by Ptitsyn.) Although the number of conformations in the collapsed state is much less than in the coil state (Dill, 1985), it is not sufficiently reduced to overcome the Levinthal paradox. Consequently, even if a hydrophobic collapse does occur, it is still necessary that some search mechanism be available to find a final structure. The molten globule model, proposed by Ptitsyn and co-workers, assumes that a compact state with the correct secondary structure is a folding intermediate (Ptitsyn et al., 1990). It has been suggested (Finkelstein and Ptitsyn, 1976) that the secondary structure, which is transiently present in the denatured state, is similar to that in folded state. This could help to overcome the search problem since many fewer degrees of freedom would be left, and the structure could form by coalescence of blocks of secondary structure. If such coalescence is described by the diffusion-collision model, numeric estimates of characteristic times are possible (see Section 4.2).

It remains unclear (and this is a key question of the protein folding problem) to what extent transient secondary structure in the unfolded protein resembles that in the native state and what degree of similarity is sufficient to overcome the search problem. The existing models do not give a definite answer to this question, and additional experimental data for the denatured state would be of interest in this regard.

4.2   Quantitative Aspects of the Diffusion-Collision Model

To illustrate one possible approach to obtaining a simplified yet quantitative model for protein folding, we briefly outline the calculational details of the diffusion-collision model (Karplus and Weaver, 1979; Bashford et al., 1984, 1987). As already mentioned, the diffusion-collision model considers the protein molecule to be divided into several parts (microdomains), each short enough for all conformations to be searched through rapidly. Consequently, several (two or more) of the microdomains have to diffuse together and to collide in order to coalesce into a structural entity that is stable. The process of folding the entire protein to the native state then involves a series of such diffusion-collision steps. These might have to follow a unique order to yield the native structure (single correct pathway); alternatively and more likely, different sequences of diffusion-collision steps might be possible. Furthermore, particularly in larger proteins, there may be several regions (domains) that individually attain their native structure by a set of diffusion-collision steps and finally come together to form the native protein molecule; for certain proteins it has been shown that a separated domain can fold by itself. (See Chapter 9.)

To illustrate the factors determining the kinetics of the diffusion-collision model, we outline an example involving two microdomains (Karplus and Weaver, 1976, 1979). Each of the microdomains is considered to be in equilibrium between the native secondary structure and the unfolded random-coil structure, which includes the rest of the available conformation space. If the microdomains form helices in the native state (as in myoglobin), the helix to random-coil transition would be involved. It is assumed that coalescence can occur only if both partners have at least partly the native structure (e.g., in the contact region) when they collide and if the collision takes place with the approximately correct orientation. If the two microdomains are regarded as diffusing together from a finite distance (due to the intervening backbone) to form a stabilized entity with the native conformation, the order of magnitude of the time t required for coalescence can be obtained by considering the radial diffusion of spherical particles. The result (Karplus and Weaver, 1976) is

$$\tau = \frac{1}{\beta} \frac{\lambda \, \Delta V}{DA} \tag{4-30}$$

where ΔV is the volume of the finite diffusion space (the space between two concentric spherical shells), A is the target surface area (the area of the inner shell) whose radius is determined by the sizes of the peptide units, D is the diffusion coefficient of a microdomain, and λ is a characteristic length (the average of the inner- and outer-shell radii, the latter being determined by the maximum distance of separation of the two units that are part of the same protein chain). The quantity β accounts for the fact that only a fraction of the microdomains can coalesce when they collide; it can involve, for example, a helix-coil equilibrium and/or an activation barrier. In a detailed theory (Lee et al., 1987), the various parameters involved (i.e., $\Delta V, A, \lambda$) would have a direct physical significance, but in this heuristic discussion they are phenomenological in character and only order-of-magnitude values can be given. With $D = \sim 1 \times 10^{-6}$ cm^2/sec, a value appropriate for spherical particles with the dimensions of protein microdomains, and with the outer-shell twice the inner radius and equal to 2×10^{-7} cm (20 Å), Equation (4–30) yields $\tau = \sim \beta^{-1}$ (10^{-7} to 10^{-8}) sec. To evaluate β, an equilibrium, two-state model can be used for simplicity; i.e., $\beta = K_1 K_2 / (1 + K_1)(1 + K_2)$, where K_1 and K_2 are the (native)/(random-coil) equilibrium constants for the two separated units. This expression ignores the possibilities that β could vary as the partners approach and that details of the kinetics might modify the equilibrium result. From the preceding formulae for τ and β, it follows that for values K_1 and K_2 equal to 10^{-2}, the folding time for a pair of units is of the order of 10^{-3} to 10^{-4} sec. Calculations have shown that the diffusion-collision dynamics of a multimicrodomain protein reduces to a network of two microdomain steps of the type we have described. For example, two units would coalesce to form a slightly more stable entity, which in turn would collide with a third entity, and so on. If this mechanism applied, the rate of the process would increase as the folding advanced and, with an appropriate set of kinetic constants, a cooperative transition could result; i.e., the coalescence of the initial microdomains could be the nucleation event.

The overall folding kinetics can be approximated by solving kinetic equations that couple the elementary steps. An example is provided by application of the diffusion-collision model to the operator-binding domain of the λ-repressor (Bashford et al., 1984). From the crystal structure (Pabo and Lewis, 1982), the operator-binding domain consists of four helices that form a well-defined globule and a fifth that interacts primarily with an equivalent helix in the other subunit of a dimer. With the calculation restricted to the four helices of a monomer, the parameters for the coalescence reaction rate were estimated from Equation (4–30). Only the correct helix contacts were allowed to form, and the coupled rate equations were solved; the possibility of dissociation of helices that had coalesced was included. Of the 65 possible intermediates, only a very small number contribute significantly. Although the exact details of the distributions depend upon the choice of parameters, as does the absolute folding time, the general behavior is typical of what is expected from the diffusion-collision model. Corresponding applications have been made to apomyoglobin (Bashford et al., 1988) and cytochrome c (Bashford et al., 1990).

One caveat with respect to quantitative calculations based on the diffusion-collision model concerns the nature of the rate-determining step in protein folding. It is possible, for example, that the diffusion-collision model describes the fast (submicrosecond) early phase in protein folding in which the approximately correct secondary structure is stabilized by loose contacts in a relatively compact globule. If this were true, the diffusion-collision step would essentially solve the Levinthal paradox; i.e., the remaining conformational search would be in the highly restricted space required for small adjustments in the secondary structure and for the relative reorientation of the secondary structural elements, coupled with the close packing of side chains. If the folding process has a final barrier near the native conformation as the "rate-determining step," as has been suggested by Creighton (1988), a calculation of the diffusion-collision type would not be germane to the overall reaction rate, but would be of interest for evaluating the time scale of earlier steps in the folding process. Evidence for this possibility comes from experiments for lysozyme (Sugihara and Segawa, 1984) that show a solvent-dependence for the rate of folding (i.e, the relative energy of the denatured and transition states depends on the solvent) but not for unfolding (the transition state is similar to the native state). The model for the thermodynamics of proteins proposed by Shakhnovich and Finkelstein (see Section 3.1.3) implies also that the main transition state on the folding pathway is close to the native state. A viscosity-dependence of the folding rate, however, as is expected for a diffusion process even if activated, has been observed in a number of proteins (Tsong, 1982; Hurle et al., 1987). Thus, additional experiments are required to resolve this question.

4.3   Atom-Based Simulation Studies

An approach that can be used to obtain information concerning the detailed kinetics of protein folding is to study peptide fragments that can be treated by detailed simulations. A number of such studies have been made, and more are in progress. To illustrate the possibilities, we shall describe several of them here and outline the more important results. Tirado-River and Jorgensen (1991) used a molecular dynamics simulation to study a 15-residue RNaseA S-peptide analogue at $5°C$ and $75°C$ in the presence of an explicit water model. They found that the peptide was stable for 300 psec at $5°C$, while it unfolded in less than 500 psec at $75°C$. Although experiments show that the peptide is more stable at low temperature (in agreement with the calculation), it is clear that equilibrium was not reached in the dynamics, since the observed helix-coil equilibrium constant is near unity at $5°C$. In the two higher-temperature simulations, unfolding occurred; in one, the transition took about 100 psec and in the other, about 350 psec. Examination of the simulation suggested that an important element of the unfolding transition is the replacement of an α-helical (*i* to *i*+4) hydrogen bond by water hydrogen bonds through an intermediate involving a 3_{10} (*i* to *i*+3) or reverse turn hydrogen bond. This is in accord with the X-ray studies of Sundaralingam and Sekhardadu (1989), who observed that when a water molecule inserts into an α-helix, a reverse turn is frequently formed. Details of the dihedral angle transitions are not given in the paper, so no conclusion can be drawn about their rates. Also, it is important to recognize that in a simulation of helix unfolding, nothing is learned about the rate-determining helix-initiation step. In related work, Case and co-workers (personal communication) have simulated the unfolding of the isolated helix H (residues 132 to 149) of myoglobin. They also found that tight (*i*, *i*+3) turns (3_{10} helices) were intermediates in the unfolding and often appeared prior to aqueous solvation. Insertion of H_2O occurred in some cases, but not all. The transitions were found to have diffusive elements with a time scale on the order of 100 psec.

Czerminski and Elber (1989, 1990) did vacuum simulations of a blocked alanine tripeptide, which is the shortest unit that can form an *i* to *i*+4 helical hydrogen bond. The simplicity of this system and the absence of solvent permitted them to make a full analysis of the multiminimum nature of the potential surface. Minima were found with the three peptide units in the neighborhood of the local minima for (ϕ, ψ) equal to (−60, −60; α-helix), (−120, 120; β sheet), (−60, 60), and (60, −60). A total of 138 minima were found in the range 0 to 6.3 kcal/mol relative to the minimum. This is to be compared with $4^{3.5}$, or 128, combinations of the most probable (ϕ, ψ) values. The 6,216 reaction paths between the minima were investigated. For direct paths connecting two minima without an intermediate, the barriers were found to be in the range of 0.5 to 5.5 kcal/mol, with the higher energies most common. Most of the direct paths involved changes in only one dihedral angle, with two thirds of the "flips" involving ψ and one third involving ϕ. Whether indirect paths led to significantly lower barriers between pairs of minima was not discussed. A master equation approach was used to study the dynamics, with the kinetic constants estimated from transition state theory. This led to lifetimes in the picosecond to nanosecond range at 400°K for the various minima.

The "folding/unfolding" transition of a blocked alanine dipeptide was investigated recently (Lazardis et al., 1991). The reaction path between the "folded" configuration (ϕ_1, ψ_1, ϕ_2, ψ_2) equal to (−72, −57, −81, −31) and an "extended" conformation (−83, 129; −83, 129) was determined in vacuum by a free energy simulation method with a dielectric constant of 50 to mimic the effect of an aqueous solvent. Two paths were found that are essentially mirror images of each other with ψ_1 and ψ_2 undergoing successive changes (ϕ_1 change very little); an intermediate was found where either ψ_1 or ψ_2

had undergone a transition. The calculated barriers were in the range of 1.5 to 2 kcal/mol. In a free energy simulation including an explicit model for the water molecules, similar results were obtained. The major difference in the latter is that the extended configuration with hydrogen bonds to water is about 3.4 kcal/mol more stable than the "turn" (which is similar to an earlier value [Tobias et al., 1990]); in the model with a dielectric constant of 50, the extended conformer was less stable by about 1 kcal/mol. The form of the potential of mean force in the presence of water is similar to that in the absence of water with an intermediate and with barriers of 2.3 and 2.8 kcal/mol.

Case (personal communication) has studied the dynamics of tetra-peptides for which there are NMR data indicating that the turn confor-mation makes a significant contribution to the equilibrium population in aqueous solution. In a 5 nsec simulation of the peptide Ala-Pro-Gly-Asp, they found transitions between "turns" with 1-3 and 1-4 bifurcated $C=O\cdots H-N$ hydrogen bonds and essentially extended conformations on a time scale of about 500 psec. It was noted that the transitions had diffusive elements, suggesting that a simple transition state description might not be valid. This corresponds to the results of stochastic dynamics simulations of the β-strand to coil transition (Yapa et al., 1992).

A study (Robert and Karplus, to be submitted) of the folding of the peptide $(Ala)_2(Asp)(Ala)_6$ from an extended chain to an α-helix in a vacuum ($\varepsilon = 1$) and in aqueous solution suggest similar folding mechanisms, but significantly different time scales. In vacuum, a structure formed rapidly (3 psec), in which the carboxyl group of the Asp interacted with three NH groups. This remained stable for about 80 psec, when a fourth NH group moved in and this initiated rapid (psec) helix formation. The simulation in aqueous solution required about 200 psec to form the three NH/Asp configurations, and helix initiation required on the order of 600 psec.

This type of detailed simulation of peptide fragments is clearly only in its infancy. It does demonstrate, however, that the nature of folding pathways for the fragments can be analyzed in detail and that information about the expected transition rates from one conformer to another range can be estimated; the time scales appear to be on the order of 100 psec to 1 nsec for elementary steps. It is important now to determine how these fragment studies are related to the behavior of larger systems.

A number of simulations of protein fragments have reached longer time scales than those possible in the all-atom plus solvent simulations described above. The method uses a potential of mean force surface that implicitly includes the solvent and employs stochastic dynamics to intro-duce the dynamic effects of the solvent. (See Section 2 and ref. 1.) To further simplify the problem, each residue is represented by a single interaction center ("atom") located at the centroid of the corresponding side chain, and the centers are linked by "virtual" bonds (Levitt, 1976; McCammon et al., 1980). For the potential energy of interaction between the residues, assumed to be Val in an α-helical simulation (McCammon et al., 1980), a set of energy parameters obtained by averaging over the side chain orientations was used (Levitt, 1976). Terms that approximate solvation and the stabilization energy of helix formation were included. The diffusive motion of the chain "atoms" expected in water was simulated by using a stochastic dynamics algorithm based upon Brownian dynamics. Starting from an all-helical conformation, the dynamics of several residues at the end of a 15-residue chain was monitored in a number of independent 12.5-nsec simulations at 298°K. The mobility of the terminal residue was quite large, with an approximate rate constant of 10^9 sec^{-1} for the transition between coil and helix states. This mobility decreased for residues further into the chain. Unwinding of an interior residue required simultaneous diplacements of several residues in the coil, so larger solvent frictional forces are involved. The coil region did not move as a rigid body. Instead, the torsional motions of the chain were correlated so as to minimize dissipative effects. Such concerted behavior is not consistent with the conventional idea that successive transitions occur independently. Analysis of the chain diffusion tensor showed that the frequent occurrence of correlated transitions resulted from the relatively small frictional forces associated with such motions (Pear et al., 1981). Further, the correlated nature of the torsional transition suggests that unwinding occurs in a relatively localized fashion and that a limiting value of about 10^7 sec^{-1} would be reached for the interior of the helix.

A similar model has been used recently to study β-sheet to coil transitions in a β-hairpin (Yapa et al., 1992). From an analysis of ten 90-nsec simulations, it was shown that rate constants for the transition between the coil and strand state are on the order of 10^{10} to 10^{11} sec^{-1}. This high rate occurs even though the adiabatic potential has barriers on the order of 3 kcal/mol. Unlike the α-helix results, the transition rate constant decreases only slowly as one goes from the end of the strand toward the interior. Also, nonterminal residues are sometimes found in the coil state, while the end residues still form a regular sheet. This behavior may be related to the occurrence of β-bulges.

A corresponding approach has been applied to the behavior of structural motifs of proteins. An example is a simulation of two α-helices connected by a coil segment (Lee et al., 1987). This simulation served to examine a possible elementary step in protein folding, as described by the diffusion-collision model, i.e., the coalescence of a pair of helices. The system considered in the simulations is a 24-residue peptide in which the first and last eight residues formed an α-helix and the intervening eight residues are initially in a random-coil conformation. Twelve trajectories were generated with a total time of 820 nsec. The exact length of the individual trajectories was not important because the parameters were chosen so that a stable coalesced structure does not form. Instead, the system folds and unfolds many times during the simulation, and rate constants for the coalescence and dissociation reactions could be determined. The values obtained are on the order of 10^8 sec^{-1}. For the unfolded system, there is a strong bias in the connecting loop toward shorter distances, with a maximum in the radial distribution function near 27 Å, even though the fully extended conformation has a length of 45 Å. This is due primarily to the entropic contribution, since the intervening chain has many more allowed conformations at intermediate lengths. Such a trend has been observed experimentally in a study of the end-to-end distances in a series of oligopeptides (Haas et al., 1978). This model calculation of coalescence used stable helices to reduce the time required. In a real system, helix to coil transitions would be coupled to the collisional association events. Given the estimates of the helix-coil transition rates, such a study should be possible by combining a stochastic dynamics simulation with a kinetic model for the helix-coil transition.

The stochastic dynamics simulation can be compared with the results of the simplified analytical treatment used in diffusion-collision, where the helices are treated as spheres (Karplus and Weaver, 1979; Bashford, 1984). The coalescence rate constant obtained from such a model is 10^8 sec^{-1}, in good agreement with the more detailed calculation. This agreement is accidental, in part, because it arises from a cancellation of two additional effects that are present in the more detailed model; one is the hydrodynamic interaction between the helices, which slows down the rate, and the other is the use of a realistic coil model, whose potential of mean force leads to closer distances and an increase in the rate. Whether this fortunate coincidence holds for more general problems will have to be determined by additional simulations.

Simulations of folding for a complete protein represent a much more complex problem than the fragment studies just described. Early work in this area (Warshel and Levitt, 1975) used a C_α-type model for the protein and coupled energy minimization with thermalization in a vacuum in an attempt to fold bovine pancreatic trypsin inhibitor (BPTI), a 58-residue protein. Conceptually such an approach is of interest: if the potential surface, or better potential of mean force surface of a protein, were such that a relatively simple procedure can find the minimum, there would be no Levinthal paradox. The limiting case would be a single potential well connecting the extended coil conformations to the native structure. Clearly that is not what is found for real proteins, although an empirical potential can be modified to obtain a relatively well-behaved surface that leads to a folded structure. In the potential used for the BPTI folding, for example, biases for turns and extended strands were introduced in the regions that have such conformations in the native structure (Warshel and Levitt, 1975; Hagler and Honig, 1978), similar to the biases introduced in recent Monte Carlo simulations (Skolnick and Kolinski, 1990a; Skolnick et al., 1990). With such a potential, one out of five runs that began with the terminal α-helix present and involved 600 cycles of minimization and thermalization had features of the native form, although the rms difference was about 6 Å, as compared with the 3.4 Å, the best value possible for the simplified model.

Molecular dynamics simulations of folding with a simplified model of 46 monomer chain represented by C_α atoms were presented in a recent study by Honeycutt and Thirumalai (1992). The simulations lacked reproducibility, and the various runs terminated with somewhat different conformations. The authors of the study suggest that this may model the metastability of the native state of real proteins. It is not clear, however, whether their result reflects metastability or is due to insufficient simulation time.

Another way of modifying the potential to achieve rapid folding is to introduce distance constraints. These have been used to simplify the energy minimization problem and, more recently, as a way of employing molecular dynamics with a simulated annealing protocol for structure determination by NMR (Brünger and Karplus, 1991). In a trial study of the protein crambin with NOE interproton distance constraints (Brünger et al., 1986), it was found that the native structure is achieved from an extended conformation in a few picoseconds of high-temperature molecular dynamics, a time on the order of 10^{12} times faster than that observed in solution. This decrease in time appears to involve two aspects of the constraints. The first is that stable secondary structural elements are formed in 1 psec, instead of approximately 10 μsec, and the second is that the need to search a large portion of the conformation space is eliminated by the long-range constraints (in terms of distance along the chain rather than physical distance). It was shown that if the secondary structure is not formed before collapse, misfolded structures that are local minima can result. This is consistent with the diffusion-collision and related phenomenological description of protein folding (Karplus and Weaver, 1979; Kim and Baldwin, 1990).

Vacuum molecular dynamics simulations with a polar hydrogen model for reduced BPTI were performed by heating the system from 300 to 1000°K over a time of 450 psec (Brady et al., to be published). Such a simulation of denaturation is in accord with the thermodynamic analysis of Shakhnovich and Finkelstein (1989) in which the first step of denaturation is regarded as a "vacuum" process and the protein expands to a molten globule–like state with a transition temperature of the order of 500°K (Section 3.1.3). A relatively sharp transition in the rms deviation from the native structure was found at about 500°K; the rms increased from about 2 Å to 4 Å. After this, the rms increased slowly until there was a second transition close to 1000°K, in which the rms reached about 7 Å. The first transition corresponds to a loss of tertiary structure with preservation of secondary structure; i.e., the β-sheet and α-helix are present but reoriented relative to each other. In the second transition all the secondary structure is lost, with the α-helix disappearing before the β-sheet. The loss of the secondary structures is accompanied by the formation of new hydrogen bonds, including "C5 ring"-like structures, a local minimum on the vacuum dipeptide map. The final system is a random coil–like globule, whose radius of gyration is only slightly larger than that of native BPTI.

A BPTI fragment, that contains the 30,51 S-S bond and has a large portion of the β-sheet and the C-terminal α-helix has been studied experimentally by Oas and Kim (1988); they have shown that the system is marginally stable with a structure close to that of native BPTI. When this construct was simulated at 500°K (Robert and Karplus, to be submitted), similar results were obtained as for complete BPTI, except that denaturation occurred more rapidly. A corresponding simulation at 500°K with the explicit inclusion of solvent (Robert and Karplus, to be submitted) showed the beginnings of denaturation over a time scale of 200 psec. The C-terminal α-helix began to unfold and water molecules replaced the terminal hydrogen bonds of the β-sheet. The behavior corresponds to the fragment studies cited previously, in terms of the slowing of the time scale relative to vacuum simulations and the explicit role of water-protein interactions at a certain stage of denaturation.

4.4 Lattice Models of Folding

Another approach to the dynamics of protein folding uses lattice models which involve the much less detailed description of the polymer chain and the reduction of the conformation space already described in Section 3.2.2. Such lattice models permit sampling of a much larger range of conformations and time scales than the more detailed off-lattice models. Formally, such lattices reduce the dynamics to a set of master equations with the transition rates between conformers as input to the calculations; in most cases, no effort has been made to relate the rates to physical estimates. The solution to the master equation is usually determined by Monte Carlo procedures.

Skolnick and co-workers (Skolnick et al., 1990; Skolnick and Kolinski, 1990a, b; Sikorski and Skolnick, 1990) have used tetrahedral (diamond) lattices, as well as more complex lattices, to represent the protein. Most of their work on the folding process has been concerned with model systems that form β-barrels or α-helical bundles (Skolnick and Kolinski, 1990a; Sikorski and Skolnick, 1990); they have also examined a more realistic model for plastocyanin (Skolnick and Kolinski, 1990b). Although the earlier simulations concerned with the equilibrium properties of the folding transition (Skolnick et al., 1990) used an extended Monte Carlo algorithm including both local and larger-scale transitions, the simulations of the folding pathways of six stranded β-barrels (Skolnick and Kolinski, 1990a) were restricted to local moves (three- and four-bond flips) to avoid "producing a distorted time scale." Secondary structure propensities (β-strand, β-turn) and distributions of hydrophobic and hydrophilic residues for a 74-amino-acid chain were used that, in previous equilibrium studies (Skolnick et al., 1990), had been shown to yield an all-or-none transition to the native state (although full enumeration to find the lowest energy conformation and determine its uniqueness was not possible). The results of a number of folding simulations showed no initial hydrophobic collapse and folding began at one or another of the predetermined β-turns of the native structure. "Collapse" took place simultaneously with the formation of tertiary structure. An intermediate was formed that contained part of the native structure (e.g., four out of six correctly positioned strands with 50 out of 74 native contacts). In the early stages of folding (i.e., from the initiation step to the formation of the intermediate), several pathways were found that led to nearly the same intermediate. Although it is stated that the intermediate to native transition is the rate-determining step, it appears from the results in Skolnick and Kolinski (1990a) that the initiation step, which involves formation of a turn that is native-like and retained through the folding transition, takes much longer than the final transition. As an example, we consider a trajectory for which some details are given (Figure 10 in Skolnick and Kolinski, 1990a). The turn that initiates folding is formed at $\tau = 675,500$; the intermediate is formed at $\tau = 681,500$; and folding is complete at $\tau = 694,750$, where τ corresponds to the number of Monte Carlo steps. From the time the intermediate is formed to the fully folded state takes only 13,250 Monte Carlo steps, while formation of the intermediate takes 681,500 steps. Estimates of the free energy as a function of the folding process do not show a clear barrier. Thus, these Monte Carlo simulations appear not to have a well-defined rate-determining step near the native structure.

The authors emphasize that the β-sheets are formed by "on-site" construction rather than diffusion-collision but that is implicit in the model since strand diffusion was not allowed in the local move Monte Carlo algorithm (Skolnick and Kolinski, 1990a). In a subsequent paper on four-helix bundles (Sikorski and Skolnick, 1990) that uses analogously biased sequences (i.e., helix and turn regions plus choices of hydrophobic and hydrophilic residues that yield all-or-none transitions to the four-helix bundle as the native state in equilibrium simulations), they introduce helix translation and rotation steps so as to provide a more realistic test of the diffusion-collision model versus the on-site assembly model. The α-helical bundle simulation results are described as being similar to those forming β-barrels. It is stated that there is a near native (three helices correctly positioned) transition state, but it is not clear from the figure in Sikorski and Skolnick (1990) that this is the transition state. Again, the initiation step takes much longer than the remainder of the trajectory. On-site construction is found to be the dominant mechanism of helix bundle formation. However, the probability of the translation and rotation steps was made to be so small (by the choice of parameters in the Monte Carlo algorithm), relative to the local moves, that such a result is not surprising. More generally, it should be noted that it is very difficult to draw conclusions from this type of simulation concerning the details of the folding mechanism because no attempt is made to determine the relative probability of different Monte Carlo moves in accord with the type of protein motions that they represent. (See Section 4.) Also, the results depend strongly on the interactions that stabilize interresidue contacts. In the simulations discussed, individual residues have a tendency to coalesce, independent of whether the chain has the correct secondary structure. This contrasts with the diffusion-collision model, in which the correct secondary structure in the coalescence region has to exist before a stable contact can be formed. Thus, although the simulations of Skolnick et al. are of interest as a model of protein folding, they do not provide a "test" of the diffusion-collision model.

A very different approach to the mechanism of protein folding (Shakhnovich et al., 1991) considered a lattice model in which the native conformations are determined only by nonlocal interactions, without any intrinsic propensities for local order and/or secondary structure. Further, no attempt was made to introduce specific structural motifs. The main idea is to find what sequences can fold by using a model where all conformations can be enumerated and therefore the "native" conformation (global minimum of energy) is known precisely. This makes it possible to address the basic questions concerned with whether and, if so, how the folding process reaches the global minimum. A 27-monomer chain on a simple cubic lattice with nearest-neighbor interactions was used (Shakhnovich and Gutin, 1990a). The conformation space of this chain included all extended conformations, as well as the subset of fully CSA conformations that completely fill a 3x3x3 fragment of the lattice. Since the total number of conformations on a cubic lattice for a 27-mer is $\sim 10^{18}$, the Levinthal paradox clearly exists in this system; in the CSA subset, there existed $\sim 10^5$ conformations and these have been enumerated (Shakhnovich and Gutin, 1990b).

The energy of a chain with N monomers is taken to have the form

$$E = B_0 \sum_{ij}^{N} \Delta (r_i - r_j) + \sum_{ij}^{N} B_{ij} \Delta (r_i - r_j) + D_2 \sum_{ij}^{N} \delta (r_i - r_j) + D_3 \sum_{ijk}^{N} \delta (r_i - r_j) \delta (r_i - r_k).$$

(4-31)

Here Δ equals 1 if $r_i - r_j = 1$ and is 0 otherwise; δ equals 1 if $r_i - r_j = 0$ and is 0 otherwise. The first term in Equation (4–31) represents a mean attraction ($B_0 < 0$) between monomers occupying neighboring sites that leads to a CSA conformation as the global energy minimum. The second term is "sequence specific"; a given sequence was generated by selecting the B_{ij} from a random distribution (Equation (4–20)) with a mean of zero. As described in Section 3.2.1, the thermodynamic behavior of heteropolymers with such interactions corresponds to the random energy model and can be well characterized analytically. The last two terms in Equation (4–31) introduce energetic penalties for conformations in which two and three links occupy the same site. Thus, although such conformations are allowed in the kinetics of folding, the values of D_2 and D_3 lead to very low thermodynamic probabilities for them, and the ground state is a CSA configuration. The dynamics of a chain was simulated by the standard Metropolis algorithm (Metropolis et al., 1973) with the kinetic scheme of Verdier (1973).

Thirty sequences corresponding to different sets of B_{ij} values were studied. The most striking result is that, of the 30 sequences, only three were able to fold to the conformation of the known global minimum and only one exhibited stable and rapid folding to this state, independent of the initial coil configuration. The one "folding" sequence required between 5×10^5 and 5×10^7 Monte Carlo steps to find the global minimum; test runs extending up to 5×10^8 steps were unsuccessful for the other sequences.

This result raises the intriguing questions of the difference between the one "folding" sequence and the other sequences that are not able to fold. Although a full understanding requires further analysis, one special feature of the "folding" sequence is that the configuration space is unusually "smooth" for a random heteropolymer. In Section 3.2, it was shown that a good measure of ruggedness of the configurational space is provided by the function $P(Q)$ defined in Equation (4–22); $P(Q)$ shows how structurally similar the low-energy conformations are. It has been demonstrated analytically (Shakhnovich and Gutin, 1989b) and numerically (Shakhnovich and Gutin, 1990a) that for a "typical" random heteropolymer the configuration space is rugged and $P(Q)$ tends to be bimodal (Fig. 4–13). The "folding" sequence is atypical in that $P(Q)$ is not bimodal (Fig. 4–14). Physically, this means that conformations that are close to the ground (native) state in energy are also close to it in structure. For the folding sequence, the conformation nearest to the native structure in energy (the difference was less than $2k_BT$) had 75% of its contacts in common with the native state.

Folding kinetics for this sequence were found to involve two stages. In the first stage, which usually required $<10^5$ Monte Carlo steps, rapid compactization took place as judged by number of interresidue contacts (Fig. 4–15A). The subsequent process of searching out the native structure required many more ($\sim 10^7$) Monte Carlo steps and involved transitions

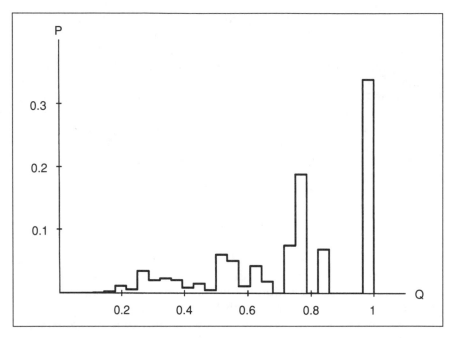

FIGURE 4–14. $P(Q)$ *for a "folding" sequence. (See text.)*

over the main activation barrier of the folding process (Fig. 4–15B). This is in accord with the result of an analytical theory (Shakhnovich and Gutin, 1989a), which predicts a free energy barrier between compact but nonnative conformations and the native conformation. The transition state has about half of the native contacts. Furthermore, an intermediate, which had the second lowest energy structure already mentioned, was present prior to the transition to the native state in most of the simulations.

The results of this simple model, if applicable to proteins, suggest that even if the native conformation is a free-energy minimum, there may be a nontrivial requirement that it be accessible kinetically in a reasonable time. In addition to selecting sequences in terms of the thermodynamic requirements for a unique global minimum, evolution may also have to select sequences in terms of the kinetic requirements for folding. An analytic analysis (Bryngelson and Wolynes, 1989) suggests that a "minimal frustration" principle (Bryngelson and Wolynes, 1987) can serve as a possible mechanism for avoiding kinetic traps under certain conditions. Further investigations are in progress to clarify the key question of what is necessary and sufficient to differentiate sequences that can fold kinetically (i.e., are able to overcome the Levinthal paradox) from the much larger set of sequences that have a unique stable structure.

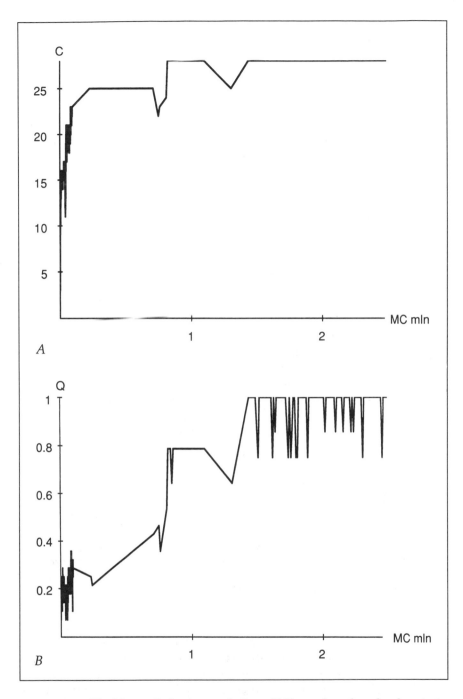

FIGURE 4–15. *The Monte Carlo time evolution of (A) number of nonlocal contacts in the chain and (B) overlap with the ground state for a typical run. The number of nonlocal contacts in CSA is 28. MC min: million Monte Carlo moves.*

5 CONCLUSION

This chapter has reviewed theoretical and computational attempts to address some fundamental problems in the thermodynamics and dynamics of protein folding. For most of these studies, essential simplifications to the potential of mean force surface were introduced, which involve the neglect of structural details of the protein chain (e.g., only the α backbone is included), the nature of the interactions (e.g., only hydrophobic/hydrophilic terms for spatially near neighbors are included), and discretization of the conformation space (e.g., only positions on a cubic or other lattice are allowed). These simplifications make it possible to treat a sufficient portion of the conformation space of a polypeptide chain to examine the large-scale changes involved in the transition between the native and denatured forms.

Formation of the native structure of a protein structure is cooperative in that the transition between the native and denatured state is of the first-order (all-or-none) type at equilibrium; i.e., only the native and denatured forms are present to a significant extent and there is a barrier between the two. A theory of protein thermodynamics must provide an explanation for this phenomenon, which differentiates proteins from ordinary heteropolymers; other aspects of the protein phase diagram are also of interest, although perhaps less fundamental.

Models based on polymer theory, when applied to ordinary homopolymers or heteropolymers, do not lead to a first-order globule-coil transition. This failure suggests that a valid theory of protein thermodynamics must be based on properties that distinguish proteins from ordinary heteropolymers. Several models, analytic and numerical, have introduced special elements of the polypeptide chain and examined the nature of the resulting folding/unfolding transition. It is clear that the only way to obtain an all-or-none transition is to have long-range interactions (i.e., distant along the polypeptide chain but possibly close in space) playing a significant role in the stability of the native structure. Chain stiffness can mimic such long-range effects but is unrealistic for proteins. One type of model focuses on special aspects of the protein as the source of cooperativity; i.e., that only the specific long-range contacts that exist in the native structure are attractive and/or that the main chain dihedral angles have propensities corresponding to the angles found in the native structure. If only the spatially nearest neighbor contacts in the native structure are attractive, a cooperative transition is found in lattice simulations, independent of other chain constraints. Also, a division into attractive and repulsive residues (e.g., hydrophobic/hydrophilic) can introduce an all-or-none transition in lattice models if the main chain propensities are chosen to correspond to the known secondary structural elements and turn regions. Local main chain propensities by themselves do not yield a first-order transition, in accord with classic studies of helix-coil equilibria. This type of model implies that the presence of cooperative folding in proteins depends upon the details of the evolutionary process and leads to a sequence-specific difference from ordinary heteropolymers (including random polypeptide chains). This is not impossible since there is survival value in the existence of cooperativity. Because the free energy stabilizing the native structure is small, large local fluctuations would destroy activity if the amino acid motions were independent. Such fluctuations are prevented by a cooperative transition.

An alternative model focuses on the presence of side chains as the property that distinguishes proteins from other heteropolymers. It shows that the requirement for tight packing in the native structure leads to an all-or-none transition. This approach is much less demanding as to the evolutionary process and implies that virtually any compact polypeptide with tight packing of side chains in the interior will undergo an all-or-none transition.

Although it is not known which, if either, of these two types of models is correct, it is important to note that they lead to different predictions concerning the structure of the protein backbone in the denatured state. In the backbone/contact and main chain propensity models, denaturation involves complete destruction of the correct fold of the protein backbone; i.e., the denatured protein can have virtually no elements of native structure. It can be either a coil or a globule without any preferential chain fold or, if it does have preferential structure, it must be different from the native one. (It is called "misfolded frozen" by Bryngelson and Wolynes (1987).) The "side chain melting" model, in contrast, assumes that the main source of the entropy in going from the compact to the denatured state is the liberation of the degrees of freedom associated with side chain rotational isomerization, so that denaturation does not require a significant rearrangement of the backbone. It is possible, therefore, that there exists a denatured globule with a relatively well-defined backbone conformation similar to that of the native structure. Because of this difference in the state of the protein after denaturation, an experimental test of the different models might be possible. It is, of course, not excluded that some elements of the two types of models contribute and that their importance varies from protein to protein.

Another aspect of proteins that distinguishes them from other heteropolymers is that they have a unique structure that is determined by their amino acid sequence. The existence of such a unique structure implies that there is one configuration, the native state, that is sufficiently low in free energy, relative to all other states, that only it is significantly populated at physiological temperatures (i.e., below the denaturation transition). A theoretical model and lattice simulations based on a C_α backbone description of the polypeptide chain suggest that a random sequence of amino acids (defined in terms of their interactions with close neighbors) can have such an "energy spectrum" and that, surprisingly, a significant fraction of all such random sequences would have a unique ground state (e.g., 10% with reasonable parameters).

Theories and experiments concerned with the dynamics of protein folding are in a more primitive state than those for protein thermodynamics. Although considerable progress has been made in detailed studies of the folding dynamics of protein fragments (e.g., α-helices, β-turns), the most basic question (the Levinthal paradox: How does the polypeptide chain search through the conformation space to find the lowest energy minima corresponding to the native state in a finite time?) is not solved. It is clear that the search must be simplified by dividing the folding process into parts, which could involve kinetic intermediates. This may be an essential difference between folding and unfolding, which could have no intermediates. Phenomenological models with intermediates (e.g., the diffusion-collision model and the closely related framework-type models) exist for the folding of the backbone. They provide one way of solving the search problem, but it is not clear that they are correct.

A number of other insights have come from folding simulations. It has been found in Monte Carlo simulations on lattices that the introduction of turn and secondary structure probabilities that lead to a thermodynamically stable unique structure (β-barrel, four α-helix bundle) can permit folding on a reasonable time scale. This appears not to be a general result, however; the existence of a unique ground state is not sufficient for the polypeptide chain to be able to find it. In fact, lattice simulations suggest that conformation space must have rather special properties for a chain to fold in a reasonable time. Although thermodynamic models have shown that the existence of a unique structure requires sequences that lead to a "rough" energy landscape, folding studies suggest there are limits on the permitted "roughness"; i.e., low-energy structures that are not too different from the native state may be required for efficient folding. In addition to selecting sequences in terms of the thermodynamic requirements for a unique global minimum, evolution may also have to select sequences in terms of the kinetic requirements for folding.

In most of the models that have been considered, folding is described as a multistage process with the formation of different features of the structure occurring at different stages. (See Figure 4–16.) It is assumed explicitly or implicitly that protein structure formation is hierarchical with specific features of the unique backbone conformation (based on side chain secondary structure propensities and effective interaction potentials) being separated from specific features of tight packing of the side chains. It is important to determine whether such a hierarchy exists in the actual folding process, particularly because it suggests essential simplifications in its analysis.

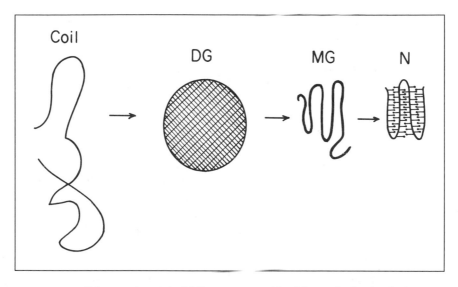

FIGURE 4–16. *Scheme of protein folding suggested by Monte Carlo simulations. The transition from the coil state to disordered globule, DG (which is like a collapsed homopolymer) has no energy barrier (second-order). Formation of the unique structure takes place as a transition within the compact state from the disordered globule to the molten globule intermediate (MG); this intermediate is likely to have some secondary structure and may have a native-like conformation of the backbone. Finally, formation of the native state (N) with a tightly packed hydrophobic core takes place. This is a first-order transition, and the main barrier to folding is between MG and N so that it involves the rate-limiting step of the overall folding pathway.*

In addition to the work described in this chapter, progress made with predictive algorithms appears to support such a hierarchical folding process. It has been shown (Finkelstein and Reva, 1991) that crude features of protein conformations, such as the folding patterns in some β-proteins, can be predicted with a simplified model that ignores many of the stereochemical details and does not consider side chains specifically. Also, there has been some success (Lee and Subbiah, 1991) in developing a Monte Carlo–based algorithm for the prediction of side chain conformations of a given fold. Such a separation of backbone folding and detailed side chain packing would have kinetic and evolutionary advantages. It is possible, however, that this separation is not valid and that the native fold is determined by numerous fine details of side chain stereochemistry and their mutual packing. This would make the protein folding problem much more complex.

To proceed further it will be necessary to have experimental knowledge of the properties of the various species that occur in the folding process and of any transition states between them. An early stage in the folding transition appears to be a rapid "collapse" of the mainchain, usually associated with the hydrophobic interaction. Although such a collapsed state does not provide a solution to the search problem, per se, the properties of the collapsed state may well hold the key to the folding problem. Is the correct secondary structure formed prior to collapse of the chain or simultaneously with the collapse or only when the unique three-dimensional structure is formed? Related to this is the question of what features of a molten globule state, if it exists, correspond to the native structure and what features of native protein are lost in molten globule. Does the cooperative denaturation transition correspond to a complete loss of the tertiary and secondary structures of the native protein or only of the tight packing of the side chains?

A question about all of the simplified models is whether the inclusion of more specific details of the protein structure and interactions would significantly modify the conclusions; i.e., do the results obtained apply to real proteins or only to the simplified models? It is essential to understand how different forces (hydrophobic, electrostatic, van der Waals) when treated by atom-based models contribute to the formation of the unique structure corresponding to the protein sequence. A knowledge of the relative role of the various interactions and the level of detail required is of practical significance, as well. It would make it possible to deal with simplified models and focus, for example, on hydrophobicity alone or on hydrophobicity and electrostatics. Also, it is of interest for evolutionary discussions whether hydrophobicity/hydrophilicity is sufficient to obtain the essential properties of proteins or whether a more complete set of amino acid properties is required.

In the field of protein folding, the present situation is in some ways analogous to that of chemical dynamics in the early 1960's (Herschbach, 1987). At that time there began a revolution in our understanding of elementary chemical reactions (e.g., $D+H_2 \rightarrow DH+H$). In both the experimental and theoretical areas, new ideas and new technologies played decisive roles in making possible the rapid advances that took place. In the protein field new experimental and theoretical techniques are now being developed. In the experimental area, stopped-flow NMR, CD, mutational, and time-resolved X-ray techniques are providing new insights. Limitations on the time scales of these approaches to milliseconds is too long, however, to provide the time resolution required for a full deconvolution of the protein folding process, and, in particular, to answer the key question as to whether the correct secondary structure is formed prior to hydrophobic collapse. The missing element is a rapid method for triggering folding or unfolding events; laser-induced pH jumps or local heating may be a possibility in this regard. In the theoretical area, the combination of new analytical methods and simplified models, as well as the application of macromolecular dynamics with realistic models of a protein, is providing much useful information on the folding process. There is again, however, a problem of time scales in the simulations; i.e., longer time simulations are needed and may become possible with a judicious combination of the appropriate simplifications and utilization of new computer technology.

We hope that this chapter, as well as the remainder of this volume, will help to set the stage for new developments toward solving the protein folding problem by a combination of theory, simulations, and experiment.

Acknowledgments

M. K. wishes to thank David Weaver for helpful discussions of protein folding and for use of some unpublished material. E. S. is grateful to A. V. Finkelstein, A. M. Gutin, O. B. Ptitsyn, and P. G. Wolynes for fruitful discussions. The work of preparing this chapter and some of the unpublished results from Harvard that are included were supported in part by a grant from the National Institutes of Health.

REFERENCES

Adelman, S. A. (1983) *Adv. Chem. Phys. 53*, 61.
Alber, T. (1989) *Ann. Rev. Biochem. 58*, 765.
Allen, M. P., and Tildesley, D. J. (1987) *Computer Simulations of Liquids*, Clarendon Press, Oxford, England.
Alonso, D., and Dill, K. A. (1991) *Biochemistry 30*, 5974.
Alonso, D., Dill, K. A., and Hutchinson, K. (1989) *Biochemistry 28*, 5439.
Alonso, D., Stigter, D., and Dill, K. A. (1991) *Proc. Natl. Acad. Sci. USA 88*, 4176.
Anfinsen, C. B. (1973) *Science 181*, 223.
Austin, R. H., Beeson, K. W., Eisenstein, L., Frauenfelder, H., and Gunsalus, I. C. (1975) *Biochemistry 14*, 5355–5373.
Baldwin, R. L. (1989) *Trends Biochem. Sci. 14*, 291.
Baldwin, R. (1990) *Nature 346*, 409.
Bashford, D. (1984) Ph.D. thesis, Tufts University, Medford, Mass.
Bashford, D., Cohen, F. E., Karplus, M., Kuntz, I. D., and Weaver, D. L. (1988) *Proteins: Struct. Funct. Genet. 4*, 211.
Bashford, D., Karplus, M., and Weaver, D. L. (1990) in *Protein Folding*, (L. M. Gierasch and J. Kuz, eds.), AAAS, Washington, D. C.
Bashford, D., Weaver, D. L., and Karplus, M. (1984) *J. Biomol. Struct. Dyn. 1*, 1243.
Baum, J., Dobson, C. M., Evans, P. A., and Hanley, C. (1989) *Biochemistry 28*, 7–13.
Binder, K., and Young, A. P. (1986) *Rev. Mod. Phys. 58*, 801.
Birstein, T. M., and Pryamitsyn, V. A. (1991) *Macromolecules 24*, 1554–1560.

Birstein, T. M., Kuznetsov, D. V., and Grosberg, A. Yu. (1987) *Vysokomol. Soed. (USSR) B29*, 951–964.

Brady, J., Ha, S., and Karplus, M. (to be published).

Brooks, C. L. III, Balk, M., and Adelman, S. (1987) *J. Chem. Phys. 19*, 784.

Brooks, C. L. III, Karplus, M., and Petitt, B. M. (1988) *Proteins: A Theoretical Perspective of Dynamics, Structure and Thermodynamics*, in *Adv. Chem. Phys. LXXI*, John Wiley & Sons, New York.

Bruccoleri, R. E., Karplus, M., and McCammon, J. A. (1986) *Biopolymers 25*, 1767–1802.

Brünger, A. T., Clore, G. M., Gronenborn, A. M., and Karplus, M. (1986) *Proc. Natl. Acad. Sci. USA 83*, 3801.

Brünger, A. T., and Karplus, M. (1991) *Acc. Chem. Res. 24*, 54–61.

Bryngelson, J. D., and Wolynes, P. G. (1987) *Proc. Natl. Acad. Sci. USA 84*, 7524–7528.

Bryngelson, J. D., and Wolynes, P. (1989) *J. Phys. Chem. 93*, 6902.

Bryngelson, J. D., and Wolynes, P. G. (1990) *Biopolymers 30*, 177.

Case, D. A., and Karplus, M. (1979) *J. Mol. Biol. 132*, 343–368.

Chan, H. S., and Dill, K. (1990a) *Proc. Natl. Acad. Sci. USA, 87*, 6388.

Chan, H. S., and Dill, K. A. (1990b) *J. Chem. Phys. 90*, 492.

Chan, H. S., and Dill, K. (1991a) *Ann. Rev. Biophys. Biophys. Chem. 20*, 447.

Chan, H. S., and Dill, K. (1991b) *J. Chem. Phys. 95*, 3775.

Chandler, D. (1978) *J. Chem. Phys. 68*, 2959.

Chick, H., and Martin, C. J. (1911) *J. Physiol. 43*, 1.

Covell, D. G., and Jernigan, R. L. (1990) *Biochemistry 29*, 3287.

Creighton, T. E. (1988) *Proc. Natl. Acad. Sci. USA 85*, 5082.

Creighton, T. E. (1990) in *Protein Folding* (L. M. Gierasch and J. Kuz, eds.) AAAS, Washington, D. C.

Czerminski, R., and Elber, R. (1989) *Proc. Natl. Acad. Sci. 86*, 6963.

Czerminski, R., and Elber, R. (1990) *J. Chem. Phys. 92*, 5580.

De Gennes, P. G. (1975) *J. Phys. Letters 36*, L55–L57.

De Gennes, P. G. (1979) *Scaling Concepts in Polymer Physics*, Cornell University Press, Ithaca, New York.

Deng, Y., Glimm, J., and Sharp, D. H. (1990) Los Alamos Report LA-UR-90-4340.

Derrida, B. (1981) *Phys. Rev. B24*, 2613–2624.

Derrida, B., and Toulouse, G. (1985) *J. Phys. Letters 46*, L223.

Dill, K. (1985) *Biochemistry 24*, 1501–1509.

Dorit, R. L., Schoenbach, L., and Gilbert, W. (1990) *Science 250*, 1377.

Elber, R., and Karplus, M. (1987) *Science 235*, 318–321.

Elber, R., and Karplus, M. (1990) *J. Am. Chem. Soc. 112*, 9161–9175.

Fasman, G. D., ed. (1988) *Prediction of Protein Structure and the Principles of Protein Conformation*, Plenum, New York.

Finkelstein, A. V., and Ptitsyn, O. B. (1976) *J. Mol. Biol. 103*, 15.

Finkelstein, A. V., and Reva, B. A. (1991) *Nature 351*, 497–499.

Finkelstein, A. V., and Shakhnovich, E. I. (1989) *Biopolymers 28*, 1682.

Flory, P. J. (1949) *J. Chem. Phys. 17*, 303.

Flory, P. J. (1953) *Principles of Polymer Chemistry*, Cornell University Press, Ithaca, New York.

Flory, P. J. (1969) *Statistical Mechanics of Chain Molecules*, Wiley, New York.

Garel, T., and Orland, H. (1988) *Europhys. Lett. 6*, 307, 597.

Gelin, B. R., and Karplus, M. (1979) *Biochemistry 18*, 1256.

Gelin, B. R., Lee, A. W. -M., and Karplus, M. (1983) *J. Mol. Biol. 171*, 489–559.

Gō, N. (1975) *Int. J. Peptide Protein Res. 7*, 313.

Gō, N., and Abe, H. (1981) *Biopolymers 20*, 991.

Gō, N., and Scheraga, H. A. (1969) *J. Chem. Phys. 51*, 4751.

Goto, Y., and Fink, A. (1990) *J. Mol. Biol. 214*, 803–805.

Gregoret, L., and Cohen, F. (1991) *J. Mol. Biol. 219*, 109.

Haas, E., Katchalski-Katzir, E., and Steinberg, I. Z. (1978) *Biopolymers 17*, 11.

Hagler, A. T., and Honig, B. (1978) *Proc. Natl. Acad. Sci. USA 75*, 554.

Harrison, S. C., and Durbin, R. (1985) *Proc. Natl. Acad. Sci. 82*, 4028.

Helfand, E., Wasserman, Z. R., and Weber, T. A. (1980) *Macromolecules 13*, 526.

Honeycutt, G. D., Thirumalai, D. (1992) *Biopolymers*, (in press).

Herschbach, D. R. (1987) *Angew. Chem. Int. Ed., Engl. 26*, 1221.

Huggins, M. L. (1942) *J. Phys. Chem. 46*, 151.

Hurle, M. H., Michelotti, G. A., Crisanti, M. M., and Matthews, C. R. (1987) *Proteins: Struct. Funct. Genet. 3*, 254.

Hynes, J. T. (1983) in *The Theory of Chemical Reactions*, vol. 4 (M. Baer, ed.), CRC Press, Boca Barton.

Jaenicke, R. (1987) *Prog. Biophys. Mol. Biol. 49*, 117.

Kanehisa, M. I., and Tsong, T. Y. (1978) *J. Mol. Biol. 124*, 177; (1979) *J. Mol. Biol. 133*, 279.

Karplus, M., Ichiye, T., and Petitt, B. M. (1987) *Biophys. J. 52*, 1083–1085.

Karplus, M., and Weaver, D. L. (1976) *Nature 260*, 404.

Karplus, M., and Weaver, D. L. (1979) *Biopolymers 18*, 1421.

Kauzmann, W. (1959) *Adv. Protein Chem. 3*, 1.

Kim, P., and Baldwin, R. (1982) *Ann. Rev. Biochem. 51*, 459.

Kim, P. S., and Baldwin, R. L. (1990) *Ann. Rev. Biochem. 59*, 631.

Kron, A., Ptitsyn, O. B., and Eisner, Yu. (1968) *J. Polymer Sci. 16*, 3509.

Lau, K. F., and Dill, K. (1989) *Macromolecules 22*, 3986.

Lazaridis, T., Tobias, D. J., Brooks, C. L. III, and Paulaitis, M. E. (1991) *J. Chem. Phys. 55*, 7612.

Lee, C., and Subbiah, S. (1991) *J. Mol. Biol. 193*, 373–388.

Lee, S., Karplus, M., Bashford, D., and Weaver, D. L. (1987) *Biopolymers 26*, 481.

Levinthal, C. (1966) *Sci. Am. 214*, 42.

Levinthal, C. (1968) *J. Chim. Phys. 65*, 44. It should be mentioned that no discussion of the type attributed to Levinthal is given in the published paper; perhaps it was given in the original lecture.

Levinthal, C. (1969) in *Mossbauer Spectroscopy in Biological Systems* (P. DeGennes et al., eds). Urbana, Illinois: University of Illinois Press. This reference does discuss the question of folding times.

Levitt, M. (1976) *J. Mol. Biol. 104*, 59.

Levitt, M. (1991) *Curr. Opinion Struct. Biol. 1*, 224.

Lifshitz, I. M., Grosberg, A. Yu., and Khohlov, A. R. (1978) *Rev. Mod. Phys. 50*, 683–713.

McCammon, J. A., Northrup, S. H., Karplus, M., and Levy, R. M. (1980) *Biopolymers 19*, 2033.

McCammon, J. A., and Harvey, S. (1987) *Dynamics of Proteins and Nucleic Acids*, Cambridge University Press, Cambridge, England.

McQuarrie, D. A. (1976) *Statistical Mechanics,* Harper & Row, New York, p. 641.

Metropolis, N., et al. (1953) *J. Chem. Phys. 21,* 1087.

Myazawa, S., and Jernigan, R. (1985) *Macromolecules 18,* 534.

Northrup, S. H., Pear, M. R., Lee, C. Y., McCammon, J. A., and Karplus, M. (1982) *Proc. Natl. Acad. Sci. USA 79,* 4035.

Oas, T. G., and Kim, P. S. (1988) *Nature 335,* 694.

Orland, H., Itzykson, C., De Dominicis, C. (1985) *J. Phys. Letters 46,* L-353.

Pabo, C. A., and Lewis, M. (1982) *Nature 298,* 443.

Parisi, G., Mezard, M., and Virasoro, M. (1985) *J. Phys. Letters 46,* L217.

Parisi, G., Mezard, M., and Virasoro, M. (1988) *Spin Glass Theory and Beyond,* World Sci., Singapore.

Pear, M. R., Northrup, S. H., McCammon, J. A., Karplus, M., and Levy, R. M. (1981) *Biopolymers 20,* 629.

Pechukas, P. (1976) *Dynamics of Molecular Collisions,* Part B (W. H. Miller, ed.), Plenum, New York, p. 269.

Perahia, D., Levy, R. M., and Karplus, M. (1990) *Biopolymers 29,* 645–677.

Perutz, M. F. (1989) *Quart. Rev. Biol. 22,* 139.

Perutz, M. F., Kendrew, J. C., and Watson, H. C. (1965) *J. Mol. Biol. 13,* 669.

Peticolas, W. L., and Kurtz, B. (1980) *Biopolymers 19,* 1153.

Petrich, J. W., Lambry, J. -C., Kuczera, K., Karplus, M., Poyart, C., and Martin, J. -L. (1991) *Biochemistry 30,* 3975–3987.

Post, C. B., and Zimm, B. H. (1979) *Biopolymers 18,* 1487.

Privalov, P. L. (1979) *Adv. Protein Chem. 33,* 167.

Privalov, P. L. (1982) *Adv. Protein Chem. 36,* 1.

Privalov, P. L., Bendsko, P., Pfeil, W., and Tiktopulo, E. I. (1988) *Biophys. Chem. 29,* 301–307.

Privalov, P. L., and Khechinashvili, N. N. (1974) *J. Mol. Biol. 86,* 665.

Privalov, P. L., et al. (1989) *J. Mol. Biol. 205,* 737–750.

Ptitsyn, O. B. (1987) *J. Protein Chem. 6,* 273–293.

Ptitsyn, O. B., and Rashin, A. (1975) *Biophys. Chem. 3,* 1.

Ptitsyn, O. B., et al. (1990) *FEBS Letters 262,* 20–24.

Reidhaar-Olson, J. F., and Sauer, R. T. (1988) *Science 241,* 53.

Richards, F. M. (1977) *Ann. Rev. Biophys. Bioeng. 6,* 151.

Robert, B., and Karplus, M. (to be submitted).

Roder, H., Elove, G. A., and Englander, S. W. (1988) *Nature 335,* 700.

Sanchez, I. C. (1970) *Macromolecules 12,* 980.

Schroedinger, E. (1945) *What is Life?* Cambridge University Press, Cambridge, England.

Schulz, G. E., and Schirmer, R. H. (1979) *Principles of Protein Structure,* Springer Verlag, New York.

Semisotnov, G. V., Rodionova, N. A., Kubyshenko, V. P., Ebert, B., Blanck, J., and Ptitsyn, O. B. (1987) *FEBS Letters 224,* 9–13.

Shakhnovich, E. I., Farztdinov, G. M., Gutin, A. M., and Karplus, M. (1991) *Phys. Rev. Lett. 67,* 1665.

Shakhnovich, E. I., and Finkelstein, A. V. (1982) *Dokl Acad. Nauk SSSR, 243,* 1247.

Shakhnovich, E. I., and Finkelstein, A. V. (1989) *Biopolymers 28,* 1667.

Shakhnovich, E. I., and Gutin, A. M. (1989) *Studia Biophysica 132,* 47.

Shakhnovich, E. I., and Gutin, A. M. (1989a) *Biophys. Chem. 34,* 187.

Shakhnovich, E. I., and Gutin, A. M. (1989b) *J. Phys. A22*, 1647.
Shakhnovich, E. I., and Gutin, A. M. (1990a) *Nature 346*, 773–775.
Shakhnovich, E. I., and Gutin, A. M. (1990b) *J. Chem. Phys. 93*, 5967.
Shakhnovich, E. I., and Gutin, A. M. (1991) *J. Theor. Biol. 149*, 537.
Sikorski, A., and Skolnick, J. (1990) *J. Mol. Biol. 212*, 819–836.
Skolnick, J., and Kolinski, A. (1990a) *J. Mol. Biol. 212*, 787–817.
Skolnick, J., and Kolinski, A. (1990b) *Science 250*, 1121.
Skolnick, J., Kolinski, A., and Yaris, R. (1990) *Comments Mol. Cell Biophys. 6*, 223.
Smith, J., Kuczera, K., and Karplus, M. (1990) *Proc. Natl. Acad. Sci. USA 87*, 1601–1605.
Stillinger, F. H., and Weber, T. A. (1982) *Phys. Rev. A25*, 978.
Sugihara, M., and Segawa, S. -I. (1984) *Biopolymers 23*, 2473.
Sundaralingham, M., and Sekharudu, Y. C. (1989) *Science 244*, 1331,
Tanford, C. (1968) *Adv. Protein Chem. 23*, 121–275.
Tanford, C. (1970) *Adv. Protein Chem. 24*, 1–95.
Tidor, B., and Karplus, M. (1991) *Biochemistry 30*, 3217–3228.
Tirado-Rives, J., and Jorgenson, W. C. (1991) *Biochemistry 30*, 3864.
Tobias, D. J., Sneddon, S. F., and Brooks, C. L. III (1990) *J. Mol. Biol. 216*, 783.
Tsong, T. Y. (1982) *Biochemistry 21*, 1493.
Tsong, T. Y., Baldwin, R. L., and McPie, P. J. (1972) *J. Mol. Biol. 63*, 453.
Ueda, Y., Taketomi, H., and Gō, N. (1975) *Int. J. Peptide Protein Res. 7*, 445.
Verdier, P. H. (1973) *J. Chem. Phys. 59*, 611.
Warshel, A., and Levitt, M. (1975) *Nature 253*, 694.
Wetlaufer, D. B. (1973) *Proc. Natl. Acad. Sci. USA 70*, 697.
Williams, D., Harding, M., and Woolfson, D. (1991) *Biochemistry 30*, 3120–3128.
Wu, H. (1929) *Am. J. Physiol. 90*, 562.
Yapa, K., Weaver, D. L., and Karplus, M. (1992) *Proteins: Struct. Funct. Genet. 12*, 237.
Zheng, C., Wong, C. F., McCammon, J. A., and Wolynes, P. G. (1988) *Nature 334*, 726.
Zwauzig, R., Szaba, A., and Bagchi, B. (1992) *Proc. Natl. Acad. Sci. USA 89*, 20.

5

Kinetics of Unfolding and Refolding of Single-Domain Proteins

FRANZ X. SCHMID

1 INTRODUCTION

The folding of a polypeptide chain under native conditions is a spontaneous and reversible process that is directed by the information encoded in the amino acid sequence. The inherent relationship between sequence and folding is presently not understood at the molecular level. The establishment of kinetic mechanisms for unfolding and refolding of a particular protein are first steps in the elucidation of its folding pathway. In subsequent steps, kinetically detected folding intermediates and the transition states between them have to be characterized in terms of their structure, stability, and position on the folding pathway. In principle, a folding pathway is understood when all intermediates and activated states are arranged in their correct order and when their structures are known. In practice, the kinetic approach is confined by several limitations. (1) Only intermediates that are followed by slow steps accumulate and can be studied. (2) Structural investigations are limited by the transient nature of intermediates. (3) The cooperativity of folding, particularly of small single-domain proteins, limits the number of steps that can be observed in the folding kinetics, and intermediates are usually not observed at equilibrium in the unfolding transitions (Chapter 3). A few exceptions do exist; they are discussed in Chapter 6.

    Folding reactions are frequently complex processes, and kinetic mechanisms are difficult to establish and to check against alternative mechanisms. The complexity of folding has two major sources: (1) The unfolded state of most proteins is heterogeneous conformationally and gives rise to multiple parallel refolding reactions. (2) Partially folded intermediates can be formed transiently in these individual refolding reactions. It is therefore mandatory to use as many different probes as possible to follow folding, to vary the conditions, to employ natural or designed variants for comparison, and also to use thermodynamic and structural information for the molecular interpretation of kinetic results.

It has been discussed for a long time, whether the native structure of a protein is under thermodynamic or kinetic control (see, e.g., Wetlaufer and Ristow, 1973), i.e., whether the native protein is the state of lowest free energy or the state that is kinetically formed most rapidly. The arguments for thermodynamic control are, at least for many small proteins, clear-cut: the unfolded and the native state exist in thermodynamic equilibrium within the unfolding transition, and this native state is by several criteria the same as the one found under physiological conditions, such as ambient temperature, neutral pH, and the absence of denaturants. On the other hand, kinetic control should be important to reach the native state in a short time. Both aspects could be satisfied by the evolution of ordered folding pathways, which, by a succession of defined intermediate states, rapidly decrease the accessible conformational space and thus allow the rapid formation of the thermodynamically most favored native state rapidly after biosynthesis. This use of kinetic control to reach the state of lowest free energy rapidly could be termed a "kinetic consistency principle" of folding (used in analogy to the consistency principle (Gō, 1983) of short-range and long-range interactions in protein folding). According to this principle, kinetic pathways of folding are consistent with, and lead to, the state of lowest free energy. Such ordered pathways would ensure that the native state is reached rapidly and efficiently, and they provide high energy barriers to other, nonproductive pathways that lead to abortive structures. Rapid folding and kinetic constraints might not be sufficient to avoid nonproductive routes. *De novo* folding in the cell may involve transient binding to other proteins ("chaperones," Goloubinoff et al., 1989; Ostermann et al., 1989). Of course, chaperones do not carry specific information to direct the folding of a protein. Rather, they help refolding chains to avoid side reactions, such as aggregation (Buchner et al., 1991). These topics are further discussed in Chapter 10.

This chapter concentrates on the *in vitro* unfolding and refolding kinetics of small, monomeric single-domain proteins. Large multidomain and multisubunit proteins are the topic of Chapter 9. The kinetics of protein folding are described in several review articles. An excellent discussion of protein folding in general, and of the interrelationship between thermodynamic and kinetic aspects, is provided by the three classic reviews by Tanford (1968a,b, 1970). The articles by Jaenicke (1987) and Creighton (1990) give general overviews on folding. The mechanism of folding, intermediates, and catalysis of folding were reviewed by Matthews (1991), Dobson (1991), and Schmid (1991). Small proteins and the role of folding intermediates are the subject of three successive reviews by Kim and Baldwin (Baldwin, 1975;

Kim and Baldwin, 1982,1990). A collection of papers on practical aspects of protein folding kinetics can be found in volume 131 of the *Methods in Enzymology* (Hirs and Timasheff, 1986). This chapter is restricted to folding reactions that do not involve the formation or breakage of disulfide bonds, which are discussed in Chapter 7. The kinetic consequences of the cooperativity of the folding of small proteins are discussed, and kinetic criteria are presented for the applicability of the most simple two-state model within the unfolding transition. In the following, the kinetic hetero-geneity of the unfolded state and one of its presumed origins, prolyl peptide bond isomerization, will be described. Then the analysis of folding in the native and unfolded baseline regions and the characterization of interme-diates and transition states on the folding pathways are discussed. Finally, two case studies of the folding of small proteins are presented.

2 THE COOPERATIVITY OF FOLDING AND ITS KINETIC IMPLICATIONS

2.1 The Two-State Approximation

The reversible unfolding of small single-domain proteins is usually a cooperative process (Chapter 3), in which only the fully native (N) and the fully unfolded (U) molecules are populated at equilibrium. Both thermal and denaturant-induced unfolding transitions are generally described adequately at equilibrium by a two-state model involving only U and N (Equation (5–1)).

$$N \rightleftharpoons U. \tag{5-1}$$

The two-state character of such protein folding transitions is a consequence of the cooperative nature of protein stabilization. A delicate balance of many relatively large energetic contributions determines the low net stability of the fully folded state. The cooperativity of these numerous interactions apparently results in partially folded conformations being less stable than either N or U under all conditions.

The primary characteristics of cooperative two-state unfolding transitions is that the same unfolding curve is given by different probes of unfolding, such as fluorescence or absorbance measurements, circular dichro-ism, and viscosity. The analysis of equilibrium unfolding transitions requires an extrapolation of the baselines for the native and for the unfolded protein into the transition region (cf. Fig. 5–1) to determine the fraction of native molecules, f_N, as a function of the unfolding parameter:

$$f_N = \frac{(A - A_U)}{(A_N - A_U)}. \tag{5-2}$$

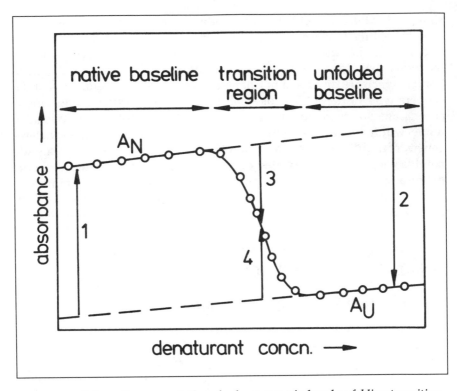

FIGURE 5–1. Schematic representation of a denaturant-induced unfolding transition of a small protein that obeys the two-state approximation. The absorbance is selected as a representative property that changes in the course of the transition. The extrapolated baselines for the native (A_N) and for the unfolded (A_U) protein are given as dashed lines. Arrow 1 represents a kinetic refolding experiment in the native baseline region, arrow 2 shows an unfolding experiment in the denatured baseline region, and arrows 3 and 4 indicate unfolding and refolding experiments within the transition region. The approximate division of the folding conditions into various regions is indicated on top of the figure. Conditions under which the native protein shows a high stability (the left end of the native baseline region) are called strongly native. Marginally native conditions are found near the transition region, where the stability of the native state is low.

A is a physical property (e.g., absorbance or fluorescence) that changes during the transition, A_N and A_U are the respective values for the native and the unfolded protein (cf. Fig. 5–1). In most cases a linear dependence of the properties A_U and A_N on the unfolding agent can be assumed for the extrapolation into the transition region (Pace, 1986). A simple procedure for fitting experimental data to a two-state transition by including the baseline regions in the fitting routine is given by Santoro and Bolen (1988).

The coincidence of normalized unfolding transitions measured by different probes is a necessary, but not a sufficient, condition for the validity of the two-state model (Equation (5–1)). A further criterion is provided by the calorimetric test. It is based on the identity of the unfolding enthalpy, measured directly in a microcalorimeter, ΔH_{cal}, with the van't Hoff enthalpy, $\Delta H_{v.H.}$, which is calculated from the temperature-dependence of the equilibrium constant under the assumption that the two-state model (Equation (5–1)) is valid. (Cf. Chapter 3.) Values close to one for the ratio $\Delta H_{cal}/\Delta H_{v.H.}$ have been found indeed for a number of small proteins. The two-state approximation and the thermodynamic measurements of protein stability are discussed in the reviews by Privalov (1979) and Privalov and Gill (1988) and in Chapter 3. Intermediates that do not differ from either N or U in enthalpy and heat capacity are not detected by the calorimetric test. The respective transitions ($N \rightleftharpoons I$ or $I \rightleftharpoons U$) would be insensitive to temperature variation and "silent" in the calorimetric experiments. Intermediates that resemble the unfolded protein in enthalpy and in heat capacity ("molten globules") were indeed detected in the transitions of a few small proteins. They are described in Chapter 6.

2.2 Kinetics of Two-State Transitions

Fully cooperative two-state transitions provide no evidence for the mechanism of folding, which requires that the partially folded intermediates that occur during folding be identified and characterized. Although intermediates may be unstable and not populated at equilibrium, they may still accumulate and be detectable during the kinetics of folding, as transient intermediates. Such intermediates will be apparent from departures from the simple kinetics expected for strictly two-state transitions. The other unique aspect of protein folding is that the fully unfolded protein can adopt so many conformations that an initial population of such molecules must be very heterogeneous conformationally. Different unfolded molecules could conceivably fold by different pathways, so there could be numerous parallel pathways rather than one single pathway as implied by a two-state transition. For protein folding to be two-state kinetically, intermediates in folding must not be populated and the structural fluctuations within the unfolded state must be rapid, relative to the rate-limiting step in folding, so that all the unfolded protein molecules can follow the same pathway. In order to identify such departures from simple two-state reactions, it is important to understand fully the kinetic properties expected of two-state reactions. The kinetics of two-state transitions have been lucidly described by Tanford (1968) and their relationship to the equilibrium unfolding transition curve are outlined in Figure 5–1. Folding and unfolding should both be single first-order reactions with the rate constants k_f and k_u, respectively:

$$N \underset{k_f}{\overset{k_u}{\rightleftharpoons}} U. \tag{5–3}$$

The values of both k_f and k_u will depend upon the conditions. The time course of unfolding in or above the transition region (for example arrows 2 and 3 in Figure 5–1) would be described by Equation (5–4):

$$A_t - A_{eq} = (A_N - A_{eq}) * \exp[-(k_f + k_u)\,t]. \tag{5–4}$$

Similarly refolding within or below the transition (arrows 1 and 4 in Figure 5–1) will follow Equation (5–5):

$$A_{eq} - A_t = (A_{eq} - A_U) * \exp[-(k_f + k_u)\,t]. \tag{5–5}$$

A_N and A_U are the absorbance values for the native and the unfolded protein, respectively. A_t is the value at time t and A_∞ is the final absorbance. First-order plots of $\ln(A_t - A_{eq})$ and of $\ln(A_{eq} - A_t)$ versus time should be linear with time under all conditions. The apparent rate constant is frequently expressed as τ^{-1}, where τ is the time constant of a reaction; it is the reciprocal of the measured rate constant. The value of τ^{-1} is equal to the sum of the individual rate constants, $k_u + k_f$. In the unfolded baseline region, $k_u \gg k_f$, in the native baseline region, $k_f \gg k_u$, so that the measured rate constants, τ_u^{-1} and τ_n^{-1}, approximate k_u and k_f, respectively.

Within the transition region, both unfolding and refolding can be followed under identical final conditions. In such experiments, the observed rate constants ($\tau^{-1} = k_u + k_f$) should be identical, irrespective of the initial conditions. Of course, the corresponding amplitudes (e.g., arrows 3 and 4 in Figure 5–1) will show different signs, reflecting the different initial states from which the reactions started. The folding kinetics should be independent of the probe used to follow the reaction. When spectral changes are employed to monitor folding, the kinetics should be independent of the wavelength used. At isosbestic, isoemissive, or isodichroic points, no signal changes should occur during the entire time course of folding and under varying final conditions. This criterion has already been pointed out by Tanford (1968b) as a good indicator for the absence of transiently populated intermediates.

These properties are helpful to evaluate the two-state character of a folding reaction, but they cannot rule out the presence of intermediates. Indeed, apparent two-state kinetics were found within the transition region for a number of proteins that slowly equilibrate in the unfolded state and/or fold via structural intermediates. Under various conditions, intermediates in a $N \rightleftharpoons I \rightleftharpoons U$ reaction are not kinetically detectable. Apparent two-state kinetic behavior will be found when the equilibrium population of the intermediate I is small, when its low steady-state concentration is rapidly established during folding, or when the signal change, e.g., from U to I in refolding is small. (See the discussions by Tanford [1968b] and by Utiyama and Baldwin [1986].) It is highly unlikely, however, that such special conditions will hold over an extended range of folding conditions, and therefore a variation of the final conditions should reveal heterogeneity in the observed kinetics that is caused by the presence of folding intermediates.

A kinetic two-state model is approximately valid only for very few proteins, such as lysozyme (Section 7.1) and a carp parvalbumin (Lin and Brandts, 1978; Kuwajima et al., 1988). These proteins are devoid of proline residues, or at least of *cis*-prolines. (See Section 3.) Even in these cases, the two-state model accounts for folding only close to or within the transition region, where partially folded species are unstable. Under native conditions structural intermediates are formed in the course of their folding. (See Sections 5 and 7.)

3 FOLDING AND PROLYL PEPTIDE BOND ISOMERIZATION

A single kinetic phase, with a single rate constant, implies that all the molecules are following the same pathway, with the same rate constant. A unique aspect of protein folding is the conformational heterogeneity of the unfolded state. For molecules to fold by the same pathway, all conformational transitions in the unfolded state must be very rapid, relative to the overall rate of folding. Perhaps not surprisingly, with hindsight, this generally is not the case. In particular, there are intrinsically slow isomerization processes in unfolded proteins that prevent the molecules from rapidly equilibrating conformationally.

3.1 Fast and Slow Folding Reactions

Early kinetic studies (Pohl, 1972) mistakenly indicated that protein unfolding could be described by the simple two-state model (Section 2). In a number of cases only single kinetic phases were observed, and the kinetic changes in signal roughly matched the respective changes in the equilibrium transitions. The application of fast mixing techniques revealed, however, additional, more rapid reactions with time constants typically in the sub-second time range (Epstein et al., 1971; Ikai and Tanford, 1971,1973; Ikai et al., 1973; Tanford et al., 1973; Tsong et al., 1971,1972; Garel and Baldwin, 1973; Garel et al., 1976; Hagerman and Baldwin, 1976). Fast and slow phases were most readily found in refolding experiments ending within the native baseline region (cf. Fig. 5–1). The fast phases were often difficult to detect within the transition region and were virtually absent in unfolding experiments under denaturing conditions. Initially, mechanisms were suggested that involved cooperative sequential steps (Tsong et al., 1972) or off-pathway intermediates (Ikai and Tanford, 1973) in order to explain the kinetic data and their dependence on the denaturant concentration. The key observation by Garel and Baldwin (1973) that both fast and slow phases of RNase A folding produced enzymatically active protein led to the suggestion that the complex kinetics were caused by the co-existence of fast-folding (U_F) and of slow-folding (U_S) species in the unfolded state:

$$U_S \underset{k_{21}}{\overset{\overset{\text{slow}}{k_{12}}}{\rightleftharpoons}} U_F \underset{k_{32}}{\overset{\overset{\text{fast}}{k_{23}}}{\rightleftharpoons}} N. \tag{5–6}$$

A reevaluation of the refolding data of lysozyme and cytochrome c on the basis of Equation (5–6) (Hagerman, 1977) indicated that these results could also be explained by assuming a co-existence of U_F and U_S species in these proteins. It should be noted that, although the initially proposed mechanisms for the folding of lysozyme and cytochrome c turned out not to be correct (since the heterogeneity of the unfolded state was not known at that time), (Ikai and Tanford, 1973; Ikai et al., 1973; Tanford et al., 1973), nevertheless, these papers represent good examples of careful experiments and critical discussions of possible kinetic mechanisms.

3.2 Prolyl Peptide Bond Isomerization in Protein Folding

The existence of slow equilibration reactions in unfolded protein chains that create a mixture of U_F and U_S species was a surprise in 1973. A plausible molecular explanation was provided by the suggestion by Brandts et al. (1975) that the U_S molecules refold slowly because they contain incorrect isomers of Xaa-Pro peptide bonds.

Peptide bonds are planar and can be either in the *trans* or in the *cis* conformation. Peptide bonds not involving proline residues are generally in the *trans* state, the *cis* conformation has not been detected in unstructured, linear oligopeptides, and the equilibrium population of the *cis* form is believed to be less than 0.1% (Brandts et al., 1975). Very few nonproline *cis* peptide bonds have been found in native proteins by X-ray crystallography (Stewart et al., 1990).

Unlike other peptide bonds, those between proline and its preceding amino acid (Xaa-Pro bonds, Figure 5–2) typically exist as a mixture of

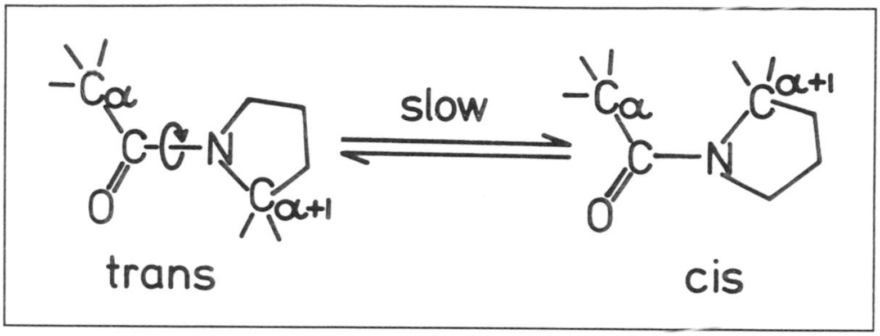

FIGURE 5–2. *Alternative isomeric states of a Pro peptide bond and its* cis ⇌ trans *isomerization.*

cis and *trans* isomers in solution, unless structural constraints, such as in folded proteins, stabilize one of the two isomers. In the absence of ordered structure and in short linear peptides, the *trans* isomer is usually favored slightly over the *cis*. A *cis* content of 10% to 30% is frequently found (Cheng and Bovey, 1977; Grathwohl and Wüthrich, 1981). It depends primarily upon the chemical nature of the flanking amino acids and upon the charge distribution around the Xaa-Pro bond. The *cis* ⇌ *trans* isomerization is an intrinsically slow reaction (time constants around 10 to 100 sec are observed at 25°C) with a high activation energy ($E_A \approx 20$ kcal/mol), since it involves the rotation about a partial double bond.

Approximately 7% of all prolyl peptide bonds in native proteins of known three-dimensional structure are *cis* (Stewart et al., 1990). The conformational state of each peptide bond is usually well defined, being either *cis* or *trans* in every molecule, depending upon the structural constraints imposed by the chain folding. Rare exceptions exist: *cis, trans* equilibria at particular Xaa-Pro bonds have been detected in native staphylococcal nuclease (Evans et al., 1987) and in calbindin (Chazin et al., 1989).

In most cases the native protein, N, has each prolyl peptide bond in a unique conformation, either *cis* or *trans*. After unfolding (N → U_F), however, these bonds become free to isomerize slowly as in small oligopeptides, thus leading to an equilibrium mixture of unfolded molecules with different prolyl isomers. The chains with the correct set of isomers usually refold rapidly; these are the U_F molecules. Chains with at least one incorrect proline (the U_S molecules) refold more slowly. The re-isomerizations of the wrong prolyl bonds are slow steps in folding. Non-native isomers do not, however, necessarily block refolding, and re-isomerization is not required to be the first step of folding, as suggested initially. Chains with certain incorrect isomers can rapidly form ordered native-like structure prior to prolyl peptide bond re-isomerization (Cook et al., 1979; Schmid and

Blaschek, 1981; Goto and Hamaguchi, 1982; Kelley et al., 1986; Kiefhaber et al., 1990b; Nall, 1990).

The general occurrence of prolyl peptide bond isomerization in protein folding is indicated by the following experimental observations.

1. The fraction of U_S molecules depends upon the number of proline residues and on their isomeric state in the native protein. In particular the presence of *cis* prolyl peptide bonds in the folded molecules leads to a high fraction of U_S, because the *cis* state is usually less populated in the absence of structural constraints. In lysozyme (Kato et al., 1981) and in cytochrome *c* (Ridge et al., 1981; Nall, 1990), which have only *trans* prolyl peptide bonds, the U_F molecules dominate in the unfolded protein. In RNase A (Garel and Baldwin, 1973) and RNase T1 (Kiefhaber et al., 1990a,b), both of which have two *cis* prolyl peptide bonds, the fraction of U_F is reduced to 0.20 and 0.04, respectively. An immunoglobulin fragment with a single *cis* prolyl peptide bond displays only 20% U_F (Goto and Hamaguchi, 1982). Lin and Brandts (1978) have studied the refolding of three different carp parvalbumins. Two of them contain one proline residue, and they show a small, slow refolding reaction. Such a reaction is not found in the folding of the third parvalbumin that lacks proline.

2. The $U_F \rightleftharpoons U_S$ reactions in unfolded proteins have properties that are characteristic of prolyl peptide bond isomerizations in small peptides. The equilibrium is independent of temperature (Schmid, 1982) and independent of the concentration of additives, such as GdmCl (Schmid and Baldwin, 1979), that strongly decrease protein stability but do not affect prolyl peptide bond isomerization. The reaction is catalyzed by strong acid and the activation energy is 20 kcal/mol (Schmid and Baldwin, 1978).

3. The refolding of the U_S molecules involves slow prolyl peptide bond-isomerization-limited steps, but coupling between folding and isomerization steps can occur. On the one hand the presence of incorrect isomers in the chain decelerates folding steps, and on the other hand, the equilibrium and kinetic properties of Xaa-Pro peptide bond isomerization are changed by the preceding folding steps. This interrelationship between structure formation and prolyl peptide bond isomerization in the course of folding will be discussed further in Section 5.

Pro peptide-bond isomerization is well established as a slow reaction in various unfolded proteins, such as lysozyme (Kato et al., 1982), the C_L fragment of the Ig light chain (Goto and Hamaguchi, 1982), thioredoxin (Kelley and Stellwagen, 1984; Kelley and Richards, 1987), yeast iso-1 and iso-2 cytochromes *c* (Ramdas et al., 1986; White et al., 1987; Wood et al., 1988), RNase T1 (Kiefhaber et al., 1990b,c), and barnase (Matouschek et al., 1990). Nevertheless, other potential sources for slow interconversion reactions in unfolded protein molecules should not be dismissed. Slow ligand exchange (Tsong, 1977), loop threading reactions of disulfide-bonded chains (Nall et al., 1978), and isomerizations of non-Pro peptide bonds have been discussed. Even though the correct *trans* state is strongly favored for non-Pro peptide bonds, the large number of such bonds in a protein molecule could nevertheless lead to a sizable portion of molecules with wrong peptide bonds (Brandts et al., 1975).

3.3 Nonessential Proline Residues

Not all Pro residues are probably equally important for protein folding. Evidence for "nonessential" prolines came from a comparison of several homologous pancreatic RNase A's (Krebs et al., 1983,1985) and cyto-chromes c (Babul et al., 1978; Nall, 1990) that differ in the number of Pro residues. It is difficult to discriminate between Pro residues that are nonessential, either because they do not interfere with folding, or because they do not isomerize after unfolding. A determination of the *cis,trans* equilibria at specific Xaa-Pro bonds in denatured proteins was attempted by various methods, such as isomer-specific proteolysis (Lin and Brandts, 1983a,b,c, 1984) or NMR (Adler and Scheraga, 1990).

Other evidence for different classes of Pro residues came from energy calculations (Levitt, 1981; Ihara and Ooi, 1985), in which the destabilization of the native state was calculated when one Pro at a time was built into the protein in its correct isomeric state. Levitt (1981) suggested that three classes of Pro residues exist. Type I residues destabilize the native state only to a small extent when in the incorrect isomeric state; consequently they should barely affect folding. Type II Pro residues are intermediate, in that they should allow folding in the presence of the incorrect isomer, although at a reduced rate. Nonnative isomers of Type III Pro residues are expected to destabilize the native state entirely, and therefore they should block refolding. This classification is somewhat arbitrary, but very useful for illustrating the varying impact of incorrect isomers on folding. Strictly "nonessential" Pro residues should retain their isomer distribution during folding. Thermodynamic linkage relation requires that Pro residues that do not affect folding (e.g., by modifying the stabilities of intermediates) should reciprocally also not be affected in their isomerization rates and equilibria by folding events. Pro43 of calbindin may be an example of such a nonessential Pro. Its *cis,trans* equilibrium is apparently unaffected by the folded state of the protein (Kördel et al., 1990).

3.4 Catalysis by Prolyl Isomerase

In 1984 Fischer and co-workers discovered peptidyl-prolyl *cis,trans* isomerase (PPI), an enzyme that accelerates the *cis,trans* isomerization of Xaa-Pro peptide bonds in oligopeptides, as well as the slow folding of several proteins. Not all Pro isomerization-limited folding steps, however, are catalyzed by PPI (Lang et al., 1987; Lin et al., 1988). Apparently, the rapid formation of partially folded structure can block access for the isomerase in some cases. The evidence is now good that PPI does in fact accelerate Pro isomerization-limited steps in protein refolding. It is thus a valuable diagnostic tool to locate Pro peptide bond isomerization steps in complex protein folding reactions. Only a positive result, however, is meaningful. When catalysis by PPI is not observed, Pro peptide bond isomerization might not be a rate-limiting step in the particular folding reaction, but this can also result from a rapid formation of ordered structure that renders the Pro peptide bond inaccessible to the isomerase. Recently, another class of PPI (FK506 binding proteins) has been discovered, which also catalyze slow steps of protein folding (Tropschug et al., 1990). The cellular functions of any PPI are not yet known (Chapter 10).

4 KINETIC PROPERTIES OF FOLDING REACTIONS

This section gives an overview of the kinetic properties of protein folding transitions. Unfolding is generally the simplest and is described first. Slow isomerizations that take place in the unfolded protein are then discussed, followed by the observations of protein refolding under native conditions, and by unfolding and refolding within the transition region.

Multiphasic kinetics can generally be represented as a sum of exponential contributions according to Equation (5–7):

$$A(t) = A_\infty - \Sigma A_i * \exp^{(-t/\tau_i)}. \tag{5–7}$$

$A(t)$ is the change in signal that is followed during folding, A_∞ is the value of A at the end of the reaction. A_i is the amplitude of phase i with the apparent time constant τ_i. Sigmoidal kinetics can be observed when consecutive reactions are involved in the folding process. The time constants τ_i and the amplitudes A_i measured under different folding and unfolding conditions provide the data to establish a kinetic mechanism, to exclude alternative mechanisms, to draw conclusions about the molecular nature of individual folding steps and folding intermediates, and to design additional experiments to test the models. Practical aspects of how to measure and analyze the kinetics of folding have been discussed by Utiyama and Baldwin (1986).

4.1 Unfolding Kinetics

The kinetics of unfolding of a homogeneous native protein under unfolding conditions is generally a single first-order reaction. Therefore, all the molecules are unfolding with the same rate constant, which is not implausible with an initially homogeneous population of folded molecules. The unfolding amplitude generally accounts for the entire change in signal as measured in the corresponding equilibrium transition, which implies that no partial unfolding has preceded the rate-limiting step in complete unfolding. The rate of unfolding increases as the conditions are made more strongly unfolding (Fig. 5–1), where the folded protein is less stable. These observations have suggested that the highest energy barrier to unfolding occurs in a form of the protein that is relatively close to the native conformation (Goldenberg and Creighton, 1985; Segawa and Sugihara, 1984a,b). A schematic energy profile for folding is given in Figure 5–3.

The rate of unfolding of a protein is generally proportional inversely to the stability of the protein, determined either by the conditions or by mutation of the protein (Chapter 8). Mutations that decrease the stability of the folded state (cf. Fig. 5–3) usually increase the rate of unfolding. This relationship can be used for a rapid qualitative assessment of changes in the stability of protein variants, such as those produced by directed mutagenesis. This correlation is not rigorous, however, for it holds only when the mutations have no, or smaller, effects on the transition state for unfolding and on the unfolded protein. Mutations have been found that affect the free energies of both the unfolded and the transition state

FIGURE 5–3. *Schematic energy profile for a protein folding reaction. It is assumed that the activated state ($\ddagger$) is close to the native state (N) in structure. U is the unfolded protein; I_x and I_y are partially folded intermediates.*

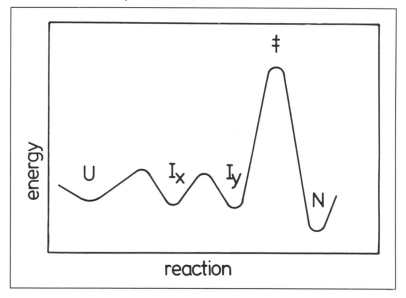

(Matthews,1987) and that change the cooperativity of unfolding transitions (Shortle and Meeker, 1986; Shortle et al., 1989).

Many small, single-domain proteins show monophasic unfolding; however, complex unfolding kinetics can be observed when the native state is heterogeneous and the individual N_i species unfold in parallel reactions. The fractional amplitudes reflect the relative population of the N_i species, if the N_i isomers of the native state are identical in their physical properties. A multiplicity of native states can arise from differences in the covalent structure, but also from slow conformational equilibria in the folded protein. A notable example is the *cis,trans* isomerization of Pro117 in native staphylococcal nuclease (Evans et al., 1987; Kuwajima et al., 1991).

4.2 Slow Isomerizations in the Unfolded State

In general, slow prolyl peptide bond isomerizations in the unfolded chains ($U_F \rightleftharpoons U_S^i$) cannot be measured directly, because the various U species are usually identical in their (e.g., spectral) properties. Invisible slow isomerizations can be detected in a two-step technique to measure the formation of the U_S species after unfolding (Brandts et al., 1975, Nall et al., 1978; Schmid and Baldwin, 1978; Schmid, 1986). In this "double jump" procedure, aliquots are withdrawn at varying time intervals after unfolding ($N \rightarrow U_F$) and transferred to native conditions, and the amplitude(s) of the resulting slow refolding reaction(s) are determined. The increase in these amplitudes with increasing time after unfolding reflects the kinetics of the underlying isomerization(s) that occur in the denatured protein. This procedure requires a knowledge of the refolding kinetics of the protein to select appropriate conditions for the refolding assays. The criteria to examine whether the observed equilibration reactions could be due to Pro peptide bond isomerizations are summarized in Section 3.3. As a consequence of its high activation energy, Pro peptide bond isomerizations become extremely slow at low temperatures ($\tau \approx 1000$ sec near 0°C). Thus, unfolding pulses at low temperature can be used to populate the U_F state for an extended time and allow the study of its properties and refolding in the virtual absence of the slow-folding molecules (Fink et al., 1988). A discussion of practical aspects of the "double jump" technique and of the data analysis is provided by Schmid (1986). The kinetic heterogeneity of unfolded proteins can also be detected by urea gradient gel electrophoresis (Creighton, 1986).

Near or within the unfolding transition, the rapid $N \rightleftharpoons U_F$ reaction is coupled with the slow $U_F \rightleftharpoons U_S$ equilibration, and thus the slow phase is correlated with a net change in signal. This coupling disappears when the conditions are shifted to "more strongly unfolding," where the $N \rightleftharpoons U_F$ equilibrium is far on the right side. Slow $U_F \rightleftharpoons U_S$ reactions contribute to the measured unfolding kinetics also when a particular isomerization is monitored by a local reporter group that is close to the isomerizing prolyl peptide bond. Such a local effect was found in unfolded RNase A, in which the isomeric state of Pro93 (which is *cis* in the native protein) affects the fluorescence emission of the neighboring Tyr92.

4.3 Multiple Equilibria in the Unfolded Protein

In unfolded polypeptide chains, isomerizations of the various Pro peptide bonds occur independently of one other. A kinetic description for the unfolding and isomerization of a protein with two Pro residues is provided by a kinetic five-species model, as in Scheme I. The independence of the two isomerizations is reflected in identical rate constants on the parallel limbs of Scheme I. The distribution of unfolded species at equilibrium is determined by the *cis,trans* equilibria at the two Xaa-Pro bonds. When the isomerization rates of the two bonds are very different, species with one incorrect isomer only accumulate transiently. For example, when the vertical isomerization in Scheme I is much faster than that of the horizontal process, the unfolded species with one incorrect isomer only (U_w^c, in the lower left corner of Scheme I) will be populated transiently to a high extent.

Such a case was encountered in the folding of porcine RNase A, and refolding experiments after various times of unfolding could be used to characterize the two isomerizations. By indirect evidence they were tentatively assigned to the Tyr92-Pro93 bond and to an isomerization in the Asn113-Pro114-Pro115 region. These two processes show different rates in the denatured protein, with time constants at 10°C of 250 sec for Pro93 and 1300 sec for Pro114-Pro115. The strong difference in isomerization rates leads to transient accumulation of a U_S species with a single incorrect isomer at Pro93 only (Fig. 5–4). It folds more rapidly than, and is thus

SCHEME I: *Kinetic model for the unfolding and isomerization of a protein with two Pro residues in either the correct (c) or incorrect (w) isomeric state (after Lin and Brandts, 1983b). This model is valid for unfolding only. The two isomerizations are independent of each other; therefore the scheme is symmetric with identical rate constants in the horizontal and the vertical directions, respectively.*

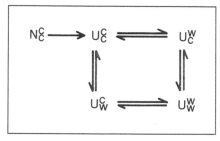

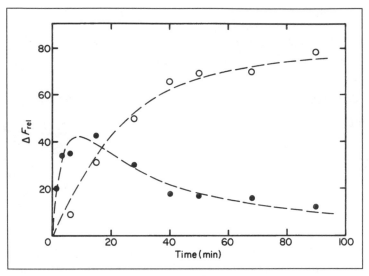

FIGURE 5–4. *Formation of the two slow refolding species of porcine RNase after N → U$_F$ unfolding. At* t = 0 *the protein is rapidly unfolded by a sixfold dilution of native RNase A to unfolding conditions of 35 μM protein in 0.1 M glycine (pH 2.0), 5.0 M GdmCl at 10°C. At various times after unfolding, samples were withdrawn and refolding assays were carried out by a 12.5-fold dilution to folding conditions of 0.4 M GdmCl (pH 6.0) at 11°C. Under these conditions, species with an incorrect Pro93 peptide bond isomer only (●) refold with τ = 72 s, whereas species with an incorrect isomer around Pro114-Pro115 (○) refold with τ = 260 s. The amplitudes of the respective refolding reactions are shown as a function of the time of incubation under unfolding conditions. (Taken from Grafl et al., 1986)*

easily discriminated from, the other U$_S$ molecules with an incorrect isomer at Pro114-Pro115. The latter species dominate in the unfolded protein at equilibrium. A change in the amplitudes of refolding with the duration of unfolding, as in Figure 5–4, provides good evidence for multiple equilibria between unfolded states and supports kinetic mechanisms like those of Scheme I. Unfortunately, extremely different isomerization rates in unfolded proteins are exceptional. In most cases the rates are similar and consequently the kinetic mechanisms are more difficult to evaluate.

Unlike the situation in unfolded polypeptide chains, the individual peptide bond isomerizations are no longer independent of each other in refolding protein molecules. Their kinetic properties can become mutually interdependent and dependent on the formation of structural intermediates (cf. Section 5).

4.4 Refolding Kinetics

To establish a kinetic mechanism for the refolding of a protein, it is necessary to discriminate sequential steps that occur on the same pathway from parallel reactions that originate from the different unfolded species that are present at the beginning of folding. The magnitude of the problem is illustrated by a few simple considerations for a protein with only two proline residues. According to Scheme I it can exist in four different unfolded states: a U_F species with only correct isomers, two U_S species with one incorrect isomer each, and one U_S species with both peptide bonds in the non-native conformation. Refolding could consequently be composed of four different pathways, and each pathway could involve the formation of transient species with properties intermediate between those of the unfolded and the native states. Folding pathways could merge at common intermediates and then share the same final steps of renaturation. It is therefore difficult to establish a comprehensive kinetic mechanism that encompasses all possible steps of folding, that is valid for all unfolded molecules under any folding and unfolding conditions, and that, furthermore, is experimentally testable against alternative mechanisms. Rather, it is necessary to use simplified models to describe folding under a restricted set of conditions only (e.g., under strongly native conditions) and/or concentrate on a subset of the unfolded species (usually the major species that dominate folding). With such a "reductionist" approach, experimentally testable models can be developed. Such models are approximate, however, and valid only for certain conditions.

Frequently, folding is most rapid under "strongly native" conditions (Fig. 5–1), and the folding rate decreases as the transition region is approached. Partially folded intermediates have a low stability and are populated only under favorable solvent conditions. Very few exceptions exist in which intermediates are also detected within the unfolding transition (cf. Chapter 6).

It is now clear that folding is not a nucleation-controlled reaction, in which a slow nucleation event is followed by rapid folding to the native state. In such a model, structural intermediates would not be observable in refolding. Limited regions of the protein chain may, however, form ordered structure very early in folding. Such "nucleation regions" have been termed *kernels* (Kim and Baldwin, 1982) to avoid potential confusion when the term *nucleation* is not used in the strict kinetic sense. The different meanings of *nucleation* in protein folding have been discussed by Wetlaufer (1990).

4.5 Parallel and Sequential Steps in Refolding

Discrimination between sequential and parallel processes is a major step in the establishment of a kinetic folding mechanism. The experimental tests for such a discrimination are related to two basic aspects of the folding mechanism: (1) kinetic phases that originate from the presence of slowly interconverting unfolded states should depend upon the duration of unfolding, because the $U_F \rightleftharpoons U_S^i$ equilibria are reached only slowly in the unfolded chains, and (2) phases that originate from the formation of intermediates should depend upon the refolding conditions, because, as mentioned before, partially folded molecules are usually stable only under strongly native folding conditions.

4.5.1 Kinetic Consequences of the Heterogeneity of the Unfolded State for Refolding

Several tests are available to identify the contributions of slowly interconverting unfolded species to the measured refolding kinetics. The relative amplitudes of refolding reactions of different U_i species depend in a characteristic manner on the duration of unfolding. All reactions that originate from U_F decrease in amplitude as the time of unfolding increases, and all refolding reactions that originate from U_S species increase, because the U_S molecules are formed from U_F slowly in the unfolded state. When multiple isomerizations occur in the unfolded chains at different rates, the amplitude profiles become more complex, but contain additional information on the mechanism of folding. (See the results with porcine RNase A in Figure 5–4.)

The equilibrium distribution of U_F and U_S species remains constant under fixed unfolding conditions. Hence, the relative amplitudes of refolding remain constant when the refolding conditions are varied, but the unfolding conditions are held constant. In a further test, both unfolding and refolding conditions are fixed, and the folding kinetics are followed by different probes. When the observed relative amplitudes are independent of the probe used, this suggests that the respective kinetic phases are caused by the parallel folding of multiple unfolded species to a common native state, N.

Such kinetic tests are only valid under conditions where the refolding reactions of the individual U_i species are much faster than their interconversion. This is most likely the case under strongly native conditions, where folding is most rapid.

The formation of partially folded intermediates during refolding can complicate such a kinetic analysis. In the refolding of the constant fragment of the immunoglobulin light chain (C_L), the fast refolding phase increased in relative amplitude from 10% at 1.0 M GdmCl to 70% at 0.1 M GdmCl. This increase originates apparently from the rapid formation of a partially folded intermediate on the slow refolding pathway under strongly native conditions. It is similar in rate to the $U_F \rightarrow N$ reaction (10% amplitude) and thus contributes to the fast folding phase under conditions where the intermediate is found (≤ 1.0 M GdmCl).

4.5.2 Detection of Sequential Steps in Refolding

The most diagnostic aspect of sequential steps in a reaction, involving one or more intermediates, rather than parallel steps, is that sequential steps will cause a lag in the appearance of the product. For example, in the reaction

$$U \xrightarrow{\ k_{UI}\ } I \xrightarrow{\ k_{IN}\ } N \qquad\qquad (5\text{–}8)$$

the lag period is caused by the need to produce the intermediate I before N will be formed at the maximum rate. The magnitude of the lag period depends upon the ratio of the two rate constants.

Spectroscopic techniques are usually not well suited to detect such lag phases, however, since intermediates on a hypothetical sequential pathway, such as in Equation (5–8), are frequently intermediate in their spectral properties between U and N. The lag that would be observed in the I → N reaction is compensated by the contribution of the U → I reaction and thus is not visible in the kinetics. For example, a model calculation based on Equation (5–8), assuming equal amplitudes for both steps and $k_{UI} = 3k_{IN}$, shows that the observed kinetics approximate a single first-order reaction with an apparent rate constant that is close to k_{IN}. A good parameter to detect a lag in the formation of native protein should ideally be sensitive only to the appearance of fully folded molecules. The regain of the (e.g., enzymatic) function of a protein can be such a parameter; it has been found in several instances, however, that folding intermediates can exhibit high catalytic activity.

The formation of native molecules during refolding can be measured selectively by a technique that is related to the energetic profile of folding (Fig. 5–3). The completely folded species, N, is separated from all other unfolded or partially folded species by a high activation barrier. Therefore, only N molecules unfold slowly, whereas all intermediates that have not yet crossed this final barrier unfold rapidly. Accordingly, the formation of native protein can be measured by unfolding assays. Folding is initiated in these experiments by dilution to the desired refolding conditions at time $t = 0$ in a test tube. Then, after various time intervals of refolding, samples are withdrawn and are transferred to unfolding conditions. The amplitude of the resulting unfolding reaction (monitored, e.g., by absorbance or fluorescence) is a measure of the amount of completely refolded molecules that were present at the time when folding was interrupted. The maximum value of the unfolding amplitude is obtained after folding in the first step of the assay has gone to completion and should correspond to the amplitude that is observed when native protein is unfolded under the same conditions. The principle of the assay is outlined in Figure 5–5. In a modified form, it can also be used to monitor the formation of intermediates. (See Section

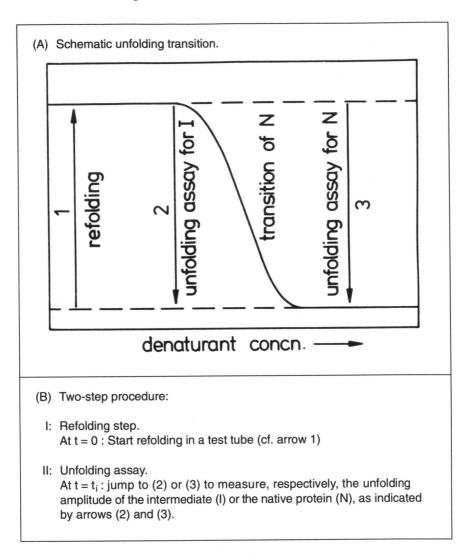

(A) Schematic unfolding transition.

refolding 1

unfolding assay for I 2

transition of N

unfolding assay for N 3

denaturant concn. ⟶

(B) Two-step procedure:

I: Refolding step.
At t = 0 : Start refolding in a test tube (cf. arrow 1)

II: Unfolding assay.
At $t = t_i$: jump to (2) or (3) to measure, respectively, the unfolding amplitude of the intermediate (I) or the native protein (N), as indicated by arrows (2) and (3).

FIGURE 5–5. *Schematic outline of the method of using unfolding assays to measure the formation of intermediates or of native protein during refolding.*

5.1.) For example, the formation of native protein during RNase A refolding is shown in Figure 5–6. The conditions for the refolding step were selected as those where partially folded intermediates are most likely to accumulate, and the assays were carried out under conditions where unfolding was complete in a convenient time range. The results in Figure 5–6 reveal several aspects of the folding mechanism of RNase A. (1) After very short

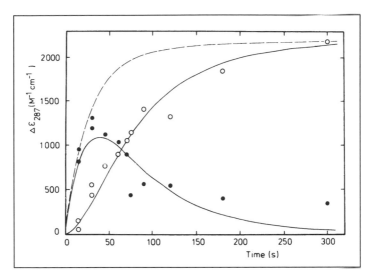

FIGURE 5–6. *Slow refolding kinetics of bovine RNase A. Concentration versus time curves for the native-like intermediate I_N (●) and for the native state N (○) during refolding in the presence of 0.5 M GdmCl (pH 6.0) at 10°C. The unfolding assays for I_N and N were carried out by subjecting aliquots from the refolding solution to a tenfold dilution to give final unfolding conditions of 4.6 M GdmCl (pH 6.0) at 10°C. The amplitudes of the resulting unfolding reaction (expressed as molar changes in Tyr absorbance at 287 nm) are shown as a function of the time of sample withdrawal from the refolding solution. A value of 560 M^{-1} cm^{-1} has been subtracted from all N values to account for the 20% N molecules formed rapidly in the $U_F \rightarrow N$ reaction. The continuous lines represent curves calculated on the basis of the following equations:*

$$U_S^{II} \xrightarrow{\tau = 25\ s} I_N \xrightarrow{\tau = 60\ s} N\ (70\%)$$

$$U_S^{I} \xrightarrow{\tau = 150\ s} I_N \xrightarrow{\tau = 60\ s} N\ (30\%)$$

For the calculations, it was assumed that I_N is an intermediate on both slow folding pathways. The dashed line is the time course of absorbance (287nm) detected folding. (Taken from Schmid, 1983)

times of refolding, about 20% of native molecules are already present; they were produced by the fast folding of the 20% U_F molecules. (2) Formation of fully folded (N) molecules in the time range of slow folding is characterized by an initial lag, which points to a sequential reaction. Indeed, the experimental curve can be approximated by a two-step mechanism according to Equation (5–8), with time constants $\tau_{UI} = 25$ s and $\tau_{IN} = 60$ s. The transient intermediate (termed I_N) was found to be native-like and enzymatically active, but it still contains at least one incorrect peptide bond isomer, probably a *trans* bond at Pro93. The final step ($I_N \to N$) is the re-isomerization of this incorrect bond. Unfolding assays as a probe of refolding yield threefold information on the folding mechanism. (1) Extrapolation of the early time points to $t = 0$ gives the amount of fast-folding U_F molecules. The method can detect even a small percentage of U_F species. It can complement or replace stopped-flow experiments, which are otherwise required to measure the fast refolding reaction. Only the amplitude of this step can be determined, however, no rate information is obtained, unless fast double-mixing techniques are employed. (2) The folding kinetics that are monitored by "conventional" techniques, such as spectroscopy or enzymatic activity, are frequently complex reactions. Comparison of these "direct" folding kinetics with the two-step unfolding assays is a powerful method to discriminate reactions that lead to the native state from others that lead to folding intermediates. (3) Coincidence of the results found by unfolding assays and in simple unfolding experiments is a sensitive test for the reversibility of folding.

4.6 Folding in the Transition Region

In general, rates of folding reactions show a V-shaped dependence on the denaturant concentration with a minimum in the transition region (Matthews, 1987). The folding of α-lactalbumin is given as an example in Figure 5–7. This protein is strongly stabilized by Ca^{2+} ions; therefore the midpoint of the unfolding transition and the minimum of the V are shifted to a higher concentration of denaturant in the presence of Ca^{2+}. In the native baseline region, both the fast and the slow refolding reactions can be monitored separately. Within the transition region they approach each other in rate and give rise to apparent two-state behavior. Pro peptide bond–isomerization-controlled steps are usually less dependent on the conditions, even when they are coupled with structure formation. The folding kinetics inside the transition region were discussed in detail by Hagerman and Baldwin (1976; Hagerman, 1977). They developed kinetic criteria to discriminate the mechanism in Equation (5–6) from other mechanisms, based on an amplitude analysis under conditions where the rates of folding and of isomerization approach each other. The most significant result was that a decrease in the amplitude of the faster phase is expected (and observed) when the two rate constants approach each other.

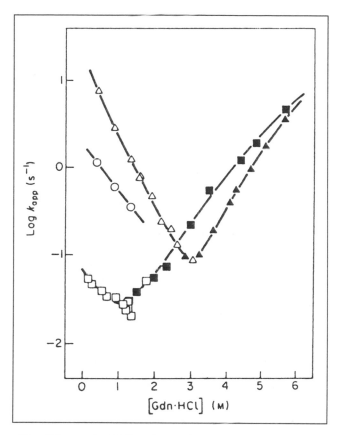

FIGURE 5–7. *Unfolding and refolding kinetics of α-lactalbumin. The dependence of log k_{app} of refolding (open symbol) and of unfolding (filled symbols) on the concentration of GdmCl at pH 7.0 and 25°C. ($\bigcirc$, $\triangle$, $\blacktriangle$) data for the holo-protein in the presence of 1 mM excess Ca^{2+}; ($\square$, $\blacksquare$) data for the apo-protein in the absence of Ca^{2+}. The open circles refer to the slow phase in biphasic refolding reactions. The minima of the V-shaped curves are near the midpoints of the respective equilibrium unfolding transitions. (Taken from Kuwajima et al., 1989)*

5 FOLDING INTERMEDIATES

The ultimate goal of kinetic folding studies is the molecular elucidation of the successive structure formation that takes place during folding. Hence, the detection and structural characterization of folding intermediates are central problems. Considerable progress has been achieved in the past few years, and detailed structural information was obtained for a number of folding intermediates. In this section, I first describe the kinetic detection of intermediates, followed by a description of "late" and of "early" folding intermediates. *Late* and *early* are used, respectively, as synonyms for "preceding the slowest, rate-limiting step" (typically peptide bond isomerization), and for "formed rapidly" (typically in the time range of a few milliseconds). Finally, a consensus model for folding of small, single-domain proteins will be discussed. An excellent account of the involvement of intermediates in folding is provided by the four reviews by Baldwin (Baldwin, 1975,1991; Kim and Baldwin, 1982,1990).

5.1 Kinetic Detection of Intermediates

Partially folded kinetic intermediates are usually apparent from the measurement of folding kinetics by different probes and under varying final conditions. In the absence of partially folded intermediates, the relative amplitudes of the different kinetic phases should be independent of the probe and the conditions and should reflect the initial distribution of unfolded species. Deviation from this behavior constitutes definite evidence for the transient presence of folding intermediates. It is also important to examine whether the kinetically resolved amplitudes correlate well with the entire change in signal as observed in the corresponding equilibrium transition. This constitutes a test for rapidly formed structural intermediates. Circular dichroism in the amide region is particularly well suited for such experiments. (See discussion following.) Discrimination of sequential steps from parallel fast $U_F \rightarrow N$ reactions is possible by "double jump" experiments. Fast phases that originate from $U_F \rightarrow N$ decrease with increasing time of unfolding, whereas the rapid formation of intermediates on $U_S \rightarrow N$ pathways gains in amplitude with increasing formation of U_S species after unfolding.

Kinetic intermediates are transient species; those in refolding should be formed early, their concentration should pass through a maximum and then decrease as the native protein is formed. It is difficult to trace the time course for an intermediate from spectral measurements, since both formation of I and formation of N (Equation (5–8)) usually contribute to the signal changes that are observed during folding. Intermediates can be measured selectively in the presence of native and unfolded molecules by a method that is similar to the unfolding assays for the detection of native protein. (See Section 4.5.2.) To apply this method, it is necessary to find conditions under which the unfolding kinetics of the intermediate can be followed independently of the presence of fully folded molecules. These can be either conditions where N, but not the intermediate, is still stably folded (close to the onset of the transition of the native protein, cf. Figure 5–5) or, alternatively, conditions that are unfolding for both I and N, but under which I unfolds much more rapidly than N. The time course of intermediate formation and decay can then be traced by unfolding assays in which samples are withdrawn after various time intervals of refolding, transferred to the assay conditions, and the amplitude of the resulting unfolding reaction of I determined (cf. Fig. 5–5). It is a measure of the concentration of intermediate that was present at the time of sample transfer. When both unfolding of I and of N can be measured under the same assay conditions, the kinetics for the two species can be determined simultaneously, as shown in Figure 5–6 for RNase A.

Fink and co-workers (1988) investigated refolding at subzero temperatures to stabilize intermediates and to retard the kinetics of their formation and decay. This method requires low-temperature equipment and the use of cryosolvents, which necessarily influence the protein structure. Another technique was developed by Adler and Scheraga (1988). They used a continuously recycling flow reactor that consisted of a refolding vessel, an observation chamber (a spectrometer cell or a NMR tube), and an unfolding vessel. The "age" of the refolding protein could be adjusted by varying the flow rate and the length of tubing between the refolding vessel and the observation chamber. An NMR spectrum of a folding intermediate of RNase A could be obtained this way. The method is restricted to proteins that withstand numerous cycles of thermal unfolding and refolding without significant aggregation or other modifications. Goldberg and co-workers used monoclonal antibodies, raised against the native protein, to follow the reappearance of native-like structure in defined regions of the folding polypeptide chain (Blond-Elguindy and Goldberg, 1990). This work is further discussed in Chapter 9.

5.2 Late Intermediates

"Late" intermediates accumulate prior to the rate-limiting step of folding, usually under strongly native conditions. They can be well populated and are readily detected when overall folding is slow, such as in $U_S \rightarrow N$ reactions. Much experimental work has concentrated on the characterization of intermediates on slow folding pathways.

In the refolding of U_S molecules, folding and isomerization steps are interrelated. The extent of structure formation before re-isomerization depends upon the folding conditions and upon the number and location of the incorrect isomers in the polypeptide chain. The presence of incorrect isomers frequently decelerates folding, which facilitates the kinetic characterization of intermediate steps. Pro peptide bond isomerization thus can act as a "kinetic trap" for intermediates. The following concentrates upon late intermediates that accumulate on $U_S \rightarrow N$ folding pathways prior to the rate-limiting re-isomerization of incorrect isomers.

The major slow refolding reaction of RNase A, $U_S^{II} \rightarrow N$, originates from a species with at least one incorrect peptide bond isomer. About 60%

of all unfolded RNase A molecules are U_S^{II}; in addition, 20% refold very slowly (U_S^{I}), while the remaining 20% are U_F molecules. Under strongly native conditions, U_S^{II} refolds on a sequential $U_S^{II} \rightarrow I_1 \rightarrow I_N \rightarrow N$ pathway. The intermediate I_1 is formed rapidly and contains elements of secondary structure. It is described in Section 5.3. The I_N intermediate is a native-like species, but with one or more incorrect Pro isomers, and accumulates prior to the final slow step. It resembles native RNase A in amide circular dichroism, tyrosine absorbance, inhibitor binding, and enzymatic activity. Indeed, in the early work on RNase A folding it was believed that the product of the folding detected by tyrosine absorbance was native protein.

The I_N intermediate unfolds much faster than completely folded RNase A. This was demonstrated by a two-step refolding/unfolding experiment (Schmid, 1983). Figure 5–8A shows the folding kinetics monitored by the increase in tyrosine absorbance. When folding is interrupted after 40 sec (as indicated by the arrow in Figure 5–8A), and the sample is transferred to unfolding conditions (Fig. 5–8B, C), only 40% of the folded molecules unfold as slowly (with a time constant of 260 sec) as native RNase A. The majority of the molecules (60%) are in the I_N state and unfold much more rapidly, with a time constant of 19 sec. The control in Figure 5–8D, E shows that, after 1 hr of folding, all molecules had attained the N state and unfolded slowly; I_N was no longer detectable.

The distinct unfolding kinetics of the I_N and N species were then used in unfolding assays (see Section 5.1 and Figure 5–5 for an outline of the method) to monitor simultaneously the time courses for appearance of I_N and of N during refolding. The results (Fig. 5–6) are well explained by a sequential mechanism, in which I_N is an obligatory intermediate on a sequential $U_S^{II} \rightarrow I_N \rightarrow N$ pathway. As expected, formation of N occurs with

FIGURE 5–8. (opposite) Refolding of ribonuclease A. (A) Slow refolding reaction at 1°C monitored by the change in Tyr absorbance. Initial unfolding conditions: 1.04 mM protein in 4.0 M GdmCl (pH 2.0). Refolding was initiated by a twentyfold dilution to 52 µM protein in 0.2 M GdmCl, 0.4 M ammonium sulfate (pH 6.1) at 1°C. (B) Unfolding assay carried out after 40 sec of folding (as indicated by the arrow in panel (A) by sample transfer to unfolding conditions of eventually 16.8 µM protein in 4.6 M GdmCl (pH 6.0) at 10°C. The kinetics of the resulting decrease in absorbance during unfolding are shown. (C) Semilogarithmic plot of the data of (B). I_N unfolds with a time constant of 19 sec (50% amplitude), N unfolds with a time constant of 260 sec (36% amplitude). (D) Refolding under the conditions of (A) was allowed to go to completion, and then the sample was transferred to unfolding conditions as in (B). (E) Semilogarithmic plot of curve (D). Only the slow unfolding of the N molecules can be detected after complete refolding. (Taken from Schmid, 1983)

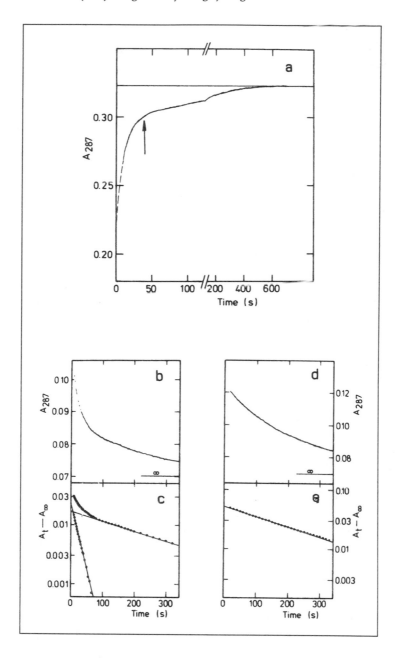

a lag period that is explained by the transient formation of I_N. The formation of I_N shows properties that are characteristic of folding reactions: major changes in the spectroscopic properties occur in this step, and the active site is formed. I_N still has incorrect peptide bond isomer(s); consequently, the product of the rapid unfolding of I_N is U_S, not U_F (Cook et al., 1979; Schmid et al., 1986). Peptide bond re-isomerization in the final $I_N \rightarrow N$ step is more rapid than is this reaction in the unfolded protein. The formation of ordered structure in I_N probably causes this acceleration.

I_N is less stable than the native protein. When populated in an interrupted folding experiment like that of Figure 5–8A and then exposed to 3.0 M GdmCl (cf. Figure 5–5), I_N unfolds rapidly, although native RNase A remains folded under these conditions. A rigorous determination of the stability of an intermediate, such as I_N, is not feasible, however, since kinetic intermediates are not stable at equilibrium. Despite the good kinetic and structural characterization of the $U_S^{II} \rightarrow N$ folding pathway of RNase A, it is still unclear which incorrect peptide bond isomers are present in U_S^{II} and re-isomerize in the final $I_N \rightarrow N$ step. Part of this uncertainty originates from the multiplicity of the folding pathways of RNase A, which complicates the assignment of kinetic phases. Site-directed mutagenesis experiments are in progress to investigate the role of individual Pro residues for the folding of this protein.

Native-like folded intermediates with one or more incorrect Pro peptide bond isomers are found in the folding of U_S species of several small proteins, such as the immunoglobulin C_L domain (Goto and Hamaguchi, 1982) and thioredoxin (Wilson et al., 1986; Shalongo et al., 1987). The major organization of the folded state has already occurred in the reaction leading to I_N. One or more chain segments are, however, "locked" in a non-native conformation by incorrect peptide bond isomers. Regain of the native state in the last step of folding is limited in rate by the re-isomerization of the incorrect isomer(s). I_N-like folding intermediates have low stability; therefore, their folding is slow and they accumulate only under strongly native conditions. Under marginally native conditions (near the transition region), they are destabilized, or their rate of formation is decreased, and alternative folding pathways are used, such as a $U_S \rightarrow U_F \rightarrow N$ mechanism, in which re-isomerization precedes folding.

5.3 Early Intermediates

Rapidly formed "early" intermediates have been detected in both fast and slow refolding reactions, either preceding the rate-limiting formation of N molecules in a $U_F \rightarrow I_F \rightarrow N$ pathway (as in lysozyme, Section 7.1), or else preceding the formation of I_N on a $U_S \rightarrow I_S \rightarrow I_N \rightarrow N$ pathway (as in RNase A).

Indirect evidence for rapidly formed intermediates came from the frequent finding that some absorbance or fluorescence changes occurred during refolding within the dead-time of stopped flow experiments (Kato et al., 1981), from multiphasic refolding kinetics of proteins without Pro residues (Lin and Brandts, 1978; Kuwajima et al., 1988; Ropson et al., 1990), and from rapid changes in amide circular dichroism that precede the rate-limiting step in the folding of proteins such as lysozyme and α-lactalbumin, which have 90% U_F species (Kuwajima et al., 1985). Kuwajima et al. (1987) have developed a stopped-flow method for the rapid measurement of changes in amide circular dichroism and have demonstrated that significant formation of secondary structure occurs within 20 ms during the refolding of β-lactoglobulin, cytochrome *c*, and carp parvalbumin (Kuwajima et al., 1988).

Early intermediates are difficult to characterize, because they are very short lived under ambient conditions. Detailed information about the structure of early folding intermediates might be obtained by NMR spectroscopy, but unfortunately NMR is a "slow" technique that is not directly appropriate for rapid kinetic experiments. A powerful method for the detection and characterization by NMR of early intermediates is based on the trapping and subsequent identification of exchangeable amide protons during folding. Shielding from the solvent and/or hydrogen bonding lead to retardation of exchange with water of amide NH protons. Udgaonkar ad Baldwin (1988) and Roder et al. (1988) have developed a technique that combines rapid pulse labeling in a stopped-flow multimixing apparatus with identification of the protected NH groups by two-dimensional NMR. Protection from exchange during refolding can be measured for individual amide positions in the millisecond time range. It is assumed that protection occurs predominantly by hydrogen bonding of NH protons and that the method monitors the rapid formation and/or stabilization of secondary structure during folding. Only those NH positions that are highly protected in the native protein can be investigated, however, since the NMR analysis is carried out after folding has gone to completion, and the fully folded conformation is necessary to inhibit further exchange. This procedure has been used to probe the refolding of RNase A, of cytochrome *c*, and of barnase. In all cases, rapid protection of many amide protons from exchange was observed under strongly native conditions. The method can be used to follow the kinetics of formation and/or stabilization of presumed individual elements of secondary structure during folding. By varying both the folding time prior to the short exchange pulse and the exchange conditions, a "protection factor" (relative to the exchange of solvent-accessible NH positions) can be measured for individual amide protons. These factors are related to "stability constants" for the protecting structure at these individual sites; they typically increase strongly with the duration of folding prior to the pulse (Udgaonkar and Baldwin, 1990, Jeng et al., 1990; Baldwin, 1991). In the U_S^{II} species of RNase A the entire central β-sheet of the native protein was concluded to be formed rapidly and cooperatively within 100 msec. This structure is sufficiently stable to provide a 10- to 100-fold protection for the involved amide protons. The protection factor increases with time of folding and exceeds 1,000 after 400 msec of folding (Udgaonkar and Baldwin, 1990).

For cytochrome *c*, protection against exchange of the amide protons (Fig. 5–9) occurred in three different stages. Amide positions in the N-terminal and C-terminal helices were protected very rapidly, in the 20 msec range, and most of the other NH protons were protected after about 200 msec. The protection of some amides that participate in tertiary interactions was observed only in the slowest step of cytochrome *c* folding, with a time constant of about 10 sec. The formation of the N- and C-terminal helices, and of specific interactions between them, are apparently very early correlated events in folding. Intriguingly, these interacting helices are also the most stable structural elements of the native protein. This implies that formation of native-like elements of structure in distant chain regions, and their stabilization by specific tertiary contacts, are among the earliest steps in folding of this protein that can be resolved. Both ends of the polypeptide chain are already necessary at this stage, arguing against the hypothesis that folding starts at the N-terminus and then proceeds to the C-terminus, as does protein biosynthesis on the ribosome (Chapter 10).

Similar results have been obtained for barnase (Bycroft et al., 1990). Helices and sheets were concluded to form with time constants in the 60 msec range, about threefold faster than the overall rate of folding. Unlike RNase A, most unfolded barnase molecules contain correct peptide bond isomers. It should thus be a good system to investigate folding in the absence of Pro peptide bond isomerization.

Early kinetic folding intermediates are very difficult to characterize, because they are formed and stabilize very rapidly and are very short lived. On the other hand, several small single-domain proteins can exist in a partially folded form at equilibrium under mildly denaturing conditions. These equilibrium intermediates are now generally called "molten globules" and are discussed in detail in Chapter 6. Unlike the elusive kinetic intermediates, the molten globule intermediates can be studied at leisure in equilibrium experiments, and it has been suggested that the molten globule is a good model for kinetic folding intermediates. Ikeguchi et al. (1986) compared a kinetic intermediate in the refolding of α-lactalbumin with the well-characterized molten globule equilibrium intermediate of this protein. The transient intermediate shares several properties with the equilibrium one, and it is hoped therefore that the latter is a valid model for the transiently formed structure. In a stable molten globule intermediate

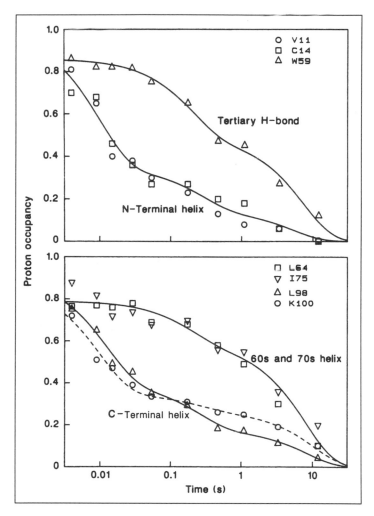

FIGURE 5–9. *Refolding of cytochrome c. Time course of protection against hydrogen exchange with the ^2H_2O solvent for a representative set of backbone NH sites and one side chain NH (of Trp59) during refolding. The proton occupancies are plotted as a function of the refolding time on a logarithmic scale. The curves represent the sums of two (top curve in each panel) or three (bottom curve in each panel) exponential terms. (Taken from Roder et al., 1988)*

of apomyoglobin, three out of five α-helices of the native conformation are already stabilized; they do not, however, interact specifically with each other by interdigitation of side chains in the helix-pairing regions as in the native protein (Hughson et al., 1991). Such a species without specific interactions could conceivably also occur as a kinetic intermediate early in refolding. Unfortunately, equilibrium intermediates are rarely observed for small proteins; also, the different conditions (native conditions for kinetic folding experiments and weakly unfolding conditions for the equilibrium experiments) could affect the respective intermediates.

All of these results indicate that elements of secondary structure are formed rapidly during folding. The results on cytochrome *c* indicate that docking reactions of local structural elements can occur very early in folding, and they are required to select and to stabilize the "correct" elements of local structure. This agrees very well with the model of Gō (1983), in which the tertiary contacts, with their "high resolving power," are used early in folding to select the correct local structures by stabilizing them.

6 TRANSITION STATE FOR FOLDING

The complete description of a kinetic pathway requires characterization of both intermediates and activated states. Application of transition state theory is dependent upon the concept of well-ordered sequential pathways in protein folding (cf. the discussion by Kim and Baldwin, 1990). Distinct transition states would not exist in statistical (e.g., jigsaw) models for protein folding, in which many different paths exist to reach the folded state (Harrison and Durbin, 1985). The activated state of folding can be characterized only with difficulty, and when certain requirements are met. (See below.) Unlike folding intermediates, which are populated at least transiently under suitable conditions, the activated state is by definition never populated, since it represents the point of highest free energy along the folding pathway. Its properties can be inferred only indirectly by a characterization of the rates of the forward and back reactions across the transition region and by comparison with the initial and final states of folding.

The folding of proteins with incorrect peptide bond isomers is not considered here. The chemical nature of the rate-limiting step in their refolding is known. It is the re-isomerization of non-native peptide bond isomers that, depending upon the conditions, is modified by the presence of ordered structure in intermediates. When peptide bond isomerization is not involved, the kinetic analysis of the rate-limiting step of folding should help to elucidate the structural properties of the transition state.

Analysis of the activated state of folding is restricted by several limitations. As pointed out, folding should not couple with other reactions, such as isomerizations or formation of partially folded intermediates. In special cases, the rapid formation of (equilibrium) intermediates can be accounted for by correcting the observed kinetic data, as was done in the case of α-lactalbumin (Kuwajima et al., 1989; Kuwajima, 1989). To characterize the activated state, refolding and unfolding kinetics have to be measured under identical conditions, where the condition of microscopic reversibility holds. The apparent rate constant $\tau^{-1} = k_u + k_f$ can be decomposed into the individual rate constants for the forward and the reverse reaction, k_u and k_f, only when the two-state approximation holds and the equilibrium constant $K_{eq} = [N]/[U] = k_f/k_u$ is known. In the experimental approach, the dependences of the rate constants k_u and k_f on variables such as denaturant type and concentration, or upon temperature are compared with the respective variation of the equilibrium constant, K_{NU}, to gain structural information about the activated state. A two-state equilibrium transition is a necessary, but not a sufficient, criterion for such an analysis, because intermediates may be involved in the kinetic pathway that are not populated in equilibrium unfolding experiments. (Cf. Section 2.2.) The presence of such intermediates between the unfolded and the activated state would make invalid an analysis that characterized the transition state relative to the native and the completely unfolded state, instead of the transient intermediate. In the presence of *kinetic* folding intermediates, a transition state analysis should not be used, unless the intermediate is also populated at equilibrium (as in the case of α-lactalbumin).

Characterization of the activated state, relative to the states before and after it, requires variation of the folding conditions. The accumulation of intermediates and the mechanism of folding, however, are known to be strongly dependent upon the conditions. This was clearly shown for lysozyme (Section 8.1). Unlike the situation in reactions of small organic molecules, the structure of the transition state of folding may depend upon the conditions. In such a case, the activated state itself (and the folding mechanism) would change, and it would be impossible to elucidate its structure by variation of the conditions, even when the respective properties of the N and U states are known.

These considerations limit the investigation of the transition state to a small number of protein folding reactions and to a narrow range of conditions. Slow isomerizations should be absent in the unfolded protein; the equilibrium transitions should conform to the two-state approximation by all available criteria; and the analysis should be restricted to the transition region, where both equilibrium and kinetics can be measured and where partially folded intermediates are less likely to occur. A general problem follows from these limitations. The activated state of folding and any kinetic folding intermediates usually cannot be characterized under identical folding conditions. Analysis of the activated state is restricted to the transition region but partially folded intermediates accumulate usually only under strongly native conditions.

Very few proteins meet the preceding criteria. They should not contain *cis* peptide bonds (barnase, α-lactalbumin, or lysozyme) or not have Pro residues at all (carp parvalbumin). The folding of lysozyme and the characterization of its activated state are described as an example in Section 7. The activated state of barnase (Matouschek et al., 1989) has been investigated by the use of designed mutants (Chapter 8).

Kuwajima and co-workers have used α-lactalbumin as an experimental system to elucidate the structure of the activated state for folding (Kuwajima et al., 1989; Kuwajima, 1989). α-Lactalbumin is an excellent protein for folding studies, because an intermediate (the "A" state) is populated at equilibrium in the unfolding transition; hence, the properties of the intermediate and of the activated state can be elucidated under identical conditions in the transition region for this protein. Furthermore, α-lactalbumin contains only 10% slow-folding species. The authors present two alternative models for the activated state of folding: the critical distortion model and the critical substructure model. In the first model the activated state is a distorted, high-energy form of the native state that resembles structurally the native state. In the second model, folding is viewed as a growing process, in which correctly ordered chain segments increase until at some point the free energy of the structure is maximal, i.e., it has reached the critical substructure in the transition state. The folding of α-lactalbumin and of carp parvalbumin (Kuwajima et al., 1988) are best explained by the critical substructure model. In particular, their Ca^{2+} binding sites are already present in the transition state, albeit with lower affinity than in the folded protein (Kuwajima et al., 1989). The folding of lysozyme, however, is more adequately described by the critical distortion model (see Section 7.1). The nature of the activated state may depend strongly upon the folding conditions, since it represents the least stable structure of highest energy along the folding pathway. Therefore it is possible that the adequacy of folding models may depend upon the protein as well as upon the folding conditions.

7 CASE STUDIES

The folding kinetics of two small proteins, lysozyme and RNase T1, are described in the following. RNase T1 is a good example to demonstrate the mutual interdependence of structure formation and Pro peptide bond isomerization during refolding. In lysozyme, most unfolded molecules contain correct peptide bond isomers, and their folding can be studied in the absence of isomerization. The experimental work on lysozyme has concentrated on the investigation of early intermediates and on the characterization of the transition state for folding.

7.1 Lysozyme

Unfolding of lysozyme is reversible under a wide variety of conditions, and its equilibrium unfolding reaction conforms to the two-state model (Privalov and Khechinasvili, 1974). The initial stopped-flow kinetic results on lysozyme folding were analyzed in terms of models with only single unfolded states (Tanford et al., 1973). A reevaluation of Hagerman (1977) indicated that these data could alternatively be explained by the extended two-state model (Equation (5–6)). Kato et al. (1981, 1982) showed that unfolded lysozyme consists indeed of fast-folding U_F (90%) and slow-folding U_S (10%) species. Lysozyme contains only two Pro residues, both with peptide bonds that are *trans* in the native protein. The high content of U_F molecules led Kato et al. (1981) to suggest that one of the Pro residues might be nonessential for folding. It is unknown, however, whether, or to what extent, both Pro residues isomerize after unfolding.

Lysozyme is a good model to study folding in the absence of Pro peptide bond isomerization, because 90% of all unfolded molecules do not contain incorrect isomers. First evidence for a structural folding intermediate came from the finding that the sum of the kinetic changes in absorbance observed during refolding was smaller than that expected from the absorbance difference observed in the equilibrium transition, which indicated that a folding intermediate is formed within the dead-time of stopped-flow mixing (Kato et al., 1981). This transient intermediate shows extensive secondary structure, as indicated by the rapid regain of the CD bands in the amide region (Kuwajima et al., 1985). Low pH and low temperature (pH 1.5 and 4.5°C) were used in this work to decelerate the rate-limiting step of $U_F \rightarrow N$ folding to $\tau = 15$ sec and to provide sufficient time to characterize the rapidly formed intermediate. It still resembles the unfolded protein in its aromatic CD spectrum, but is close to the native protein in the amide CD (Fig. 5–10). The transient folding intermediate of lysozyme is similar to a kinetic intermediate in the folding of α-lactalbumin and to an equilibrium intermediate of the molten globule type that is found at neutral pH (Ikeguchi et al., 1986). α-Lactalbumin and lysozyme are homologous and have very similar three-dimensional structures. Although the unfolding equilibrium of lysozyme is well described by the two-state approximation, a kinetic intermediate accumulates during folding in the native region.

The two structural lobes of lysozyme appear to be folding units. A competition experiment between refolding and solvent exchange of amide protons, followed by NMR analysis, showed that the formation of the intermediate protects many NH positions from exchange with the solvent (Miranker et al., 1991). The extent of protection was different for the two lobes under the employed conditions (≈80% for the region of residues 36 to 84 and ≈30% for the discontinuous region of residues 4 to 35 and 88 to 125).

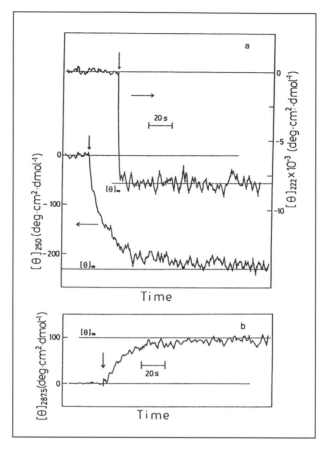

FIGURE 5–10. *Refolding kinetics of lysozyme. Kinetic progress curves of refolding measured by circular dichroism (A) at 250nm and at 222nm, (B) at 287.5 nm, all at pH 1.58 and at 4.5°C. The vertical arrows indicate the zero time at which the refolding was initiated. [θ]∞ denotes the ellipticity at equilibrium. (Taken from Kuwajima et al., 1985)*

Lysozyme meets most of the criteria for the applicability of transition state theory to protein folding (cf. Section 6). Its equilibrium transitions are two-state, and the U_F species dominates in the unfolded protein. Thus, the nature of the activated complex can possibly be inferred by analysis of the folding kinetics inside the transition region, where intermediates are not present. (See preceding section.) Such an analysis was carried out by Segawa and Sugihara (1984a,b). The unfolding and refolding rate constants, k_u and k_f, were derived from the measured rate constant $\tau^{-1} = k_u + k_f$ and from $K_{eq} = k_u/k_f$ (Equation(5–3)). The neglect of 10% U_S molecules that co-exist with

U_F in the unfolded protein leads to a corresponding small error (10%) in the refolding rate constant, k_f. Segawa and Sugihara concluded from their analysis that the activated state for folding of lysozyme is similar to the native state. Both states are virtually identical in heat capacity, which indicates that the activated state is compact, with the hydrophobic side chains buried as in the native protein. An extra cross-link introduced between Glu35 and Trp108 stabilized the protein predominantly by increasing the rate constant for folding k_f. Apparently, the covalent cross-link has only a minor effect on the activated state, which must be compact in the presence as well as in the absence of such a cross-link. Similar results were obtained from a kinetic analysis of the GdmCl-dependent unfolding transition of lysozyme. It indicates that the activated state is "70% native-like" with respect to the interaction with GdmCl (Tanford, 1970). A different result was obtained when formation of the active site (located between the two lobes of the protein) was probed by binding of an inhibitor. The inhibitor, an N-acetylglucosamine trimer, binds only to the native state and did not stabilize the transition state. Possibly, the two lobes of lysozyme are not yet correctly positioned relative to each other in the transition state of folding.

7.2 Ribonuclease T1

Ribonuclease T1 (RNase T1) is a small single-domain protein of 104 amino acids with an extended α-helix of 4.5 turns and two antiparallel β-sheets (Koepke et al., 1989). Two disulfide bonds form small (2–10) and large (6–103) covalently linked loops. RNase T1 contains four Pro peptide bonds; two are *trans* (Trp59-Pro60 and Ser72-Pro73) and the other two are *cis* (Tyr38-Pro39 and Ser54-Pro55) in the native protein. Unfolding is reversible under a wide variety of conditions (Pace, 1990; Pace et al., 1990) and is well described by the two-state approximation (Kiefhaber et al., 1990d). RNase T1 is strongly stabilized by NaCl (Pace and Grimsley, 1988), and it can fold to a native-like, catalytically active form in the absence of the disulfide bonds (Pace et al., 1988).

The refolding kinetics of RNase T1 are largely determined by two slow processes. Both originate from unfolded species that are formed by Pro peptide bond isomerizations in the denatured protein. All slow refolding reactions are catalyzed by prolyl isomerase, PPI (Kiefhaber et al., 1990b), albeit with varying effectiveness. Several peculiar features of the refolding of RNase T1 are revealed when the formation of native molecules is monitored by unfolding assays. (See Section 4.5.2 for a description of the technique.) Folding apparently occurs on three parallel pathways (Fig. 5–11). About 3.5% U_F molecules are present that refold to N within the dead-time of the experiment. The slow refolding kinetics are composed of a sequential

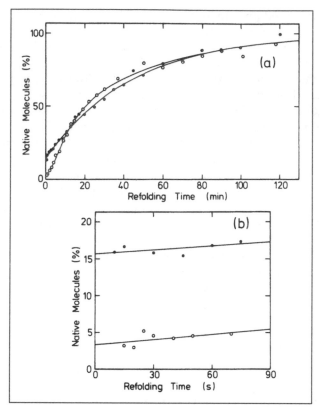

FIGURE 5–11. *Slow refolding of RNase T1. Time course of formation of native protein during refolding of wild-type RNase T1 (○) and of the S54G,P55N mutant (●) at pH 8 and 10°C. (A) Entire time course. The percentage of native molecules was measured by unfolding assays in the presence of 5.5 M (wild-type) and 5.2 M (mutant) GdmCl at pH 1.9 and 10°C. The theoretical curve for the wild-type protein was calculated by using the kinetic scheme (Scheme II) and the time constants given therein. The corresponding curve for the mutant was calculated by assuming that 15% of the molecules fold rapidly and that all slow-folding molecules (85%) regain the native state in a single slow reaction with a time constant of 2,700 sec. (B) Early region of panel (A) to determine the amount of fast-folding species by extrapolation to time zero. (Taken from Kiefhaber et al., 1990c)*

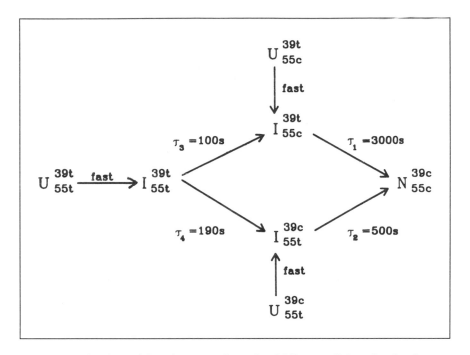

SCHEME II: *Kinetic model under strongly native folding conditions for the slow refolding reactions of RNase T1. U stands for unfolded species, I indicates intermediates of refolding, and N is the native protein. The superscript and the subscript indicate the isomeric states of Pro residues 39 and 55, respectively, in the correct, native-like cis (c) or in the incorrect, non-native trans (t) isomeric state. For example, I_{55c}^{39t} represents an intermediate with Pro55 in the correct cis state and Pro39 in the incorrect trans state. The time constants given for the individual steps refer to folding conditions of 0.15 M GdmCl, pH 8.0, at 10°C. About 3.5% of all unfolded molecules have both correct isomers and refold rapidly. They are not shown in the two schemes here. (Taken from Kiefhaber et al., 1990c)*

two-step reaction (with time constants of 190 sec and 500 sec, Scheme II) and a very slow reaction with $\tau = 3{,}000$ sec (Kiefhaber et al., 1990a,b). Structural intermediates are formed much more rapidly than these slow steps occur, as indicated by the regain of a native-like CD spectrum in the amide region within the dead-time of mixing (20 sec). The transient formation of these largely folded intermediates could be monitored by unfolding assays (cf. Section 5.1) that are based upon the reduced stability of folding intermediates relative to the native protein.

The kinetic mechanism for RNase T1 refolding is shown in Scheme II. It is based on the assumption that two slow Pro peptide bond isomerizations occur in the unfolded protein and lead to four distinguishable species (cf. Scheme I). About 3.5% of all molecules contain the correct isomers and fold rapidly ($U_F \rightarrow N$). In addition, three slow-folding species exist: two with one incorrect Pro isomer each and another, dominant species with both of these peptide bonds in the incorrect isomeric state. In the refolding mechanism (Scheme II) it is proposed that all slow-folding molecules can regain rapidly most of their secondary structure and presumably part of their tertiary structure (the $U_i \rightarrow I_i$ steps). The slow, rate-limiting steps of folding are caused by the re-isomerizations of the incorrect peptide bond isomers. The major unfolded species with two incorrect isomers can enter two alternative folding pathways (the upper or the lower pathway in Scheme II), depending upon which isomerization occurs first. This choice is determined by the relative rates of re-isomerization at the stage of the rapidly formed intermediate. It is changed in the presence of PPI, which catalyzes these two isomerizations with different efficiencies (Kiefhaber et al., 1990b). The slow steps have been tentatively assigned to the isomerization of the two *cis* prolines of RNase T1, Pro39, and Pro55, because they should be largely in the incorrect *trans* state in the unfolded protein. This could be confirmed for the Ser54-Pro55 bond. When this bond is replaced by a "normal" Gly54-Asn55 bond, the sequential folding pathway (the lower branch in Scheme II) is absent, and the amount of U_F molecules increases about fourfold from 3.5% to 15.5% (Kiefhaber et al., 1990c) (cf. Fig. 5–11). RNase T1 variants with replacements of Pro39 have not yet been obtained.

In summary, the folding kinetics of RNase T1 are dominated by slow Pro peptide bond isomerization reactions. They lead to the formation of U_S molecules after $N \rightarrow U_F$ unfolding according to Scheme I, and they control the rates of slow steps in refolding. Folding and re-isomerization are interrelated. Under strongly native conditions, intermediates with native-like secondary structure are formed rapidly. These intermediates can tolerate the presence of nonnative peptide bond isomers. Their stabilities, however, are lowered by the presence of the incorrect isomers, and they are not populated under marginally native conditions. Correct isomers are required for the final events of refolding, and hence these steps are limited by re-isomerization. The rapid formation of ordered structure affects the isomerization kinetics. Unlike the situation in RNase A, where an acceleration of isomerization was found, one isomerization in RNase T1 (presumably at Pro39) is strongly decelerated in a folding intermediate. The folding mechanism in Scheme II is complex, but is still a simplification. It is valid only under strongly native conditions, where intermediates are populated and reverse reactions are not significant. In addition, contributions from minor species with incorrect isomers at the *trans* Pro residues 60 and 73 were not considered.

8 FOLDING MECHANISM OF SMALL PROTEINS: PERSPECTIVES

The stage is now set to use the currently available kinetic results to construct a "consensus pathway" for the folding of small monomeric proteins. It is clear that folding does not proceed by a trial and error method or by a mechanism that resembles the solution of a jigsaw puzzle (Harrison and Durbin, 1985). Rather, folding follows ordered pathways that are characterized by a sequence of steps in which structure is successively formed and/or stabilized. An excellent discussion of folding pathways is found in the article by Baldwin (1991).

Folding is presumably initiated by formation of isolated elements of secondary structure. These elements are formed very rapidly, typically within the dead-time of stopped-flow spectrometers. They may have fairly low stabilities, and they rapidly interconvert with each other. The tendency of model peptides to form specific ordered structures of low stability in water is good evidence for the existence of such early events in folding.

In the next step these local elements rapidly diffuse together, possibly to form an intermediate of the molten-globule type (Chapter 6). The high surface tension of water would tend to minimize the protein-solvent surface and probably stabilizes such molten globules. The collapse also stabilizes the (native-like) secondary structure. It is still unclear whether this step is accompanied by the formation of the first specific tertiary contacts, to generate a framework intermediate. Alternatively, formation of the molten globule and of the specific framework contacts may be separated in time. In any case, both processes lead to a significant stabilization and to a corresponding large increase in the protection factors of amide protons, as observed in the hydrogen-deuterium exchange experiments. These early folding events can occur in chains with both correct and incorrect Pro peptide bond isomers. Whether or not a hydrophobic core exists at this stage of folding is a matter of debate. Certainly, some hydrophobic regions have to exist to help to dock structural elements and to stabilize the hydrogen-bonded secondary structure. On the other hand, aromatic residues are still largely exposed to the solvent, and hydrophobic patches exist at the surface that bind hydrophobic dyes (Ptitsyn et al., 1990). The nature of the driving forces for these early steps is also not yet clear. Possibly, early hydrogen bonding cooperates with the formation of limited hydrophobic regions to direct the initial collapse in a way that promotes correct further folding.

In the rate-limiting step of folding, major changes in the absorbance of the aromatic residues are observed, suggesting that water molecules are removed from the protein interior at this stage. Presumably, the tight packing of the protein interior and the desolvation of hydrophobic regions occur cooperatively in this step.

In the case of polypeptide chains with incorrect Pro peptide bond isomers, re-isomerization can occur in any of the folding steps outlined here, depending upon the location of the Pro residues in the sequence and upon the folding conditions. Under unfavorable conditions, the stabilities of folding intermediates are low, and correct Pro isomers are already required for the early steps in folding. Under strongly native conditions, on the other hand, close packing of the protein chain can occur before the incorrect isomers are reversed, and native-like intermediates are formed that show several properties of the native state, but are decreased in their stability. In that case, folding is possible in the absence as well as in the presence of incorrect isomers, and peptide bond re-isomerization is the last and rate-limiting step of folding.

REFERENCES

Adler, M., and Scheraga, H. A. (1988) *Biochemistry 27*, 2471–2480.
Adler, M., and Scheraga, H. A. (1990) *Biochemistry 29*, 8211–8216.
Babul, J., Nakagawa, A., and Stellwagen, E. (1978) *J. Mol. Biol. 126*, 117–121.
Baldwin, R. L. (1975) *Ann. Rev. Biochem. 44*, 453–475.
Baldwin, R. L. (1991) in *CIBA Foundation Symposium: "Protein Conformation,"* vol. 161, Wiley Press, London, 190–205.
Blond-Elguindy, S., and Goldberg, M. E. (1990) *Biochemistry 29*, 2409–2417.
Brandts, J. F., Halvorson, H. R., and Brennan, M. (1975) *Biochemistry 14*, 4953–4963.
Buchner, J., Schmidt, M., Fuchs, M., Jaenicke, R., Rudolph, R., Schmid, F. X., and Kiefhaber, T. (1991) *Biochemistry 30*, 1586–1591.
Bycroft, M., Matouschek, A., Kellis, J. T., Serrano, L., and Fersht, A. R. (1990) *Nature 346*, 488–490.
Chazin, W. J., Kördel, J., Drakenberg, T., Thulin, E., Brodin, P., Grundström, T., and Forsén, S. (1989) *Proc. Natl. Acad. Sci. USA 86*, 2195–2198.
Cheng, H. N., and Bovey, F. A. (1977) *Biopolymers 16*, 1465–1472.
Cook, K. H., Schmid, F. X., and Baldwin, R. L. (1979) *Proc. Natl. Acad. Sci. USA 76*, 6157–6161.
Creighton, T. E. (1986) *Methods Enzymol. 131*, 156–172.
Creighton, T. E. (1990) *Biochem. J. 270*, 1–16.
Dobson, C. M. (1991) *Curr. Opinion Struct. Biol. 1*, 22–27.
Epstein, H. F., Schechter, A. N., Chen, R. F., and Anfinsen, C. B. (1971) *J. Mol. Biol. 60*, 499–508.
Evans, P. A., Dobson, C. M., Kautz, R. A., Hatfull, G., and Fox, R. O. (1987) *Nature 329*, 266–268.
Fink, A. L., Anderson, W. D., and Antonino, L. (1988) *FEBS Letters 229*, 123–126.
Fischer, G., Bang, H., and Mech, C. (1984) *Biomed. Biochim. Acta 10*, 1101–1111.
Garel, J. R., and Baldwin, R. L. (1973) *Proc. Natl. Acad. Sci. USA 70*, 3347–3351.
Garel, J. R., Nall, B. T., and Baldwin, R. L. (1976) *Proc. Natl. Acad. Sci. USA 73*, 1853–1857.
Gō, N. (1983) *Ann. Rev. Biophys. Bioeng. 12*, 183–210.

Goldenberg, D. P., and Creighton, T. E. (1985) *Biopolymers 24*, 167–182.

Goloubinoff, P., Christeller, J. T., Gatenby, A. A., and Lorimer, G. H. (1989) *Nature 342*, 884–889.

Goto, Y., and Hamaguchi, K. (1982) *J. Mol. Biol. 156*, 891–910.

Grafl, R., Lang, K., Wrba, A., and Schmid, F. X. (1986) *J. Mol. Biol. 191*, 281–293.

Grathwahl, C., and Wüthrich, K. (1981) *Biopolymers 20*, 2623–2633.

Hagerman, P. J. (1977) *Biopolymers 16*, 731–747.

Hagerman, P. J., and Baldwin, R. L. (1976) *Biochemistry 15*, 1462–1473.

Harrison, S. C., and Durbin, R. (1985) *Proc. Natl. Acad. Sci. USA 82*, 4028–4030.

Hirs, C. H. W., and Timasheff, S. N., eds. (1986) *Methods Enzymol. 131*.

Hughson, F. M., Barrick, D., and Baldwin, R. L. (1991) *Biochemistry*, 4113–4118.

Ihara, S., and Ooi, T. (1985) *Biochim. Biophys. Acta 830*, 109–112.

Ikai, A., Fish, W. W., and Tanford, C. (1973) *J. Mol. Biol. 73*, 165–184.

Ikai, A., and Tanford, C. (1971) *Nature 230*, 100–102.

Ikai, A., and Tanford, C. (1973) *J. Mol. Biol. 73*, 145–163.

Ikeguchi, M., Kuwajima, K., Mitani, M., and Sugai, S. (1986) *Biochemistry 25*, 6965–6972.

Jaenicke, R. (1987) *Prog. Biophys. Mol. Biol. 49*, 117–237.

Jeng, M.-F., Englander, S. W., Elöve, G. A., Wand, A. J., and Rodei, H. (1990) *Biochemistry 29*, 10433–10437.

Kato, S., Okamura, M., Shimamoto, N., and Utiyama, H. (1981) *Biochemistry 20*, 1080–1085.

Kato, S., Shimamoto, N., and Utiyama, H. (1982) *Biochemistry 21*, 38–43.

Kelley, R. F., and Richards, F. M. (1987) *Biochemistry 26*, 6765–6774.

Kelley, R. F., and Stellwagen, E. (1984) *Biochemistry 23*, 5095–5102.

Kelley, R. F., Wilson, J., Bryant, C., and Stellwagen, E. (1986) *Biochemistry 25*, 728–732.

Kiefhaber, T., Quaas, R., Hahn, U., and Schmid, F. X. (1990a) *Biochemistry 29*, 3053–3060.

Kiefhaber, T., Quaas, R., Hahn, U., and Schmid, F. X. (1990b) *Biochemistry 29*, 3061–3070.

Kiefhaber, T., Grunert, H.-P., Hahn, U., and Schmid, F. X. (1990c) *Biochemistry 29*, 6475–6479.

Kiefhaber, T., Schmid, F. X., Renner, M., Hinz, H. -J., Hahn, U., and Quaas, R. (1990d) *Biochemistry 29*, 8250–8257.

Kim, P. S., and Baldwin, R. L. (1982) *Ann. Rev. Biochem. 51*, 459–489.

Kim, P. S., and Baldwin, R. L. (1990) *Ann. Rev. Biochem. 59*, 631–660.

Koepke, J., Maslowska, M., Heinemann, U., and Saenger, W. (1989) *J. Mol. Biol. 206*, 475–488.

Kördel, J., Forsén, S., Drakenberg, T., and Chazin, W. J. (1990) *Biochemistry 29*, 4400–4409.

Krebs, H., Schmid, F. X., and Jaenicke, R. (1983) *J. Mol. Biol. 169*, 619–633.

Krebs, H., Schmid, F. X., and Jaenicke, R. (1985) *Biochemistry 24*, 3846–3852.

Kuwajima, K. (1989) *Proteins: Struct. Funct. Genet. 6*, 87–103.

Kuwajima, K., Hiraoka, Y., Ikeguchi, M., and Sugai, S. (1985) *Biochemistry 24*, 874–881.

Kuwajima, K., Mitani, M., and Sugai, S. (1989) *J. Mol. Biol. 206*, 547–561.

Kuwajima, K., Okayama, N., Yamamoto, K., Ishihara, T., and Sugai, S. (1991) *FEBS Letters 290*, 135–138.

Kuwajima, K., Sakuraoka, A., Fueki, S., Yoneyama, M., and Sugai, S. (1988) *Biochemistry 27*, 7419–7428.

Kuwajima, K., Yamaya, H., Miwa, S., Sugai, S., and Nagamura, T. (1987) *FEBS Letters* *221*, 115–118.

Lang, K., and Schmid, F. X. (1990) *J. Mol. Biol.* *212*, 185–196.

Lang, K., Schmid, F. X., and Fischer, G. (1987) *Nature* *329*, 268–270.

Levitt, M. (1981) *J. Mol. Biol.* *145*, 251–263.

Lin, L.-N., and Brandts, J. F. (1978) *Biochemistry* *17*, 4102–4110.

Lin, L.-N., and Brandts, J. F. (1983a) *Biochemistry* *22*, 559–563.

Lin, L.-N., and Brandts, J. F. (1983b) *Biochemistry* *22*, 564–573.

Lin, L.-N., and Brandts, J. F. (1983c) *Biochemistry* *22*, 573–580.

Lin, L.-N., and Brandts, J. F. (1984) *Biochemistry* *23*, 5713–5723.

Lin, L.-N., Hasumi, H., and Brandts, J. F. (1988) *Biochim. Biophys. Acta* *956*, 256–266.

Matouschek, A., Kellis, J. T., Serrano, L., Bycroft, M., and Fersht, A. R. (1989) *Nature* *340*, 122–126.

Matouschek, A., Kellis, J. T., Serrano, L., Bycroft, M., and Fersht, A. R. (1990) *Nature* *346*, 440–445.

Matthews, C. R. (1987) *Methods Enzymol.* *154*, 498–511.

Matthews, C. R. (1991) *Curr. Opinion Struct. Biol.* *1*, 28–35.

Miranker, A., Radford, S. E., Karplus, M., and Dobson, C. M. (1991) *Nature* *349*, 633–636.

Nall, B. T. (1990) in *Protein Folding* (L. M. Gierasch and J. King, eds.), AAAS Press, Washington, D. C., pp. 198–207.

Nall, B. T., Garel, J. -R., and Baldwin, R. L. (1978) *J. Mol. Biol.* *118*, 317–330.

Ostermann, J., Horwich, A. L., Neupert, W., and Hartl, F. -U. (1989) *Nature* *341*, 125–130.

Pace, C. N. (1986) *Methods Enzymol.* *131*, 266–280.

Pace, C. N. (1990) *Trends Biochem. Sci.* *15*, 14–17.

Pace, C. N., and Grimsley, G. R. (1988) *Biochemistry* *27*, 3242–3246.

Pace, C. N., Grimsley, G. R., Thomson, J. A., and Barnett, B. J. (1988) *J. Biol. Chem.* *263*, 11820–11825.

Pace, C. N., Laurents, D. V., and Thomson, J. A. (1990) *Biochemistry* *29*, 2564–2572.

Pohl, F. (1972) *Angew. Chemie* *84*, 931–944.

Privalov, P. L. (1979) *Adv. Protein Chem.* *33*, 167–241.

Privalov, P. L., and Gill, S. J. (1988) *Adv. Protein Chem.* *39*, 191–234.

Privalov, P. L., and Khechinasvili, N. N. (1974) *J. Mol. Biol.* *86*, 665–684.

Ptitsyn, O. B., Pain, R. H., Semisotnov, G. V., Zerovnik, E., and Razgulyaev, O. I. (1990) *FEBS Letters* *262*, 20–24.

Ramdas, L., Sherman, F., and Nall, B. T. (1986) *Biochemistry* *25*, 6952–6958.

Rehage, A., and Schmid, F. X. (1982) *Biochemistry* *21*, 1499–1505.

Ridge, J. A., Baldwin, R. L., and Labhardt, A. M. (1981) *Biochemistry* *20*, 1622–1630.

Roder, H., Elöve, G. A., and Englander, S. W. (1988) *Nature* *335*, 700–704.

Ropson, I. J., Gordon, J. I., and Frieden, C. (1990) *Biochemistry* *29*, 9591–9599.

Santoro, M. M., and Bolen, D. W. (1988) *Biochemistry* *27*, 8063–8068.

Schmid, F. X. (1982) *Eur. J. Biochem.* *128*, 77–80.

Schmid, F. X. (1983) *Biochemistry* *22*, 4690–4696.

Schmid, F. X. (1986) *Methods Enzymol.* *131*, 70–82.

Schmid, F. X. (1991) *Curr. Opinion Struct. Biol.* *1*, 36–41.

Schmid, F. X., and Baldwin, R. L. (1978) *Proc. Natl. Acad. Sci. USA* *75*, 4764–4768.

Schmid, F. X., and Baldwin, R. L. (1979) *J. Mol. Biol.* *133*, 285–287.

Schmid, F. X., and Blaschek, H. (1981) *Eur. J. Biochem. 114*, 111–117.

Schmid, F. X., Grafl, R., Wrba, A., and Beintema, J. J. (1986) *Proc. Natl. Acad. Sci. USA 83*, 872–876.

Segawa, S., and Sugihara, M. (1984a) *Biopolymers 23*, 2473–2488.

Segawa, S., and Sugihara, M. (1984b) *Biopolymers 23*, 2489–2498.

Shalongo, W., Ledger, R., Jagannadham, M. V., and Stellwagen, E. (1987) *Biochemistry 26*, 3135–3141.

Shortle, D., and Meeker, A. K. (1986) *Proteins: Struct. Funct. Genet. 1*, 81–89.

Shortle, D., Meeker, A. K., and Gerring, S. L. (1989) *Arch. Biochem. Biophys. 272*, 103–113.

Stewart, D. E., Sarkar, A., and Wampler, J. E. (1990) *J. Mol. Biol. 214*, 253–260.

Tanford, C. (1968a) *Adv. Protein Chem. 23*, 121–217.

Tanford, C. (1968b) *Adv. Protein Chem. 23*, 218–282.

Tanford, C. (1970) *Adv. Protein Chem. 24*, 1–95.

Tanford, C., Aune, K. C., and Ikai, A. (1973) *J. Mol. Biol. 73*, 185–197.

Tropschug, M., Wachter, E., Mayer, S. Schönbrunner, E. R., and Schmid, F. X. (1990) *Nature 346*, 674–677.

Tsong, T. Y. (1977) *J. Biol. Chem. 252*, 8778–8780.

Tsong, T. Y., Baldwin, R. L., and Elson, E. L. (1971) *Proc. Natl. Acad. Sci. USA 68*, 2712–2715.

Tsong, T. Y., Elson, E. L., and Baldwin, R. L. (1972) *Proc. Natl. Acad. Sci. USA 69*, 1809–1812.

Udgaonkar, J. B., and Baldwin, R. L. (1988) *Nature 335*, 694–699.

Udgaonkar, J. B., and Baldwin, R. L. (1990) *Proc. Natl. Acad. Sci. USA 87*, 8197–8201.

Utiyama, H., and Baldwin, R. L. (1986) *Methods Enzymol. 131*, 51–71.

Wetlaufer, D. B. (1990) *Trends Biochem. Sci. 15*, 414–415.

Wetlaufer, D. B., and Ristow, S. (1973) *Ann. Rev. Biochem. 42*, 135–158.

White, T. B., Berget, P. B., and Nall, B. T. (1987) *Biochemistry 26*, 4358–4366.

Wilson, J., Kelley, R. F., Shalongo, W., Lowery, D., and Stellwagen, E. (1986) *Biochemistry 25*, 7560–7566.

Wood, L. C., Muthukrishnan, K., White, T. B., Ramdas, L., and Nall, B. T. (1988) *Biochemistry 27*, 8554–8561.

6

The Molten Globule State

OLEG B. PTITSYN

1 INTRODUCTION

There are two main problems in the physics of proteins: how amino acid sequences encode the three-dimensional (3D) structures of native proteins and how protein chains fold into their native structures in spite of the astronomically large number of alternatives.

It is very likely that the key to both of these problems is the hierarchy of structural levels in native proteins. The most important generalization that follows from the 30 years of X-ray studies of protein structures is probably that they can be divided into three levels: *secondary structure* (the positions of α-helices and β-strands along the chain), *"protein fold"* or *"folding pattern"* (the mutual positions of the α-helices and β-strands in 3D-space), and *tertiary structure* (a set of coordinates of the protein atoms). It is extremely important that only the third level is really unique for each protein, while the first and second ones can be similar or even identical for many related proteins. In fact, there are typical folds ("globin fold," "immunoglobulin fold," "trypsin fold," and so on) that are common for protein families, in spite of large differences in their amino acid sequences. This suggests that only the third level is determined by all the details of the amino acid sequence, while the codes for the first and second levels are highly degenerate. If this is the case, the problem of the "physical code" that connects protein 3D structures with their amino acid sequences can be solved in two steps: a general physical theory to predict the secondary structure and the tertiary fold, followed by conformational analysis to predict the atomic coordinates (tertiary structure) within the framework of this tertiary fold (Ptitsyn and Finkelstein, 1980a, b; Ptitsyn, 1985; Finkelstein and Ptitsyn, 1987; Ptitsyn et al., 1989).

The best experimental approach to study the levels of protein structure may be to "switch off" part of the interactions in the native protein in order to understand which type of structure can be preserved by the remaining interactions. To this end it is useful to study the physical states of a protein under different *denaturing conditions*, to identify and to investigate possible *equilibrium intermediates* between the native and the completely unfolded states.[1] A related approach is to investigate the *kinetics of protein folding*, to identify and to study *kinetic intermediates* in protein folding, and to compare them with the equilibrium intermediates.

The discovery of the equilibrium molten globule state of protein molecules (Dolgikh et al., 1981,1983,1985) and the identification of this state as a general kinetic intermediate in protein folding (Dolgikh et al., 1984; Semisotnov et al., 1987; Ptitsyn et al., 1990) should be helpful in this respect. This chapter is devoted to the description of the molten globule state, which is also reviewed by Ptitsyn (1987), Kuwajima (1989), and Dill and Shortle (1991). Section 2 of this chapter outlines briefly the history of the idea of protein equilibrium intermediates. Section 3 describes the properties of the equilibrium molten globule state. A model of the molten globule that attempts to comprise all the available experimental data is presented in Section 4. Section 5 describes the transitions between three main conformational states of protein molecules: native, molten globule, and unfolded. Section 6 contains a list of proteins and the conditions under which something like the molten globule state has been observed. Section 7 describes the role of the molten globule state in protein folding and the properties of the kinetic molten globule state, which are similar to those of the equilibrium molten globule. In Section 8, the possible role of the molten globule state in the living cell is discussed briefly.

2 BACKGROUND

More than 20 years ago, Tanford (1968) summarized all available data on nonnative (or denatured) physical states of protein molecules. He came to the conclusion that at high concentrations of strong denaturants (GdmCl or urea) proteins are more or less completely unfolded, although even in these cases some fluctuating structures cannot be excluded. (See also Dill and Shortle, 1991.) Other denaturing conditions often lead to only "partly unfolded" states. Denaturation of a number of proteins by high temperatures

---

[1]The term *native* is used here for the rigid state of a protein under physiological conditions in which the majority of the atoms are more or less fixed. The term *completely unfolded* is less definite, because an unfolded, statistical coil molecule can have a large number of conformations that depend upon the solvent and the temperature (Flory, 1953). The term *completely unfolded* will be used for all these states to distinguish them from *partly unfolded* states in which some residual structure can be observed.

and/or low pH (as well as by adding different salts) leads to remarkably small changes in their hydrodynamic and optical parameters, as compared to their denaturation by high concentrations of GdmCl.

Some of the differences between proteins denatured by different agents can be explained simply by the solvent- or temperature-dependencies of the physical properties of unfolded molecules, but many of them clearly need other explanations. For example, the denaturation of serum albumin by moderate pH (3.5) leads to a very small increase in its intrinsic viscosity, from 3.7 to 4.5 cm^3/g (Foster, 1960), which suggests that the protein remains globular under these conditions. The most definitive evidence for the presence of residual structure in acid- and temperature-denatured proteins is that they can undergo another cooperative transition under the influence of GdmCl (Aune et al., 1967) or urea (Brandts and Hunt, 1967). Thus, the existence of "partly folded" denatured states of proteins was well established as early as the late 1960's, although their nature was not known at that time.

Later, the existence of residual structure in temperature- and pH-denatured proteins was questioned by Privalov (1979) by a very convincing argument that the changes of enthalpy, entropy, and heat capacity upon denaturation are basically the same for temperature-, pH-, or GdmCl-induced denaturations. These data led him to the conclusion that temperature- and pH-denatured proteins do not contain stable residual structure and are nearly as unfolded as GdmCl- or urea-denatured proteins (Privalov, 1979). Especially informative is the change in heat capacity of denaturation, which is due mainly to the exposure of nonpolar side chains to water. (See Privalov, 1979, Chapter 3.) The careful measurements of Spolar et al. (1989) and especially of Privalov and Makhatadze (1990) have shown that the changes in heat capacity upon temperature- and acid-denaturation of apomyoglobin, cytochrome c, RNase A, and lysozyme are those expected if all their nonpolar groups are exposed to water, as would be the case for completely unfolded chains (Chapter 3).

The temperature melting of small proteins (Privalov, 1979) and protein domains (Privalov, 1982) is well known from the work of Privalov and his collaborators to be an "all-or-none" transition, i.e., a transition without any intermediate states apparent (Chapter 3). Together with the evidence for temperature-denatured proteins being unfolded, it suggested that there are no stable states of protein molecules that are intermediate between the native and the unfolded states. This point of view became popular in spite of hydrodynamic and optical data that showed that temperature-denatured proteins are far from being completely unfolded and that they can undergo another cooperative transition when GdmCl or urea is added.

Strong evidence for the existence of equilibrium intermediates between the native and unfolded states came from the studies of protein unfolding by GdmCl, especially on the unfolding of bovine (Kuwajima et al., 1976) and human (Nozaka et al., 1978) α-lactalbumins. These proteins undergo two different conformational transitions when GdmCl is added. The first transition (at lower concentrations of the denaturant) causes a drastic decrease of CD in the "aromatic" (near-UV) region, i.e., destruction of the rigid environment of the aromatic groups in the fully folded state. The second transition (at higher concentrations of the denaturant) produced a drastic decrease of CD in the "peptide" (far-UV) region, i.e., destruction of the secondary structure. Similar data had been obtained even earlier for bovine carbonic anhydrase B (Wong and Tanford, 1973), and for growth hormones (Holladay et al., 1974). These observations demonstrated the existence of one or more equilibrium intermediate states that differ from the native state by the absence of a rigid environment of aromatic side chains, but differ from the unfolded state by the presence of secondary structure. Moreover, the optical properties of bovine growth hormone (Burger et al., 1966), bovine carbonic anhydrase B (Wong and Hamlin, 1974), and bovine (Kuwajima et al., 1976) and human (Nozaka et al., 1978) α-lactalbumins at acid pH are similar to the states observed at intermediate concentrations of GdmCl. The absence of rigid tertiary structure and the presence of secondary structure in these intermediates lead Kuwajima (1977) to model these intermediates as unfolded, noncompact molecules with local secondary structure.

The equilibrium intermediates described in those papers seemed to be of key importance for the whole problem of protein structure and protein folding. Therefore, we have studied the intermediate forms of bovine and human α-lactalbumins (Dolgikh et al., 1981,1985; Gilmanshin et al., 1982; Ptitsyn et al., 1983,1986; Pfeil et al., 1986; Damaschun et al., 1986; Gast et al., 1986; Timchenko et al., 1986; Bychkova et al., 1990; Semisotnov et al., 1991a) and bovine carbonic anhydrase B (Dolgikh et al., 1983; Ptitsyn et al., 1983; Brazhnikov et al., 1985; Rodionova et al., 1989; Semisotnov et al., 1989,1991a) by a wide variety of physical methods. These studies revealed a novel physical state of protein molecules that has been named the "molten globule" by Ohgushi and Wada (1983). Similar states have been observed later in a number of other proteins. (See Section 6.)

3 PROPERTIES OF THE MOLTEN GLOBULE

The physical properties of the molten globule state will be described using bovine and human α-lactalbumins and bovine carbonic anhydrase B as examples, because these three proteins have been studied more carefully than have others. Other proteins for which data are not as complete will be described briefly in Section 6.

3.1 Compactness

One of the most important physical properties of the molten globule state is that it is almost as compact as the native protein. Table 6–1 summarizes the corresponding data for α-lactalbumins obtained from their diffusion and sedimentation coefficients. The very good coincidence between three different methods, after correction for small amounts of aggregation, indicates that the hydrodynamic radius of α-lactalbumin increases in the molten globule state by $14 \pm 2\%$ as compared to the native state; this corresponds to a volume increase of $50 \pm 8\%$. For comparison the hydrodynamic radii of these proteins in the unfolded state (with intact S-S bridges) are increased by $49 \pm 5\%$, which corresponds to a volume 3.3 ± 0.3 times larger than that of the native molecule. The large difference between the volume expansion in the molten globule state (1.5) and in the unfolded state (3.3) clearly shows that the molten globule state is relatively compact.

Data on the intrinsic viscosities of α-lactalbumins (Dolgikh et al., 1981,1985) and of carbonic anhydrase (Wong and Hamlin, 1974; Dolgikh et al., 1983), as well as on the radius of gyration of bovine α-lactalbumin (Dolgikh et al., 1981,1985; Izumi et al., 1983; Timchenko et al., 1986), are also consistent with a small expansion of protein molecules in the molten globule state.

3.2 Internal Water

The approximately 50% increase in a protein's volume in the molten globule state suggests that water can penetrate inside; indeed, there must be several hundred such molecules in the case of α-lactalbumin. Several experimental data support this point of view:

Partial specific volume. If the molten globule were "empty" (i.e., did not contain water) its partial specific volume would be much greater (~50%) than that of the native (or of the unfolded) protein. Although no careful comparison of the partial specific volumes of molten globule, unfolded, and native proteins has been carried out, there is little doubt that this is not the case.

Heat capacity. As mentioned in Section 2, the change in protein heat capacity upon denaturation reflects mainly the exposure to aqueous solvent of nonpolar side chains. No careful comparison of the heat capacities of molten globule and other denatured proteins has been performed. There is no doubt, however, that the heat capacity of a molten globule is larger than that of native proteins, which suggests that some nonpolar groups buried in the native protein become exposed to water in the molten globule state.

TABLE 6-1. Diffusion and Sedimentation Constants of α-lactalbumins in Different States

	State	Condition	$\dfrac{M_{app}}{M}$ [a]	$D_{20,W} \times 10^7$, cm^2/sec [c] QLS	$D_{20,W} \times 10^7$, cm^2/sec [c] PI	$s_{20,W}$, s [e]	R/R_N [f] $D_{20,W}$ QLS	R/R_N [f] $D_{20,W}$ PI	R/R_N [f] $s_{20,W}$
Bovine α-lactalbumin	Native	pH 7.5; 20°C	1.05	—	11.8 (11.6)[d]	1.63 (1.67)	—	—	—
	Molten globule	pH 2.0; 20°C	0.96	—	10.1	1.40	—	1.17	1.17
	Unfolded	6.4 M GdmCl pH 7.5; 20°C	—	—	7.7	—	—	1.53	—
Human α-lactalbumin	Native	pH 7.5; 20°C	1.04	12.3 (12.1)	12.5 (12.3)	1.62 (1.64)	—	—	—
	Molten globule	pH 2.0; 20°C	1.16	10.8 (11.0)	11.1 (10.5)	1.40 (1.56)	1.14	1.13	1.11
	Molten globule	Ca^{2+}-free (10 mM EDTA) pH 7.5; 50°C	1.40[b]	10.9 (9.2)	—	—	1.13	—	—
	Unfolded	6.0 M (QLS) or 6.4 M (PI) GdmCl; pH 7.5; 20°C	—	8.6	8.3	—	1.43	1.51	—

[a] M_{app}: apparent molecular weights measured by equilibrium sedimentation (Gilmanshin et al., 1982); M: molecular weight calculated from amino acid content.

[b] Evaluated from M_{app} of Ca^{2+}-free form at 20°C and temperature-dependence of light scattering (Gast et al., 1986).

[c] Diffusion coefficients measured by quasi-elastic light scattering QLS (Gast et al., 1986) and by polarization interferometer PI (Bychkova et al., 1990).

[d] Values of diffusion and sedimentation coefficients have been corrected for slight aggregation, using a method analogous to that published earlier (Gast et al., 1986). Noncorrected values are shown in brackets. No corrections have been made for unfolded proteins in 6.0–6.4 M GdmCl, because proteins do not aggregate under these conditions.

[e] Sedimentation coefficients measured by Gilmanshin et al. (1982).

[f] R/R_N — ratio of the hydrodynamic radii to its value in the nature (N) state obtained from diffusion (D) and sedimentation (s) coefficients as $R_D/(R_D)_N = D_N/D$ and $R_s/(R_s)_N = s_N/s$. The correction for a probable change of asymmetry of α-lactalbumins in the molten globule state (Timchenko et al., 1986) is small and does not change the results significantly.

3.3 Core

There is clear evidence, on the other hand, that the molten globule has a relatively dense core. In native proteins, many atom pairs are in van der Waals contact at a distance of ~4.5 Å; consequently, diffuse X-ray scattering by these proteins has a maximum corresponding to this Bragg distance (Echols and Anderegg, 1960; Grigoryev et al., 1971; Gernat et al., 1986). This maximum is due in part to van der Waals contacts in secondary structure but is mainly due to long-range contacts in the tertiary structure (Fedorov and Ptitsyn, 1977); it is absent or weak in unfolded or helical polypeptides (Damaschun et al., 1986; Gernat et al., 1986).

Figure 6–1 demonstrates that this maximum is present also in the acid molten globule state of human α-lactalbumin (Damaschun et al., 1986). The only difference is that it is shifted from 4.50 to 4.65 Å, i.e., by 3% to 4%. This small shift is similar to the greater contact distances that are present in typical liquids, as compared to crystals. A similar result has been observed in the heat-denatured molten globule state of this protein (Ptitsyn et al., 1986; Ptitsyn, 1987). This indicates that the nonpolar groups in the protein core remain in contact in the molten globule, although their packing is not as tight as in the native protein. There is also NMR evidence for the presence of a cluster of aromatic groups in the molten globule state of guinea pig α-lactalbumin (Baum et al., 1989). (See Section 3.5.)

3.4 Secondary Structure

Circular dichroism spectra of α-lactalbumins (Kuwajima et al., 1976; Nozaka et al., 1978; Dolgikh et al., 1981, 1985) and carbonic anhydrases (Wong and Tanford, 1973; Wong and Hamlin, 1974; Jagannadham and Balasubramanian, 1985; Bolotina, 1987; Rodionova et al., 1989) in the far-UV region are very pronounced in various molten globule states (acid pH, high temperature, or ~2 M GdmCl) and suggest a high content of secondary structure. The CD spectra are not the same as those of the native proteins, but they usually are even more pronounced. For example, in human carbonic anhydrase B the negative molar ellipticity at 210 nm is nearly four times greater in the acid molten globule than in the native state (Jagannadham and Balasubramanian, 1985). This does not necessarily mean a change in secondary structure, however, as far-UV CD spectra can be influenced by aromatic side chains. These side chains can contribute to CD spectra not only in the near-UV, but also in the far-UV region (Sears and Beychok, 1973; Manning and Woody, 1989). This contribution may be especially large for clusters of aromatic groups.

Aromatic groups lose their rigid environment in the molten globule state (see Section 3.6), so the contribution of these groups to far-UV CD must

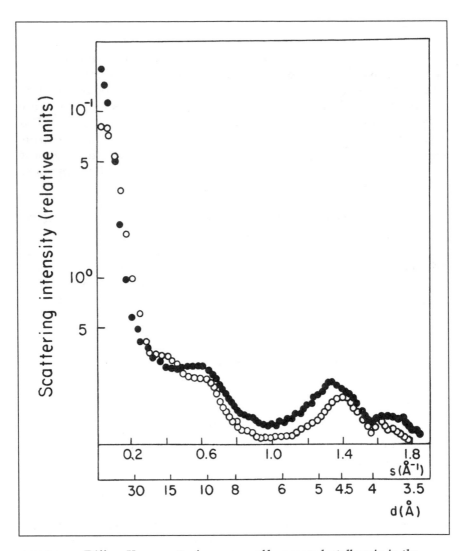

FIGURE 6–1. *Diffuse X-ray scattering curves of human α-lactalbumin in the native (○) and in the acid-denatured, molten globule (●) states, measured at protein concentrations of ~10% (w/v). s = (4π/λ) sinΘ is the value of scattering vector (wavelength, λ = 1.54 Å, 2Θ-scattering angle), d = 2π/s is the Bragg distance. The shift of the maximum from s = 1.40 Å⁻¹ for the native protein to s = 1.35 Å⁻¹ for the molten globule corresponds to an increase of d from 4.49 to 4.65 Å. (Adapted from Damaschun et al., 1986)*

vanish or at least be greatly reduced, which can lead to the large changes of far-UV optical properties (Kronman et al., 1966). It is worthwhile to note that aromatic groups can have both positive and negative bands in the far-UV region (Brahms and Brahms, 1980; Manning and Woody, 1989); therefore, far-UV CD spectra of the molten globule state may be either more or less pronounced than those of the native protein.

An approximate method for the decomposition of far-UV CD spectra into the contributions of peptide and aromatic groups (Bolotina and Lugauskas, 1985) suggests that these spectra of the molten globule states of α-lactalbumins and carbonic anhydrase differ from those of the native proteins mainly by the contribution of aromatic groups (Bolotina, 1987).

More precise information can be obtained from infrared spectra, as in this case the contribution of side chains can be measured and subtracted from the total absorption (Chirgadze et al., 1975; Venyaminov and Kalnin, 1991). Figure 6–2 shows that the infrared spectra of bovine α-lactalbumins in the amide I region are almost identical for the native and molten globule states (Dolgikh et al., 1985). Very small differences between these spectra exclude any changes in content of α-helices and β-structure of greater than

FIGURE 6–2. Infrared spectra of bovine α-lactalbumin in the amide I region of the native (———) and acid-denatured molten globule (– – –) states (after the subtraction of the side chain contributions). (Adapted from Dolgikh et al., 1985)

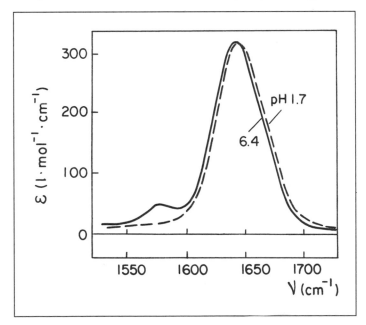

10% and 5%, respectively. Infrared spectra of native and of acid-denatured bovine carbonic anhydrase B have different shapes, but their decomposition into the contributions of β- and unfolded structures gives nearly identical β-content: 39% for the native form and 37% for the molten globule state (Dolgikh et al., 1983; see also Brazhnikov et al., 1985; Ptitsyn, 1987). The difference in the shapes of the spectra can be due to the broadening of the main β-band, which suggests some disordering of the β-structure.

3.5 Native-like Structural Organization

Definitive evidence that at least some α-helices of α-lactalbumin are located in their native positions along the polypeptide chain in the molten globule came from NMR studies of hydrogen exchange (Baum et al., 1989; Dobson et al., 1991). In these experiments, the protein was first allowed to exchange in the molten globule state for a given time; it was then transformed into the native state, and 2D-NMR was used to identify those NH protons that were protected from exchange in the molten globule state. In this way, Dobson and his collaborators (Baum et al., 1989; Dobson et al., 1991) have shown that the NH-protons of the B- and C-helices in guinea pig α-lactalbumin (see Figure 6–3) are protected in the acid molten globule state. It is almost certain that protected protons are involved in intramolecular hydrogen bonds and are shielded from solvent (e.g., Udgaonkar and Baldwin, 1988). Therefore, at least two α-helices of native α-lactalbumin (B, residues 23 to 34, and C, residues 86 to 99) are present also in the molten globule state. These α-helices contribute to a common nonpolar core in the native protein (Acharya et al., 1989), and there is the intriguing possibility that this may also be the case in the molten globule (Dobson et al., 1991). On the other hand, no evidence for a native-like structure has been found thus far for the other subdomain of this protein, which has predominantly β-structure in the native state.

In a similar way, Baldwin's group has shown the existence of at least three native α-helices in the molten globule state of sperm whale apomyoglobin (Hughson et al., 1990). They have measured individual exchange rates for many protons and have shown that the NH-protons of helices A, G, and H exchange 5 to 200 times more slowly than expected for the unfolded state (although much faster than in the native state). The NH-protons of helix B exchange only 2 to 10 times more slowly than in the unfolded protein, which suggests that this helix is only partly folded or is unstable. No traces of protection have been observed for helix E, which therefore seems to be unfolded. The NH-protons of helix F and of the short helices C and D could not be studied. These observations lead to the attractive idea that the compact unit formed by helices A, G, and H in native myoglobin may remain in the molten globule state (Hughson et al., 1990).

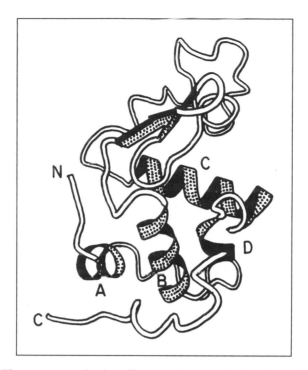

FIGURE 6–3. *The structure of α-lactalbumin, showing the locations of the major α-helices and β-strands. (Adapted from Acharya et al., 1989)*

Similar results have been obtained for cytochrome *c* (Jeng et al., 1990). In this protein, the main parts of all three large native α-helices (residues 6 to 14, 60 to 69, and 87 to 102) (Bushnell et al., 1990) remain protected from hydrogen exchange in the molten globule state at acid pH and high ionic strength.[2] This protection is, of course, much weaker than in the native state. On the other hand, many NH groups that are involved in hydrogen bonds in reverse turns and in the tertiary structure of the native protein are not protected substantially in the molten globule state.

A direct approach to the study of the tertiary fold of the molten globule would be by 2D-NMR spectra. NMR spectra of the molten globule state are very poorly resolved (see Section 3.6), however, and cannot be

---

[2]The existence of the molten globule state of cytochrome *c* at acid pH and high ionic strength was first claimed by Ohgushi and Wada (1983). Later it was suggested (Ptitsyn, 1987) that this state is native-like (differing only by the change of the heme state) as it melts cooperatively upon heating (Potekhin and Pfeil, 1989). However, recent experimental data presented by Jeng et al. (1990) and by Goto et al. (1990a) suggest that the acid form of cytochrome *c* at high ionic strength really has many properties typical of the molten globule state.

assigned by conventional methods. In Dobson's group, such assignments have been partly achieved by magnetization transfer. If the rate of interconversion between the native and the molten globule state in the transition region is faster than the nuclear relaxation rates, magnetic saturation of resonances of a given residue in the native state can be transferred to the resonances of the same residue in the molten globule. This can permit the "transfer" of the assignment of the NMR spectrum of the native state to that of the molten globule. Using this approach, Dobson's group (Baum et al., 1989; Dobson et al., 1991) has succeeded in correlating the strongly perturbed aromatic resonances of guinea pig α-lactalbumin in the acid molten globule state of those in the native state. The result was that the resonances of four aromatic residues, which are assigned tentatively to Trp 26, Phe 31, Tyr 103, and Trp 104, have the largest NMR chemical shifts in both the native and the molten globule states. The corresponding residues (two of them belong to α-helix B, residues 23 to 34, and two others are between α-helices C, residues 86 to 99, and D, residues 105 to 109; see Figure 6–3) are close together in the native structure of baboon α-lactalbumin (Acharya et al., 1989). It is very likely that this cluster of aromatic groups exists also in the molten globule state.

Recently, the molten globule state has been obtained for ubiquitin in 60% methanol at pH 2 (Harding et al., 1991). In this case, the NMR spectrum of the molten globule state was relatively sharp and permitted assignment of a number of resonances. As a result, it was shown that three β-strands of the β-sheet of the native protein are also present in the molten globule state, both in their positions in the polypeptide chain and in their mutual positions in space. The single α-helix of ubiquitin is also present and probably has a native-like position relative to the β-sheet. The other parts of the native conformation of this small protein, including two small additional β-strands, could not be detected in the molten globule state.

3.6 Fluctuating Environment of Side Chains

The data just presented may create an impression that a protein in the molten globule state is "almost native." The other properties of the molten globule state, however, are quite different from those of the native protein and are similar to those of the unfolded state.

^{1}H-NMR spectra of acid- and temperature-denatured forms of bovine (Ptitsyn et al., 1983; Dolgikh et al., 1985; Ikeguchi et al., 1986; Kuwajima et al., 1986) and guinea pig (Baum et al., 1989; Dobson et al., 1991) α-lactalbumins are quite different from those of the native proteins and are more similar to the spectra of the unfolded proteins. This is illustrated by Figure 6–4, which shows that the NMR spectrum of the molten globule state is much simpler than that of the native protein; the number of perturbed

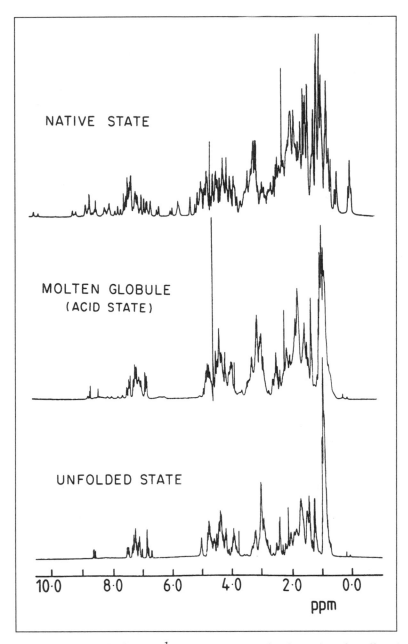

FIGURE 6–4. *Five hundred–MHz* ^{1}*H-NMR spectra of guinea pig α-lactalbumin in the native (pH 5.4), acid (pH 2.0), and unfolded (in 9-M urea) states recorded at 52°C. (Adapted from Baum et al., 1989)*

resonances is much smaller, and the overall picture is much more similar to the unfolded state. Note particularly that some pronounced high field resonances (at <1 ppm) of the native protein almost completely disappear in the molten globule state. This shows that the rigid mutual arrangement of aromatic and aliphatic side chains in the protein core is largely destroyed. On the other hand, a number of resonances are still remarkably perturbed, which reflects traces of structure and distinguishes the NMR spectrum of the molten globule state from that of the unfolded state. These are the perturbed resonances that have been used by Baum et al., (1989) for a tentative assignment of the molten globule NMR spectrum and for the conclusion of the existence of the native-like aromatic cluster in the molten globule state. (See Section 3.5.)

^{1}H-NMR spectra of the acid molten globule states of bovine carbonic anhydrase B (Rodionova et al., 1989; Semisotnov et al., 1989; Ptitsyn, 1987) and human retinol-binding protein (Bychkova et al., 1992) are also much more similar to those of the unfolded than of the native state.

In a similar way, the near-UV CD of α-lactalbumins (Kuwajima et al., 1976; Nozaka et al., 1978; Dolgikh et al., 1981,1985), carbonic anhydrase (Wong and Tanford, 1973; Wong and Hamlin, 1974) and other proteins is practically absent or greatly reduced in the molten globule state, suggesting that the local environment of the aromatic side chains is much more flexible.

3.7 Side Chain Movements

NMR and near-UV CD measurements show that the environment of many side chains is much less rigid in the molten globule than in the native state. This means that the fluctuations of side chains in the molten globule state are greatly increased as compared with the native state. This conclusion has been confirmed by direct experiments on spin-lattice and spin-spin relaxation times (T_1 and T_2) for bovine carbonic anhydrase B in the native, molten globule, and unfolded states (Semisotnov et al., 1989).

The spin-spin relaxation time decreases with an increase in the ratio of molecular movements. Figure 6–5A illustrates that the spin-spin relaxation on time T_2 of methyl groups in the molten globule state coincides with that of the unfolded state (0.085 sec) while it is quite different for the native state (0.021 sec).[3]

The molecular volumes of the molten globule and native states are similar (Wong and Hamlin, 1974), so the main reason for the difference in their relaxation times must be intramolecular movements. The similarity of the spin-spin relaxation times for the molten globule and for the unfolded

---

[3]Spin-spin relaxation time decreases with the increase of times of molecular movements.

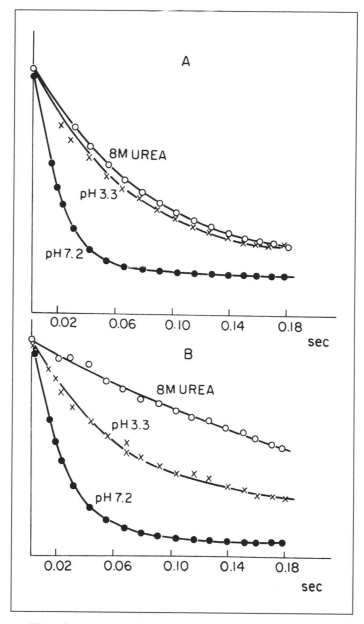

FIGURE 6–5. *Time-dependence of the spin echo amplitudes for protons of the methyl (A) and aromatic (B) groups of bovine carbonic anhydrase B in the native (pH 7.2), molten globule (pH 3.3), and unfolded (in 8 M urea) states. Spin echo of the methyl groups was measured at 0.89 ppm and of the aromatic groups at 7.1 ppm. (Adapted from Semisotnov et al., 1989)*

states suggests that the intramolecular movements for methyl nonpolar groups in the molten globule and in the unfolded states are practically the same.

On the other hand, spin echo curves for aromatic groups in the molten globule state are intermediate between those in the native and in the unfolded states (Fig. 6–5B), and the spin-spin relaxation time for these protons in the molten globule state (0.062 sec) is much closer to those in the native state (0.031 sec) than those in the unfolded protein (0.22 sec). Therefore, intramolecular movements of aromatic side chains (unlike aliphatic ones) are much more hindered in the molten globule than in the unfolded state, although not as hindered as in the native state. It appears that there is sufficient space inside the molten globule state for the "free" movements of small and symmetric aliphatic groups, but not enough for the movements of larger and less symmetric aromatic groups (see also Shakhnovich and Finkelstein, 1989).

Hindering of the motion of aromatic side chains in the molten globule state is confirmed by the study of urea denaturation of bovine carbonic anhydrase B (Rodionova et al., 1989). The proton resonances of the aliphatic side chains change at the first stage of denaturation (together with near-UV CD), i.e., at the transition from the native to the molten globule state. The proton resonances of aromatic side chains change at the second stage of denaturation (together with the decrease in far-UV CD and compactness), i.e., at the transition from the molten globule to the unfolded state. (See Figure 6–7B.)

Polarization of luminescence of Trp residues also shows that intramolecular mobilities of the Trp indole rings in the molten globule states of bovine and human α-lactalbumins (Dolgikh et al., 1981,1985) and in bovine carbonic anhydrase (Rodionova et al., 1989) are nearly as restricted as in the native state, while the restriction in the unfolded state is much less.

3.8 Large-scale Movements

Restricted large-scale fluctuations in the molten globule state are reflected in the field-dependent broadening of individual resonances in ^{1}H-NMR spectra. This broadening is especially apparent for aromatic groups and has been observed in the acid molten globule state of guinea pig α-lactalbumin (Baum et al., 1989). This is consistent with interconversion of different local conformations of the molten globule at rates slower than $\sim 10^3$ sec^{-1}.

Another approach to the study of large-scale fluctuations is the accessibility of the internal parts of protein molecules to the solvent or to other molecules. The most common technique used in these studies is hydrogen-deuterium exchange. Measurements of the overall rate of this exchange for the NH-protons of the backbone of bovine α-lactalbumin (Dolgikh et al., 1981,1985) and of bovine carbonic anhydrase B (Dolgikh et al., 1983) have shown that deuterium exchange is much faster in the molten globule than in the native protein and is closer to that in unfolded chains. More detailed studies of deuterium exchange using 2D-NMR (see Section 3.5) have shown that NH groups of many α-helices are still protected in the molten globule state (although much less than in the native state), while the NH groups of loops are not protected (Baum et al., 1989; Hughson et al., 1990; Jeng et al., 1990; Dobson et al., 1991; Baldwin, 1991).

The accessibility of a protein molecule to proteases also increases in the molten globule state. For example, digestion of channel peptide of colicin E1 by papain and bromelain is much faster in the molten globule state than in the native one (Merrill et al., 1990).

3.9 Stability

The small increase in protein volume in the molten globule should be sufficient to destroy the tight packing of side chains, which should lead to a large decrease in their van der Waals attractions. Even though being packed more loosely ("liquid-like"), nonpolar groups still preserve their attraction in an aqueous environment. It was suggested (Shakhnovich and Finkelstein, 1982; Gilmanshin et al., 1982; Dolgikh et al., 1985), therefore, that the molten globule state is stabilized mainly by the liquid-like interactions of nonpolar groups in water that are often called hydrophobic interactions.

This idea has been confirmed recently by site-directed mutagenesis of apomyoglobins (Hughson and Baldwin, 1989; Hughson et al., 1991; Baldwin, 1991). These experiments have shown that an increase in side chain hydrophobicity (replacements of Cys $\rightarrow$ Leu, Ala $\rightarrow$ Leu, Phe $\rightarrow$ Trp, Ser $\rightarrow$ Leu, Ser $\rightarrow$ Phe) stabilizes the molten globule state against unfolding, while the same mutations almost always destabilize the native state. This suggests that the molten globule state is stabilized mainly by less specific hydrophobic interactions (i.e., liquid-like interactions of nonpolar groups), while specific tight packing is important for the native state.

It is important to note that although the molten globule state is already "molten" (as it has lost the tight packing of side chains—see also Sections 4 and 5), it still preserves many elements of structure that can be melted by the influence of stronger denaturants. First, α-helices and other types of secondary structure certainly are preserved in the molten globule (see Sections 3.4 and 3.5) and the enthalpy of the helix-coil transition in water is as large as ~1 kcal/mol (Finkelstein et al., 1991; Baldwin, 1991). Second, the enthalpy of hydrophobic interactions is small at 20°C, but greatly increases with an increase in temperature and thus can provide an important contribution to heat effects of protein unfolding (Baldwin, 1986). It is not possible, therefore, to exclude that unfolding of the molten globule state may be accompanied by measurable heat effects, especially at high temperatures.

Unfortunately the available experimental data do not provide much information on this important point. In all proteins that have been studied, the molten globule state can be unfolded by GdmCl or urea in an apparently cooperative transition. (See Section 6, Table 6–2.) In one case —for human α-lactalbumin—this unfolding at 40°C is not accompanied by a large cooperative heat absorption (Pfeil et al., 1986). This suggests that the enthalpy of the molten globule state in this protein and at this temperature does not differ very much from the enthalpy of the unfolded state.

Unfolding of the molten globule state can also be caused by a further increase in electrostatic interactions. For example, the acid molten globule state of carbonic anhydrase can be unfolded by further decrease in the pH (Wong and Hamlin, 1974) and the acid molten state of cytochrome c by a decrease in the ionic strength (Ohgushi and Wada, 1983). Interesting effects occur when the ionic strength is varied by an increase of an acid (e.g., HCl) rather than a salt (e.g., KCl). Fink and his collaborators have shown that some proteins (e.g., β-lactamase, cytochrome c, and RNase A) first become fully or partly unfolded at acid pH (in the absence of salt), but with further decrease in pH they change into the molten globule state (Goto and Fink, 1989,1990; Goto et al., 1991a,b; Fink et al., 1990,1991). The explanation is simple: after deionization of all (or the majority of) the acid side groups, further decrease of the pH leads, not to an increase, but to a decrease of electrostatic interactions, due to the increase in the concentration of ions.

Data on the temperature melting of the molten globule state are controversial: The acid molten globule states of bovine (Dolgikh et al., 1985) and human (Pfeil et al., 1986) α-lactalbumins do not melt cooperatively upon heating. This can be expected because these proteins have similar molten globule states at low pH and at high temperatures (Dolgikh et al., 1981,1985; Gast et al., 1986). Microcalorimetric data on the acid form of retinol-binding protein (Bychkova et al., 1992) are consistent with a cooperative melting of the molten globule, with a small enthalpy change. Cytochrome c is in a state similar to the molten globule at low pH and high ionic strength (see Section 3.5) and melts upon heating in a manner similar to the native protein (Potekhin and Pfeil, 1989). It is unknown to what extent these proteins are unfolded at high temperatures.

4 MODEL OF THE MOLTEN GLOBULE

4.1 Initial Model

As early as 1973, the author predicted the molten globule state as a kinetic intermediate on the protein folding pathway (Ptitsyn, 1973). This state was specified as an "intermediate compact structure which is still different from the unique native protein structure and which formation is determined mainly by non-specific interactions of amino acid residues with their environment (water and a hydrophobic core of a forming globule)," i.e., by hydrophobic interactions. This hypothesis implied that the intermediate compact structure is formed "by 'the merging' of pre-existing embryos of chain regions with secondary structure and that the mutual arrangement of chain regions in this structure satisfying the crude criterion of the maximal screening of hydrophobic side groups from contacts with water must be close to their arrangement stabilized by specific long-range interactions between space-neighboring residues in the final compact structure" (Ptitsyn, 1973). According to this hypothesis, therefore, the predicted intermediate structure is compact, has native-like secondary structure and a native-like tertiary fold, but differs from the native state by the lack of specific interactions between neighboring residues, i.e., by the lack of a rigid tertiary structure.

Section 3 of this chapter shows that the equilibrium "molten globule" state is compact, has native-like secondary structure, and differs from the native state by the lack of the rigid tertiary structure. There is some preliminary indirect evidence that the molten globule state also has a native-like tertiary fold. The molten globule appears therefore to be consistent with the model that was proposed in 1973.

There are some contradictions, however, between experimental data that need to be explained. (See, e.g., Baldwin, 1990.) One group of experimental data—on specific partial volume and heat capacity (see Section 3.2)—suggests that the molten globule must have a lot of water inside and is something like "an amorphous squeezed coil of fluctuating helices" (Griko et al., 1988). In more rigorous terms, these data are consistent with an unfolded chain below the Θ-temperature[4] (see Flory, 1953); i.e., a chain with a predominance of intramolecular attractions that stabilize its squeezed state by occasional contacts between attracting groups (Finkelstein and Shakhnovich, 1989).

Another group of experiments, however, strongly contradicts this model. Large angle diffuse X-ray scattering shows the existence of a protein core with many van der Waals contacts at distances only ~3% to 4% larger than in the native state. (See Section 3.3.) Moreover, there is evidence that at least some of the contacts of native structure remain also in the molten globule state. (See Section 3.5.)

The most plausible explanation for this contradiction is the assumption on a nonuniform expansion of a protein molecule in the molten globule state. This assumption was proposed several years ago (Ptitsyn et al., 1986; Damaschun et al., 1986) and will be presented in this section.

---

[4]The Θ temperature is that at which interactions between nonadjacent atoms of a polymer compensate for interactions between the polymer and the solvent, so that the polymer has the average dimensions expected for a random coil.

4.2 "Nonuniform" Model

The hypothesis of nonuniform expansion implies that the "frame" of the native protein is essentially preserved in the molten globule state, while other parts of the molecule can be more or less unfolded (Fig. 6–6). The frame includes central parts of the main α-helices and β-sheets, i.e., those parts of these regions that carry nonpolar side chains contributing to the hydrophobic core. The rest of the molecule can be unfolded and includes loops, ends of α- and β-regions, and those α- and β-regions that are not part of the frame of the protein.

According to this hypothesis, a protein core formed mainly by nonpolar groups of α- and β-regions must expand only slightly in the molten globule state, so that water molecules may not penetrate inside this core to any considerable extent. The packing of nonpolar side chains in the core becomes looser (as in usual crystal-liquid transitions), but at least a substantial part of these contacts is preserved. The looser packing of the side chains permits them to increase the amplitudes of their rotational vibrations; moreover, aliphatic side chains probably can even jump from one rotational isomer into another (Shakhnovich and Finkelstein, 1982, 1989). Therefore, this expansion will lead not only to an increase in the van der Waals energy of the core, but also to an increase in its entropy. In this (and only in this!) sense, the molten globule is liquid-like, as the native-like mutual arrangement

FIGURE 6–6. Schematic representation of the native and the molten globule states of protein molecules. Nonpolar side chains are hatched.

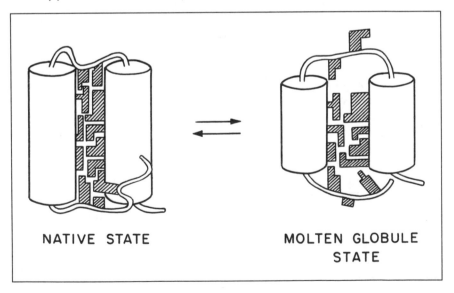

NATIVE STATE   MOLTEN GLOBULE
                                       STATE

of α- and β-regions may be stabilized by liquid-like hydrophobic contacts between their nonpolar side chains. The point is that these side chains belong to rigid α- and β-regions, so their interactions are cooperative (Ptitsyn and Finkelstein, 1980b); this permits preservation of the native-like mutual arrangement of the α- and β-regions. "Liquid crystal" is probably the best way to describe this situation in usual molecular physics terms.

The important point of this hypothesis is that the stability of the molten globule state is due to "liquid-like" (hydrophobic) interactions in a relatively small part of the protein molecule. This possibility is justified by the fact that only a relatively small number of side chains are completely buried in globular proteins; e.g., in globins the number of these side chains is 33 out of approximately 150, or only ~20% (Perutz et al., 1965). As a rule, all, or almost all, of these residues are conserved as nonpolar in all members of the given protein family, e.g., in globins (Perutz et al., 1965). It is natural to assume that these 20% of the side chains (plus probably some nonpolar side chains that are only partly buried in the native state but enter or belong to a hydrophobic core) preserve their contacts in the molten globule state. The energies of these contacts, even when reduced due to the expansion of the core, seem sufficient to stabilize the α- and β-regions and their native-like arrangement.

The hypothesis of a nonuniform expansion probably can help to reconcile the apparently contradictory experimental data mentioned previously. On the one hand, the hypothesis implies preservation of native-like core, which is consistent with the large-angle X-ray scattering and the NMR data. On the other hand, it implies that a large part of the molecule is unfolded and therefore that water penetrates into a thick shell of the molecule, "washing" its nonpolar core, which may be consistent with the data on the partial specific volume and the heat capacity.

The unfolded part of a molecule must have the partial specific volume of an unfolded protein, which is very near to that of a native one (e.g., Bendzko et al., 1988). Therefore, the difference between the partial specific volumes of the molten globule and the native states is reduced to the small expansion of a small part of a molecule that remains impenetrable to the solvent (i.e., to a few percent). In a similar way, the penetration of water inside the molecular shell can lead to a large increase in the heat capacity, even in the case when the small part of a molecule remains screened from water.

Of course, this is no more than a crude draft of a possible explanation for this contradiction. Precise experiments are needed to check the existence of expected differences in partial specific volumes and heat capacities between the molten globule and unfolded proteins. We need quantitative measurements of the partial specific volume, the heat capacity, and the intramolecular contacts in the molten globule state to understand why it preserves many intramolecular contacts, while having a lot of water inside.

Unfolding of the loops of a molecule make them susceptible to rapid hydrogen exchange. This is consistent with the evidence that those NH-protons that are involved in native hydrogen bonds of reverse turns or tertiary structure are not protected from exchange in the molten globule state. (See Section 3.5.) On the other hand, parts of a molecule that are involved in regular secondary structure become much more accessible to water, which facilitates their "breathing" and thus dramatically decreases the protection from exchange of even these fixed parts.

One of the consequences of this "double nature" of the nonuniform expanded molten globule is its high affinity to hydrophobic probes like 8-anilinonapthalene-1-sulfonate (ANS), which is bound by the molten globule state much stronger than by unfolded chains and by the majority of native proteins (Semisotnov et al., 1987,1991a; Rodionova et al., 1989; Ptitsyn et al., 1990). This probe binds to solvent-accessible clusters of nonpolar atoms (Stryer, 1965), which are absent in unfolded chains and relatively rare in native proteins. A molten globule in which the nonpolar core is preserved, but becomes less tightly packed and accessible to the solvent, is the ideal case for this binding.

4.3 Terminology

It is worthwhile now to discuss some terminology questions. As was mentioned in the beginning of this section, the molten globule state was predicted as a compact state with native-like secondary structure and native-like tertiary fold, but without rigid tertiary structure. When this state was detected experimentally (Dolgikh et al., 1981) it was called a "compact globule with native-like secondary structure and with slowly fluctuating tertiary structure." The term *molten globule* was introduced later by Ohgushi and Wada (1983).

When designating the molten globule as a state with a "fluctuating tertiary structure," the term *tertiary structure* was used in its modern sense, i.e., for the *detailed* protein structure at atomic resolution, to distinguish it from the *crude* protein structure (or tertiary fold). It led later, however, to some misunderstandings. For example, Kim and Baldwin (1990) have described "Ptitsyn's model" as "the rapidly fluctuating liquid-like structure" which "is incompatible with a structure determined by fixed tertiary interactions" (p. 642). In fact, my model implied just the opposite picture— a fixed native-like tertiary fold with greatly increased fluctuations of side chains and irregular regions.

Kim and Baldwin (1990) proposed distinguishing between "the term 'molten globule' for Ptitsyn's model and the term 'collapsed form' for the experimentally observed intermediate." It is a good idea to distance the model (which implies the native-like tertiary fold) from the existing experimental data. The term *molten globule* has been widely used, however, as an operational term for a compact state with pronounced secondary structure but with unknown location of secondary structure elements in the sequence and in space, and it is probably too late to change its meaning.

Therefore, I propose to retain the general term *molten globule* for this state of our knowledge and to introduce the terms *native-like molten globule* and *disordered molten globule* for compact states with and without a native-like tertiary fold. Thus, we shall designate the "molten globule" state of a given protein as either the "native-like" or "disordered" one when we learn whether it does or does not have a native-like tertiary fold. If it happens that the globularization of a protein molecule during folding occurs before formation of the native-like tertiary fold, we can speak of fast formation of a "disordered molten globule" and of its slow transition to a "native-like molten globule." The advantage of this terminology is that it avoids any misunderstanding when reading papers published since the early 1980's on this subject. It is of course too early now to speculate on the possible properties of "disordered" molten globules and particularly on their relation to squeezed coils.

5 TRANSITIONS BETWEEN CONFORMATIONAL STATES

5.1 Denaturation as an All-or-None Transition

Many years ago it was suggested that denaturation of globular proteins is an all-or-none process, i.e., that in the transition region there are only native and fully denatured molecules, without any measurable amount of intermediate states. The first evidence supporting this point of view was the coincidence of denaturation curves measured by different methods (Lumry et al., 1966; Tanford, 1968). The definitive evidence, however, was obtained by Privalov and his collaborators (Privalov and Khechinashvili, 1974; Privalov, 1979; Chapter 3) after they had developed the technique of precise microcalorimetric measurements (Privalov, 1974; Privalov et al., 1975). These measurements have shown that for small, single-domain proteins, the calorimetric enthalpy of denaturation per protein molecule is practically equal to the enthalpy per cooperative unit calculated from the temperature interval of melting. Therefore, the cooperative unit coincides with a protein molecule, i.e., a single-domain protein denatures as a whole without any visible intermediate states. For larger proteins consisting of two or more domains, a cooperative unit usually coincides with a domain (Privalov, 1982). The all-or-none mechanism of protein denaturation has been strictly established only for temperature denaturation. A number of less definitive data, however, strongly suggest that it is true also for other types of denaturation.

The all-or-none mechanism of protein denaturation is very important for the stability of protein function. If proteins could denature non-cooperatively (i.e., in small pieces), thermal motions would destroy a protein structure at all temperatures. As a result, proteins would not have rigid active centers, and this would inhibit their biological activities. Only the high cooperativity of denaturation makes a protein structure resistant to thermal motions unless and until the temperature becomes large enough to destroy the structure as a whole.

Thus, proteins are unique examples of molecules that can possess an *intramolecular all-or-none transition*—a fact that requires a full and convincing explanation.

The first attempt to explain the all-or-none character of protein denaturation (Ptitsyn and Eizner, 1965; Ptitsyn et al., 1968) was based on the "liquid drop" model of a protein molecule and describes protein denaturation as a globule-coil transition. Two approximate versions of the theory have been proposed; the first one used the virial expansion of the free energy of long-range intramolecular interactions (limited by the interactions of pairs and triplets of residues), while the second one used the approximate expression of this free energy similar to the van der Waals expression for real gases. Both of these lines have been followed afterwards—the first one by De Gennes (1975) and others and the second one by Sanchez (1979); for further references, see the review by Chan and Dill (1991). Birshtein and Pryamitsyn (1987,1991) have improved this treatment by replacing distribution functions for the end-to-end distance by that of the radius of gyration, which better represents the linear dimensions of squeezed coils. A more strict and sophisticated, but physically equivalent treatment has been elaborated by Lifshitz et al. (1978).

All of these versions of the theory give the same result—they predict a cooperative but not an all-or-none globule-coil transition for flexible homopolymers. Formally, an all-or-none transition is predicted for rigid chains, but even in this case the barrier between native and denatured states is approximately of magnitude $k_B T$ (A. V. Finkelstein, private communication) and therefore intermediate states can also be populated. The drastic difference between a phase liquid-gas transition and a smooth globule-coil transition is due to the fact that monomers in polymer molecules are connected in a chain; therefore, each monomer has a limited freedom that does not depend upon the dimensions of the entire molecule.

The existence of cooperative coil-globule transitions in a homopolymer has been confirmed experimentally (Anufrieva et al., 1972) by polarized fluorescence, which permits the use of small concentrations of polymer and avoids it aggregation. Neither this nor the subsequent studies (see, e.g., Sun et al., 1980), however, give a clear answer to the character of this transition. See reviews by Chan and Dill (1991), Anufrieva

(1982), and Anufrieva and Krakovyak (1987) for other references. Although there is some evidence against the all-or-none character of such transitions (Anufrieva et al., 1972), the problem still needs careful investigation. Monte Carlo computer experiments on the collapse of a nonintersecting chain on a cubic lattice (Kron et al., 1967) also did not show an all-or-none transition.

The theory of collapse of heteropolymers has also been developed (Dill, 1985; Grosberg and Shakhnovich, 1986a,b; Dill et al., 1989). Grosberg and Shakhnovich (1986a,b) extended the approach of Lifshitz et al. (1978) to heteropolymers and have shown that the coil-globule transition in a very long heteropolymer must be similar to that of a homopolymer.

A liquid drop is, of course, not the best model for a protein molecule, which has rigid α- or β-regions. There are no detailed theories that take this important fact into account. Grosberg (1984) has considered a simple model of a polymer chain in which collapse (coil-globule transition) is coupled with the helix-coil transition. He has shown that this coupling may dramatically increase the cooperativity of the transition, transforming it into an all-or-none transition.

Another approach to this problem (Ptitsyn, 1975) is based on the fact that proteins (or protein domains) consist of only a few "structural blocks" (α-helices and β-strands). If these blocks are unstable in the unfolded state, there are only few states that can be intermediate between the completely unfolded states. In this system, even relatively small cooperativity (e.g., the unfavorable initiation and / or favorable termination of a folded structure) will lead to a small total statistical weight of all the intermediates. This will lead to a transition that practically is all-or-none not because intermediates are forbidden but because there are only a few slightly unstable intermediates. The difference between this case and the real all-or-none transition is the height of the free energy maximum between the folded and unfolded states—for the "quasi all-or-none" transition this maximum must be relatively low.

5.2 Native–Molten Globule Transition

The problem of protein denaturation dramatically changed after it was shown that all-or-none temperature transitions occur in bovine (Dolgikh et al., 1981,1985) and human (Dolgikh et al., 1981; Pfeil et al., 1986) α-lactalbumins. The temperature-denatured states of these proteins are compact (Dolgikh et al., 1981,1985; Gast et al., 1986) and have pronounced secondary structure (Dolgikh et al., 1981,1985), i.e., they are in the molten globule state. Therefore, at least for these proteins, the all-or-none transition is *not* a globule-coil transition, i.e., is not connected with the unfolding of a protein chain or with the destruction of its secondary structure. The single common feature of all denatured proteins—from molten globules to unfolded chains—is the absence of the tight packing of side chains. This suggests that the common reason for the all-or-none character of protein denaturation is the destruction of the tight packing.

This idea was the basis of the novel theory of protein denaturation proposed by Shakhnovich and Finkelstein (1989). Their theory describes protein denaturation as melting of a crystal rather than as the evaporation of a liquid drop. It explains the all-or-none transition by using the model of tightly packed side chains attached to rigid regions (α-helices and β-strands). Due to this attachment, the movement of side chains can be released only cooperatively, because it is impossible to increase the distance between a few of them, while leaving the other distances unaltered. In the native state, side chain movements typically are limited to their rotational vibrations with relatively small amplitudes. Therefore, the entropy of this state is low. However, its energy also is low (due to large van der Waals attractions, which stabilize this state). If the distances between side chains increase, amplitudes of rotational vibrations increase, but the corresponding gain in entropy is too small to compensate for the increase in energy. At some threshold distance, however, a new type of aliphatic side chain movement is released—these groups begin to jump from one rotational isomer to another. The additional increase in entropy compensates for the increase in energy, and the denatured state becomes stable. As side chain–side chain distances can change only simultaneously, they can be described by a single parameter—the molecular volume. This means that a protein molecule as a whole can have two stable states: an energetically stable native state with a small volume and an entropically stable denatured state with an increased volume. The states with intermediate volumes have high energy and low entropy and therefore are unstable.

According to this theory the "fate" of the denatured protein molecule depends upon the solvent and upon the temperature. Sometimes it can be the molten globule state with a moderately increased volume and release of some new movements. In other conditions (e.g., at high concentrations of strong denaturants), however, it can be the completely unfolded state, with a large increase of the volume and the release of all movements.

Shakhnovich-Finkelstein's theory is a very important step in our understanding of the all-or-none character of protein denaturation. It emphasizes two essential points: the role of the destruction of tight packing of the side chains and the release of new types of movements at some threshold value of the volume. This theory, however, refers to the "uniform" model of the molten globule, i.e., it does not distinguish between the protein core and the protein shell.

According to the nonuniform model of the molten globule state (see Section 4), the main difference between the molten globule and the native protein is the state of the protein shell rather than that of the protein core; loops forming the protein shell are more or less unfolded, while the protein core still exists, becoming only a little less compact. In native proteins, loops are attached to the hydrophobic core by their nonpolar side chains. The small increase in volume of this core will release these side chains and this process can be cooperative if all side chains need the same threshold value of core volume to be released. This threshold value can be smaller than the value that permits side chains to rotate inside a core, as it is easier to remove a side chain than to rotate it in a dense environment.

The cooperative release of nonpolar side chains belonging to loops may lead to the cooperative unfolding of these loops, because their native structure may become unstable without this support. This unfolding will lead to a large increase in entropy, which can compensate for the increase in the core volume.

The release of a number of nonpolar side chains from the core will permit other side chains to increase the amplitudes of their rotational vibrations and probably even to jump from one rotational isomer to another. Therefore, this interpretation probably supplements the Shakhnovich-Finkelstein theory rather than contradicts it. The main point is that "new" movements that must be switched on at the threshold volume can be not only the rotational isomerization of core side chains, but also, or even predominantly, the unfolding of loops (of course, together with rotational isomerization of their side chains). According to this point of view, the cooperative increase of a core volume may be more the trigger than the reason for the all-or-none transition.

5.3 Globule–Coil Transition

Does the existence of an all-or-none transition between the native and the molten globule states mean that the molten globule state is thermo-dynamically equivalent to the unfolded state? In other words, is the molten globule just a limiting case of the unfolded state, with maximum compactness and secondary structure, or is it divided from the unfolded state by another all-or-none transition?

GdmCl- and urea-induced molten globule–coil transitions appear to be cooperative, in that they occur within a relatively narrow interval of concentrations of strong denaturants. These intervals are determined by the differences of the number of denaturant molecules "bound" to a cooperative unit in the unfolded and in the molten globule state. We cannot measure the numbers of denaturant molecules bound to one residue in the unfolded and in the molten globule states, so it is practically impossible to evaluate the number of residues in a cooperative unit and to compare it with the number of residues in a protein or in a protein domain. Thus, the cooperativity of the molten globule–coil transition does not give us any information as to whether this transition is all-or-none.

On the other hand, the small enthalpy of the molten globule–coil transition in human α-lactalbumin (Pfeil et al., 1986) does not exclude the existence of an all-or-none transition, as the solvent-induced transition may be connected with a jump in the number of bound denaturant molecules rather than with a jump of enthalpy.

A common, but not definitive, criterion for all-or-none transitions is to compare the transition curves obtained by different methods. If all these curves coincide, it is likely to be an all-or-none transition; if not, this type of transition is excluded. It was this method that had been used in the "precalorimetric" era to suggest that protein denaturation is an all-or-none transition. This method has been used to study the urea-induced molten globule–coil transition in bovine carbonic anhydrase B (Rodionova et al., 1989). The result was that the unfolding of the molten globule state of this protein monitored by the change of its compactness (intrinsic viscosity, polarization, and spectrum of fluorescence) coincided with the unfolding monitored by the change of its secondary structure, i.e., by the ellipticity at 220 nm (Fig. 6–7A). This suggests that in the transition region there are no stable compact intermediates without secondary structure, as well as no noncompact intermediates with secondary structure.

This result is confirmed in Figure 6–7B, which presents the urea-induced unfolding of native carbonic anhydrase. This unfolding process is biphasic. The first phase involves the destruction of the rigid tertiary structure, while the second is the unfolding of the molten globule state. Again the curves monitored by compactness (polarization and spectrum of fluorescence) and by secondary structure coincide, which shows the absence of intermediates between the molten globule and the unfolded states. Similar results have been obtained for GdmCl-induced unfolding of native human carbonic anhydrase B. In addition, the changes of the NMR signals shown in Figure 6–7B demonstrate that the mobility of the aliphatic side chains increases at the first stage of denaturation (at the native–molten globule transition), while the mobility of the aromatic side chains changes only at the second stage (at the molten globule–coil transition).

The results of this study suggest that the transition from the molten globule to the unfolded chain may also be an all-or-none transition. The most direct method to prove this would be to demonstrate a bimodal distribution of protein molecules between the two states in the transition region. Recently, we have used this approach (Uversky et al., 1992) to study the unfolding of the molten globule states of two proteins by FPLC size-exclusion chromatography. The main difficulty with these experiments is that it is possible to separate only slow-exchanging conformational states of protein molecules. We have found that this exchange is slow enough only at a low temperature (4°C) and for GdmCl-induced (not urea-induced!) unfolding. Under these conditions, two peaks in the

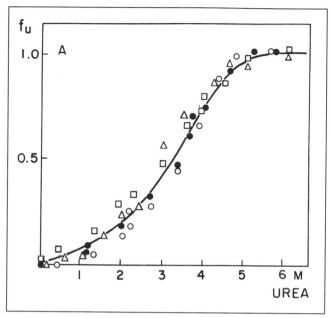

FIGURE 6–7. *Urea-induced unfolding of bovine carbonic anhydrase B at pH 3.6 (A) and at pH 7.5 (B). At pH 3.6 in the absence of urea, the protein is in the molten globule state and unfolds by a one-step process. At pH 7.5, the protein is in the native state and unfolds in two steps (native–molten globule and molten globule–unfolded transitions) without additional intermediates. The fraction of unfolding is given by $f_u = (x - x_o)/(x_u - x_o)$, where x is the value of a measured parameter, x_o is its value in the absence of urea, and x_u is its value at high concentration of urea. (A) ●, the increase of intrinsic viscosity [η], □, the decrease of I_{320}/I_{360} (I is the intensity of tryptophan fluorescence at the given wavelength), ○, the increase of 1/P (P is the polarization of tryptophan fluorescence), △, the decrease of the negative ellipticity at 220 nm. Intrinsic viscosity and the spectrum and polarization of tryptophan fluorescence both reflect the compactness of the molecule, while the ellipticity at 220 nm reflects its secondary structure. (Adapted from Rodionova et al., 1989)*

elution curves have been observed within the equilibrium molten globule–coil transition for bovine carbonic anhydrase B and for β-lactamase from *Staphylococcus aureus*. The first peak is close to that of the native protein (reflecting the almost native compactness of the molten globule state) while the second has a smaller elution volume, demonstrating that this state is more expanded than the first. When the GdmCl concentration increases in the transition zone, the more compact molten globule peak decreases in magnitude, while the unfolded peak increases; this

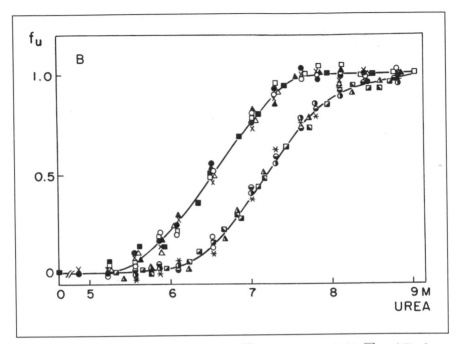

FIGURE 6.7. (B) △, *the increase of* I_{320}/I_{360}; ◣, *the increase of* $1/P$; ◢ *and* ■, *the decrease of negative ellipticity at 220 nm (◢) and 270 nm (■); ▲, the decrease of the area under high-field NMR resonances; other signs mean the increase of signal intensities of protons of aliphatic side chains at 3.17 (×), 2.97 (○), 2.00 (●), 1.38 (□), and 0.86 (△) ppm as well as of aromatic side chains at 7.5 (◖), 7.10 (◓), and 6.79 (∗) ppm.* I_{320}/I_{360} *and* P *reflect the compactness of the molecule, ellipticity at 220 nm reflects its secondary structure, while the ellipticity at 270 mm and the area under the high-field NMR resonances reflect its rigid tertiary structure. (Adapted from Rodionova et al., 1989)*

demonstrates the all-or-none transition of proteins from the molten globule to the more expanded state.

This analysis of the molten globule–coil transition leads us back to the general problem of the cooperativity of globule-coil transitions. (See Section 5.1.) As has been already mentioned, the globule-coil transition may be an all-or-none transition of the first order if it is coupled with a change of linear order, e.g., with the helix-coil transition (Grosberg, 1984). Therefore, unfolding of the molten globule, with the simultaneous destruction of its secondary structure, can be similar to the melting of a liquid crystal and may be an all-or-none transition.

It is pertinent to mention that we also know an example of a smooth (not all-or-none) molten globule–coil transition. The first experimental evidence for the molten globule state was actually obtained for random copolymers of glutamic acid with leucine (Anufrieva et al., 1975; Bychkova et al., 1980; Semisotnov et al., 1981). Upon deionization of the glutamic acid residues, these copolymers first became helical and then collapsed to a compact helical state. The transition of these copolymers from the noncompact helical state to the collapsed helical state clearly is not an all-or-none transition, because the relaxation time of depolarization of fluorescence goes through a maximum. The difference between the pH-induced globule-coil transition in these copolymers and the urea- or GdmCl-induced transitions in proteins is that the pH-induced transition occurs between collapsed helical and noncompact helical polymer, while the influence of urea and GdmCl may transform a protein from a condensed state with secondary structure to an unfolded state without secondary structure. Therefore, this difference suggests that the coupling between condensation and linear structure of a chain may be very important for an all-or-none transition.

On the other hand, the molten globule, unlike globules from random copolymers, may have a unique tertiary fold. (See Section 3.5.) Although a globule-coil transition for a globule with a unique tertiary fold has not been extensively studied theoretically, the possibility cannot be excluded that this transition may be an all-or-none transition (Shakhnovich and Gutin, 1990). Much more experimental and theoretical work is needed to clarify the physical reasons for this unusual phenomenon, the all-or-none globule-coil transitions of denatured proteins.

Depending upon the stability of the molten globule state, it can unfold at different concentrations of strong denaturants (as well as at different pH values and temperatures). Therefore, the unfolded protein at the end of this transition can be expanded to different extents, depending upon the properties of the solvent (Damaschun et al., 1991). One cannot exclude the possibility that very unstable native or molten globule proteins can unfold into "squeezed coils," i.e., coils below the Θ-temperature. (See footnote on p. 261.) These coils can be transformed into Gaussian ones and then into expanded coils by a further increase in the concentration of the strong denaturant. This transition is *not* an all-or-none transition, rather it may be a phase transition of the second order (Lifshitz et al., 1978).

6 OCCURRENCE OF THE MOLTEN GLOBULE

After the molten globule state was first described for bovine and human α-lactalbumins, a number of other proteins have been reported to have similar properties. These are collected and classified in Table 6–2. (See also Bychkova and Ptitsyn, 1992.) Table 6–2A collects proteins for which all three main features of the molten globule state (compactness, the presence of secondary structure, and the absence of rigid tertiary structure) have been

TABLE 6-2. *Molten Globule States in Proteins*

Protein	Conditions	Compactness[a]	Secondary Structure[b]	Tertiary Structure[c]	Cooperative Unfolding[d]	Binding to Hydrophobic Probe	References
		A. WELL-ESTABLISHED MOLTEN GLOBULE STATES					
Bovine α-lactalbumin	pH 2; 0.05 M KCl	R_D=21.1; N: 18.4; U: 32.3 [η]=3.1; N:3.4; U:6.1 R_g=16.8; 15.7; N: 15.6; 15.5 $λ_{max}$=338	10,000 N: 9,500 IR	−20 (270 nm) N: −320 NMR	—	ANS[e], phospholipids	Robbins and Holms, 1970; Kuwajima et al., 1976; Dolgikh et al., 1981,1985; Gilmanshin et al., 1982; Izumi et al., 1983; Kim and Kim, 1986; Bychkova et al., 1990; Semisotnov et al., 1991a
	pH 7	—	—	—	2.0 and 2.6[f] GdmCl	—	
	pH 7; 50°C; 0.05 M KCl Ca²⁺-free	[η]=3.0	7,000	−50; NMR	—	—	
	pH 11.3; 0.15 M KCl	R_g=16.8; U: 24.4	$[Θ]_{208}$=14,500 N: 10,500	−100; N: −300	—	—	
Human α-lactalbumin	pH 2; 0.05 M KCl	[η]=4.2;N:3.1; U:6.6 R_D=19.9; N:17.7; U: 24.7 $λ_{max}$=331	14,000; N: 9,000	−60 (270 nm) N: −240; U: −50	—	ANS, melittin	Nozaka et al., 1978; Dolgikh et al., 1981; Gilmanshin et al., 1982; Gast et al., 1986; Bychkova et al., 1990; Permyakov et al., 1991
	pH 7.4; 1.85 M GdmCl	$λ_{max}$=336	9,000	−60	3.5 GdmCl	—	
	pH 7; 50°C; Ca²⁺-free	R_D=19.6	Similar to pH 2	Similar to pH 2	—	—	

Protein	Conditions					ANS	References
Bovine carbonic anhydrase B	pH 3.6	$[\eta]=4.1$; N:3.7; U:29 P=0.09; N:0.085; U:0.05 $\lambda_{max}=340$	4,600 N: 3,500 IR	+5 (270 nm) N: -100; U: 0	3.5 urea	ANS	Wong and Tanford, 1973; Wong and Hamlin, 1974; Dolgikh et al., 1983; Rodionova et al., 1989; Semisotnov et al., 1991a; Uversky et al., 1992
	pH7	—	—	—	6.5 and 7.2 urea[f]	—	
	pH7; 4°C; 1.3 M GdmCl	$R_S \cong (R_S)_N$	—	—	1.5 GdmCl	ANS	
	pH7; 1.9 M GdmCl	$R_S \cong (R_S)_N$	Similar to N	Similar to U	2.3 GdmCl	ANS	
Human carbonic anhydrase B	pH3	—	5,800; N: 600; U: 1,800	-40; N: -180	—	—	Jagannadham and Balasubramanian, 1985; Rodionova et al., 1989
	pH 8; 1.7 M GdmCl	η_{sp}/c similar to N	5,000; N: 970; U: 1,500	10% from N	2.5 GdmCl	ANS	
β-Lactamase *Staphylococcus aureus*	pH 7; 4°C 0.5 M GdmCl	$R_S \cong (R_S)_N$	$-[m]_{234}=3,000$ N: 3,600 (1 M GdmCl at 20°C)	Nonactive	0.75 GdmCl	—	Robson and Pain, 1976a,b; Uversky et al., 1992
Ribonuclease A	pH 3; 3 M LiClO$_4$	$[\eta] = 5.5$ N: 3.4; U: 7.5	8,000 N: 9,000	-20 N: -250	2.5 urea	—	Ahmad and Bigelow, 1979; Denton et al., 1982
	pH <2 (or pH 2 at high KCl)	$R_S \cong (R_S)_N$	Similar to N	Similar to U	—	—	Fink et al., 1990
T4 Lysozyme	pH 7.4; 2.0 M GdmCl	$\lambda_{max}=(\lambda_{max})_N$	60% from N	—	2.2 GdmCl	—	Desmadril and Yon, 1981
Bovine pancreatic trypsin inhibitor, reduced	0.5 M GdmCl	Specific energy transfer P=0.18; N: 0.18[h]	13,250 (200 nm)[g] N: 19,800 (202.5 nm)	0 (275 nm) N: -520	—	—	Kosen et al., 1981, 1983; Amir and Haas, 1988

continued

Protein	Conditions	Compactness[a]	Secondary Structure[b]	Tertiary Structure[c]	Cooperative Unfolding[d]	Binding to Hydrophobic Probe	References
Horse cytochrome c	pH 2.2; 0.5 M NaCl	$[\eta]$=3.1; N: 2.8; U: 12.7; R_g=20.1 N: 19.8	11,300; N: 10,600	Weak near UV circular dichroism. Small protection against H ⇄ D exchange (at 1.5 M NaCl)	—	—	Ohgushi and Wada, 1983; Jeng et al., 1990
Bovine growth hormone, reduced	pH 9.1; 4.5 M urea	$R_S \cong (R_S)_N$[h]	8,500 N: 15,000	ε_{290}=7,300 N: 9,600	7.8 urea	—	Holzman et al., 1986
Fragment F_c of IgG	pH 4.5; 64°C	s=4.1; N: 4.1[i]	3,500 N: 3,300	−80 (270 nm) N: −120 Fluorescence	—	—	Vonderviszt et al., 1987
Apomyoglobin	pH 4.3; 27°C	$[\eta]$=5.0; N: 3.5; U: 20.7	11,500 N[i]: 18,300	0; N: +20	—	ANS, phospholipids	Griko et al., 1988; Lee and Kim, 1988
Diphtheria toxin, fragment A	pH 3–5; 0.15 M NaCl	λ_{max}=336 N: 328; U: 350	3,000 N[i]: 3,000 IR	Proteolysis	0.5 urea	Liposomes, TID[j], membranes	Zhao and London, 1988; Dumont and Richards, 1988; Cabiaux et al., 1989
Colicin A (pore-forming domain)	pH 4; 0.1 M KCl	s=3.66; N[i]: 3.66	14,000 N: 14,500	Proteolysis	—	Membranes	Cavard et al., 1988
Colicin E1 (channel peptide)	pH 3.5; 0.1 M NaCl	R_S=20.7; N[i]: 22.7; U: 84.8	Similar to N[i]	Proteolysis	—	Membranes	Merrill et al., 1990
Staphylococcal nuclease, fragment 1-128	pH 7.0; 0.05 M NaCl	R_S=21; N[h]: 21 (0.2 M NaCl)	8,000 N: 13,000	Nonactive (though includes active center)	0.3 GdmCl	—	Shortle and Meeker, 1989

Rhodanese	pH 7.4; 2 M GdmCl	λ_{max}=348; N: 335; U: 356	90% from N	Much less than N; proteolysis	3.3 GuHCl	ANS	Tandon and Horowitz, 1989
	pH 7.4; 4 M urea	λ_{max}=345	80% from N	-SH groups are easily oxidized	5.0 urea	—	Horowitz and Criscimagna, 1990
β-Lactamase *Bacillus cereus*	pH 12; 1.5 M KCl	R_D=31; N: 23; U: 55; λ_{max}=332	10,000 N: 12,000	0; N: -100	—	ANS	Goto and Fink, 1989
	pH 2; 0.5 M KCl	R_S=26.5; N:24; U:51; λ_{max}=332	11,000	+20	—	ANS	Goto et al., 1990a
Aspartate amino transferase (monomer)	0.9–1.1 M GdmCl	s=3.1; N:3.4[k];U:2.2 (6°C) λ_{max}=339 N: 335; U: 355	50% from native dimer	Fluorescence	1.7 GuHCl	—	Herold and Kirschner, 1990
Retinol binding protein (RBP) apo-form	pH 2	η_{sp}/c=9.7[l]; N[l]: 4.4; U: 24.5 λ_{max}=336	7,500 (207 nm) N: 5,000	+5 (285 nm) N: -130; NMR	—	ANS	Bychkova et al., 1992

B. POSSIBLE MOLTEN GLOBULE STATES

Bovine growth hormone	pH 2; 0.1 M NaCl	η_{sp}/c=10.5; N: 3.0	b_o= -225; N: -225	ε_{290} is changed	—	—	Burger et al., 1966
	pH 8.5; 3.3 M GdmCl	R_S=26; N: 18; U: 37; λ_{max}=342	11,000 N: 15,000	ε_{290}=7,400; N: 9,800; U: 7,400	3.7 GdmCl	—	Brems et al., 1985; Brems and Havel, 1989
	pH 4.0; 4.5 M urea	λ_{max}=$(\lambda_{max})_N$	8,000 (50% N)	ε_{290}=7,000 N: 9,000	5.3 urea	—	Holzman et al., 1990
β-lactoglobulin	pH 2; 2.7 M GdmCl	—	6,500 N: 6,000	-15 N: -90	3.0 GdmCl	—	Ananthanarayanan et al., 1977
	pH 2; 90°C 0.1 M KCl	η_{sp}/c=11.5 U: [η]=19.1	11,000 N: 5,500	-10	2 GdmCl	—	Ananthanarayanan and Ahmad, 1977

continued

Protein	Conditions	Compactness[a]	Secondary Structure[b]	Tertiary Structure[c]	Cooperative Unfolding[d]	Binding to Hydrophobic Probe	References
Parvalbumin, carp, apo-form	pH 7; 42°C	—	$-[m']_{233}=6,500$ N: 7,500	NMR	—	—	Cave et al., 1979
apo-form	pH 2.5–2.8; 0.1=M NaCl; 4.5°C	—	9,000 N: 14,000	−20 (267 nm) N:-40; U:-15	0.7 GdmCl	—	Kuwajima et al., 1988
Human α-interferon	pH 1.5	—	9,000 N: 15,000	0 (292 nm) N: -150	—	—	Bewley et al., 1982
Human γ-interferon	pH 3.5; 0.05 M NaCl	—	12,000 N: 15,000	0 N: +120	—	—	Arakawa et al., 1987
	pH 2.0; 0.05 M NaCl	—	10,000	+10	—	—	Arakawa et al., 1987

[a] R_s and R_D: Stokes radii (in Å) from FPLC and from diffusion coefficient, respectively; R_g, radius of gyration from diffuse X-ray and neutron scattering data; $[\eta]$ and η_{sp}/c, intrinsic and reduced viscosity (in $cm^3 \cdot g^{-1}$), respectively; s, sedimentation coefficient in Svedberg units (10^{-13} sec^{-1}); λ_{max} wavelength of the maximum of tryptophan fluorescence spectrum (in nm). Data for native (N) and unfolded (U) states (if available) are shown for comparison. λ_{max} for N-state is usually equal to 325 to 330 nm and for U-state 350 to 355 nm.

[b] $-[\Theta]_{222}$ or $-[\Theta]_{220}$ (in deg.cm^2.dmol^{-1}) unless otherwise stated ([m'] in deg.cm^2.dmol^{-1} or b$_o$ in degrees). Data for the N-state are shown for comparison. "IR" means that infrared spectrum shows pronounced secondary structure.

[c] $[\Theta]_{280}$ (in deg.cm^2.dmol^{-1}) if other (molar absorption ϵ_{290} in 1.mol^{-1}.cm^{-1}) is not stated. Data for N-state are shown for comparison. "NMR" means near (U-like) NMR spectrum. "Proteolysis" means an increase in susceptibility to proteases as compared to the native state. "Fluorescence" means an increase of the intensity of tryptophan fluorescence as compared to the native state.

[d] Molar concentrations of urea or GdmCl corresponding to the midpoint of the transition to the unfolded state.

[e] ANS: 8-anilino-1-naphthalene sulfonate.

[f] In these cases two overlapping urea- or GdmCl-induced transitions have been observed by $[\Theta]_{222}$ and $[\Theta]_{280}$, which indicates the existence of an intermediate state.

[g] For BPTI, the contribution of aromatic side chains in far- UV CD can be very large (Manning and Woody, 1989).

[h] "Native" values are the values for an intact protein (i.e., for a protein with intact disulfide bridges or to a protein of wild-type) at native conditions.

(continued)

established. Therefore, it is fairly certain that these proteins adopt the molten globule state. Table 6–2B includes proteins for which only the presence of secondary structure and the absence of tertiary structure have been established; the data on compactness are either absent or doubtful. These proteins may be assumed to be in the molten globule state, although additional information is necessary to be certain.

Four conclusions can be drawn from this table:

1. About 20 to 25 different proteins have properties similar to that of the molten globule state. This means that the molten globule state is not a rare exception. Rather, it is probably typical for many proteins under mild denaturation conditions.
2. Proteins can be transformed into the molten globule state by low or high pH, by high temperature, by moderate concentrations of GdmCl or urea, and by the influence of $LiClO_4$ and other salts, i.e., by quite different mild denaturation conditions.
3. Proteins can be transformed into the molten globule state without a change of the environment (i.e., almost under physiological conditions) by small alterations of their chemical structure. Some examples include:
   a. Staphylococcal nuclease with 21 C-terminal residues removed (Shortle and Meeker, 1989);
   b. Point mutants of λ repressor (W. A. Lim and R. T. Sauer, personal communication);
   c. Bovine pancreatic trypsin inhibitor with reduced disulfide bridges (Amir and Haas, 1988);
   d. Bovine and human α-lactalbumins after removal of the Ca^{2+} ion (Kuwajima, 1989).
4. Among proteins that possess a pH-induced molten globule state are a number of membrane-binding proteins (colicins, retinol-binding protein, growth hormone, α- and γ-interferons, diptheria toxin). The question arises as to whether the electrostatic field of the membrane (McLaughlin, 1989; Stegmann et al., 1989) can produce similar transitions in these proteins. This is part of the very important question of the possible role of molten globules in a living cell. (See Section 8.)

Table 6–2 footnotes, continued

[i] "Native" values are the values at pH 7.

[j] TID: 3-(trifluoromethyl)-3-(m-[^{125}I] iodophenyl)-diazirine.

[k] Calculated from data for dimeric protein.

[l] η_{sp}/c at acid pH and low ionic strength may be increased relative to the native state due to electrostatic repulsion of ionized molecules.

All data refer to room temperature, unless otherwise stated.

7 PROTEIN FOLDING

7.1 The Molten Globule and Protein Folding

The problem of the kinetics of protein folding, especially of kinetic intermediates in folding pathways, has been extensively reviewed (e.g., Kim and Baldwin, 1982,1990); Creighton, 1978,1990; Fischer and Schmid, 1990) and is the subject of Chapters 5 and 7 of this book. Therefore, this chapter does not give a full description of the problem, but it concentrates on the role of the molten globule state in protein folding and on some related questions.

The molten globule state was first predicted as a *kinetic* intermediate in protein folding pathway. (See Section 4.) That hypothesis of protein folding (Ptitsyn, 1973) suggested that folding starts with the formation of "fluctuating embryos" of regions with secondary structure (stabilized mainly by hydrogen bonds), followed by the collapse of these regions into an "intermediate compact structure" (stabilized mainly by hydrophobic interactions) and completed by the adjustment of this intermediate structure to the unique native structure (driven by van der Waals and other specific interactions). This hypothesis implies that the positions of the embryos of secondary structure in an unfolded chain must be close to their positions in the folded globular state and that the tertiary fold of the polypeptide chain in a compact intermediate must be similar to that in the final native structure. This scheme has subsequently been designated the "framework model" (Kim and Baldwin, 1982).

The first experimental evidence for the existence of a compact kinetic intermediate during protein folding was obtained using urea-gradient gel electrophoresis near $0°C$ (Creighton, 1980; Creighton and Pain, 1980). It has also been shown that the folding of β-lactamase (Robson and Pain, 1976b) and carbonic anhydrase (McCoy et al., 1980) is accompanied by the accumulation of kinetic intermediates that have pronounced secondary structure (monitored by far-UV CD) but no rigid tertiary structure (monitored by near-UV CD). Similar evidence has been obtained using amide hydrogen exchange and ^{3}H labeling for RNase A (Schmid and Baldwin, 1979; Kim and Baldwin, 1980). Finally, Dolgikh et al. (1984) have shown that the early kinetic intermediate of bovine carbonic anhydrase B has viscosity and far-UV ellipticity similar to those of the native state, but has no rigid tertiary structure and no enzyme activity. Therefore, it was concluded that the molten globule state accumulates during protein folding.

All of these studies have not answered, however, the question of the time scale for the formation of the molten globule state and particularly for the time scales of the formation of secondary structure and of the molecular globularization. These questions have been answered only after the development of methods for the rapid monitoring of these processes (Gilmanshin, 1985,1988; Kuwajima et al., 1987; Gilmanshin and Ptitsyn, 1987; Semisotnov et al., 1987,1991a; Ebert et al., 1990).

7.2 Secondary Structure

By combining far-UV CD and the stopped-flow technique, Kuwajima et al. (1987), Gilmanshin and Ptitsyn (1987), Goldberg et al. (1990), and Semisotnov and Kuwajima (unpublished data reviewed by Ptitsyn and Semisotnov, 1991) have shown that the far-UV ellipticity of proteins changes drastically at a *very early stage* of their folding (within the dead time of these experiments, which is approximately 0.01 sec). The negative far-UV ellipticity that is reached after 0.01 sec can be smaller, equal to, or even greater than the native value, but in all cases it is substantially different from the value for the unfolded protein. This suggests the very fast formation of pronounced secondary structure at an early stage of protein folding.

If the far-UV CD ellipticity value reached after 0.01 sec is not equal to the native value, is usually approaches this value relatively slowly, with a rate comparable to the rate of restoration of the rigid tertiary structure monitored by near-UV ellipticity (Kuwajima et al., 1987,1988; Gilmanshin and Ptitsyn, 1987; Goldberg et al., 1990); this is illustrated in Figure 6–8.

The interpretation of these data must take into account that the far-UV CD is very likely influenced by the contribution of aromatic side chains, which can be much greater in the native, rigid state than in nonrigid intermediates. (See Section 3.4.) Therefore, only the change of the far-UV ellipticity that occurs *before* the change in the near-UV can be unambiguously interpreted, while the slow changes can also reflect the increase in the aromatic side chain contribution. Even an absence of slow changes may not indicate that the secondary structure content is constant, as it could be due to changes of secondary structure and the aromatic contribution compensating each other.

7.3 Globularization

To study the next step of protein folding—the collapse of the protein molecule—energy transfer from Trp residues to fluorescent dansyl labels has been used (Semisotnov et al., 1987; Ptitsyn and Semisotnov, 1991). If the Trp residues are close in space to dansyl labels, part of their excitation energy can migrate to the dansyl groups and be emitted as dansyl fluorescence, even if it is the Trp residues that are excited. In proteins with several Trp residues and/or several dansyl labels, energy transfer is averaged over all the Trp-dansyl pairs; it therefore must reflect the overall dimensions of the

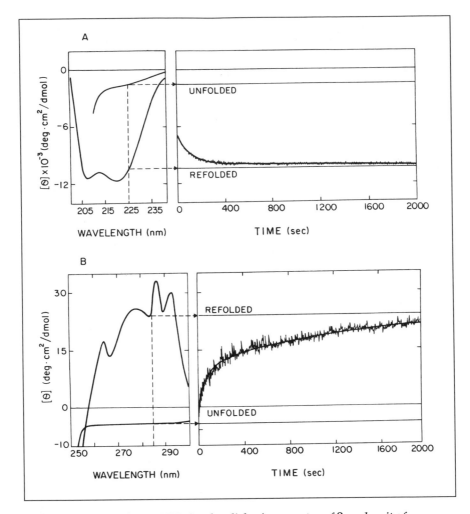

FIGURE 6–8. *Far and near UV circular dichroism spectra of* β_2 *subunit of tryptophan synthase in the unfolded and folded states and time-dependence of molar ellipticities at 225 nm (A) and at 285 nm (B) upon of its refolding from 5.5 to 0.5 M urea at 12°C. (Adapted from Goldberg et al., 1990)*

protein molecule. In addition, the intensities of electron spin resonance signals of labels covalently attached to a protein also can be used to monitor the compactness of a chain (Semisotnov et al., 1987; Ebert et al., 1990), because these intensities depend upon the environment of labels. Special experiments on protein equilibrium unfolding (Rodionova, 1990; see also Ptitsyn and Semisotnov, 1991) have shown that the energy transfer and ESR signals can be used to monitor the kinetics of the molecular collapse of protein folding.

Figure 6–9 illustrates the kinetics of the restoration of the native level of energy transfer, measured by the intensity of dansyl fluorescence at Trp excitation, for bovine carbonic anhydrase B, and for β-lactoglobulin,

FIGURE 6–9. Kinetics of energy transfer increase, monitored by the increase of dansyl fluorescence, during (A) carbonic anhydrase refolding from 8.5 to 4.2 M urea at 23°C and (B) β-lactoglobulin refolding from 4 to 0.4 M GdmCl at 4.5°C. (Adapted from Ptitsyn and Semisotnov, 1991)

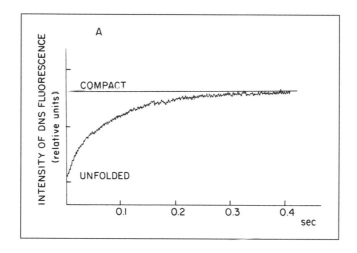

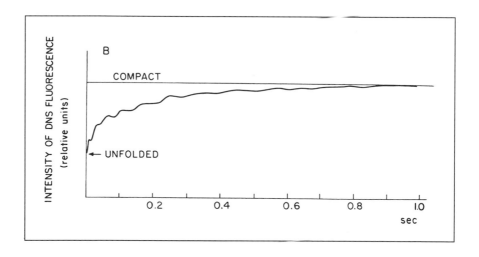

which reflects the collapse of the protein molecules. In both proteins, the collapse occurs much more slowly than does the formation of a substantial part of secondary structure; in both proteins, the CD molar ellipticity in the far-UV region changes drastically within the first 0.01 sec.

As has been mentioned already (Section 4), a sensitive test for the molten globule state is the binding of a hydrophobic probe, such as ANS (Semisotnov et al., 1987,1991a; Rodionova et al., 1989). The kinetics of ANS binding in carbonic anhydrase and in β-lactoglobulin, as measured by the increase in ANS fluorescence, consist of two phases (Semisotnov et al., 1991a; Ptitsyn and Semisotnov, 1991). The first phase occurs within the dead time of the experiment (about 0.01 sec), while the second practically coincides with the kinetics of restoration of energy transfer (Semisotnov et al., 1987; Ptitsyn and Semisotnov, 1991). The first phase may reflect ANS binding to the forming secondary structure and/or the first stage of molecular collapse, which is not sufficient to be monitored by energy transfer (Ptitsyn and Semisotnov, 1991). On the other hand, the coincidence of the second stage of ANS binding with the kinetics of restoration of energy transfer suggests that the globularization of a protein molecule leads to the formation of a nonpolar core (Semisotnov et al., 1987,1991a; Ptitsyn and Semisotnov, 1991). Therefore, the increase in ANS fluorescence can be used to monitor the formation of the molten globule state.

The left part of Figure 6–10 illustrates the increase of ANS binding that occurs upon the folding of six proteins (Ptitsyn et al., 1990; Goldberg et al., 1990). These data show that the globularization of a protein molecule, with the formation of a solvent-accessible nonpolar core, occurs in a time that varies from 0.05 sec (for β-lactamase at 23°C) to a few seconds (for the β_2 subunit of tryptophan synthase; Goldberg et al., 1990). As the time required for ANS binding to the molten globule state is less than 0.002 sec (Semisotnov et al., 1991a), these times reflect the rates of formation during protein folding of a kinetic intermediate with strong affinity for ANS.

7.4 Tertiary Structure

The right part of Figure 6–10 shows the decrease in ANS fluorescence that occurs during folding. This release of ANS from the protein suggests that the non- polar core of the protein is being screened from the solvent. At least in one case, for carbonic anhydrase (Semisotnov et al., 1987), it was shown that high field ^{1}H-NMR resonances appear simultaneously with ANS desorption. This suggests that the screening of the nonpolar core occurs simultaneously with its tight packing. The restoration of near-UV CD ellipticity and of protein activity in a number of proteins also occurs simultaneously with ANS desorption (Goldberg et al., 1990; Semisotnov and Kuwajima, unpublished data). In some cases, however, as with carbonic anhydrase (Semisotnov et

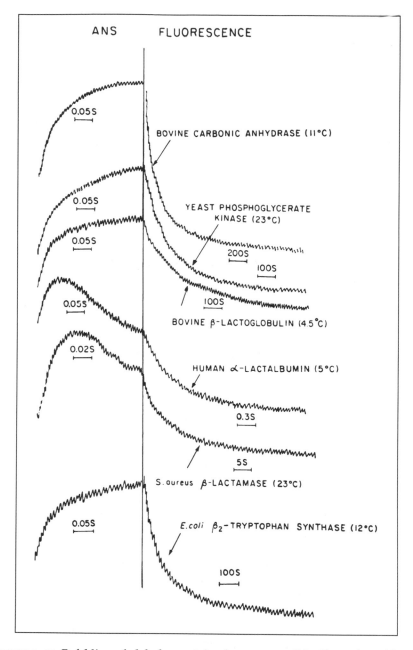

FIGURE 6–10. *Refolding of globular proteins from urea or GdmCl monitored by ANS fluorescence. Different time scales are used for the left and the right parts of the figure. The final concentration of denaturant was from 1.95 to 4.0 M urea or from 0.07 to 0.7 M GdmCl.*

al., 1987), the restoration of near-UV ellipticity and of enzymatic activity occurs more slowly than does ANS desorption and the appearance of high field ^{1}H-NMR resonance; this suggests that the solidification of a protein core can precede formation of the full native tertiary structure.

Slow phases of protein folding, which can be as slow as 500 to 2,500 sec at 25°C, are often due to *cis,trans* isomerization of peptide bonds preceding Pro residues (Kim and Baldwin, 1982,1990; Chapter 5) as described by the proposal of Brandts et al. (1975) which was extended by Semisotnov et al. (1990). This is not always the case, however; in horse (Betton et al., 1985), pig, and yeast (Semisotnov et al., 1991b) phosphoglycerate kinases, slow refolding stages do not depend upon the time of preincubation of the protein in the unfolded state and therefore are not due to Pro peptide bond isomerization.

7.5   The Molten Globule as a General Kinetic Intermediate

All proteins studies thus far have early kinetic intermediates with strong affinity for ANS, i.e., with a solvent-accessible nonpolar core. The compactness of these intermediates has been confirmed for some proteins by more direct methods: urea gradient electrophoresis, energy transfer, and viscosity measurements. (See above.) All of these kinetic intermediates have pronounced secondary structure, as indicated by their far-UV CD ellipticities (Kuwajima et al., 1987; Goldberg et al., 1990; Ptitsyn and Semisotnov, 1991).

On the other hand, these kinetic intermediates have no rigid tertiary structures. For example, ^{1}H-NMR spectra of kinetic intermediates of carbonic anhydrase (Semisotnov et al., 1987; Ptitsyn and Semisotnov, 1991) and of β-lactamase (Ptitsyn et al., 1990) are much simpler than are the spectra of the native proteins and are similar to the spectra of thermodynamically stable molten globules. Near-UV CD ellipticities of these kinetic intermediates are virtually absent (Robson and Pain, 1976a,b; McCoy et al., 1980; Dolgikh et al., 1984; Kuwajima et al., 1987,1988; Gilmanshin and Ptitsyn, 1987; Goldberg et al., 1990; Ptitsyn and Semisotnov, 1991). In addition, these intermediates are totally inactive enzymatically (Ikai et al., 1978; McCoy et al., 1980; Dolgikh et al., 1984; Mitchinson and Pain, 1985; Semisotnov et al., 1987; Ptitsyn and Semisotnov, 1991).

All of these data taken together leave little doubt that *a protein molecule while folding passes through an early kinetic intermediate that is very similar to the equilibrium molten globule state.*

The list of proteins for which these kinetic intermediates have been observed includes carbonic anhydrase, α-lactalbumin, β-lactoglobulin, β-lactamase, phosphoglycerate kinase (Ptitsyn et al., 1990; Ptitsyn and Semisotnov, 1991), and the β$_2$ subunit of tryptophan synthase (Goldberg et al.,

1990). In addition, strong evidence for a molten globule kinetic intermediate has been obtained by Roder et al. (1988) for cytochrome *c* and by Matouschek et al. (1989,1990) and Bycroft et al. (1990) for barnase (Section 7.6). These proteins belong to different structural types and include both single- and multidomain proteins, as well as proteins with and without disulfide bridges. Therefore, we can conclude that *the molten globule state is a general intermediate in protein folding* (Ptitsyn et al., 1990; Ptitsyn and Semisotnov, 1991).

In addition, there are a number of proteins including RNase A (Schmid and Baldwin, 1979; Kim and Baldwin, 1980), lysozyme (Kuwajima et al., 1985; Ikeguchi et al., 1986), the α subunit of tryptophan synthase (Beasty and Matthews, 1985), growth hormone (Brems et al., 1987) and parvalbumin (Kuwajima et al., 1988) for which it has been shown that their early kinetic intermediates have secondary structure but do not have rigid tertiary structures although the compactness of these intermediates has not been demonstrated.

7.6 Native-like Features of Structural Organization

Now we know (Ptitsyn et al., 1990; Ptitsyn and Semisotnov, 1991) that protein folding includes at least three main stages:

1. Formation of a substantial part of the secondary structure ($<10^{-2}$ sec);
2. Formation of a compact (molten globule) intermediate (≤ 1 sec);
3. Formation of a rigid tertiary structure (1 to 10^3 sec).

These three stages coincide with the predictions made by Ptitsyn (1973) and occur in the folding of all proteins studied by his group. Sometimes, the last stage can split into two or more stages—the formation of the rigid structure of a protein core can precede the formation of the full tertiary structure described previously. Table 6–3 illustrates these three stages for carbonic anhydrase.

These main stages of protein folding, monitored by methods sensitive to the structure of a molecule as a whole, give a "time frame" for the consideration of other, more specific events that can occur during protein folding and reflect the formation of some native-like features of structural organization.

A powerful method of looking for these specific events is to use hydrogen exchange to label protected NH groups of kinetic intermediates, with subsequent identification of the protected groups in the refolded protein by 2D-NMR (Udgaonkar and Baldwin, 1988; Roder et al., 1988). These studies identify NH groups that have been protected from exchange, and therefore were involved in some structure, at different stages of folding.

TABLE 6–3. Folding Stages of Bovine Carbonic Anhydrase B (4.2 M urea, 23°C)

Formation of	Techniques	Half-time $t_{1/2}$ (sec)[a]	References
Secondary structure	Ellipticity at 222 nm[b]	$\ll 0.01$	Semisotnov and Kuwajima, unpublished results
Molten globule (compact state)	Energy transfer	0.04	Semisotnov et al., 1987
	Spin-label immobilization	0.03	Semisotnov et al., 1987; Ebert et al., 1990
Molten globule (solvent-accessible nonpolar core)	ANS binding	0.04	Semisotnov et al., 1987,1991a
Screened and tightly packed nonpolar core	ANS desorption	140	Semisotnov et al., 1987,1991a
	High-field NMR	140	
Native tertiary structure	Ellipticity at 270 nm	580	Semisotnov et al., 1987
	Esterase activity	530	

[a] For monoexponential process with rate constant k, the half-time is given by $t_{1/2} = (\ln 2)/k$.
[b] Ellipticity at 222 nm can also be influenced by the contribution of aromatic side chains.

Interpretation of these data is based on the assumption that "most highly protected amide protons in native proteins are both hydrogen-bonded and inaccessible to solvent" (Udgaonkar and Baldwin, 1988). A good illustration is that in the molten globule state only NH groups involved in α-helices are protected but their protection is much weaker than in the native state. (See Section 3.5.) This suggests that both the presence of secondary structure and its screening from solvent are important for substantial protection from exchange.

Using this method, it has been shown that the NH groups of residues that are part of β-sheets and of the main α-helices in the native conformations of RNase A (Udgaonkar and Baldwin, 1988,1990) and of barnase (Bycroft et al., 1990) are protected at early stages of folding, while NH groups involved in the final tertiary structure are protected later. These data suggest that the secondary structure that is formed at the beginning of the folding process is present in the same positions in the polypeptide chain as in the native conformation, in accordance with the framework model (Udgaonkar and Baldwin, 1988).

Even more interesting results have been obtained for cytochrome *c* (Roder et al., 1988), where the NH groups of the N- and C-terminal α-helices became protected very early in folding, with a half-time of approximately 10^{-2} sec. That these α-helices are protected simultaneously (Roder et al., 1988) and that their protection follows the same pH-dependence

(Roder, private communication) suggests that protection reflects their docking, rather than their independent formation. This is extremely important evidence that the basic feature of the tertiary fold of this simple protein is formed at a very early stage of its folding pathway. It is especially interesting that the rate of this packing is close to the rate of the first stage of fluorescence quenching of the single Trp residue, number 59, as a result of its proximity to the heme group, which suggests that "association of chain termini may be accompanied by a general condensation of the initially unfolded chain" (Roder et al., 1988). In this case also, NH groups involved in the tertiary structure are protected later.

It is worthwhile to mention that the NH groups of some elements of secondary structure, for example, the α-helices of residues 60 to 69 and 71 to 75 in cytochrome c (Roder et al., 1988) and the N-terminal α-helix in RNase A (Udgaonkar and Baldwin, 1990), become protected only at the same time as restoration of the tertiary structure, and much later than the other elements of secondary structure.

The protein for which the general time frame of folding has been most carefully compared with specific events is the β_2 subunit of tryptophan synthase, which has been extensively studied using energy transfer (Blond and Goldberg, 1986) and monoclonal antibodies specific for the native protein (Blond and Goldberg, 1987; Murray-Brelier and Goldberg, 1988; Blond-Elguindi and Goldberg, 1990) to monitor specific events of protein folding, such as the approach of two pairs of residues and the formation of an antigenic determinant. These specific events have been compared to the restoration of the overall characteristics of protein structure (Goldberg et al., 1990). The results presented in Table 6–4 show that an antigenic determinant is formed within the molten globule state, before formation of the rigid tertiary structure, while the close interaction of Trp177 with Lys87, which are well separated along the polypeptide chain, occurs simultaneously with the first stage in formation of the rigid tertiary structure.

A new interesting approach to the study of the role of various interactions in protein folding has been proposed by Matouschek et al. (1989,1990). By changing different residues by site-specific mutagenesis, they have determined the contribution of these residues to the free energies of three stages of protein folding: a kinetic intermediate (before the free energy barrier), the transition state (at the barrier), and the fully folded state (after the barrier). Applying this approach to barnase, and combining it with NMR data, they came to the following conclusions:

1. All three α-helices and all β-strands of barnase are already formed in the early kinetic intermediate;
2. The C-terminus of the first α-helix (residues 6 to 18) also is stabilized in this intermediate;

TABLE 6-4. *Folding Stages of β₂ Subunit of E.coli Tryptophan Synthase (0.5 M urea; 12°C)*

		Time Interval									
		<0.01 sec		0–5 sec		0–50 sec		0–200 sec		0–3500 sec	
Formation of:	Technique	$t_{1/2}$ (sec)[a]	A(%)[b]	$t_{1/2}$ (sec)	A(%)	$t_{1/2}$ (sec)	A(%)	$t_{1/2}$ (sec)	A(%)	$t_{1/2}$ (sec)	A(%)	References
Secondary structure	Ellipticity at 225 nm[c]	<0.007	57	—	—	5	7	70	29	700	7	Goldberg et al., 1990
Molten globule (solvent-accessible nonpolar core)	ANS binding	—	—	0.03 0.2 1	40 28 32	—	—	—	—	—	—	Goldberg et al., 1990
Native-like antigenic determinant	Monoclonal antibody binding	—	—	—	—	12	100	—	—	—	—	Murry-Brelier and Goldberg, 1988; Blond-Elguindi and Goldberg, 1990
Closing Trp 177 with Lys 87	Energy transfer	—	—	—	—	—	—	35	100	—	—	Blond and Goldberg, 1986
Closing Trp 177 with Cys 170	Energy transfer	—	—	—	—	—	—	90	100	—	—	Blond and Goldberg, 1986
Screened nonpolar core	ANS desorption	—	—	—	—	—	—	35	30	700	70	Goldberg et al., 1990
Native tertiary structure	Ellipticity at 285 nm	—	—	—	—	—	—	35	43	700	57	Goldberg et al., 1990

[a] Half-time of the process.
[b] Amplitude of the process.
[c] Slow change of ellipticity at 225 nm may be determined or at least influenced by contribution of aromatic side chains.

3. The hydrophobic core of barnase is present in the early intermediate, becomes more condensed in the transition state, and reaches its final tight packing in the folded state;
4. The N-termini of the first (residues 6 to 18) and the second (residues 26 to 34) α-helices become stabilized only in the final folded state.

These data are consistent with the kinetic intermediate in barnase being the molten globule state and with the packing of nonpolar groups in the protein core increasing in the sequence molten globule → transition state → folded state, as predicted by the theory of protein denaturation of Shakhnovich and Finkelstein (1989). They contain much additional information, as they present perhaps a detailed picture of what happens at different stages of protein folding.

These data indicate that the native features of protein structure may form at different stages of protein folding. Some features are formed simultaneously with the molten globule, such as the docking of the N- and C-terminal helices in cytochrome *c*, others are formed within the molten globule state such as the antigenic determinant in β₂ subunit of tryptophan synthase, and some (such as the N-termini of α-helices in barnase) occur only in the final native state. Many more experiments are needed to clarify this complicated picture.

8 POSSIBLE ROLES OF THE MOLTEN GLOBULE

Is the molten globule state described in this chapter only a curious feature of proteins *in vitro* or is it involved with proteins in a living cell? Trying to answer this question, it must be remembered that a dilute solution of one protein in a "native" buffer is far from being the ideal model for proteins in a cell. In fact, proteins in a living cell can be subjected to different mild denaturing influences, such as high temperatures, low pH, membranes, and so forth. Therefore, it is not meaningless to consider the possible roles of "slightly denatured" proteins, including the molten globule state, in a living cell.

1. The typical time for biosynthesis of a polypeptide chain is approximately 10 to 100 sec. It is greater than the time for formation of the molten globule state (about 1 sec), but much less than the time for maturation of rigid native structure in some proteins, which can be as long as 10^3 sec. (See Section 7.4.) Therefore, a nascent polypeptide chain during and immediately after its biosynthesis may be in the molten globule state. Cells may even want to keep a nascent protein in this state, as it is sufficiently flexible to adjust to subsequent events,

such as oligomerization, transmembrane transport, and so forth. In fact, it is known that:

a. Heat-shock GroEL protein binds and stabilizes a non-native state of a nascent polypeptide chain (Bochkareva et al., 1988; Rothman, 1989);

b. Some heat-shock proteins bind inactive subunits of oligomeric proteins to prevent them from nonspecific aggregation and/or to transport them to the place of their assembly (Pelham, 1986; Ellis, 1987,1990; Schlesinger, 1990);

c. Some heat-shock proteins bind secretory and imported proteins to prevent them from premature tight folding or aggregation (Deshaies et al., 1988) and are involved in transmembrane protein translocation (Goloubinoff et al., 1989; Scherer et al., 1990; Lazdunski and Benedetti, 1990).

Therefore, it is reasonable to assume that *some heat-shock proteins, particularly GroEL, recognize and bind a nascent protein in the molten globule state* (Bychkova et al., 1988). Recently, this assumption has been confirmed by Semisotnov et al. (1992), who have shown that GroEL retards protein folding, both when it is added at the time when protein is completely unfolded and when it is added 1 to 2 sec later, when the protein is already folded into the molten globule state.

2. There is numerous evidence (Vestweber and Schatz, 1988; Vestweber et al., 1989; Lazdunski and Benedetti, 1990) that proteins cannot be translocated through membranes in the native state, but are competent for translocation when in a non-native state in which they are susceptible to proteolysis. This has given rise to the idea (Bychkova et al., 1988) that the molten globule state might be involved in protein translocation through membranes. It is confirmed by the evidence that proteins can bind to membranes or micelles at low pH (e.g., Zhao and London, 1986; Kim and Kim, 1986), when at least some of them might be in the denatured, molten globule state. It is possible that the interactions of proteins with the membrane surface trigger their transition to the molten globule state, in which they can easily adapt, perhaps by partial unfolding, to various environments, such as water, membrane surface, and the inner part of a membrane. There is evidence (Rassow et al., 1990; Neupert et al., 1990) that the part of the protein chain that is transversing the contact site of mitochondrial membranes is in an extended conformation.

3. It is possible that protein degradation in lysosomes, at acid pH, or by ATP-dependent proteosomes, as in the ubiquitin-dependent system, can be facilitated by prior denaturation, which

may be the transition to the molten globule state. It is interesting that a heat-shock protein is involved in lysosomal protein degradation (Chiang et al., 1989).

Although very little is known about non-native protein states in a living cell, the possibility that they might play a role in cell processes is intriguing.

9 CONCLUSION

In summarizing this chapter, it should be emphasized that there is now no doubt that the molten globule state actually exists as a separate equilibrium state of protein molecules, that it occurs frequently in denatured proteins, and that it plays an important role in protein folding. The main question that has to be answered is to what extent the equilibrium molten globule state possesses the native tertiary fold of the corresponding protein and at what stage in the formation or maturation of the molten globule–like kinetic intermediate it achieves the native tertiary fold. A very intriguing question, of course, is the possible role of the molten globule state in protein life in a living cell.

Acknowledgments

Sections 6 and 8 of this chapter were written together with Dr. V. E. Bychkova, who also prepared Table 6–2. Drs. A. V. Finkelstein and G. V. Semisotnov carefully read the manuscript and made many valuable comments. My wife, I. G. Ptitsyna, and M. S. Shelestova have rendered important assistance in preparing the manuscript. I am extremely grateful to all of them for their generous help.

Note

Recently two important papers have been published that directly confirmed the suggestion (Bychkova et al., 1988) that the molten globule may be physiologically important. J. Martin et al. (1991, *Nature 352*, 36–42) have shown that dihydrofolate reductase and rhodanese being in the complexes with GroEL are in the molten globule state. They came to the conclusion that "a molten globule-like pre-folded state is a physiological, early intermediate of chaperonin-mediate folding." F. G. van der Goot et al. (1991, *Nature 354*, 408–410) have shown that pH-dependence of the membrane insertion of the pore-forming domain of colicin A correlates with its transition into the molten globule state. This paper documents a local pH decrease near negatively charged membrane surface (see p. 282), which has been measured and taken into account. Thus the idea of the physiological role of the molten globule state may be meaningful.

REFERENCES

Acharya, K. R., Stuart, D. I., Walker, N. P. C., Lewis, M., and Phillips, D. C. (1989) *J. Mol. Biol. 208*, 99–127.

Ahmad, F., and Bigelow, C. C. (1979) *J. Mol. Biol. 131*, 607–617.

Amir, D., and Haas, E. (1988) *Biochemistry 27*, 8889–8893.

Ananthanarayanan, V. S., and Ahmad, F. (1977) *Can. J. Biochem. 55*, 239–243.

Ananthanarayanan, V. S., Ahmad, F., and Bigelow, C. C. (1977) *Biochim. Biophys. Acta 492*, 194–203.

Anufrieva, E. V. (1982) *Pure Appl. Chem. 54*, 533–584.

Anufrieva, E. V., Bychkova, V. E., Krakovyak, M. G., Pautov, V. D., and Ptitsyn, O. B. (1975) *FEBS Letters 55*, 46–49.

Anufrieva, E. V., and Krakovyak, M. G. (1987) *Vysokomol. Soed. (USSR) A29*, 211–222.

Anufrieva, E. V., Volkenstein, M. V., Gotlib, Yu. Ya., Krakovyak, M. G., Pautov, V. D., Stepanov, V. V., and Skorokhodov, S. S. (1972) *Dokl. Akad. Nauk SSSR 207*, 1379–1382.

Arakawa, T., Hsu, Y. -R., Yphantis, D. A. (1987) *Biochemistry 26*, 5428–5432.

Aune, K. C., Salahuddin, A., Zarlengo, M. H., and Tanford, C. (1967) *J. Biol. Chem. 242*, 4486–4489.

Baldwin, R. L. (1986) *Proc. Natl. Acad. Sci. USA 83*, 8069–8072.

Baldwin, R. L. (1990) *Nature 346*, 409–410.

Baldwin, R. L. (1991) in CIBA Foundation Symposium 161, *"Protein conformation,"* London, January 22–24, 1991.

Baum, J., Dobson, C. M., Evans, P. A., and Hanly, C. (1989) *Biochemistry 28*, 7–13.

Beasty, A. M., and Matthews, C. R. (1985) *Biochemistry 24*, 3547–3553.

Bendzko, P. I., Pfeil, W. A., Privalov, P. L., and Tiktopulo, E. I. (1988) *Biophys. Chem. 29*, 301–307.

Betton, J.-M., Desmadril, M., Mitraki, A., and Yon, J. M. (1985) *Biochemistry 24*, 4570–4577.

Bewley, T. A., Levine, H. L., and Wetzel, R. (1982) *Int. J. Peptide Protein Res. 20*, 93–96.

Birshtein, T. M., and Pryamitsyn, V. A. (1987) *Vysokomol. Soed. (USSR) A29*, 1858–1864.

Birshtein, T. M., and Pryamitsyn, V. A. (1991) *Macromolecules 24*, 1554–1560.

Blond, S., and Goldberg, M. E. (1986) *Proteins: Struct. Funct. Genet. 1*, 247–255.

Blond, S., and Goldberg, M. E. (1987) *Proc. Natl. Acad. Sci. USA 84*, 1147–1151.

Blond-Elguindi, S., and Goldberg, M. E. (1990) *Biochemistry 29*, 2409–2417.

Bochkareva, E. S., Lissin, N. M., and Girshovich, A. S. (1988) *Nature 336*, 254–257.

Bolotina, I. A. (1987) *Mol. Biol. (USSR) 21*, 1625–1635.

Bolotina, I. A., and Lugauskas, V. Yu. (1985) *Mol. Biol. (USSR) 19*, 1409–1421.

Brahms, S., and Brahms, J. (1980) *J. Mol. Biol. 138*, 149–178.

Brandts, J. F., Halvorson, H. R., and Brennan, M. (1975) *Biochemistry 14*, 4953–4963.

Brandts, J. F., and Hunt, L. (1967) *J. Am. Chem. Soc. 89*, 4826–4838.

Brazhnikov, E. V., Chirgadze, Yu. N., Dolgikh, D. A., and Ptitsyn, O. B. (1985) *Biopolymers 24*, 1899–1907.

Brems, D. N., and Havel, H. A. (1989) *Proteins: Struct. Funct. Genet. 5*, 93–95.

Brems, D. N., Plaisted, S. M., Dougherty, J. J. Jr., and Holzman, T. F. (1987) *J. Biol. Chem. 262*, 2590–2596.

Brems, D. N., Plaisted, S. M., Havel, H. A., Kauffman, E. W., Stodola, J. D., Eaton, L. C., and White, R. D. (1985) *Biochemistry 24*, 7662–7668.

Burger, H. G., Edelhoch, H., and Condliffe, P. G. (1966) *J. Biol. Chem. 241*, 449–457.

Bushnell, G. W., Louie, G. V., and Brayer, G. D. (1990) *J. Mol. Biol. 214*, 585–595.

Bychkova, V. E., Bartoshevich, S. F., and Klenin, S. I. (1990) *Biofizika (USSR) 35*, 242–248.

Bychkova, V. E., Berni, R., Rossi, G. L., Kutyshenko, V. P., and Ptitsyn, O. B. (1992) *Biochemistry*, in press.

Bychkova, V. E., Pain, R. H., and Ptitsyn, O. B. (1988) *FEBS Letters 238*, 231–234.

Bychkova, V. E., and Ptitsyn, O. B. (1992) *Biofizika (USSR)*, in press.

Bychkova, V. E., Semisotnov, G. V., Ptitsyn, O. B., Gudkova, O. V., Mitin, Yu. V., and Anufrieva, E. V. (1980) *Mol. Biol. (USSR) 14*, 278–286.

Bycroft, M., Matouschek, A., Kellis, J. T. Jr., Serrano, L., and Fersht, A. R. (1990) *Nature 346*, 488–490.

Cabiaux, V., Brasseur, R., Wattiez, R., Falmagne, P., Ruysschaert, J. -M., and Goormaghtigh, E. (1989) *J. Biol. Chem. 264*, 4928–4938.

Cavard, D., Sauve, P., Heitz, F., Pattus, F., Martinez, C., Dijkman, R., and Lazdunski, C. (1988) *Eur. J. Biochem. 172*, 507–512.

Cave, A., Pages, M., Morin, P., and Dobson, C. M. (1979) *Biochimie 61*, 607–613.

Chan, H. S., and Dill, K. A. (1991) *Ann. Rev. Biophys. Biophys. Chem. 20*, 447–490.

Chiang, H.-L., Terlecky, S. R., Plant, C. P., and Dice, J. F. (1989) *Science 246*, 382–385.

Chirgadze, Yu. N., Fedorov, O. V., and Trushina, N. P. (1975) *Biopolymers 14*, 679–694.

Creighton, T. E. (1978) *Progr. Biophys. Mol. Biol. 33*, 231–297.

Creighton, T. E. (1980) *J. Mol. Biol. 137*, 61–80.

Creighton, T. E. (1990) *Biochem. J. 270*, 1–16.

Creighton, T. E., and Pain, R. H. (1980) *J. Mol. Biol. 137*, 431–436.

Damaschun, G., Damaschun, H., Gast, K., Zizwer, D., and Bychkova, V. E. (1991) *Int. J. Biol. Macromol. 13*, 217–221.

Damaschun, G., Gernat, C., Damaschun, H., Bychkova, V. E., and Ptitsyn, O. B. (1986) *Int. J. Biol. Macromol. 8*, 226–230.

De Gennes, P.-G. (1975) *J. Phys. Letters 36*, 55–57.

Denton, J. B., Konishi, Y., and Scheraga, H. A. (1982) *Biochemistry 21*, 5155–5163.

Deshaies, R. J., Koch, B. D., Werner-Washburne, M., Craig, E. A., and Schekman, R. (1988) *Nature 332*, 800–805.

Desmadril, M., and Yon, J. M. (1981) *Biochem. Biophys. Res. Commun. 101*, 563–569.

Dill, K. A. (1985) *Biochemistry 24*, 1501–1509.

Dill, K. A., Alonso, D. O. U., and Hutchinson, K. (1989) *Biochemistry 28*, 5439–5449.

Dill, K. A., and Shortle, D. (1991) *Ann. Rev. Biochem. 60*, 795–825.

Dobson, C. M., Hanley, C., Radford, S. E., Baum, J. A., and Evans, P. A. (1991) in *Conformations and Forces in Protein Folding* (B. T. Nall and K. A. Dill, eds.), AAAS, Washington, D.C., pp. 175–181.

Dolgikh, D. A., Abaturov, L. V., Bolotina, I. A., Brazhnikov, E. V., Bychkova, V. E., Bushuev, V. N., Gilmanshin, R. I., Lebedev, Yu. O., Semisotnov, G. V., Tiktopulo, E. I., and Ptitsyn, O. B. (1985) *Eur. Biophys. J. 13*, 109–121.

Dolgikh, D. A., Abaturov, L. V., Brazhnikov, E. V., Lebedev, Yu. O., Chirgadze, Yu. N., and Ptitsyn, O. B. (1983) *Dokl. Akad. Nauk SSSR 272*, 1481–1484.

Dolgikh, D. A., Gilmanshin, R. I., Brazhnikov, E. V., Bychkova, V. E., Semisotnov, G. V., Venyaminov, S. Yu., and Ptitsyn, O. B. (1981) *FEBS Letters 136*, 311–315.

Dolgikh, D. A., Kolomiets, A. P., Bolotina, I. A., and Ptitsyn, O. B. (1984) *FEBS Letters* *165*, 88–92.

Dumont, M. E., and Richards, F. M. (1988) *J. Biol. Chem. 263*, 2087–2097.

Ebert, B., Semisotnov, G. V., and Rodionova, N. A. (1990) *Studia Biophys. 137*, 125–131.

Echols, G. H., and Anderegg, J. W. (1960) *J. Am. Chem. Soc. 82*, 5085–5093.

Ellis, R. J. (1987) *Nature 328*, 378–379.

Ellis, R. J. (1990) *Science 250*, 954–959.

Fedorov, B. A., and Ptitsyn, O. B. (1977) *Dokl. Akad. Nauk SSSR 233*, 716–718.

Fink, A. L., Calciano, L. J., Goto, Y., and Palleros, D. R. (1990) in *Current Research in Protein Chemistry* (J. Villafranca, ed.), Academic Press, New York, pp. 417–424.

Fink, A. L., Calciano, L. J., Goto, Y., and Palleros, D. R. (1991) in *Conformations and Forces in Protein Folding* (B. T. Nall and K. A. Dill, eds.), AAAS, Washington, D. C., pp. 169–174.

Finkelstein, A. V., Badretdinov, A. Ya., and Ptitsyn, O. B. (1991) *Proteins: Struct. Funct. Genet. 10*, 287–299.

Finkelstein, A. V., and Ptitsyn, O. B. (1987) *Progr. Biophys. Mol. Biol. 50*, 171–190.

Finkelstein, A. V., and Shakhnovich, E. I. (1989) *Biopolymers 28*, 1681–1694.

Fischer, G., and Schmid, F. X. (1990) *Biochemistry 29*, 2205–2212.

Flory, P. J. (1953) *Principles of Polymer Chemistry*, Cornell University Press, Ithaca, New York.

Foster, J. F. (1960) in *Plasma Proteins*, vol. 1, Academic Press, New York, pp. 179–239.

Gast, K., Zirwer, D., Welfle, H., Bychkova, V. E., and Ptitsyn, O. B. (1986) *Int. J. Biol. Macromol. 8*, 231–236.

Gernat, C., Damaschun, G., Kröber, R., Bychkova, V. E., and Ptitsyn, O. B. (1986) *Studia Biophys. 112*, 213–219.

Gilmanshin, R. I. (1985) *Biofizika (USSR) 30*, 581–587.

Gilmanshin, R. I. (1988) *Biofizika (USSR) 33*, 27–30.

Gilmanshin, R. I., Dolgikh, D. A., Ptitsyn, O. B., Finkelstein, A. V., and Shakhnovich, E. I. (1982) *Biofizika (USSR) 27*, 1005–1016.

Gilmanshin, R. I., and Ptitsyn, O. B. (1987) *FEBS Letters 223*, 327–329.

Goldberg, M. E., Semisotnov, G. V., Friguet, B., Kuwajima, K., Ptitsyn, O. B., and Sugai, S. (1990) *FEBS Letters 263*, 51–56.

Goloubinoff, P., Christeller, J. T., Gatenby, A. A., and Lorimer, G. H. (1989) *Nature 342*, 884–889.

Goto, Y., Calciano, L. J., and Fink, A. L. (1990a) *Proc. Natl. Acad. Sci. USA 87*, 573–577.

Goto, Y., and Fink, A. L. (1989) *Biochemistry 28*, 945–952.

Goto, Y., and Fink, A. L. (1990) *J. Mol. Biol. 214*, 803–805.

Goto, Y., Takahashi, N., and Fink, A. L. (1990b) *Biochemistry 29*, 3480–3488.

Grigoryev, A. I., Volkova, L. A., and Ptitsyn, O. B. (1971) *FEBS Letters 15*, 217–219.

Griko, Yu. V., Privalov, P. L., Venyaminov, S. Yu., and Kutyshenko, V. P. (1988) *J. Mol. Biol. 202*, 127–138.

Grosberg, A. Yu. (1984) *Biofizika (USSR) 29*, 569–573.

Grosberg, A. Yu., and Shakhnovich, E. I. (1986a) *Zh. Exp. Theor. Fiz. (USSR) 91*, 2159–2170.

Grosberg, A. Yu., and Shakhnovich, E. I. (1986b) *Biofizika (USSR) 31*, 1054–1057.

Harding, M. M., Williams, D. H., and Woolfson, D. N. (1991) *Biochemistry 30*, 3120–3128.

Herold, M., and Kirschner, K. (1990) *Biochemistry 29*, 1907–1913.

Holladay, A., Hammonds, R. G., Jr., and Puett, D. (1974) *Biochemistry 13*, 1653–1661.

Holzman, T. F., Brems, D. N., and Dougherty, J. J., Jr. (1986) *Biochemistry 25*, 6907–6917.

Holzman, T. F., Dougherty, J. J., Jr., Brems, D. N., and MacKenzie, N. E. (1990) *Biochemistry 29*, 1255–1261.

Horowitz, P. M., and Criscimagna, N. L. (1990) *J. Biol. Chem. 265*, 2576–2583.

Hughson, F. M., and Baldwin, R. L. (1989) *Biochemistry 28*, 4415–4422.

Hughson, F. M., Barrik, D., and Baldwin, R. L. (1991) *Biochemistry 30*, 4113–4118.

Hughson, F. M., Wright, P. E., and Baldwin, R. L. (1990) *Science 249*, 1544–1548.

Ikai, A., Tanaka, S., and Noda, H. (1978) *Arch. Biochem. Biophys. 190*, 39–45.

Ikeguchi, M., Kuwajima, K., Mitani, M., and Sugai, S. (1986) *Biochemistry 25*, 6965–6972.

Izumi, Y., Miyake, Y., Kuwajima, K., Sugai, S., Inoue, K., Iizumi, M., and Katano, S. (1983) *Physica 120 B*, 444–448.

Jagannadham, M. V., and Balasubramanian, D. (1985) *FEBS Letters 188*, 326–330.

Jeng, M. F., Englander, S. W., Elöve, G. A., Wang, A. J., and Roder, H. (1990) *Biochemistry 29*, 10433–10437.

Kim, J., and Kim, H. (1986) *Biochemistry 25*, 7867–7874.

Kim, P. S., and Baldwin, R. L. (1980) *Biochemistry 19*, 6124–6129.

Kim, P. S., and Baldwin, R. L. (1982) *Ann. Rev. Biochem. 51*, 459–489.

Kim, P. S., and Baldwin, R. L. (1990) *Ann. Rev. Biochem. 59*, 631–660.

Kosen, P. A., Creighton, T. E., and Blout, E. R. (1981) *Biochemistry 20*, 5744–5754.

Kosen, P. A., Creighton, T. E., and Blout, E. R. (1983) *Biochemistry 22*, 2433–2440.

Kron, A. K., Ptitsyn, O. B., Skvortsov, A. M., and Fedorov, A. K. (1967) *Mol. Biol. (USSR) 1*, 576–582.

Kronman, M. J., Blum, R., and Holmes, L. G. (1966) *Biochemistry 5*, 1970–1978.

Kuwajima, K. (1977) *J. Mol. Biol. 114*, 241–258.

Kuwajima, K. (1989) *Proteins: Struct. Funct. Genet. 6*, 87–103.

Kuwajima, K., Harushima, Y., and Sugai, S. (1986) *Int. J. Peptide Protein Res. 27*, 18–27.

Kuwajima, K., Hiraoka, Y., Ikeguchi, M., and Sugai, S. (1985) *Biochemistry 24*, 874–881.

Kuwajima, K., Nitta, K., Yoneyama, M., and Sugai, S. (1976) *J. Mol. Biol. 106*, 359–373.

Kuwajima, K., Sakuraoka, A., Fueki, S., Yoneyama, M., and Sugai, S. (1988) *Biochemistry 27*, 7419–7428.

Kuwajima, K., Yamaya, H., Miwa, S., Sugai, S., and Nagamura, T. (1987) *FEBS Letters 221*, 115–118.

Lazdunski, C. J., and Benedetti, H. (1990) *FEBS Letters 268*, 408–414.

Lee, J. W., and Kim, H. (1988) *FEBS Letters 241*, 181–184.

Lifshitz, I. M., Grosberg, A. Yu., and Khokhlov, A. R. (1978) *Rev. Modern Phys. 50*, 683–713.

Lumry, R. T., Biltonen, R., and Brandts, J. F. (1966) *Biopolymers 4*, 917–944.

Manning, M. C., and Woody, R. W. (1989) *Biochemistry 28*, 8609–8613.

Matouschek, A., Kellis, J. T., Jr., Serrano, L., Bycroft, M., and Fersht, A. R. (1990) *Nature 346*, 440–445.

Matouschek, A., Kellis, J. T., Jr., Serrano, L., and Fersht, A. R. (1989) *Nature 340*, 122–126.

McCoy, L. F., Rowe, E. S., and Wong, K. -P. (1980) *Biochemistry 19*, 4738–4743.

McLaughlin, S. (1989) *Ann. Rev. Biophys. Biophys. Chem. 18*, 113–136.

Merrill, A. R., Cohen, F. S., and Cramer, W. A. (1990) *Biochemistry 29*, 5829–5836.

Mitchinson, C., and Pain, R. H. (1985) *J. Mol. Biol. 184*, 331–342.

Murry-Brelier, A., and Goldberg, M. E. (1988) *Biochemistry 27*, 7633–7640.

Neupert, W., Harte, F.-U., Craig, E. A., and Pfanner, N. (1990) *Cell 63*, 447–450.

Nozaka, M., Kuwajima, K., Nitta, K., and Sugai, S. (1978) *Biochemistry 17*, 3753–3758.

Ohgushi, M., and Wada, A. (1983) *FEBS Letters 164*, 21–24.

Pelham, H. R. B. (1986) *Cell 46*, 959–961.

Permyakov, E. A., Grishchenko, V. M., Kalinichenko, L. P., Orlov, N. Y., Kuwajima, K., and Sugai, S. (1991) *Biophys. Chem. 39*, 111–117.

Perutz, M. F., Kendrew, J. C., and Watson, H. C. (1965) *J. Mol. Biol. 13*, 669–678.

Pfeil, W., Bychkova, V. E., and Ptitsyn, O. B. (1986) *FEBS Letters 198*, 287–291.

Potekhin, S. A., and Pfeil, W. (1989) *Biophys. Chem. 34*, 55–62.

Privalov, P. L. (1974) *FEBS Letters 40*, S140–S153.

Privalov, P. L. (1979) *Adv. Protein Chem. 33*, 167–241.

Privalov, P. L. (1982) *Adv. Protein Chem. 35*, 1–104.

Privalov, P. L., and Khechinashvili, N. N. (1974) *J. Mol. Biol. 86*, 665–684.

Privalov, P. L., and Makhatadze, G. I. (1990) *J. Mol. Biol. 213*, 385–391.

Privalov, P. L., Plotnikov, V. V., and Filimonov, V. V. (1975) *J. Chem. Thermodyn. 7*, 41–47.

Ptitsyn, O. B. (1973) *Dokl. Akad. Nauk SSSR 210*, 1213–1215.

Ptitsyn, O. B. (1975) *Dokl. Akad. Nauk SSSR 223*, 1253–1255.

Ptitsyn, O. B. (1985) *Suppl. J. Biosci.* (Proc. Int. Symp. Biomol. Struct. Interactions) *8*, 1–13.

Ptitsyn, O. B. (1987) *J. Protein Chem. 6*, 273–293.

Ptitsyn, O. B., Damaschun, G., Gernat, C., Damaschun, H., and Bychkova, V. E. (1986) *Studia Biophys. 112*, 207–211.

Ptitsyn, O. B., Dolgikh, D. A., Gilmanshin, R. I., Shakhnovich, E. I., and Finkelstein, A. V. (1983) *Mol. Biol. (USSR) 17*, 569–576.

Ptitsyn, O. B., and Eizner, Yu. Ye. (1965) *Biofizika (USSR) 10*, 3–6.

Ptitsyn, O. B., and Finkelstein, A. V. (1980a) *Quart. Rev. Biophys. 13*, 339–386.

Ptitsyn, O. B., and Finkelstein, A. V. (1980b) in *Protein Folding* (R. Jaenicke, ed.), Elsevier/North Holland Biomedical Press, Amsterdam-New York, pp. 101–115.

Ptitsyn, O. B., Kron, A. K., and Eizner, Yu. Ye. (1968) *J. Polymer Sci. Pt C 16*, 3509–3517.

Ptitsyn, O. B., Pain, R. H., Semisotnov, G. V., Zerovnik, E., and Razgulyaev, O. I. (1990) *FEBS Letters 262*, 20–24.

Ptitsyn, O. B., Reva, B. A., and Finkelstein, A. V. (1989) *Highlights of Modern Biochemistry* (Proc. 14th Intern. Congress of Biochemistry, Prague, Czechoslovakia, July 10–15, 1988), vol. 1 (A. Kotyk, J. Škoda, V. Pačes, and V. Kostka, eds.), VSP, Utrecht-Tokyo, pp. 11–17.

Ptitsyn, O. B., and Semisotnov, G. V. (1991) in *Conformations and Forces in Protein Folding* (B. T. Nall and K. A. Dill, eds.), AAAS, Washington, D. C., pp. 155–168.

Rassow, J., Hartl, F.-U., Guiard, B., Pfanner, N., and Neupert, W. (1990) *FEBS Letters 275*, 190–194.

Robbins, F. M., and Holmes, L. G. (1970) *Biochim. Biophys. Acta 221*, 234–240.

Robson, B., and Pain, R. H. (1976a) *Biochem. J. 155*, 325–330.

Robson, B., and Pain, R. H. (1976b) *Biochem. J. 155*, 331–334.

Roder, H., Elöve, G. A., and Englander, S. W. (1988) *Nature 335*, 700–704.

Rodionova, N. A. (1990) The study of compactization of globular proteins. Ph.D. thesis, Moscow Physico-Technical Institute.

Rodionova, N. A., Semisotnov, G. V., Kutyshenko, V. P., Uversky, V. N., Bolotina, I. A., Bychkova, V. E., and Ptitsyn, O. B. (1989) *Mol. Biol. (USSR) 23*, 683–692.

Rothman, J. E. (1989) *Cell 59*, 591–601.

Sanchez, I. C. (1979) *Macromolecules 12*, 980–988.

Scherer, P. E., Krieg, U. C., Hwang, S. T., Vestweber, D., and Schatz, G. (1990) *EMBO J. 9*, 4315–4322.

Schlesinger, M. J. (1990) *J. Biol. Chem. 265*, 12111–12114.

Schmid, F. X., and Baldwin, R. L. (1979) *J. Mol. Biol. 135*, 199–215.

Sears, D. W., and Beychok, S. (1973) in *Circular Dichroism in Physical Properties and Techniques of Protein Chemistry*, part C (S. J. Leach, ed.), Academic Press, New York, pp. 445–593.

Semisotnov, G. V., Kutyshenko, V. P., and Ptitsyn, O. B. (1989) *Mol. Biol. (USSR) 23*, 808–815.

Semisotnov, G. V., Rodionova, N. A., Kutyshenko, V. P., Ebert, B., Blank, J., and Ptitsyn, O. B. (1987) *FEBS Letters 224*, 9–13.

Semisotnov, G. V., Rodionova, N. A., Razgulyaev, O. I., Uversky, V. N., Gripas, A. F., and Gilmanshin, R. I. (1991a) *Biopolymers 31*, 119–128.

Semisotnov, G. V., Sokolovsky, I. V., Bochkareva, E. S., and Girshovich, A. S. (1992), in preparation.

Semisotnov, G. V., Uversky, V. N., Sokolovsky, I. V., Gutin, A. M., Razgulyaev, O. I., and Rodionova, N. A. (1990) *J. Mol. Biol. 213*, 561–568.

Semisotnov, G. V., Vas, M., Chemeris, V. V., Kashparova, N. J., Kotova, N. V., Razgulyaev, O. I., and Sinev, M. A. (1991b) *Eur. J. Biochem.*, in press.

Semisotnov, G. V., Zikherman, K. Kh., Kasatkin, S. B., Ptitsyn, O. B., and Anufrieva, E. V. (1981) *Biopolymers 20*, 2287–2309.

Shakhnovich, E. I., and Finkelstein, A. V. (1982) *Dokl. Akad. Nauk SSSR 267*, 1247–1250.

Shakhnovich, E. I., and Finkelstein, A. V. (1989) *Biopolymers 28*, 1667–1680.

Shakhnovich, E. I., and Gutin, A. M. (1990) *Nature 346*, 773–775.

Shortle, D., and Meeker, A. K. (1989) *Biochemistry 28*, 936–944.

Spolar, R. S., Ha, J.-H., and Record, M. T., Jr., (1989) *Proc. Natl. Acad. Sci. USA 86*, 8382–8385.

Stegmann, T., Doms, R. W., and Helenius, A. (1989) *Ann. Rev. Biophys. Biophys. Chem. 18*, 187–211.

Stryer, L. (1965) *J. Mol. Biol. 13*, 482–495.

Sun, S. T., Nishio, I., Swislow, G., and Tanaka, T. (1980) *J. Chem. Phys. 73*, 5971–5975.

Tandon, S., and Horowitz, P. M. (1989) *J. Biol. Chem. 264*, 9859–9866.

Tanford, C. (1968) *Adv. Protein Chem. 23*, 121–282.

Timchenko, A. A., Dolgikh, D. A., Damaschun, H., and Damaschun, G. (1986) *Studia Biophys. 112*, 201–206.

Udgaonkar, J. B., and Baldwin, R. L. (1988) *Nature 335*, 694–699.

Udgaonkar, J. B., and Baldwin, R. L. (1990) *Proc. Natl. Acad. Sci. USA 87*, 8197–8201.

Uversky, V. N., Semisotnov, G. V., Pain, R. H., and Ptitsyn, O. B. (1992), in preparation.

Venyaminov, S. Yu., and Kalnin, N. N. (1991) *Biopolymers*, in press.

Vestweber, D., Brunner, J., Baker, A., and Schatz, G. (1989) *Nature 341*, 205–209.

Vestweber, D., and Schatz, G. (1988) *J. Cell Biol.* 107, 2037–2043.

Vonderviszt, F., Lakatos, S., Gál, P., Sárvári, M., and Zavodszki, P. (1987) *Biochem. Biophys. Res. Commun.* 148, 92–98.

Wong, K.-P., and Hamlin, L. M. (1974) *Biochemistry* 13, 2678–2683.

Wong, K.-P., and Tanford, C. (1973) *J. Biol. Chem.* 248, 8518–8523.

Zhao, J.-M., and London, E. (1986) *Proc. Natl. Acad. Sci. USA* 83, 2002–2006.

Zhao, J.-M., and London, E. (1988) *J. Biol. Chem.* 263, 15369–15377.

—7—

Folding Pathways Determined Using Disulfide Bonds

THOMAS E. CREIGHTON

1 INTRODUCTION

The complexity of protein structures, described in Chapter 2, and the cooperativity of protein folding transitions, described in Chapter 3, are the major factors that have hampered experimental elucidation of protein folding pathways. The partially folded conformations of the type that would define the folding pathway of a small single-domain protein are inherently unstable thermodynamically, relative to the fully folded or unfolded states. Consequently, such partially folded conformations are not populated significantly at equilibrium. They might accumulate transiently as kinetic intermediates during unfolding or refolding (Chapter 5), but only if they precede the rate-limiting step on the pathway and if, under the particular folding conditions, their free energy is lower than or similar to that of the starting protein. None of the other intermediate states that must occur along a complex folding pathway will accumulate substantially, and they will not be detectable directly. Those intermediates that do accumulate kinetically are likely to be very transient and elusive; moreover, their kinetic roles in the folding process can be very difficult to ascertain.

The greatest exceptions are the molten globule states that are stable in certain proteins under certain conditions and are described in Chapter 6. They appear to correspond to an intermediate state in refolding of a number of proteins, but the relevance of the equilibrium state that can be studied in detail needs to be confirmed by comparison with the kinetic state of the protein during folding.

The difficulties of identifying the intermediates in protein folding transitions could be alleviated if it were possible to control the stability of any type of interactions within the protein. This would make it possible to manipulate the strength of the interaction, to trap intermediate conformations with intermediate numbers of such interactions, and to determine which interactions were present in the trapped intermediates. It is easy to imagine how a protein folding pathway could be determined if intramolecular protein hydrogen bond formation, for example, could be manipulated in this way. If the fully unfolded protein contained no intramolecular hydrogen bonds and if the rate of forming each hydrogen bond in the refolding protein could be made to be rate-limiting and constant, protein molecules with 1, 2, 3, 4, ... internal hydrogen bonds would accumulate transiently to significant levels during refolding. If protein hydrogen bond formation, breakage, and rearrangement could be quenched rapidly at any time, the intermediates with various numbers and identities of hydrogen bonds would be trapped. These trapped intermediates could then be identified in terms of their hydrogen bonds. Their kinetic roles in the folding transition would need to be determined, but this would be feasible if the rates of hydrogen bond formation and breakage could be manipulated.

No such method is available for hydrogen bonds or for van der Waals and electrostatic interactions, but it is possible with another type of interaction that is present in some proteins, namely the disulfide bond that can be formed between the thiol groups of Cys residues (Creighton, 1978, 1984,1986):

$$2 -SH \rightarrow -S-S- + 2H^+ + 2e^-. \tag{7-1}$$

The reduction/oxidation nature of this interaction means that its formation or breakage requires the presence of an appropriate electron acceptor or donor, respectively, which is under experimental control. Consequently, the rates of protein disulfide bond formation and breakage can be controlled experimentally, as can the intrinsic stability of the disulfide bond interaction. As a result, all disulfide bonds present in a protein at any time may be trapped in a stable form, and the disulfides may be identified chemically (Fig. 7–1). The presence of a disulfide bond between two Cys residues requires that they be within a few angstroms of each other, so disulfides serve as probes of the conformational properties of proteins. The disulfide bonds present in the structures of folded proteins have been studied extensively (Thornton, 1981; Srinivasan et al., 1990), and no evidence has been found to indicate that the disulfide interaction between two Cys residues to form a disulfide bond is intrinsically different from the other interactions that stabilize proteins. The most unique aspects of disulfides are the strict stereochemical requirements for forming the covalent disulfide bond and the ability of the disulfide to be retained in the unfolded state.

Using disulfide bond formation to study protein folding necessitates that the protein folded conformation require one or more disulfides for its stability. The reduced protein must be unfolded under the conditions used

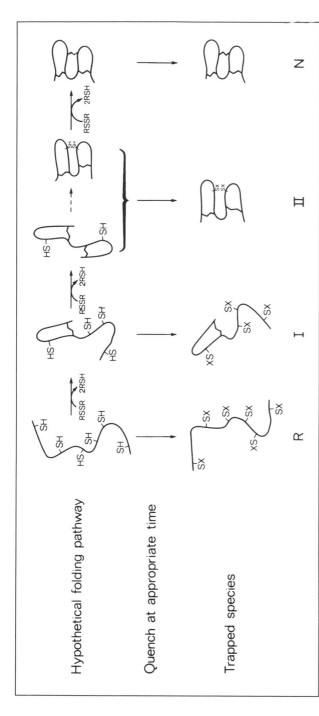

FIGURE 7-1. *Illustration of the rationale of the experimental approach to using disulfide bonds to trap intermediates in protein folding. A hypothetical folding pathway for a protein with six cysteine residues is illustrated at the top, using an intermolecular disulfide reagent, RSSR, to form the disulfide bonds; the mixed-disulfide intermediate involved in each step is not depicted. The reduced protein must be unfolded under the conditions used, but the intermediates are expected to adopt nonrandom conformations if certain disulfide bonds predominate. An intramolecular conformation change of possible kinetic significance is illustrated in the two-disulfide intermediate; this will be trapped in whichever conformation is the more stable. The quenching reaction is one that blocks each thiol group with a moiety "X"; this can be a proton or irreversible adducts introduced by alkylation with reagents such as iodoacetate or iodoacetamide. (Creighton, 1978)*

for folding and in the absence of any denaturant. Reduced proteins without disulfides tend to be unfolded because they lack the stabilizing influence of the disulfide bonds, but this depends upon the intrinsic stability of the folded conformation and upon the stabilizing contributions of the disulfide bonds. Disulfide bonds stabilize protein folded conformations by decreasing the conformational entropy of the unfolded state (Pace et al., 1988), but they also contribute directly to the folded conformation (Goldenberg, 1985; Creighton, 1988a). It is also unfavorable energetically to bury two Cys -SH groups in the interior of a protein designed for just the two sulfur atoms of a disulfide bond, especially if the thiol groups are ionized. The intrinsic pK_a values of Cys thiol groups are generally in the region of 8 to 9, so it is best to carry out disulfide folding studies at alkaline pH, where the ionized thiols of free Cys residues tend to remain accessible to solvent and to keep the protein with incomplete disulfides unfolded. In this case, folding is linked to disulfide bond formation; no denaturants of uncertain action are required, the equilibrium between folded and unfolded conformations can be altered simply by changing only the thiol/disulfide intrinsic stability, and the ways that the different redox conditions affect the rates of disulfide bond formation and breakage are known. Therefore, the kinetics of unfolding and refolding can be determined unambiguously under all unfolding and refolding conditions.

Proteins that adopt folded conformations in the absence of disulfides are not suitable for these studies, for disulfide bond formation is then a matter primarily of the accessibilities and reactivities of protein thiol groups within the folded conformation. On the other hand, much useful information can be gained in such circumstances about the stabilities of the folded state with and without the disulfides (Goto and Hamaguchi, 1981).

2 DISULFIDE BOND FORMATION AS A PROBE OF PROTEIN CONFORMATIONAL TRANSITIONS

2.1 Making and Breaking Protein Disulfide Bonds

The rates of forming or breaking protein disulfide bonds are pertinent for protein folding only if they reflect the protein conformational transitions that occur, not simply the reduction oxidation chemistry that must be involved. A great number of different ways of forming and breaking protein disulfides are known, but those found most suitable for studies of protein folding use thiol-disulfide exchange between the protein and a small molecule reagent in either the thiol or disulfide form (Fig. 7–1). The intrinsic reaction between a thiol and a disulfide

$$R_1\text{-SH} + R_2\text{-S-S-}R_2 \overset{k_{ex}}{\longleftrightarrow} R_2\text{-SH} + R_1\text{-S-S-}R_2 \qquad (7\text{-}2)$$

is rapid, reversible, extremely specific, free of side reactions, and predictable in rate (Creighton, 1975b; Shaked et al., 1980; Szajewski and Whitesides, 1980; Gilbert, 1990). The thiol must actually be in the ionized -S⁻ form, but the reaction will be depicted here with protonated thiols. Equilibrium constants of reactions such as Equation (7–2) give the relative stabilities and free energies of the different disulfide bonds. The relative simplicity of this chemical reaction makes it possible to interpret observed rates of disulfide bond formation and breakage in terms of the protein conformational transitions involved. Such interpretations clearly are not warranted when the disulfide formed is highly constrained, due to the small size of the covalent loop formed by it (Zhang and Snyder, 1989; Burns and Whitesides, 1990), but such strained systems are not expected to be encountered in protein folding reactions, where the folded conformation stabilized by a disulfide bond is only slightly stable. Experience with disulfide bond formation and breakage during protein folding, to be described here, and in studying conformational equilibria in small peptides (e.g., Lin and Kim, 1989; Altmann and Scheraga, 1990) thus far indicates that interpretations of the equilibria and rates of disulfide bond formation and breakage are relatively straightforward.

Protein disulfides are formed or broken in two sequential thiol-disulfide exchange reactions with the reagent, passing through an intermediate mixed disulfide between the protein and the reagent (Fig. 7–2). Both steps are reversible, so there are four rate constants involved in making and breaking each protein disulfide bond. Two fundamentally different types of reagent are useful. In one, the disulfide links two molecules that are separated when the disulfide is reduced, such as in oxidized glutathione (GSSG) and in mercaptoethanol disulfide; for that reason, this type of disulfide is considered to be "intermolecular." In the other type, the disulfide is intramolecular between two thiols on the same molecule, as in oxidized dithiothreitol (DTT_S^S):

$$\begin{array}{c} \nearrow CH_2 \searrow \\ HOCH \qquad\quad S \\ | \qqu\qquad\quad | \\ HCOH \qquad\quad S \\ \searrow \qquad\quad \nearrow \\ CH_2 \end{array} \qquad (7\text{-}3)$$

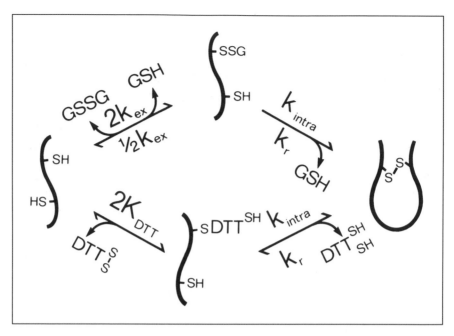

FIGURE 7–2. *Protein disulfide bond formation and breakage by thiol-disulfide interchange with reagents that form either intermolecular (top) or intramolecular (bottom) disulfide bonds; the reagents illustrated are glutathione and dithiothreitol. The polypeptide chain of the protein with two Cys residues is indicated by the solid line. The two Cys thiol groups are assumed to be accessible and normally reactive with either reagent, and there are assumed to be no other interactions between the protein and the reagents. Consequently, the values of* k_{intra} *and* k_r *are assumed to be identical with both reagents. The rate constant* k_{ex} *is that for thiol-disulfide interchange between normal thiol and disulfide groups; the factors of 2 and 1/2 are statistical factors correcting for the numbers of thiol and disulfides. The equilibrium constant* K_{DTT} *is that for formation of the mixed disulfide between a Cys thiol group and dithiothreitol.*

The process of forming or breaking a protein disulfide is exactly analogous with both types of reagents (Fig. 7–2), but there are important differences that result in them providing complementary kinetic information.

2.1.1 Intermolecular Disulfide Reagents

With an intermolecular disulfide reagent, RSSR, three of the four rate constants are simply of bimolecular thiol-disulfide exchange reactions of the reagent with the protein thiols or disulfide (Fig. 7–2). The rates of each of these reactions depend upon the concentration of either the thiol or disulfide form of the reagent. With normal, accessible protein thiol groups, these reactions occur with the rate of the intrinsic thiol-disulfide exchange reaction, k_{ex} (Equation (7–2)), which may be measured with model compounds (Creighton, 1975b, 1980b). The value of this rate constant depends somewhat upon the electrostatic properties of both the thiol and disulfide molecules, but with typical Cys residues of proteins and of glutathione, the value of k_{ex} is 10 sec^{-1}M^{-1} at pH 8.7 and 25 °C. The observed value of k_{ex} changes with temperature, with an Arrhenius activation energy of 12 kcal/ mole, and with pH, reflecting the fraction of thiol groups that are ionized.

The values of the individual rate constants can be determined only when the mixed-disulfide intermediate accumulates to detectable levels, and then the possibility of both Cys residues existing as the mixed disulfide must be taken into account:

$$(7\text{–}4)$$

The forward and reverse rate constants for forming each mixed disulfide from the reduced protein and the disulfide reagent are frequently found to have their expected values, as would be expected if the reduced protein were fully unfolded. The rate constant of breaking a normal protein disulfide by the thiol reagent, k_r, also might have the value of k_{ex}, but it may differ if the protein disulfide is strained or inaccessible due to the protein conformation. The value of k_r may be determined most readily by the rate of reduction of the protein disulfide by dithiothreitol. (See below.)

The intramolecular step in which a second Cys thiol displaces the mixed disulfide to form the protein disulfide bond is the step that is relevant for protein folding, and all rates of disulfide formation mentioned here will refer to this rate. Its rate constant, k_{intra}, is the only one that involves protein conformational transitions that bring the two Cys residues of the protein into proximity to form the disulfide bond. If this step is fast, as is usually the case in the absence of conformational restrictions, the rate of forming the protein disulfide bond with a reagent RSSR is determined by the initial bimolecular reaction between a protein thiol and the disulfide reagent; the mixed disulfide does not accumulate significantly. The rate, therefore, reflects only the thiol-disulfide exchange reaction and the number of Cys thiols involved. This rate is not of interest for folding, unless it is low due to inaccessibility or unreactivity of the protein thiols. Its value, however, generally decreases upon forming successive disulfide bonds in a protein, as a result of the decreasing number of Cys thiol groups, which has the advantage of ensuring significant accumulation of intermediates of the protein with all possible numbers of disulfide bonds.

In the case where k_{intra} is large, different disulfide bonds are distinguished by their rates of breakage by the thiol reagent, because initial breakage of the protein disulfide by RSH to generate the mixed disulfide is rapidly reversed (Fig. 7–2). Complete reduction of the protein disulfide requires that the competing reaction of the mixed disulfide with a second molecule of RSH be sufficiently fast. Consequently, the rate of complete reduction of a protein disulfide generally depends upon the square of the thiol reagent concentration. The observed third-order rate constant for breaking the protein disulfide, k_{obs}, can be used to calculate the apparent value of k_{intra}:

$$k_{intra} = 1/2 \, k_{ex} \, k_r \, / \, k_{obs} \tag{7-5}$$

assuming the standard value for k_{ex} and the measured value for k_r. The more stable the protein disulfide, the slower its rate of breakage by monothiols.

The other extreme, where the value of k_{intra} is very small, generally occurs only if there are conformational restrictions preventing Cys residues from forming disulfide bonds, or if the Cys residues are very far apart in an unfolded polypeptide chain. In this case, the mixed disulfide accumulates to detectable levels, and the value of k_{intra} can be measured directly.

Whenever a protein disulfide is not formed during protein folding, the question arises as to whether this is due to unreactivity of the protein thiols or due to the conformation keeping the Cys residues apart. If inaccessibility or unreactivity of the protein thiols is the reason, formation of the mixed disulfide will be rate limiting, and the mixed disulfide will not accumulate. On the other hand, if the thiols are reactive but the conformation keeps them from forming a disulfide bond, they will accumulate as the mixed disulfide. Accumulation of the mixed disulfide is positive evidence that nonformation of a disulfide bond is due to conformational restrictions, not simply to unreactivity of the thiols (Creighton, 1981).

Protein disulfides are intramolecular, whereas that of the reagent RSSR is intermolecular. The equilibrium constant for their interconversion, K,

$$\text{(7–6)}$$

$$K = \frac{\left[P\,^S_S \right]\left[RSH \right]^2}{\left[P\,^{SH}_{SH} \right]\left[RSSR \right]} \tag{7–7}$$

compares the free energies of their disulfide bonds. As with any comparison of otherwise identical inter- and intramolecular reactions (Kirby, 1980), this gives the effective concentration of the pair of Cys thiols in the protein. This parameter reflects the tendency of the protein conformation to keep the Cys thiols in proximity in the absence of the disulfide bond, plus any conformational changes after formation of the disulfide bond. With unfolded polypeptides, effective concentrations between pairs of Cys residues of roughly 0.05 M are measured, depending upon the distance between the Cys residues in the polypeptide chain. In folded proteins, however, values of between 100 and 500,000 M have been measured (Creighton and Goldenberg, 1984). These large values are a result primarily of the smaller entropy lost in making an intramolecular interaction than in an intermolecular interaction between two independent molecules (Page and Jencks, 1971), although any energetically unfavorable aspects of the protein without the disulfide bond, such as burying Cys thiol groups in an unfavorable environment in a folded protein, will also contribute to increase the measured effective concentration.

2.1.2 Intramolecular Disulfide Reagents

Dithiothreitol (DTT^{SH}_{SH}) forms a stable, intramolecular disulfide bond because of the high effective concentration of its two thiol groups, which has been measured to be 1.3×10^4 M (Cleland, 1964) or 8.8×10^3 M (Szajewski and Whitesides, 1980) at pH 7 and 1.2×10^3 M at pH 8.7 (Creighton and Goldenberg, 1984) (see note on pp. 348–349). This causes mixed disulfides involving dithiothreitol to be very unstable, for the disulfide of DTT^{S}_{S} is formed intramolecularly with a rate constant of approximately 10^4 to 10^5 sec^{-1} under normal conditions, and the mixed disulfide never accumulates substantially. As a result, protein disulfide formation using DTT^{S}_{S} as a reagent is relatively slow, but the observed rate is proportional to the value of k_{intra}. Because of the unfavorable equilibrium for forming the mixed disulfide between a Cys thiol group and dithiothreitol, with a value K_{DTT}, the observed rate constant is given by

$$k_{obs} = 2\, K_{DTT}\, k_{intra}. \tag{7-8}$$

Knowing K_{DTT} and measuring k_{obs}, the value of k_{intra} may be calculated. In this way, large values of k_{intra} of up to 10^5 sec^{-1} may be measured. The limit is set by the requirement that k_{intra} for the protein disulfide be no greater than that for closing the ring of DTT$_{S}^{S}$.

    With DTT$_{S}^{S}$ as disulfide reagent, there is no requirement for protein Cys residues to form disulfides, either intramolecular or with the reagent. Only favorable intramolecular protein disulfides are formed at significant rates.

    The rate of reducing a protein disulfide with DTT$_{SH}^{SH}$ is almost always determined by the initial exchange, i.e., k_r, because the second step is the very rapid closure of the ring of DTT$_{S}^{S}$. The value of k_r is close to k_{ex} for normal, accessible disulfides, but is lower for those that are inaccessible to the reagent. Likewise, strained disulfides, like that of lipoic acid (Creighton, 1975b), are reduced more rapidly. The rate of reduction of a protein disulfide by DTT$_{SH}^{SH}$ gives little or no information about the disulfide's stability and rate of formation, which are determined by the value of k_{intra}, but it does depend upon the conformational properties of the protein with the disulfide bond.

2.1.3 Disulfide Rearrangements
Disulfides can rearrange spontaneously and intramolecularly within polypeptide chains containing both disulfides and free Cys thiol groups:

$$\tag{7-9}$$

$$\tag{7-10}$$

In this case, the equilibria of the rearrangements will depend upon the relative free energies of the different isomers, and the rate will depend upon the ability of a free Cys residue of the protein to come into appropriate proximity of the disulfide bond.

Whether or not protein disulfide bonds rearrange when given the opportunity is a very powerful method for analyzing protein stability. For example, native proteins retain their disulfide bond pairings under conditions where their native conformation is stable, but will rearrange their disulfides under unfolding conditions (Creighton, 1977c). In contrast, the disulfide bonds of insulin (Givol et al., 1965) and of the S-protein of RNase A (Kato and Anfinsen, 1969) were shown to rearrange under physiological conditions, indicating that the native disulfide pairings and the protein conformation were metastable in these cases. The native disulfide bonds of RNase S-protein (residues 21 to 124) are only stable in the presence of the S-peptide (residues 1 to 20), to form RNase S. The instability of the disulfides of insulin suggested that this protein must have been synthesized as a precursor in which the native disulfides and conformation were more stable (Givol et al., 1965). This precursor was subsequently found to be proinsulin.

2.1.4 The Molten Globule State of α-lactalbumin

A similar approach has been used to characterize the molten globule state of α-lactalbumin, which has four disulfide bonds and binds a single Ca^{2+} ion. A three-disulfide form of α-lactalbumin, in which the disulfide between Cys6 and Cys120 had been reduced, was found to maintain the native-like conformation when Ca^{2+} was bound (Ewbank and Creighton, 1991). Having three disulfide bonds and two thiol groups, this species has the potential for intramolecular disulfide bond rearrangements, but the three native disulfides were found to be stable so long as Ca^{2+} was bound. When the Ca^{2+} was removed, however, the protein adopted the molten globule conformation, and its disulfide bonds spontaneously rearranged intramolecularly to a large number of alternative pairings. The rearranged molecules tended to maintain the molten globule conformation; the rearranged disulfides were between Cys residues relatively far apart in the polypeptide chain, and those that tend to keep the protein molecule compact. The molten globule state of this protein is compatible with a large number of disulfide pairings.

The disulfide bonds of this molten globule state rearranged at the same rate as when the protein was totally unfolded in 8-M urea. The rearrangement products in 8-M urea were different, however, in that they tended to maximize the overall dimensions of the polypeptide chain, presumably so as to maximize its conformational entropy. The rapidity of the disulfide rearrangements in the molten globule state indicate that this conformational state is much closer to an unfolded, but collapsed, form of the protein than to have an expanded form of the native conformation that tends to maintain the native topology. (Cf. Chapter 6.)

2.2 Trapping Protein Disulfides

Whatever disulfides are present at any instant of time may be trapped in a stable form by simply blocking, rapidly and irreversibly, all thiol groups in the solution (Fig. 7–1). In the absence of other nucleophiles, disulfide bonds may no longer be made, broken, or interchanged. The quenching reaction should be as rapid as possible to ensure that it faithfully traps the species present at the time of addition of the quenching reagent. Trapping by reaction with reagents like iodoacetamide or iodoacetic acid has the advantage of being irreversible, but the reaction with a free thiol group has a half-time of 0.3 seconds with 0.2 M iodoacetate at pH 8.7 and 25 °C. Consequently, disulfide rearrangements could occur during the trapping reaction. The other difficulty with this trapping procedure is that buried thiol groups do not react at a sufficient rate. For this reason, a two-disulfide species of bovine pancreatic trypsin inhibitor (BPTI) was overlooked initially, because it adopted a stable native-like conformation in which two thiols were buried and unreactive (States et al., 1984). The two thiols could be reacted under more drastic conditions, but considerable rearrangements of the two disulfides of the protein occurred before the trapping reaction was complete. This is an inherent difficulty with this experimental approach. Nevertheless, once all free thiol groups are blocked irreversibly, the trapped species are stable indefinitely. Consequently, they may be separated and characterized in detail.

Thiol-disulfide interchange requires the thiolate anion, so acidification will also quench disulfide bond formation, breakage, and rearrangement. Acidification has the advantage that it is very rapid and is not prevented by steric accessibility, but it has the disadvantage of not being irreversible. At pH 2, thiol-disulfide interchange will still occur at 10^{-6} its rate at pH 8.7. This may be adequate for many purposes, but intramolecular disulfide rearrangements occur at rates of up to 10^5 sec^{-1} at pH 8.7, so they could occur with a half-time as short as 7 seconds at pH 2. There is also a very real possibility that any separation method used to separate the acidified species, such as their adsorption to reversed-phase HPLC columns at elevated temperatures, will induce such rearrangements. Nevertheless, acidification has been found adequate to trap the unstable 3-disulfide form of α-lactalbumin (see p. 311) for analysis by reversed phase chromatography in 0.1% trifluoroacetic acid (Ewbank and Creighton, 1991). Acid-trapped species have the advantage that they can be isolated and then returned to the usual folding conditions, to define their kinetic roles more directly.

Only the disulfide bonds are trapped by the preceding procedures, but the conformation of the protein can also be effectively trapped, due to linkage between the stabilities of the disulfide bonds and the protein conformation that favored them (Fig. 7–3). Whatever conformation stabilizes,

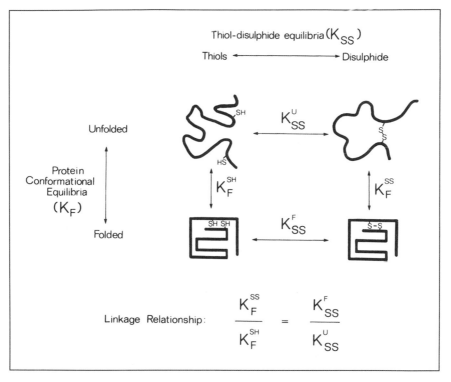

FIGURE 7–3. *Linkage between the stabilities of protein disulfides and of conformations stabilizing them. An unfolded polypeptide chain is depicted at the top, with and without the disulfide bond. The folded conformation that stabilizes the disulfide bond, and is stabilized by it, is shown at the bottom. The equilibrium constants for the conformational equilibrium and for disulfide bond formation, involving an unspecified thiol and disulfide reagent, are indicated. The linkage relationship between them shown at the very bottom results from the thermodynamic necessity that the free-energy change around such a cycle be zero. The linkage relationship states that whatever nonrandom conformation that favors formation of a specific disulfide bond must be favored to the same extent by the presence of that disulfide bond, as in the trapped intermediates. Comparable linkage relationships apply to all interactions within the folded conformation, not just disulfide bonds. (Creighton, 1986)*

a particular disulfide bond will be stabilized to the same extent by the presence of that disulfide bond, as in the trapped intermediate. Therefore, trapped intermediates with one or more disulfide bonds should tend to adopt the conformation that favored formation of those disulfides. This assumes that the trapping reaction does not alter the protein conformation and that the conformational fluctuations of the protein are rapid relative to the rates of disulfide bond formation. If the protein slowly adopts a more stable conformation after trapping, its conformational properties in the trapped form will not account for its behavior during folding. Also, the rates of forming and breaking the disulfide bond during the folding transition will not correspond to the observed equilibrium constant for that step. No such instances have been observed, except when conformational transitions of the protein are slowed by an intrinsically slow step, such as the *cis,trans* isomerization of peptide bonds preceding Pro residues (Chapter 5). This phenomenon greatly complicates the kinetic analysis of fast protein folding reactions, but it need not be rate determining if the rates of disulfide bond formation are made slow. In the other examples to be described next, the same transition state seems to be encountered in both directions under the usual experimental conditions, for the ratio of the measured rate constants agrees with the observed equilibrium constant. This implies that all of the conformational transitions are fast relative to the rate of disulfide bond formation.

2.3 Separating and Characterizing the Trapped Intermediates

Once trapped irreversibly by alkylation with iodoacetamide or iodoacetate, the protein species with different disulfide bonds are stable indefinitely and may be separated and characterized by any methods that do not permit disulfide interchange. This excludes alkaline conditions, due to transient alkaline hydrolysis of disulfide bonds, and the presence of other nucleophiles that can break disulfides, even only transiently. Acid-trapped intermediates must be kept under acidic conditions and only for relatively short periods of time.

Trapping by alkylation with a charged reagent, such as iodoacetate, has the advantage that the charged group introduces an additional heterogeneity into protein molecules with different numbers of disulfide bonds. In the case of iodoacetate, they will differ in the number of acidic carboxymethyl groups introduced and in their net charges. Iodoacetamide, in contrast, introduces very similar carboxamidomethyl groups, but without the charge.

Separation of the trapped species can take place with very many techniques, but electrophoresis in polyacrylamide gels is especially informative, for it separates on the basis of both net charge and conformation, as reflected in the hydrodynamic volume of the protein molecule (Goldenberg and Creighton, 1984a). Comparison of protein samples trapped with the neutral reagent iodoacetamide and the acidic iodoacetate is generally sufficient to determine the number of disulfide bonds in each trapped species. The electrophoretic mobility is divided into conformational and net charge factors, and only the latter factor is assumed to be different when trapped with iodoacetamide or iodoacetate. The charge factor is quantified from the difference in mobilities of the fully reduced protein, with known number of Cys thiol groups, when trapped with the two different reagents (Creighton, 1974b).

Which Cys residues are paired in intramolecular disulfide bonds, which are involved in any mixed disulfides between protein and reagent, and which have free thiols, can be determined by a number of different methods, but only after all the free thiols are blocked irreversibly. A particularly elegant procedure is diagonal electrophoresis (Fig. 7–4), which can identify the state of all the protein Cys residues of a homogeneous species (Brown and Hartley, 1966). Other more quantitative methods are desirable, however, when more than one protein species is present.

2.4 Reconstructing the Pathway of Folding and Unfolding

Determining the kinetic roles of intermediates on a complex pathway can be difficult, but it is relatively straightforward with the disulfide interaction, due to the ability to manipulate experimentally the rates of disulfide bond formation and breakage. Steps involving disulfide bond formation generally occur with a rate proportional to the concentration of disulfide reagent, whereas the rates of steps involving disulfide breakage are usually proportional to the concentration of reagent in the thiol form. The rates of intramolecular steps, including disulfide rearrangements, should be independent of both forms of the reagent. By varying the concentrations of both disulfide and thiol forms of the reagent during both unfolding and refolding, it is usually clear whether the rate-limiting steps in formation and disappearance of a protein species involve protein disulfide formation or breakage, or intramolecular steps. Where there are different alternative pathways to a species, the dependence may not be clear-cut, but there are thermodynamic restrictions on the possible values of rate constants with multiple, parallel paths, due to the necessity that there be no net-free-energy change around a cyclic pathway.

The kinetic behavior of an intermediate on a complex pathway will be affected by the rates of the steps preceding its formation or after its disappearance, so it is necessary to demonstrate that the kinetic behavior of *all* the species under *all* the folding conditions, during *both* unfolding and refolding, can be adequately simulated using a *single* set of rate constants for all of the steps. The conditions are kept constant, except for varying the concentrations of the thiol and disulfide forms of the reagent. Each rate constant for disulfide bond formation or breakage should include the dependence of the rate on the concentration of the disulfide or thiol forms

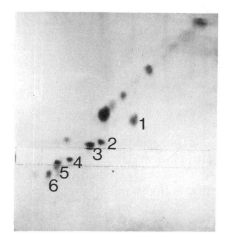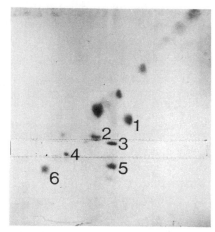

*FIGURE 7–4. Diagonal maps of fully reduced BPTI (left) and the (30-51)
intermediate trapped by alkylation with iodoacetic acid (right). The two proteins
were cleaved into peptides by hydrolysis with trypsin followed by chymotrypsin.
The peptides were separated by paper electrophoresis at pH 3.5 in the horizontal
direction (anode at the left). After exposure to performic acid vapor, to cleave the
disulfide bonds and oxidize the sulfur atoms, the electrophoresis was repeated in
the vertical direction (anode at the bottom). The peptides were stained with
ninhydrin-cadmium acetate. Peptides unaltered by the performic acid vapor
have the same mobility in both dimensions and define a "diagonal" of peptides
running from the origin to the upper right; most of these peptides have migrated
off the paper. The six numbered peptides in the left map contain the six
carboxymethyl-Cys (CMCys) residues of the protein and define the "CMCys
diagonal"; these peptides are more acidic in the second dimension as a result of
oxidation of the sulfur atoms of the CMCys residues. The peptides correspond to
the following residues of BPTI: peptide 1: residues 36 to 39; 2: 1 to 15; 3: 47 to
53; 4: 5 to 15; 5: 27 to 33; 6: 54 to 58. With intermediate (30-51), on the right,
the two peptides containing Cys 30 and Cys 51, numbered 5 and 3, respectively,
had a common mobility in the first dimension as a result of being linked by the
disulfide bond. After performic oxidation, they migrated independently. The
absence of peptides 3 and 5 from the "CMCys diagonal" indicates the
homogeneity of the (30-51) intermediate. (Creighton, 1947b)*

of the reagent, that is, be either second- or third-order. (See pp. 304–310.)
By systematically varying the concentrations of the two forms of the
reagent, values for the apparent rate constants for all of the steps should be
obtained by fitting the simulated kinetics to the observed behavior of all
the species. When using different reagents, the values of the multiple-order

rate constants involving the reagent may differ systematically, but the values for all the intramolecular steps should be independent of the nature of the reagent. Generally, one rate constant will be obtained for making a protein disulfide and one for reducing it, unless the mixed disulfide accumulates. The observed rate constants can generally be interpreted in terms of the principles outlined previously. (See pp. 304–310.) Ideally, both inter- and intramolecular disulfide reagents will be used; for example, both GSSG and DTT$_S^S$ were used with BPTI folding (See pp. 319–332), and the values of k_{intra} for forming each of the disulfide bonds indicated by the two reagents agreed within a factor of two (Creighton and Goldenberg, 1984). This ability to simulate the kinetics of unfolding and refolding under a wide variety of redox conditions, because only the thiol and disulfide forms of the reagent are varied, contrasts with the uncertainty of analyzing the kinetics of folding when the concentration of a denaturant, the pH, or the temperature must be varied (Chapter 5).

Knowing the disulfide bonds in the various species trapped during folding helps to infer a plausible pathway, for the pathway should provide for a stepwise proces of disulfide bond formation, breakage, and rearrangement. A crucial intermediate might not accumulate to high levels, however, for it could be rapidly converted to other species by disulfide rearrangements. In general, the levels of accumulation of an intermediate do not necessarily reflect its kinetic importance. Also, the presence of two species differing only in a single disulfide bond should not be taken as indicating that they are necessarily interconverted directly.

All protein molecules with the same disulfide bonds are assumed initially to be equivalent and to represent a homogeneous species. This need not be the case, however, for species with a partial set of disulfide bonds should be only partially folded and have flexible conformations; different molecules are likely to have different conformations at each instant of time. They can still behave kinetically as a homogeneous species, but only if all conformational interconversions are rapid on the time scale of disulfide bond formation or breakage. This need not be the case; for example, *cis,trans* isomerization of Pro peptide bonds (Chapter 5) can introduce conformational heterogeneity if disulfide bond formation occurs rapidly (Creighton, 1977b).

In many cases, the observed rate constant will be the sum of a number of different steps. For example, the observed rate of forming the first disulfide bond in a fully reduced unfolded protein is expected to be the sum of the rates of all the possible disulfides that can be formed between all the possible pairs of Cys residues. The pathway of disulfide formation and breakage can be further dissected using molecules in which one or more of the Cys residues are blocked irreversibly or replaced by other residues, so as to prevent their involvement in disulfide bonds (Creighton,

1977a). The irreversibly trapped intermediates are useful sources of such altered proteins, but they have blocking groups on the free Cys thiols. It is preferable to replace the Cys residues by Ser residues, using the techniques of protein engineering. One advantage of acid-trapped intermediates is that they may be purified and their kinetic behavior followed when they are returned to folding conditions (Creighton, 1977c).

2.5 Conformational Properties of Trapped Intermediates

The ultimate goal of using disulfide bonds to elucidate protein folding pathways is to determine the role of the protein conformation in guiding disulfide bond formation. Some aspects of conformation can be deduced from which Cys residues come into proximity to form disulfides, for the Cys residues should be probes of the conformational transitions undergone at different stages of folding. The conformations of the trapped intermediates should provide the greatest information in this regard, however, for whatever conformation favored the formation of a particular disulfide bond should be stabilized to the same extent by the presence of that disulfide bond in the trapped intermediate (Fig. 7–3). To the extent that the trapped intermediates are stable, they may be studied by a wide variety of techniques (Creighton et al., 1978; Darby et al., 1991,1992; Kosen et al., 1980,1981,1983; States et al., 1987; van Mierlo et al., 1991a,b).

The trapped intermediates have the disadvantage of having the blocking groups on their Cys residues not involved in disulfide bonds. Even those trapped with acid have their free Cys residues fully protonated, whereas the thiol groups should be at least partially ionized during folding studies. Protonated thiols are somewhat hydrophobic (e.g., Radzicka and Wolfenden, 1988), so acid-trapped intermediates are much more likely to adopt their fully folded conformations under acidic conditions (Weissman and Kim, 1991) than they would when they are folding. A better approach is to replace the free Cys residues with Ser or Ala and to prepare the analogues of the folding intermediates by protein engineering methods (Goldenberg, 1988; Darby et al., 1991). This also has the advantage that intermediates that do not accumulate to substantial levels, which can be as important, or more so, than those that do, can be prepared in quantities sufficient for structural analysis. Whether the free Cys residues are replaced by Ser or Ala will depend upon whether or not one wants the intermediate analogues to adopt folded conformations. Replacing a Cys residue by Ala is likely to encourage that residue to engage in nonpolar interactions with the rest of the protein. The hydroxyl group of a Ser residue seems to be a better mimic of a partially ionized thiol, but there appears to be no perfect replacement. (See note on pp. 348–349.)

3 FOLDING PATHWAYS ELUCIDATED USING DISULFIDE BONDS

Only a few proteins have had their disulfide folding pathways elucidated in detail. These will be described here, both to point out the strengths and weaknesses of the approach and to illustrate the most relevant principles of protein folding.

3.1 Bovine Pancreatic Trypsin Inhibitor (BPTI)

BPTI is a very small protein of only 58 amino acid residues, with a well-determined three-dimensional structure (Wlodawer et al., 1987) that is exceedingly stable to unfolding and to chemical modifications, properties that have made it very amenable to both theoretical and experimental analysis. The folded conformation consists primarily of an antiparallel β-sheet, with one substantial segment of α-helix near the C-terminus (Fig. 7–5A). The folded protein contains three disulfide bonds, linking Cys5 to Cys55, Cys14 to Cys38, and Cys30 to Cys51, which will be referred to here as the 5-55, 14-38, and 30-51 disulfides. The pathway of disulfide bond breakage during unfolding and of disulfide bond formation during refolding has been extensively studied and is the best-characterized protein folding pathway available. Folding and disulfide bond formation and breakage have been carried out normally at pH 8.7, where thiol-disulfide interchange is rapid and where the Cys thiol groups are ionized at least a significant fraction of the time, so as to discourage folding of the reduced protein. The species trapped by acid, by iodoacetate, and by iodoacetamide have been analyzed by nondenaturing polyacrylamide gel electrophoresis and ion-exchange chromatography (Fig. 7–6; Creighton, 1974b,1975a, 1977a,b,c,1986) and by reversed-phase HPLC (Weissman and Kim, 1991). The same pathway, with the same intramolecular rate constants, has been shown to be obtained with GSSG, oxidized mercaptoethanol, and DTT$^{S}_{S}$ as disulfide reagents (Creighton and Goldenberg, 1984). The current disulfide folding pathway for BPTI is illustrated in Figure 7–7, and the rate constants for the various steps are compiled in Table 7–1.

3.3.1 Reduced BPTI

Reduced BPTI is a very unfolded protein, even in the absence of denaturants and irrespective of whether or not its six free thiol groups are blocked by carboxymethyl or other groups. Its hydrodynamic volume as indicated by its mobility through nondenaturing polyacrylamide gels relative to that of the native protein is comparable to that of other unfolded proteins (Creighton, 1974a; Goldenberg and Creighton, 1984). Its six thiol groups react with iodoacetate at nearly equal rates and at the rate observed with

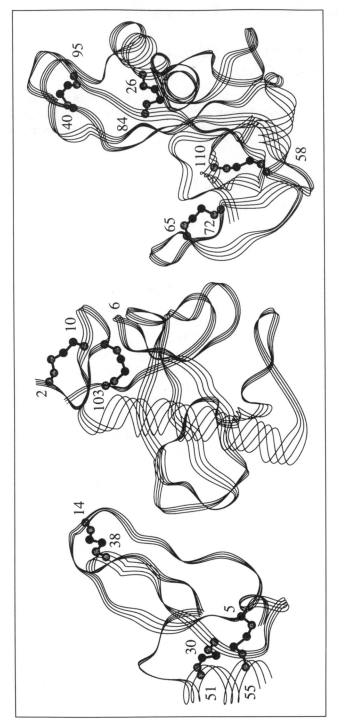

FIGURE 7-5. *Structures of BPTI, RNase T₁, and RNase A drawn on the same scale. The course of the polypeptide is traced by a ribbon. The C^α and C^β atoms the Cys side chains are depicted as stippled spheres, the S^γ atoms as solid spheres. (Kindly provided by Gerrit Vriend)*

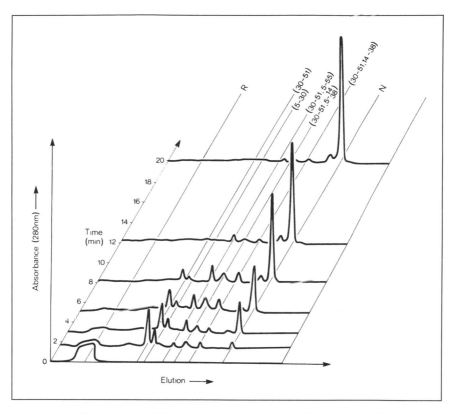

FIGURE 7–6. *Chromatographic separation of the intermediates trapped by iodoacetate during folding of reduced BPTI. Refolding was initiated by the addition to reduced BPTI at time zero of GSSG to a final concentration of 0.15 mM. At the indicated times, the species present were trapped with 0.1 M iodoacetate and separated by ion-exchange chromatography. The major species in each peak is identified. Peak N contains the quasi-native (5-55,14-38) in addition to fully refolded (5-55,14-38,30-51) BPTI. (Creighton, 1984)*

model thiols (Creighton, 1975c).[1] Its UV and ^{1}H-NMR spectral properties are comparable to those of other unfolded proteins (Kosen et al., 1980, 1981, 1983, States et al., 1987; Darby et al., 1991). Its aromatic residues are nearly fully exposed to the solvent (Kosen et al., 1980). It demonstrates negligible tendencies to cross-react with antibodies directed against the native protein

---

[1]This study measured the amount of ^{14}C label incorporated at each of the six Cys residues by reaction of one molecule of [^{14}C]-iodoacetate with one molecule of reduced BPTI, under the same conditions as used for folding. As an internal control, it used reduced BPTI in which all
(continued)

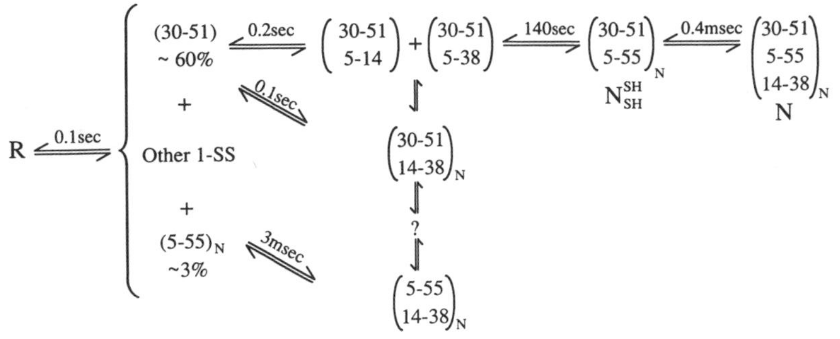

FIGURE 7–7. *Disulfide folding pathway of BPTI. The fully reduced protein is depicted as R, the other species by the disulfide bonds they contain. Species that tend to adopt the native-like conformation are depicted with the subscript N. The one-disulfide intermediates are in rapid equilibrium, as indicated by the bracket; their approximate levels under the normal conditions of folding at pH 8.7 are given. Second disulfide bonds are depicted as being formed in specific one-disulfide intermediates, but this is not certain, due to the rapid equilibrium. The + between the (30-51,5-14) and (30-51,50-38) intermediates indicates that they have comparable kinetic roles.*

The half-times for the intramolecular transitions in forming and rearranging the disulfide bonds are depicted for those steps where they are known. Those for forming the 14-38, 5-14, and 5-38 disulfides in (30-51) came from studies with both or either Cys14 and Cys38 blocked by reaction with iodoacetamide (Creighton, 1977a). The other half-times came from Creighton and Goldenberg (1984).

(Creighton et al., 1978). The β-sheet and α-helix that are present in the native conformation are barely detectable in the reduced protein and in peptide segments (Oas and Kim, 1988; Goodman and Kim, 1989). Nevertheless, there are claims that reduced BPTI adopts a compact structure, even in the presence of denaturants (Amir and Haas, 1988), and it has been included as an example of a molten globule in Table 6–2 of Chapter 6.

six Cys thiols were blocked completely with [^{3}H]-iodoacetate. Separation of the Cys-containing peptides used performic acid oxidation, which oxidized the sulfur atoms to the sulfones. This may have labilized the ^{3}H atoms of the blocking groups of the internal standard to exchange with the solvent. The ^{3}H internal standard may not, therefore, have been valid. Nevertheless, the similar levels of ^{14}C in the various Cys peptides indicate that their thiol groups had comparable reactivities in fully reduced BPTI, as previously concluded (Creighton, 1975c). Those of Cys14 and Cys38 may be somewhat greater than the others, probably due to the basic residues adjacent to them in the primary structure.

TABLE 7-1. Rate Constants for Disulfide Bond Formation, Breakage, and Rearrangement in the Disulfide Folding Transitions of BPTI and Some Related Proteins

Step	BPTI[a]	BPTI in 8 M Urea[b]	14,38 Cam₂-BPTI[c]	14 Cam-BPTI[c]	38 Cam-BPTI[c]	5,55 CM₂-BPTI[d]	Protein I[e]	Protein K[e]
R ↔ I								
forward ($s^{-1} M^{-1}$)	0.022	0.011	0.016	0.016	0.014	0.014	0.029	0.028
reverse ($s^{-1} M^{-1}$)	20	25	n.d.	n.d.	n.d.	n.d.	25	45
I ↔ II								
forward ($s^{-1} M^{-1}$)	0.030	2.6×10^{-3}	—	0.005	0.005	0.020	0.038	0.022
reverse ($s^{-1} M^{-1}$)	250	70	—	30	30	1×10^{3}	100	45
II ↔ N_{SH}^{SH}								
forward (s^{-1})	5.0×10^{-3}	3.8×10^{-3}	—	$\sim 5 \times 10^{-3f}$	$\sim 5 \times 10^{-3f}$	—	2.0×10^{-3}	3.7×10^{-2}
reverse (s^{-1})	2.0×10^{-5}	1.1×10^{-2}	—	n.d.	n.d.	—	7.0×10^{-3}	2.4×10^{-3}
I ↔ N_{SH}^{SH}[g]								
forward ($s^{-1} M^{-1}$)	8×10^{-6h}	n.d.	8×10^{-6h}	n.d.	n.d.	—	1.9×10^{-3}	3.6×10^{-3}
reverse ($s^{-1} M^{-1}$)	$\sim 5 \times 10^{-5}$	n.d.	$\sim 5 \times 10^{-5}$	n.d.	n.d.	—	0.64	5.1×10^{-2}
N_{SH}^{SH} ↔ N_{S}^{S}								
forward	5.7	3.8	—	—	—	—	1.5	3.0
reverse	30	20	—	—	—	—	40	30

The thiol and disulfide reagent was dithiothreitol in all cases. The conditions were the same in each case: 0.1 M Tris-HCl (pH 8.7), 0.2 M KCl, 1 mM EDTA, 25°C. R is fully reduced protein; I represents the one-disulfide intermediates; II, all the two disulfide intermediates other than the quasi-native (5,55,14-38) and the native-like (30-51,5-55), which is designated as N_{SH}^{SH}; N_{S}^{S} is the native-like protein; the quasi-native species (5-55,14-38) is ignored. A dash indicates that this step is not possible; n.d., not determined.

[a] Normal BPTI; from Creighton and Goldenberg (1984).
[b] Normal BPTI, but in the presence of 8 M urea; from Creighton (1977c).
[c] BPTI in which either or both of the Cys14 and Cys38 thiols were blocked by reaction with iodoacetamide; from Creighton (1977a).
[d] BPTI in which Cys5 and Cys55 thiols had been reacted with iodoacetate; from Creighton (1977a).

(continued)

3.1.2 Formation of the First Disulfide Bond

Reduced BPTI readily forms an initial disulfide bond under normal folding conditions at pH 8.7, with a collective value of k_{intra} of 2.2 sec^{-1}. All six Cys thiol groups are presumed to participate in this reaction; their equal reactivities with alkylating reagents indicate that they should react at similar rates with disulfide reagents, such as mercaptoethanol disulfide and oxidized glutathione (GSSG). Yet no mixed disulfides of the reduced protein with radioactive forms of these reagents accumulate to detectable levels (Creighton, 1975a), indicating that each Cys residue can form at least one disulfide bond rapidly, with k_{intra} of at least 1 sec^{-1}. The collective rate of forming the first disulfide bond indicates a total effective concentration of all the Cys thiol groups relative to each other of approximately 1 M. As there are 15 possible pairs of six Cys residues, the average effective concentration of 67 mM is approximately what would be expected if reduced BPTI were a random coil. The rate of forming collectively the initial disulfide bonds was not altered by replacement of residues important for the fully folded conformation, which drastically decreased the stability of the fully folded protein (Goldenberg et al., 1989).

On the other hand, the rate of disulfide bond formation did not decrease at pH 7.5 as much as expected from the diminished ionization of the Cys thiol groups (Creighton, 1980b), suggesting the possibility that the decreased ionization of the thiols may have caused the Cys groups to be in increased proximity to each other and to have greater effective concentrations. The microscopic rates of forming five of the individual disulfide bonds have been measured by mutating four of the Cys residues to Ser, so that only a single disulfide bond between the remaining two Cys residues is possible (N.J. Darby, unpublished observation). The rates measured indicate that the values of k_{intra} vary approximately as expected from the number of residues between the Cys residues in a random polypeptide chain.

3.1.3 The Single-Disulfide Intermediates

Although formation of the first disulfide bond by reduced BPTI appears to be relatively random, with all 15 possibilities contributing to the observed rate, the single-disulfide intermediates that accumulate are very nonrandom. Indistinguishable collections of intermediates were trapped by acid, by iodoacetamide, and by iodoacetate (Creighton, 1974a). Approximately 60% of the molecules contain the 30-51 disulfide bond (Creighton, 1974b). Other

---

Table 7-1 footnotes, continued

[e]Black mamba proteins I and K; in this case, intermediate I represents just (30-51); from Hollecker and Creighton (1983).

[f]Value not specified precisely by experimental data.

[g]Usually measured with the protein in which Cys14 and Cys38 had been blocked.

[h]Estimated from the approximate value for $k_{intra} = 5 \times 10^{-3} s^{-1}$.

intermediates present in quantities sufficient to be detected were (5-30), (5-51), (30-55), and (5-55). Reversed-phase HPLC analysis of the trapped intermediates confirmed these assignments, but suggested that the earlier disulfide mapping techniques had overestimated the abundance of intermediate (5-30) (Weissman and Kim, 1991). Rather than being 30% of the total one-disulfide intermediates, its concentration may not be much greater than those of the other minor intermediates.

The reason why intermediate (30-51) predominates, even though all the disulfides are formed at comparable rates, is that the disulfides formed initially are rapidly rearranged intramolecularly, so that an equilibrium spectrum is present (Creighton, 1977c). Intermediate (30-51) predominates because it has favorable conformational properties under normal folding conditions; a very different spectrum of intermediates, in which none predominates, is obtained in the presence of 6 M GdmCl or 8 M urea. This relatively random mixture of one-disulfide intermediates rearranges under normal folding conditions to the normal spectrum within five seconds.

The (30-51) intermediate seems to be crucial for folding of reduced BPTI, because all further productive intermediates retain its disulfide bond (Fig. 7–7). It has a somewhat more compact conformation than the reduced protein, as judged by its electrophoretic mobility in polyacrylamide gels, whether trapped by acid or by iodoacetamide, but is much less compact than the native protein (Creighton, 1974a). Its aromatic residues are partially protected from exposure to aqueous solvent (Kosen et al., 1980), and it binds only moderately well to antibodies against both the native and reduced forms of BPTI (Creighton et al., 1978). The 30–51 disulfide bond was noted to link the main elements of secondary structure in BPTI, suggesting their importance at an early stage of folding (Creighton, 1974b). The CD spectrum was distinct from that of both native and reduced BPTI (Kosen et al., 1983), but interpretation of the CD spectrum of BPTI in terms of secondary structure is not possible, presumably due to the contribution of aromatic residues to the far-UV spectrum (Manning and Woody, 1989; Hollecker and Larcher, 1989). NMR analysis of the iodoacetate-trapped intermediate indicated the presence of β-sheet structure similar to that of the native protein (States et al., 1987), but little evidence for the presence of the α-helix could be obtained (Kosen et al., 1983). Oas and Kim (1988) considered the possibility that the conformation stabilizing the 30-51 disulfide bond would be localized around the disulfide bond, which would include both the β-sheet and the α-helix of the fully folded conformation and its largely hydrophobic interior. Accordingly, they synthesized two peptides containing each of the Cys30 and Cys51 residues, comprising residues 20 to 33 and 43 to 58, in which Cys55 was replaced by Ala. The first segment comprises most of the β-sheet of native BPTI, whereas the second includes the major α-helix, so the peptides were designated P_β and P_α, respectively.

Individually, the two peptides had little nonrandom structure when alone or when present together in the reduced form. When linked by the 30-51 disulfide bond, however, they adopted an ordered structure that was claimed to be very like that they adopt in the fully folded conformation. These observations suggest that the particular stability of the (30-51) intermediate arose because of interactions between the major elements of secondary structure of the protein, which include most of the hydrophobic interior of the protein. Individually, the elements of secondary structure are not sufficiently stable to be populated substantially in the reduced protein or in short peptides (Goodman and Kim, 1989), but their interaction probably stabilizes each other and the 30-51 disulfide bond between them. When the disulfide bond is present, as in the trapped intermediate, the ordered conformation is further stabilized (Fig. 7–3).

That the conformation stabilizing the (30-51) intermediate is localized around the disulfide bond would be consistent with mutational analysis, which has shown that the stabilities of the one-disulfide intermediates are not altered by the drastic replacements of Tyr35 $\rightarrow$ Gly and Asn43 $\rightarrow$ Gly or Ala, residues that in the native conformation are not involved in the β-sheet/α-helix interactions, but that contribute greatly to the stability of the fully folded conformation (Goldenberg et al., 1989). A significant destabilization of the one-disulfide intermediates was caused by the replacement of Tyr23 $\rightarrow$ Leu, a residue that is part of the hydrophobic core that surrounds the 30-51 disulfide bond. Likewise, changing Met52 to more polar residues decreased somewhat the stability of (30-51) relative to the other one-disulfide intermediates (Creighton and Dyckes, 1981; N. J. Darby, unpublished observations).

The conformational properties of the entire (30-51) intermediate are being characterized by 2D NMR techniques, using the analogue of the intermediate prepared by protein engineering, with the other four Cys residues replaced by Ser (van Mierlo et al., 1992). Preliminary analysis of this analogue, designated (30-51)$_{Ser}$ suggests that its ordered structure consists of β-sheet and α-helix around the 30-51 disulfide, similar to that in native BPTI, but with the remainder of the polypeptide chain disordered or at least very flexible.

Although intermediate (30-51) predominates during folding, it is only marginally more stable than the other single-disulfide intermediates. How much stability the 30-51 disulfide bond should give to its folded conformation can be estimated, which made it possible to calculate that the same folded conformation should be present in the absence of the disulfide bond, i.e., in reduced BPTI, between 0.16% and 5% of the time (Creighton, 1988b). Therefore, the conformation that stabilizes the 30-51 disulfide bond does not preexist in the reduced protein to a substantial extent, but it is stabilized by the presence of that disulfide bond, as expected from the linkage between the two (Fig. 7–3). Whatever other conformations are present in the bulk of the reduced BPTI molecules seem to have no direct role in directing folding.

The other one-disulfide intermediates that accumulate to significant levels during folding, both in their trapped forms (Kosen et al., 1980, 1981; States et al., 1987) and as the analogues with Ser replacing the free Cys residues (Darby et al., 1991, 1992), appear to have little nonrandom conformation and to be largely unfolded. An exception is the intermediate (5-55), which accumulates to only very low levels at pH 8.7, approximately 3% of the one-disulfide intermediates. It was suspected to form the 14-38 disulfide bond very rapidly and consequently to be the precursor of a quasi-native two-disulfide species, (5-55,14-38); this suggested that it adopted a native-like conformation with Cys14 and Cys38 in proximity (Creighton and Goldenberg, 1984). A peptide model comprising peptide P_α (residues 42 to 58, with Cys51 replaced by Ala) linked by the 5-55 disulfide bond to peptide P_γ (residues 1 to 9 linked by 6 Gly residues to residues 20 to 33) was claimed to adopt a native-like conformation (Staley and Kim, 1990). The analogue of the intermediate, $(5-55)_{Ser}$, with Ser residues replacing the four free Cys residues, was found by detailed 2D NMR analysis to adopt at low temperatures a folded conformation (Darby et al., 1991) that is very close to that of native BPTI (van Mierlo et al., 1991b).

The (5-55) intermediate is a quasi-native form of BPTI that adopts the native-like conformation even in the absence of two of the native disulfide bonds. The 5-55 disulfide links the two ends of the polypeptide chain and would be expected on the basis of its greatest decrease in the conformational entropy of the unfolded conformation to have the greatest contribution to stability of the native conformation. The 5-55 disulfide is also that which contributes most to the stability of native BPTI with three disulfide bonds (Creighton and Goldenberg, 1984). The (5-55) intermediate is relatively unstable relative to the other one-disulfide intermediates at pH 8.7, and the folded conformation of $(5-55)_{Ser}$ is very unstable, with a midpoint melting temperature near 5 °C. That the native conformation of BPTI can be populated even to this extent with only a single disulfide bond, although that most stabilizing, is a reflection of the extreme stability of that conformation.

The native-like conformation of intermediate (5-55) undoubtedly means that the thiol groups of Cys30 and Cys51 are buried in the interior of the protein, where they could be inaccessible to thiol and disulfide reagents. The occurrence of such a quasi-native intermediate interferes with the disulfide approach to folding, where folding and disulfide bond formation need to be coupled (Fig. 7–3). The quasi-native intermediate (5-55) accumulates to greater extents at lower pH, where the thiols are not ionized to substantial extents (Weissman and Kim, 1991). This illustrates the importance of carrying out disulfide folding at alkaline pH, where ionization of the thiols helps to keep the protein from completing refolding until all the disulfides are formed, and substituting Ser residues, rather than Ala, for Cys in analogues of the intermediates (see note on pp. 348–349).

3.1.4 Formation of the Second Disulfide Bonds

The quasi-native (5-55) intermediate rapidly forms the 14-38 disulfide bond, with k_{intra} = 200 to 500 sec^{-1}, because its conformation brings Cys14 and Cys38 into appropriate proximity on the surface of the protein. Although much faster than forming the first disulfide bonds in the reduced protein and the other second disulfides, the rate of forming the 14-38 disulfide in (5-55) is only 10% of that in the stable native-like (30-51,5-55) (Table 7–1). This suggests that intermediate (5-55) is sufficiently unstable to have the native-like conformation only 10% of the time under the normal folding conditions of pH 8.7 and 25°C. The two-disulfide species that results, (5-55,14-38), retains the native-like conformation and is considerably more stable, as a result of the presence of the second disulfide bond (States et al., 1984; Eigenbrot et al., 1990).

The other second disulfide bonds that are formed probably occur primarily in the predominant intermediate (30-51), as the subsequent intermediates retain this disulfide. The collective rate is only slightly greater than the rate of forming the first disulfide bonds (Table 7–1), with k_{intra} = 2.5 sec^{-1}. This is further evidence that intermediate (30-51) does not have a quasi-native conformation like that of (5-55), for the 14-38 disulfide bond should then have been formed at least 10^2 times faster.

Other second disulfides are also formed, in particular 5-14 and 5-38, to generate directly intermediates (30-51,5-14) and (30-51,5-38). This was demonstrated directly using reduced BPTI in which either Cys14 or Cys38 had been blocked covalently by reaction with iodoacetamide; the 5-14 and 5-38 disulfides seem to be formed collectively at about half the rate of 14-38 (Creighton, 1977a). A revised pathway suggested by Weissman and Kim (1991) that omitted these intermediates is incompatible with these experimental observations (Creighton, 1992).

One disulfide bond that is not formed directly in intermediate (30-51) at a comparable rate is 5-55. This can be demonstrated most readily by blocking the thiol groups of both Cys14 and Cys38 irreversibly (Creighton, 1977a), or by replacing them with Ser residues (Goldenberg, 1988). In this case, a second disulfide bond is formed only very slowly, with k_{intra} = 0.005 sec^{-1} for forming the 5-55 disulfide bond in (30-51); this rate is less than 10^{-3} that of forming the second disulfides involving Cys14 and Cys38. Corroborating evidence was obtained with the normal protein, in which the mixed-disulfide of Cys55 with the disulfide reagents GSSG and oxidized mercaptoethanol accumulated in intermediate (30-51) during folding. This indicates that there is a block in forming any disulfide bond with Cys55 in (30-51) and that this was not due to inaccessibility of Cys55 to disulfide reagents, even though this has frequently been suggested by others (Goto and Hamaguchi, 1981); Oas and Kim, 1988; Staley and Kim, 1990). Instead, the kinetic block must be in the second step in which the second Cys residue cannot react with the mixed-disulfide for conformational reasons. (See p. 308.) It is most likely that this kinetic block occurs due to the high energy of the transition state in which a buried disulfide bond and the native conformation would be formed. (See pp. 345–348.)

3.1.5 The Two-Disulfide Intermediates

The quasi-native (5-55,14-38) intermediate has a conformation very close to that of native BPTI (States et al., 1984; Eigenbrot et al., 1990). This quasi-native species cannot complete refolding because Cys30 and Cys51 are buried and unreactive. It is trapped in this form unless the 14-38 disulfide bond is reduced, when the (5.55) intermediate can rearrange to the (30-51) intermediate and follow the productive pathway (Fig. 7–7).

The intermediate (30-51,14-38) also has two native-like disulfide bonds and also adopts a quasi-native conformation (Creighton et al., 1978; States et al., 1987). The conformation of the analogue (30-51,14-38)$_{Ser}$ was shown by detailed 2D NMR analysis to differ from that of native BPTI only at the ends of the polypeptide chain, where there is a disulfide bond in native BPTI between residues 5 and 55 (van Mierlo et al., 1991a). The thiol groups of these two Cys residues may consequently be somewhat restricted in their reaction with alkylating and disulfide reagents, but they are observed to react with both, except at low temperatures in the region of 5°C. This is especially relevant, because this intermediate also does not readily form the third disulfide bond between Cys5 and 55 (Creighton, 1975a). Instead, these two Cys residues accumulate as mixed disulfides with disulfide reagents like GSSG. As in the case of (30-51), this indicates that the kinetic blockage in forming the disulfide bond is in the second step (Fig. 7.2) and is not due to inaccessibility of the two Cys thiol groups.

This (30-51,14-38) intermediate slowly rearranges, with comparable rates, either to the quasi-native (5-55,14-38) or to the native-like (30-51,5-55) intermediate. The latter rearrangement occurs through either intermediate (30-51,5-14) or (30-51,5-38) with comparable frequencies. The kinetics of this rearrangement process remain to be determined in detail; until then, the three intermediates, (30-51,14-38), (30-51,5-14), and (30-51,5-38), are treated collectively for kinetic and energetic analyses (Creighton and Goldenberg, 1984). The necessary intermediate in the rearrangement of (30-51,14-38) to (5-55,14-38) has not been identified, but the 30-51 and 5-55 disulfides are quite close in the folded conformation, so several such intermediates are possible; they would be expected to be relatively unstable.

The intermediates (30-51,5-14) and (30-51,5-38) were found in lower quantities than the quasi-native (30-51,14-38) when acid-trapped (Weissman and Kim, 1991). In contrast, these three two-disulfide species were trapped by iodoacetate in roughly comparable quantities (Creighton, 1975a). Whether this discrepancy is due to disulfide rearrangements during covalent trapping or after trapping by acid remains to be determined. It is not a central question, however, for the kinetic analysis of the BPTI folding pathway has not depended upon the relative levels of these intermediates, which probably are in rapid equilibrium during folding.

Intermediates (30-51,5-14) and (30-51,5-38) are important intermediates in the productive refolding process (Fig. 7–7), and they appear to play comparable roles. That they are the predominant intermediates in folding to the native conformation was shown by the refolding of BPTI in which either Cys14 or Cys38 had been blocked irreversibly (Creighton, 1977a). These proteins cannot form the 14-38 disulfide bond, yet they refolded to the native-like (30-51,5-55) at nearly the normal rate. Both of these intermediates have partially folded conformations like that of (30-51), with the non-native disulfides linking the flexible portions of the polypeptide chain (Darby et al., 1992). There is no evidence that the non-native 5-14 and 5-38 disulfides result from or stabilize particular conformations. The random two-disulfide intermediates in 8-M urea rearrange to (30-51,5-55) at almost the normal rate, and the Tyr35 → Gly, Asn43 → Gly, or Ala, and Tyr23 → Leu replacements had only small energetic effects on the productive two-disulfide intermediates in folding (Goldenberg et al., 1989). Therefore, the 5-14 and 5-38 disulfides may not actively guide the rearrangement process. Their roles apparently are to provide stable and flexible conformations that can participate in the slow rearrangements to the native-like intermediate (30-51,5-55). These rearrangements are the rate-limiting overall step and must involve a high-energy transition state. Nevertheless, this is the energetically most favorable means of getting into and out of the native conformation.

3.1.6 Forming the Third Disulfide Bond

The intermediate (30-51,5-55) plays a vital role in folding, for only it can form rapidly the third correct disulfide bond to complete refolding. It has a conformation that is exceedingly close to that of native BPTI and is much more stable than the other two-disulfide intermediates (Stassinopoulou et al., 1984; States et al., 1987). Its stability relative to the fully reduced protein is affected by amino acid replacements in a similar way to that of the totally native protein with three disulfide bonds (Goldenberg et al., 1989). The crucial difference between this intermediate and the quasi-native intermediates with two native disulfide bonds is probably that Cys14 and Cys38 are on the surface of the folded protein, and a disulfide bond can be formed between them without perturbing the native conformation; the 14-38 disulfide is also the weakest in folded BPTI, so the energetic consequences of making and breaking it are not so great. The low energy barrier and the proximity of the two thiol groups produced by the native-like conformation, causes them to form a disulfide bond very rapidly (Table 7–1), with $k_{intra} = 500 \text{ sec}^{-1}$.

The refolded BPTI that results from this process of refolding is indistinguishable from native BPTI.

3.1.7 The Pathway of Unfolding

Native BPTI unfolds under the same conditions as used in refolding, but in the presence of excess thiol reagent so that the protein disulfide bonds become reduced. The kinetics of the unfolding process are compatible with it being the reverse of the pathway of disulfide bond formation (Table 7–1).

The first step is reduction of the 14-38 disulfide bond, to produce the intermediate (30-51,5-55), which retains the native conformation. As a consequence, the 14-38 disulfide bond can be very rapidly reformed, and the two-disulfide form can be considered as a minor variant of the native conformation. It is probably comparable to other minor variants of folded proteins with slightly altered conformations, such as with broken hydrogen bonds or slightly altered van der Waals contacts, that result from protein flexibility.

The 14-38 disulfide bond is the least stable of the three disulfide bonds of native BPTI and consequently contributes least to the stability of the native conformation (Creighton and Goldenberg, 1984); it is also the most accessible of the three disulfides to thiol reagents. Either or both considerations probably account for it being the only disulfide bond to be reduced directly.

The two remaining disulfides of (30-51,5-55) are not reduced directly at substantial rates, but the protein is reduced most readily after rearranging intramolecularly its disulfide bonds with either Cys14 or Cys38. This was demonstrated both by the rate of reduction of the second disulfide bond being independent of the concentration of thiol reagent and by it being much slower when both the Cys14 and Cys38 thiols were blocked irreversibly (Creighton, 1977a) or when these residues were replaced by Ser (Goldenberg, 1988). This intramolecular rearrangement would require that the thiol group of Cys14 and Cys38 undergo thiol-disulfide interchange with either of the two remaining disulfides, most likely 5-55. This process must involve a considerable distortion of the native conformation, for this disulfide bond is at the opposite end of the native conformation from the Cys14 and Cys38 thiols (Fig. 7–5). Probably for that reason, it is an infrequent event; nevertheless, it is the most favorable method for reducing the two buried disulfides of BPTI with concentrations of thiol reagent of less than 0.1 M.

The two-disulfide intermediates that result from this rearrangement are reduced rapidly and do not accumulate to detectable levels; neither do one-disulfide intermediates accumulate. The kinetics of appearance of fully reduced BPTI are consistent with the one- and two-disulfide intermediates being the same as those that are observed during refolding, but this has not been proven. Nevertheless, the kinetics of both disulfide bond formation and breakage are consistent with the measured equilibrium constants for each step and for the overall equilibrium between fully reduced and fully folded BPTI, suggesting that the pathways of unfolding and refolding are the reverse of each other (Creighton, 1977b).

3.2 Black Mamba Proteins I and K

Two dendrotoxins isolated from the venom of black mamba snakes, known as proteins I and K, have sequences that are respectively 33% and 40% identical to that of BPTI (Strydom, 1973). They are more closely related to each other, having 60% of their residues in common. The residues that are conserved are those that appear to play important structural roles in the folded conformation of BPTI, and it is virtually a certainty that the two black mamba proteins have essentially the same conformation as BPTI, which has been confirmed by NMR analysis in the case of protein K (Keller et al., 1983). Their far-UV CD spectra are markedly different, however, confirming that these spectra reflect more than just the secondary structure in this case (Manning and Woody, 1989; Hollecker and Larcher, 1989). These proteins must have diverged during evolution from a common ancestor of BPTI, and this then raised the question as to whether their folding pathways had been conserved during evolution, as well as their folded conformations. The pathways and kinetics of disulfide bond formation and breakage were determined for both proteins under the same conditions as used for BPTI (Table 7–1; Hollecker and Creighton, 1983).

The rates of forming the first disulfide bond in the reduced proteins were similar to that of BPTI for both proteins (Table 7–1). The predominant one-disulfide intermediate was (30-51) in both cases. The other one-disulfide intermediates were somewhat different from those of BPTI, apparently comprising mainly (30-55), (38-51), (38–55), and (14-51). The one-disulfide intermediates of proteins I and K had similar stabilities, relative to R, as those of BPTI, although they did not appear to be in as rapid equilibrium by thiol-disulfide interchange.

The major two-disulfide intermediate formed directly was (30-51, 14-38) for both proteins I and K. This species did not form readily the 5-55 disulfide bond, as it accumulated as the mixed disulfide in each case. Instead, it rearranged intramolecularly to the native-like (30-51,5-55), which then formed rapidly the 14-38 disulfide bond. These steps were similar to those in BPTI, but the intermediates involved in the rearrangement, which presumably are (30-51,5-14) and (30-51,5-38), were not trapped in substantial quantities and were not identified. With proteins I and K in which Cys14 and Cys38 were blocked irreversibly by alkylation, the direct formation of the 5-55 disulfide in intermediate (30-51) was shown to be significant in each case and much faster than in the case of BPTI, although still slower than formation of the 14-38 disulfide (Table 7.1).

The pathway of reduction was similar to that in BPTI, in that only the 14-38 disulfide bond was reduced at a significant rate, with kinetics similar to those observed in BPTI. The resulting (30-51,5-55) intermediate tended to rearrange before being reduced further, and the rearrangement product was shown in the case of protein I to be (30-51,14-38). A very substantial difference was that this rearrangement occurred some two orders of magnitude faster than in BPTI. Also, the direct reduction of the 5-55 disulfide bond in (30-51,5-55) was found in proteins K and I to be, respectively, three and four orders of magnitude faster than in BPTI.

The three-disulfide folded conformations of proteins I and K are less stable, relative to their fully reduced form, by 4.2 and 1.5 kcal/mol, respectively, than is that of BPTI. These conformations are also less stable to thermal unfolding with the disulfides intact (Keller et al., 1983), as would be expected (Fig. 7–3). This energetic difference in stabilities is not reflected in the stabilities of the first and third disulfide bonds to be formed during folding, but is reflected almost entirely in the conformational transitions at the two-disulfide stage. In particular, the rates of rearrangement of (30-51,5-55) were very much faster, as were the rates of direct reduction of its 5-55 disulfide bond. Correspondingly, the 5-55 disulfide bond was formed directly in (30-51) much more rapidly than in the case of BPTI. As a consequence, the direct pathway of disulfide bond formation (30-51 first, then 5-55, followed by 14-38) was much more significant with the less stable proteins I and K than with BPTI.

Presumably as a result of the lower stabilities of their folded conformations, the quasi-native species (5-55) and (5-55,14-38) that occur with BPTI were not apparent with proteins I and K. Intermediate (30-51, 14-38) accumulated with both intermediates, but the trapped form did not have the native conformation, as judged by its hydrodynamic volume. Whether or not it is a quasi-native species in these less stable proteins remains to be determined.

In summary, the pathways of unfolding and refolding of the homologous proteins I and K are very similar to that of BPTI, indicating that the folding pathway has been conserved during the evolutionary divergence of the genes for these proteins. There are, however, substantial differences in the energetics of the folding transitions of the different proteins, which provide valuable clues to the mechanism of folding. (See pp. 337–348.)

3.3   *Ribonuclease T₁*

Ribonuclease T_1 is a small protein of 104 residues that has only two di-sulfide bonds: one linking Cys2 to Cys10 and the other linking Cys6 to Cys103. The 6-103 disulfide bond is totally buried, whereas the 2-10 disulfide bond is about 90% accessible to solvent (Fig. 7–5). In the presence of DTT_{SH}^{SH}, the 2-10 disulfide bond of the native protein is reduced at about 0.05 the rate of a model disulfide (Fig. 7–8), presumably due to limitations on its accessibility and to the presence of acidic residues around the disul-fide (Pace and Creighton, 1986). The resulting one-disulfide species (6-103) retains the native folded conformation; its remaining disulfide tends not to be reduced directly, but only after intramolecular rearrangement of the disulfide by exchange with one of the two free thiol groups.

The fully reduced protein is unfolded under the conditions used for folding, although it will adopt the native conformation at low temper-atures and high ionic strengths (Pace et al., 1988). Upon disulfide bond formation and folding, the reduced protein formed its first disulfide bond between any pair of the residues Cys2, Cys6, and Cys10. No initial disulfides were observed involving Cys103, probably due at least in part to its distance in the primary structure from the other three Cys residues. The initial

FIGURE 7–8. *Disulfide folding pathway of RNase T_1. R is the fully reduced protein, and the other species are designated by the disulfide bonds they contain; N is the fully folded protein and the subscript N indicates that the protein adopts the native-like conformation. The initial one-disulfide intermediates are likely to be in rapid equilibrium, as indicated by the bracket. The half-times for the intramolecular steps in forming and rearranging disulfide bonds are indicated for each step. (Pace and Creighton, 1986)*

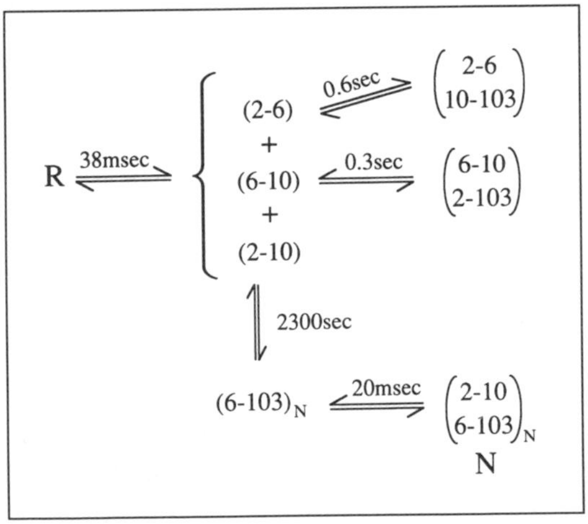

one-disulfide intermediates formed directly the nonnative pairings (2-6,10-103) and (6-10,2-103) at greatest rate. The (2-10) one-disulfide intermediate did not readily form directly the 6-103 disulfide bond to complete folding. Instead, the native conformation was regained most readily by undergoing intramolecular rearrangements of the one-disulfide intermediates to that with the 6-103 disulfide bond. This (6-103) intermediate adopts the native-like conformation, and the accessible 2-10 disulfide bond can then be formed readily.

3.4 Ribonuclease A

Ribonuclease A has 124 residues and four disulfide bonds, 26-84, 40-95, 58-110, and 65-72 (Fig. 7–5), and has been one of the proteins most thoroughly studied with regard to unfolding and refolding (Anfinsen et al., 1961; Chapter 5).

Fully reduced RNase A is well established to be a very unfolded molecule, approximating a random coil in its global properties, although with some tendencies to adopt local nonrandom conformations, especially a helical conformation near the N-terminus (Haas et al., 1988; Osterhout et al., 1989). Accordingly, its hydrodynamic volume is markedly greater than that of native RNase A (Creighton, 1977d), although no more so than in the case of BPTI. Reduced RNase A also makes first disulfides at about the same rate as does reduced BPTI, with a collective $k_{intra} = 4$ sec^{-1} (Fig. 7–9). The disulfide bonds found in the one-disulfide intermediates seem to be relatively random, and it is likely that all 28 possibilities are formed in similar quantities (Creighton, 1979). The greater number of disulfides possible to

FIGURE 7–9. Disulfide folding pathway of RNase A. R is the fully reduced protein, and N is the fully folded protein with all four native disulfides; the other species are designated collectively by the number of protein disulfides they contain. The half-times for the intramolecular steps in forming the disulfide bonds are indicated. Those for the first and second disulfides are for the collective steps in forming all the various first and second disulfides. Those for the third and fourth disulfides are the average for individual steps, estimated from the accumulation of mixed disulfides. The slowest step in both directions is formation and breakage of one or more of the four native disulfides. (Creighton, 1979)

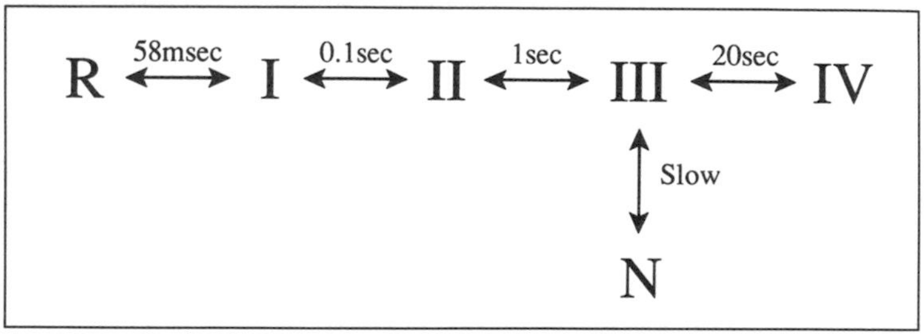

be formed in reduced RNase A than in reduced BPTI seems to be compensated by the greater separation of the Cys residues in the RNase sequence. That no single-disulfide intermediates predominate in the case of RNase A suggests that the individual disulfides contribute less to stability of local nonrandom conformation than in the case of BPTI.

Second disulfides are formed in the one-disulfide intermediates at a somewhat lower rate, but they are reduced nearly an order of magnitude more rapidly. Again, no particular disulfides predominate at this stage. Disulfide bond formation using DTT_S^S as disulfide reagent comes to a stop at this stage, due to the rate of forming third disulfides being slow and to their reduction being rapid, even though only low concentrations of DTT_{SH}^{SH} are generated by disulfide formation. No native protein was generated under these conditions, demonstrating that the rate-limiting step in refolding had not yet been encountered, in spite of claims to the contrary (Konishi et al., 1982). This was confirmed by demonstrating that the rate-limiting step in unfolding was reduction of the first disulfide bond using DTT_{SH}^{SH}. No partially reduced disulfide intermediates preceding the rate-limiting step in unfolding could be detected (Wearne and Creighton, 1988). The rate-limiting step in refolding is likely to be the reverse of this step.

Three- and four-disulfide intermediates could be generated during refolding using GSSG as disulfide reagent, but a large number of species were observed, and none predominated. Conformational restrictions on the ability of the protein to form disulfides were evident from the tendency of the disulfide intermediates to accumulate as mixed disulfides with glutathione (Creighton, 1977d, 1979; Konishi et al., 1981). All of the trapped one- and two-disulfide intermediates appear to be largely unfolded, irrespective of whether they were trapped by acidification (Konishi and Scheraga, 1980) or by alkylation (Creighton, 1977d, 1979; Galat et al., 1981). Only one quasi-native species was detected, that lacking the 40-95 disulfide bond, with the two thiol groups alkylated (Creighton, 1980a). Ignoring this species, the disulfide folding pathway of RNase A can only be depicted relatively simply, as in Figure 7–9.

Refolding of reduced RNase A is both similar to and different from that of reduced BPTI. It is similar in that the process is random initially and that the rate-limiting step seems to separate the native conformation from all other disulfide intermediates. The differences are that the folding process is much less directed in the case of RNase A, presumably due to the greater size of its polypeptide chain, and that there is no native-like species lacking one native disulfide equivalent to (30-51,5-55) in BPTI and (6-103) in RNase T_1; the latter difference probably arises from all the disulfides in native RNase A being buried. In view of the large number of disulfide intermediates, it is very likely that disulfide rearrangements are important in the case of RNase A, but this aspect has not been clarified.

4 GENERAL LESSONS FROM DISULFIDE FOLDING PATHWAYS

The disulfide folding pathways that have been elucidated thus far demonstrate some common principles and are compatible with the general observations of the folding mechanisms not involving disulfide bond formation and breakage of single-domain proteins. For example, the disulfide folding pathways are not random, but proceed through a limited number of intermediate states, as was postulated to be necessary in general, due to the very large number of protein conformations that are possible (Levinthal, 1968). In each case, the folding process starts out randomly, as a result of the unfolded nature of the fully reduced protein, in which all possible pairs of Cys residues come into proximity, probably depending primarily upon the distance in the primary structure between the Cys residues. Yet the pathways converge to a limited number of intermediates that are more stable than the other possibilities. This greater thermodynamic stability of some intermediate conformations can have kinetic consequences, as further folding from the stable conformations is favored. This is illustrated most clearly in the case of BPTI (Fig. 7–7), least clearly with RNase A (Fig. 7–9), where the convergence is much less pronounced. These intermediates are only metastable, however, and they are readily reduced to the unfolded, fully reduced form of the protein. Under typical folding conditions, there is a rapid pre-equilibrium between these intermediates and the fully unfolded protein.

These observations may be directly applicable to the folding of individual protein domains in general (Creighton, 1978, 1988c, 1990). They illustrate how all the molecules of an unfolded protein, which must be a conformationally very heterogeneous population in which every molecule probably has a unique conformation at any instant of time, can refold at the same rate (unless hindered by intrinsically slow isomerizations like *cis,trans* isomerization of the peptide bonds preceding Pro residues; Chapter 5). These observations also explain why the same rate of folding is obtained under the same folding conditions, irrespective of the nature of the starting form of the unfolded protein (Garel et al., 1976; Kato et al., 1981; Lynn et al., 1984); the rate of refolding of small proteins generally depends only upon the final folding conditions. This can be accounted for by the rapid pre-equilibrium that occurs very quickly when the protein is placed under folding conditions; the nature of the pre-equilibrium mixture will depend upon only the final folding conditions and will be independent of the initial unfolding conditions. Any favorable conformations present in the pre-equilibrium need not be present at detectable levels in the initial folded protein, and other nonrandom conformations that might be present may have no relevance for folding. The most stable partially folded conformations in the pre-equilibrium appear in the case of BPTI to involve interactions between elements of secondary structure, and there is considerable evidence for the rapid formation of some elements of secondary structure during the folding of many proteins (e.g., Ikeguchi et al., 1986).

The rate-limiting step in disulfide folding is very late in refolding, just before acquiring the native conformation (Goldenberg and Creighton, 1985). The highest free-energy barrier separates the native conformation from all the other intermediates. Where quasi-native forms are observed, the height of the energy barrier to unfolding is at least approximately proportional to the stability of the quasi-native conformation. When considering the rate and mechanism of unfolding, it becomes apparent that the overall transition state in the folding transition is a high-energy, distorted form of the native conformation. The data that are available on the transition states of protein folding not involving disulfide bonds are generally consistent with this proposal (Segawa and Sugihara, 1984a, b; Chen et al., 1989; Kuwajima et al., 1989; Matouschek et al., 1989).

The general consistency of these properties of disulfide folding transitions with other protein folding transitions are not entirely unexpected. Under the conditions developed for disulfide bond formation and breakage, the disulfide bond is not the static covalent cross-link that is so often imagined. Instead, it is a reversible interaction of variable stability that can be made, broken, and rearranged on relatively short time scales. Under these conditions, it is useful to consider the disulfide bond interaction as a special type of hydrogen bond that is possible only between the sulfur atoms of thiol groups (Fig. 7–10). But how far can the disulfide folding mechanisms be extrapolated to protein folding mechanisms in general?

FIGURE 7–10. (opposite) Possible similarity between making and breaking disulfide bonds and hydrogen bonds under the conditions used for disulfide folding. At the top is illustrated the steps involved in using the reagents DTT_S^S and DTT_{SH}^{SH} to make and break a disulfide bond that is buried in the interior of the folded protein. To reduce this disulfide bond, the folded conformation must be distorted for the reagent to obtain accessibility to the disulfide; if the same transition state is involved in formation of this disulfide bond, the transition state would be distorted by the reagent leaving. The transient formation of mixed disulfides between the Cys thiol groups of the reduced protein is indicated by the dashed lines. All thiol groups are depicted as protonated, but the thiolate anion is the reactive form. At the bottom is illustrated the energetically most favorable method of making and breaking a buried hydrogen bond, by interchanging the hydrogen bond partners with the solvent, rather than breaking them in isolation. (Creighton, 1978)

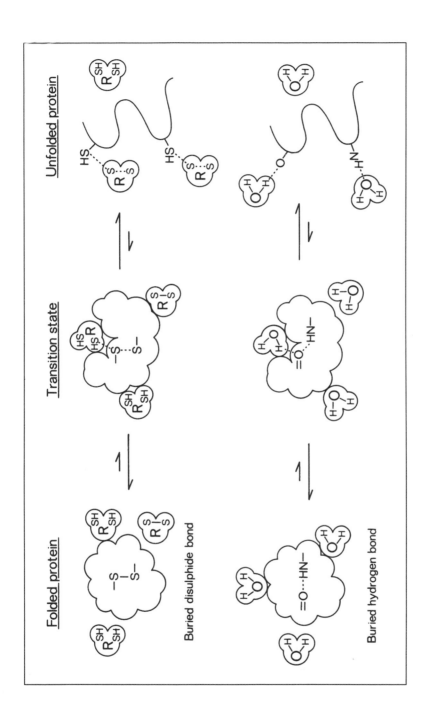

How do the unique properties of disulfide bonds influence the disulfide folding transitions? What is the relevance to protein folding in general of the disulfide rearrangements that occur so frequently? How can one analyze the energetics of disulfide bond formation and relate them to folding in the absence of this phenomenon? It is informative to consider these aspects in somewhat greater detail.

4.1   Energetics of Disulfide Bonds and of Folding

The protein disulfide bond has no fixed intrinsic stability, relative to the reduced form, but its stability depends upon the ratio of the thiol and disulfide forms of the reagent that are present, which determine the disulfide redox environment. This is analogous to the way that the difference is stability of the unfolded and folded forms of a protein that do not differ in disulfides depends upon the denaturant concentration, the temperature, the pH, and so on. In the latter cases, the net stability of the folded protein may be extrapolated to some standard conditions (see Chapter 3), usually physiological conditions of temperature, pH and salt concentration, in the absence of denaturant. Under such conditions, small monomeric proteins are generally stable by -5 to -15 kcal/mol. When disulfide formation is involved, there are no such standard disulfide redox conditions, but a standard condition can be defined as being that where the folded protein has the typical net stability relative to the fully reduced form. For example, three-disulfide native BPTI is 5.2 kcal/mol more stable than the reduced form when the ratio of DTT_S^S to DTT_{SH}^{SH} is 10^3.

Equilibrium constants under the selected redox conditions define the relative free energies of the various intermediates and final states. The free energies of the transition states can be calculated from intramolecular rate constants, using traditional transition state theory. Complications arise with the rates of disulfide bond formation and breakage, which are inter-molecular reactions. The rate of disulfide bond formation is best taken as that of the intramolecular step, k_{intra} (Fig. 7–2), which defines the free energy of the transition state involved in forming a disulfide bond. There is no intramolecular step in disulfide reduction, but the height of the free energy barrier is defined by k_{intra} and the equilibrium constant (Fig. 7–11).

From the free-energy profiles of Figure 7–11, it is apparent that the highest free-energy barrier separates all the less folded forms from the fully folded and the native-like (30-51,5-55) forms of BPTI, and of the homologous proteins I and K. The reaction coordinate is arbitrary in this case, but if it were expressed as compactness or degree of folding, the fully folded and (30-51,5-55) intermediates would be virtually indistinguishable in this respect, and the free-energy barrier would be seen to be very close to the native conformation, at a very late stage of folding. A similar free-energy

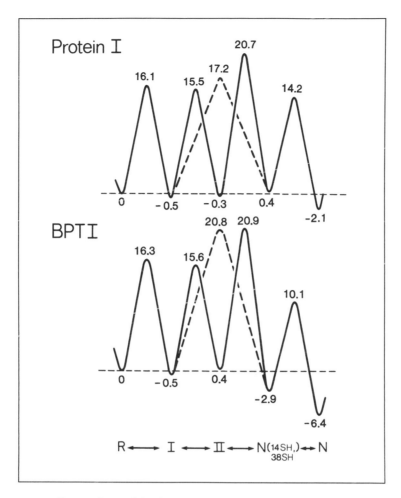

FIGURE 7–11. *Comparison of the free energy profiles for the disulfide folding transitions of black mamba protein I (top) and of BPTI (bottom). The apparent free energies (in kcal/mol) are given for the fully reduced (R), intermediate, and fully folded (N) forms of the proteins and for the transition states for the intramolecular transitions in forming or rearranging the protein disulfide bonds. The free energies are plotted along an arbitrary reaction coordinate, with the species along the pathway equally spaced. The species I includes all the one-disulfide intermediates, II includes the species (30-51,14-38), (30-51,5-14), and (30-51,5-38), while N(14SH,38SH) is the native-like intermediate (30-51,5-55). The solid line represents the disulfide rearrangement pathway (see Figure 7–7), whereas the dashed line is for formation directly of the 5-55 disulfide in intermediate (30-51). The redox conditions were arbitrarily chosen to be those where folded BPTI is 6.4 kcal/mole more stable than the reduced form. (Goldenberg and Creighton, 1985)*

profile can be constructed for RNase T$_1$ (Pace and Creighton, 1986), but all the data are not yet available for RNase A.

It is clear from energy profiles such as those of Figure 7–11, under redox conditions where the folded protein has the usual magnitude of stability, that the initial one- and two-disulfide intermediates in BPTI folding are only marginally stable relative to the reduced protein. The free-energy barriers separating them are also lower than that of the overall rate-limiting step, as are the energy barriers to the relatively rapid disulfide rearrangements within them. Consequently, there is under typical folding conditions a rapid pre-equilibrium between the reduced protein and the initial one- and two-disulfide intermediates on the left-hand side of the rate-limiting step; these intermediates unfold some 10^3 to 10^4 times more frequently than they complete refolding under these conditions. On the right-hand side, the native-like (30-51,5-55) form of BPTI is clearly of considerably higher free energy than is N, so loss of the 14-38 disulfide produces a significant increase in free energy. The energy barrier separating these two species is relatively low, however, so breakage of the 14-38 disulfide is rapidly reversed, some 10^7 times more frequently than unfolding is completed. Consequently, the (30-51,5-5) intermediate is barely populated even kinetically during unfolding under such redox conditions.

The population of only a limited number of intermediates can be shown to be important for the kinetics of refolding, by comparing the rates of refolding of BPTI in the absence and presence of 8 M urea, where essentially random intermediates are populated (Creighton, 1977c). Native BPTI with the three native disulfide bonds remains folded in 8 M urea and seems to be the three-disulfide form of lowest free energy. Consequently, refolding to native BPTI from the reduced protein occurs in the presence of 8 M urea, although more slowly than in the absence of urea (Table 7–1). The difference in overall rate is 14-fold, about the value expected if all 45 possible two-disulfide intermediates are present and unfolded in 8 M urea, rather than the three that normally comprise the kinetic species, II, and that rearrange to (30-51,5-55). The free-energy profiles in the presence and absence of urea indicate that the effect of the urea occurs primarily on the folded forms (30-51,5-55) and fully refolded BPTI, which are each destabilized relative to the reduced protein by 5.3 kcal/mole. The smaller energetic effects on the initial one- and two-disulfide intermediates are probably the entropic result of the greater number of species of each that are populated in 8 M urea.

Circularizing the polypeptide backbone of BPTI by introducing a peptide bond between its normal amino and carboxyl termini did not alter the folding pathway (Goldenberg and Creighton, 1983, 1984b). Cleaving the circular backbone after the residue that is normally number 15 did not alter the ability of the reduced protein with a circularly permuted amino acid sequence to refold to the normal native-like conformation (Goldenberg and Creighton, 1983). Clearly, protein folding pathways can usually adapt to numerous situations to reach the native conformation, although not always (Silen and Agard, 1989).

The free energies of all the one- and two-disulfide intermediates of BPTI, and of the homologues, are of higher free energy than either the fully reduced or fully folded forms, under all redox conditions. The BPTI disulfide folding transition is therefore cooperative, and it can be demonstrated experimentally that the transition is two-state at equilibrium, in that essentially only the fully folded and fully reduced forms are populated substantially at the midpoint of the disulfide transition (Creighton, 1977b). The folding transitions of RNase T_1 and of RNase A are also cooperative to similar extents. The disulfide folding transitions therefore share one of the most fundamental characteristics of protein folding transitions (Chapter 3).

4.2 The Nature of Protein Cooperativity

Cooperativity is necessary to maintain a stable folded conformation using interactions that are individually very weak, and the physical basis of cooperativity is one of the major problems of protein structure (Chapter 3, p. 123). The disulfide bond interaction illustrates how this cooperativity can arise, for unique information is available about the energetics of this interaction at each stage of folding. Whereas it is not feasible to measure directly the relative free energy of a hydrogen bond or of a van der Waals interaction in a protein, the equilibrium constants for disulfide bond formation provide this information for each disulfide bond made at each stage of folding (e.g., Equation (7–7)).

The initial disulfide bonds formed in the reduced protein are not very stable, for they are not stabilized by much nonrandom conformation in the protein, and the reduced protein gains considerable conformational entropy when the disulfide bond is broken. In other words, the effective concentrations of the various pairs of Cys thiol groups in the reduced protein are low, of the order of 10^{-2} to 10^{-1} M. The second disulfide bonds formed in reduced BPTI and RNase A also tend to be relatively unstable, for these disulfides are not favored to a great extent by the presence of the first disulfide. Consequently, the initial one- and two-disulfide intermediates tend to have comparable or higher free energies than the fully reduced form under typical redox conditions, like those in Figure 7–11.

Within the fully folded conformation, the stabilities of the disulfides are greatly increased. The folded protein conformation tends to keep the Cys residues in proximity, whether or not the disulfide bond is present. Also, having two Cys thiol groups in an environment within the folded protein designed for a disulfide can be energetically unfavorable to varying extents, which further increases the free-energy difference between the protein with and without the disulfide bond. Both factors result in the disulfide bonds within fully folded conformations being very stable, with effective concentrations of between 200 M and 5×10^5 M measured in folded BPTI (Creighton and Goldenberg, 1984). The native disulfide bonds are much more stable in the fully folded conformation than in the initial disulfide intermediates. The equilibrium between fully reduced and fully folded conformations depends upon the average stability of the disulfide bonds formed sequentially. Under conditions where the average disulfide would be present half the time, i.e., at the midpoint of the equilibrium between fully folded and fully reduced forms of the protein, the initial disulfides are less stable, and the final disulfides in the folded conformation are much more stable, so the disulfide transition is cooperative (Fig. 7–11).

The stability of each disulfide bond depends upon the conformation stabilizing it. In turn, the stability of that conformation depends upon the stability of the disulfide bond (Fig. 7–3). The two are linked functions, and the greater the stability of the conformation stabilizing the disulfide, the greater the stability of the disulfide bond. Such considerations obviously do not apply only to disulfide bonds, but they will also be pertinent to all other interactions within a folded conformation, be it a hydrogen bond, salt bridge, or van der Waals interaction. The cooperativity of protein structure can then be considered to be a result of a number of different interactions acting simultaneously to stabilize a particular conformation and, in doing so, to stabilize each other (Creighton, 1983a,b).

Being intramolecular, interactions within a folded protein can be much more stable than those between molecules. In other words, the effective concentration between two groups in a folded conformation can be much greater than is possible for feasible concentrations of molecules in a liquid (Page and Jencks, 1971; Kirby, 1980). For example, the stabilities of the disulfides of folded BPTI are much greater than are those of the intermolecular disulfide of GSSG, and the effective concentrations of the Cys thiols in the folded protein can be as high as 5×10^5 M. It is undoubtedly this factor that makes it possible for a protein like BPTI to be stable under *in vivo*–like redox conditions, where the majority of the glutathione is in the thiol form (Chapter 10; Creighton, 1983b). Similarly, this is undoubtedly the primary explanation as to how intramolecular hydrogen bonds and van der Waals interactions can stabilize a folded protein (Chapter 3), even though the unfolded protein will make otherwise similar, but intermolecular, interactions with the solvent.

4.3 Relevance of the Disulfide Rearrangements During Folding

What is the general relevance of the disulfide rearrangements that occur during the course of disulfide folding of the proteins that have been described here? It seems extremely likely that the intramolecular shuffling of disulfide bonds that occurs at the early stages of disulfide folding is analogous to the conformational flexibility that must be present in an initially unfolded polypeptide chain, which permits all the molecules to equilibrate rapidly to a finite set of conformations so as to be able subsequently to follow the same pathway. (See p. 337.) But what about the slow rearrangements that seem to occur in the rate-limiting steps of disulfide folding, as exemplified by the pathways of BPTI and RNase T_1? This phenomenon is not unique to these proteins, for there are ample indications that it is a general property of disulfide bond folding transitions (e.g., Taniyama et al., 1991). Its prevalence and importance probably accounts for the widespread occurrence of the enzyme protein disulfide isomerase and for the presence of disulfides primarily in proteins that encounter that enzyme in the endoplasmic reticulum (Chapter 10). Do comparable conformational rearrangements occur during protein folding that does not involve disulfide bonds? There are no data to rule this out. Further consideration, however, suggests that the disulfide rearrangements that occur in the rate-limiting steps, at a late stage of folding, are very exaggerated examples of the types of conformational rearrangements that normally occur.

One reason that disulfide rearrangements appear to be so prevalent in disulfide folding is that they are intramolecular; consequently, they tend to occur at much faster rates than the intermolecular steps in disulfide bond formation. This is not the explanation in the case of BPTI, however, for the rearrangements of the two-disulfide intermediates are no slower than the intramolecular rate of the alternative direct disulfide bond formation. The dashed lines in Figure 7–11 give a comparison of the energetics of the rearrangement steps with the intramolecular step in forming directly the 5-55 disulfide bond in intermediate (30-51), which avoids the rearrangements. In the case of BPTI, the two are about the same energy, as would be expected, because the same conformational transitions could take place in both, with the mixed disulfide playing the role in the direct step of the disulfide being rearranged. So the rearrangement pathway is the energetically preferred pathway in the case of BPTI. This is not the case with black mamba proteins I and K, however, where the direct pathway is energetically preferred when considered on a solely intramolecular basis. These proteins still tend to follow the rearrangement pathway during folding under the usual conditions, however, because disulfide bond formation is slow relative to intramolecular steps. The data are not available to make such comparisons in the case of RNase T_1 and RNase A.

What is the reason for the very slow rates of formation of the 5-55 disulfide bond in intermediates (30-51) and (30-51,5-55), which cause the rearrangement pathway to predominate in BPTI? Accumulation of the mixed disulfide in each case demonstrates directly that it is not due to inaccessibility or unreactivity of the thiol groups in the intermediates, but is due to a kinetic block in the second Cys residue displacing the mixed disulfide to generate the protein disulfide bond. It is a general observation that disulfide bond formation is slow when the disulfide bond that would be formed will produce a stable conformation in which the disulfide bond will be buried and inaccessible (Creighton, 1978). Moreover, the slowness of the rate of disulfide bond formation appears to be proportional to the stability of that conformation. For example, the 5-55 disulfide bond is formed in reduced BPTI in which the other Cys residues have been replaced by Ser at approximately the expected rate relative to the other one-disulfides (N. J. Darby, unpublished observation); the resulting (5-55) intermediate tends to adopt a quasi-native conformation, but this conformation is only marginally stable. The 5-55 disulfide bond is formed much more slowly in the (30-51) intermediate, and the resulting (30-51,5-55) intermediate has a very stable native-like conformation. Quantitative data are not available on the rate of forming the 5-55 disulfide in (30-51,14-38), but the Cys5 and Cys55 thiols in this intermediate accumulate to the greatest extent with mixed disulfides, suggesting the slowest rate of forming 5-55. The most stable fully folded conformation would result in this case. Finally, the 5-55 disulfide bond is formed in intermediate (30-51) much more rapidly in proteins I and K, which have much less stable folded conformations, than in BPTI. These observations indicate that the rate of forming a protein disulfide bond is governed by the tendency of the transition state to adopt a distorted form of the fully folded conformation (Creighton, 1985, 1988c, 1990).

The rates of forming each of the disulfide bonds of BPTI, of proteins I and K, and of RNase T_1 appear to be governed by the same transition state as occurs in reducing the disulfides, because the ratio of the rate constants observed in each case appears to be the same as the equilibrium constant measured directly. This need not be the case, for other steps could become rate limiting under conditions far from equilibrium. Only minor examples have been observed when disulfide bond formation is rapid, which have been attributed to *cis,trans* isomerization of Pro peptide bonds becoming rate limiting (Creighton, 1977b; Chapter 5). Otherwise, conformational changes appear to be rapid relative to disulfide bond formation. In this case, the native conformation could tend to be formed as the disulfide is being formed and the transition state in forming a native disulfide bond would be the same as that in reducing the same disulfide in the folded conformation.

The transition state in reducing a stable disulfide bond within the interior of a folded protein (Fig. 7–10) will necessarily be distorted and of high free energy. If the same high-energy transition state occurs in the reverse direction, forming the disulfide bond will also be relatively slow. The only unexplained observation is why the height of the free energy barrier to forming the disulfide varies with the stability of the resulting conformation (Fig. 7–11). Either the lower free energy of the folded conformation is also reflected in the free energies of the reduced proteins and the initial one- and two-disulfide intermediates, which seems unlikely, or the free energy of the transition state is affected more than is that of the fully folded state.

Similar considerations are likely to apply to making and breaking hydrogen bonds within a protein (Fig. 7–10), but with quantitative differences. All other considerations being comparable, rearranging hydrogen bonds, either within a protein or with the aqueous solvent, should be the energetically preferred mechanism of conformational flexibility in folded proteins, in contrast to simply separating two hydrogen bond partners in isolation. The energetic preference is, however, unlikely to be greater than 5 kcal/mol per hydrogen bond. With disulfide bonds, rearrangements either within the protein or with the reagent are a virtual necessity, for breaking a disulfide bond in isolation is energetically prohibitive. Therefore, rearrangements of disulfides are much more energetically favored than are rearrangements of hydrogen bonds. The other difference is that with hydrogen bonds there is always an alternative donor or acceptor no further away than the next amino acid residue, whereas disulfide bonds can occur only between Cys residues, which are relatively rare. Intramolecular disulfide rearrangements consequently would be expected to involve much more extensive conformational changes than would rearrangements of hydrogen bonds.

In summary, the extensive disulfide rearrangements that occur in the rate-limiting step in refolding of BPTI do not necessarily imply that comparable conformational rearrangements must occur during folding not involving disulfide bonds. It can also be argued that the disulfide rearrangement pathway of BPTI is largely due to the exceptional stability of the folded conformation of this protein, as is the tendency of quasi-native intermediates to occur. The disulfide rearrangement pathway appears to be an attempt to avoid the energetically extremely difficult path of forming the native disulfides and the native conformation directly. The less stable black mamba proteins I and K may be more representative of protein folding pathways in general. They refold most readily on an intramolecular basis by forming the disulfide bonds sequentially, first 30-51, then 5-55, followed by 14-38. The rate-limiting step is still forming the second disulfide bond, 5-55, to generate the native-like conformation, and it undoubtedly involves going through a distorted form of the native conformation.

Conformational rearrangements may also occur in this rate-limiting step, but in this case there are no Cys residues in positions to detect them as disulfide rearrangements.

The rearrangements that take place in the slow steps in disulfide protein folding may be extreme examples of the conformational rearrangements that take place during the rate-limiting step in folding transitions not involving disulfide bonds. Some such conformational rearrangements are likely in the latter case, as a result of the cooperativity of protein structure and the similarity of the overall folding transition state to the native conformation.

5 SUMMARY

The known disulfide folding pathways illustrate the known general properties of protein folding transitions (Creighton, 1988c). The rapid interchange of the initial disulfide bonds formed, and their relative instabilities, indicate how a conformationally heterogeneous population of unfolded protein molecules can rapidly equilibrate to be able subsequently to follow the same pathway. Convergence of unfolded molecules to a common pathway occurs because certain partially folded intermediates have lower free energies than others, and consequently predominate. The (30-51) intermediate of BPTI illustrates the importance at early stages of folding of native-like conformations, of mutually stabilizing interactions between unstable individual elements of secondary structure, and of nonpolar interactions. Such partially folded intermediates are much less stable than the fully folded and unfolded molecules, illustrating the importance of cooperativity in folding. The high free-energy barrier that separates the partially folded intermediates from the fully folded state reflects the need to go through a distorted form of the native conformation, which has high energy due to disruption of the cooperative fully folded conformation.

Note

M.-H. Chau and J. W. Nelson (*FEBS Letters 291*, 296–298, 1991) have measured the equilibrium constant for thiol-disulfide interchange between dithiothreitol and glutathione (Equation (7–7)) directly and have obtained values of 200 M at pH 7 and 375 M at pH 8.7. We have confirmed these measurements, but the reason for the discrepancy with the much greater values measured by Cleland (1964) and Szajewski and Whitesides (1980) is not known. This new value indicates a value for K_{DTT} of Equation (7–8) about $(187 \text{ M})^{-1}$ at pH 8.7, about threefold greater than that estimated previously (Creighton and Goldenberg, 1984). It gives values of k_{intra} using Equation (7–8) that are consistent with those measured directly using glutathione (N. J. Darby, unpublished observations). Therefore, values of k_{intra} estimated previously using Equation (7–8) should be recalculated using the new value of K_{DTT}, and the half-times for disulfide bond formation in Figures 7–7, 7–8, and 7–9 should be increased by a factor of three.

J. P. Staley and P. S. Kim (*Proc. Natl. Acad. Sci. USA 89*, 1519–1523, 1992) have shown that the quasi-native conformation of an analogue of the (5–55) intermediate of BPTI is much more stable when the free Cys residues are replaced by Ala than by Ser (van Mierlo et al., 1991b). The Ala analogue is also much more stable than the conformation of the true intermediate during folding at pH 8.7 (Section 3.1.4), illustrating that Ser residues are much better mimics of partially-ionized Cys residues than are Ala.

REFERENCES

Altmann, K.-H., and Scheraga, H. A. (1990) *J. Am. Chem. Soc. 112*, 4926–4931.

Amir, D., and Haas, E. (1988) *Biochemistry 27*, 8889–8893.

Anfinsen, C. B., Haber, E., Sela, M., and White, F. H. Jr. (1961) *Proc. Natl. Acad. Sci. USA 47*, 1309–1314.

Brown, J. R., and Hartley, B. S. (1966) *Biochem. J. 101*, 214–241.

Burns, J. A., and Whitesides, G. M. (1990) *J. Am. Chem. Soc. 112*, 6296–6303.

Chen, B.-I., Baase, W. A., and Schellman, J. A. (1989) *Biochemistry 28*, 691–699.

Cleland, W. W. (1964) *Biochemistry 3*, 480–482.

Creighton, T. E. (1974a) *J. Mol. Biol. 87*, 579–602.

Creighton, T. E. (1974b) *J. Mol. Biol. 87*, 603–624.

Creighton, T. E. (1975a) *J. Mol. Biol. 95*, 67–199.

Creighton, T. E. (1975b) *J. Mol. Biol. 96*, 767–776.

Creighton, T. E. (1975c) *J. Mol. Biol. 96*, 777–782.

Creighton, T. E. (1977a) *J. Mol. Biol. 113*, 275–293.

Creighton, T. E. (1977b) *J. Mol. Biol. 113*, 295–312.

Creighton, T. E. (1977c) *J. Mol. Biol. 113*, 313–328.

Creighton, T. E. (1977d) *J. Mol. Biol. 113*, 329–341.

Creighton, T. E. (1978) *Prog. Biophys. Mol. Biol. 33*, 231–297.

Creighton, T. E. (1979) *J. Mol. Biol. 129*, 411–431.

Creighton, T. E. (1980a) *FEBS Letters 118*, 283–288.

Creighton, T. E. (1980b) *J. Mol. Biol. 144*, 521–550.

Creighton, T. E. (1981) *J. Mol. Biol. 151*, 211–213.

Creighton, T. E. (1983a) *Biopolymers 22*, 49-58.

Creighton, T. E. (1983b) In *Functions of Glutathione: Biochemical, Physiological, Toxicological and Clinical Aspects* (A. Larsson, S. Orrenius, A. Holmgren, and B. Mannervik, eds.), Raven Press, New York, pp. 205–213.

Creighton, T. E. (1984) *Methods Enzymol. 107*, 305–329.

Creighton, T. E. (1985) *J. Phys. Chem. 89*, 2452–2459.

Creighton, T. E. (1986) *Methods Enzymol. 131*, 83–106.

Creighton, T. E. (1988a) *BioEssays 8*, 57–63.

Creighton, T. E. (1988b) *Biophys. Chem. 31*, 155–162.

Creighton, T. E. (1988c) *Proc. Natl. Acad. Sci. USA 85*, 5082–5086.

Creighton, T. E. (1990) *Biochem. J. 270*, 1–16.

Creighton, T. E. (1992) *Science 256*, 111–112.

Creighton, T. E., and Dyckes, D. F. (1981) *J. Mol. Biol. 146*, 375–387.

Creighton, T. E., and Goldenberg, D. P. (1984) *J. Mol. Biol. 179*, 497–526.

Creighton, T. E. , Kalef, E., and Arnon, R. (1978) *J. Mol. Biol. 123*, 129–147.

Darby, N. J., van Mierlo, C. P. M., and Creighton, T. E. (1991) *FEBS Letters 279*, 61–64.

Darby, N. J., van Mierlo, C. P. M., Scott, G. H. E., Neuhaus, D., and Creighton, T. E. (1992) *J. Mol. Biol. 224*, 905–911.

Eigenbrot, C., Randal, M., and Kossiakoff, A. A. (1990) *Protein Eng. 3*, 591–598.

Ewbank, J., and Creighton, T. E. (1991) *Nature 350*, 518–520.

Galat, A., Creighton, T. E., Lord, R. C., and Blout, E. R. (1981) *Biochemistry 20*, 594–601.

Garel, J.-R., Nall, B. T., and Baldwin, R. L. (1976) *Proc. Natl. Acad. Sci. USA 73*, 1853–1857.

Gilbert, H. (1990) *Adv. Enzymol. 63*, 69–172.

Givol, D., DeLorenzo, F., Goldberger, R. F., and Anfinsen, C. (1965) *Proc. Natl. Acad. Sci. USA 53*, 676–684.

Goldenberg, D. P. (1985) *J. Cell. Biochem. 29*, 321–335.

Goldenberg, D. P. (1988) *Biochemistry 27*, 2481-2489.

Goldenberg, D. P., and Creighton, T. E. (1983) *J. Mol. Biol. 165*, 407–413.

Goldenberg, D. P., and Creighton, T. E. (1984a) *Anal. Biochem. 138*, 1-18.

Goldenberg, D. P., and Creighton, T. E. (1984b) *J. Mol. Biol. 179*, 527–545.

Goldenberg, D. P., and Creighton, T. E. (1985) *Biopolymers 24*, 167–182.

Goldenberg, D. P., Frieden, R. W., Haack, J. A., and Morrison, T. B. (1989) *Nature 338*, 127–132.

Goodman, E. M., and Kim, P. S. (1989) *Biochemistry 28*, 4343–4347.

Goto, Y., and Hamaguchi, K. (1981) *J. Mol. Biol. 146*, 321–340.

Haas, E., McWherter, C. A., and Scheraga, H. A. (1988) *Biopolymers 27*, 1–21.

Hollecker, M., and Creighton, T. E. (1983) *J. Mol. Biol. 168*, 409–437.

Hollecker, M., and Larcher, D. (1989) *Eur. J. Biochem. 179*, 87–94.

Ikeguchi, M., Kuwajima, K., Mitani, M., and Sugai, S. (1986) *Biochemistry 25*, 6965–6972.

Kato, I., and Anfinsen, C. B. (1969) *J. Biol. Chem. 244*, 1004–1007.

Kato, S., Okamura, M., Shimamoto, N., and Utiyama, H. (1981) *Biochemistry 20*, 1080–1085.

Keller, R. M., Baumann, R., Hunziker-Kwik, E.-H., Joubert, F. J., and Wüthrich, K. (1983) *J. Mol. Biol. 163*, 623–646.

Kirby, A. J. (1980) *Adv. Phys. Org. Chem. 17*, 183–278.

Konishi, Y., Ooi, T., and Scheraga, H. A. (1981) *Biochemistry 20*, 3945–3955.

Konishi, Y., Ooi, T., and Scheraga, H. A. (1982) *Biochemistry 21*, 4734–4740.

Konishi, Y., and Scheraga, H. A. (1980) *Biochemistry 19*, 1308–1316.

Kosen, P. A., Creighton, T. E., and Blout, E. R. (1980) *Biochemistry 19*, 4936–4944.

Kosen, P. A., Creighton, T. E., and Blout, E. R. (1981) *Biochemistry 20*, 5744–5454.

Kosen, P. A., Creighton, T. E., and Blout, E. R. (1983) *Biochemistry 22*, 2433–2440.

Kuwajima, K., Mitani, M., and Sugai, S. (1989) *J. Mol. Biol. 206*, 547–561.

Levinthal, C. (1968) *J. Chim. Phys. 65*, 44–45.

Lin, T.-Y., and Kim, P. S. (1989) *Biochemistry 28*, 5282–5287.

Lynn, R. M., Konishi, Y., and Scheraga, H. A. (1984) *Biochemistry 23*, 2470–2477.

Manning, M. C., and Woody, R. W. (1989) *Biochemistry 28*, 8609–8613.

Matouschek, A., Kellis, J. T. Jr., Serrano, L., and Fersht, A. R. (1989) *Nature 340*, 122–126.

Oas, T. G., and Kim, P. S. (1988) *Nature 336*, 42–48.

Osterhout, J. J., Baldwin, R. L., York, E. J., Stewart, J. M., Dyson, H. J., and Wright, P. E. (1989) *Biochemistry 28*, 7059–7064.

Pace, C. N., and Creighton, T. E. (1986) *J. Mol. Biol. 188*, 477–486.

Pace, C. N., Grimsley, G. R., Thomson, J. A., and Barnett, B. J. (1988) *J. Biol. Chem. 263*, 11820–11825.

Page, M. I., and Jencks, W. P. (1971) *Proc. Natl. Acad. Sci. USA 68*, 1678–1683.

Radzicka, A., and Wolfenden, R. (1988) *Biochemistry 27*, 1664–1670.

Segawa, S.-I., and Sugihara, M. (1984a) *Biopolymers 23*, 2473–2488.

Segawa, S.-I., and Sugihara, M. (1984b) *Biopolymers 23*, 2489–2498.

Shaked, Z., Szajewski, R. P., and Whitesides, G. M. (1980) *Biochemistry 19*, 4156–4166.

Silen, J. L., and Agard, D. A. (1989) *Nature 341*, 462–464.

Srinivasan, N., Sowdhamini, R., Ramakrishnan, C., and Balaram, P. (1990) *Intl. J. Peptide Protein Res. 36*, 147–155.

Staley, J. P., and Kim, P. S. (1990) *Nature 344*, 685–688.

Stassinopoulou, C. I., Wagner, G., and Wüthrich, K. (1984) *Eur. J. Biochem. 145*, 423–430.

States, D. J., Creighton, T. E., Dobson, C. M., and Karplus, M. (1987) *J. Mol. Biol. 195*, 731–739.

States, D. J., Dobson, C. M., Karplus, M., and Creighton, T. E. (1984) *J. Mol. Biol. 174*, 411–418.

Strydom, D. J. (1973) *Nature New Biol. 243*, 88–89.

Szajewski, R. P., and Whitesides, G. M. (1980) *J. Am. Chem. Soc. 102*, 2011–2026.

Taniyama, Y., Kuroki, R., Omura, F., Seko, C., and Kikuchi, M. (1991) *J. Biol. Chem. 266*, 6456–6461.

Thornton, J. M. (1981) *J. Mol. Biol. 151*, 261–287.

van Mierlo, C. P. M., Darby, N. J., Neuhaus, D., and Creighton, T. E. (1991a) *J. Mol. Biol. 222*, 353–371.

van Mierlo, C. P. M., Darby, N. J., Neuhaus, D., and Creighton, T. E. (1991b) *J. Mol. Biol. 222*, 373–390.

van Mierlo, C. P. M., Darby, N. J., and Creighton, T. E. (1992) *Proc. Natl. Acad. Sci. USA*, in press.

Wearne, S. J., and Creighton, T. E. (1988) *Proteins: Struct. Funct. Genet. 4*, 251–261.

Weissman, J. S., and Kim, P. S. (1991) *Science 253*, 1386–1393.

Wlodawer, A., Nachman, J., Gilliland, G. L., Gallagher, W., and Woodward, C. (1987) *J. Mol. Biol. 198*, 469–480.

Zhang, R., and Snyder, G. H. (1989) *J. Biol. Chem. 264*, 18472–18479.

8

Mutational Analysis of Protein Folding and Stability

DAVID P. GOLDENBERG

1 INTRODUCTION AND OVERVIEW

Over the past ten years, the use of genetically altered proteins has become an important experimental tool for studying the physical principles that determine the thermodynamic stabilities and folding mechanisms of proteins. Although genetic and chemical methods have been utilized for several decades to prepare modified proteins for structural and functional studies, powerful new genetic engineering techniques have revolutionized our ability to manipulate the covalent structures of polypeptide chains. By comparing the thermodynamics and kinetics of folding among proteins that differ by only one or a few amino acid residues, the roles of individual residues and interactions can be tested. Numerous laboratories are now using mutations to study protein folding and stability, and several reviews, emphasizing different approaches, have appeared recently (e.g., Alber, 1989; Goldenberg, 1988; Matthews, 1987a; Pakula and Sauer, 1989; Shortle, 1989).

One of the important results to emerge from mutational studies is a qualitative picture of how different residues and interactions contribute to the stability of a folded structure. Although it might have been anticipated that a few residues of a protein would be particularly important in stabilizing the native state, it now appears that a quite large fraction (perhaps one third) of the residues make significant contributions to stability and that few residues make absolutely essential contributions. This conclusion is drawn from the finding, with several proteins, that the native state can be significantly destabilized by replacements of many different kinds of residues throughout the polypeptide chain. At the same time, the folded structures of proteins have been found to be more tolerant of replacements

353

than might have been expected. The residues that are most important to stability are usually located in the interior of the folded protein.

In addition to this qualitative picture, there is now a large body of quantitative data about the degree to which different types of amino acid substitutions destabilize the native state. To the extent that a mutation acts by simply removing a single interaction in the folded protein, these data can be used to estimate the stabilization arising from individual interactions. As a result, we now have a much better sense of how much hydrogen bonds, salt bridges, van der Waals interactions, disulfide bonds, and the hydrophobic effect contribute to stability. However, it is also apparent that the magnitude of the stabilization from an individual interaction can be very much dependent on the structure and interactions that surround it in the native protein and, possibly, in the unfolded protein. Thus, there is a great need for improved theoretical analyses of the interactions in protein structures. Data from mutational experiments may play an important role in the development and testing of such theoretical treatments.

Amino acid replacements have also been utilized to study the mechanism of protein folding. In these studies, the effects of amino acid replacements on the relative free energies of the intermediates and transition states making up the folding pathway have been examined. Replacements at different sites have been seen to have distinct effects on different stages of the pathways, providing evidence that folding takes place by mechanisms in which the structure of the native protein forms in a preferred order. Further studies of mutant proteins should allow additional details of the structures of folding intermediates and transition states to be inferred.

While mutational studies have led to rapid progress in understanding some aspects of protein folding and stability, they have also helped focus attention on aspects that are still poorly understood. For instance, in some cases it has been found that a substitution causes only very local changes to the three-dimensional structure of a protein while other changes cause more extensive perturbations. It is not yet understood why the structural changes are more localized in some cases than others, and it is not known what the energetic consequences of these changes are. Even less is known about how mutations alter the properties of the unfolded protein and its interaction with solvent. Since the stability of the native protein is usually measured relative to that of the unfolded protein, it is clear that a complete understanding of the effects of mutations requires a better understanding of the unfolded state. Finally, most studies of protein folding have examined refolding of unfolded chains *in vitro*. It is now apparent that the folding of newly synthesized chains *in vivo* is often facilitated or regulated by other proteins (see Chapter 10), and it is possible that some amino acid replacements alter the ability of the protein to correctly interact with these "molecular chaperones" and thereby influence *in vivo* folding in ways that could not be predicted from *in vitro* studies.

The sections that follow describe some of the mutational strategies and methodologies currently being used to study protein folding and discuss some of the important conclusions, as well as new questions, that have arisen from these experiments.

2 STRATEGIES AND METHODOLOGIES FOR MUTATIONAL ANALYSIS OF FOLDING

Since the complete analysis of one protein variant might take a single worker more than a year (particularly if the analysis includes high-resolution structure determination), considerable care is required in choosing the protein variants to prepare and what techniques to use in characterizing them. A thorough characterization of a single mutant protein may be required to provide detailed information about how the altered residue contributes to stability and the folding mechanism. On the other hand, a less detailed survey of a large number of substitutions, at one or many sites, may reveal general patterns that would not emerge from studies of one or a few mutants. Fortunately, different investigators have chosen to use a variety of strategies for designing or screening for mutants and have utilized different methods for measuring the effects of the substitutions.

2.1 Strategies for Generating Mutant Proteins

It is now possible to make virtually any point mutation in any DNA sequence, including a protein-encoding gene, by chemically synthesizing a short deoxy-oligonucleotide and enzymatically incorporating it into a larger cloned sequence (Smith, 1985). Provided that the altered gene can be expressed in a bacterial or other easily grown cell type, variants with nearly any desired amino acid replacement can be produced (although the protein may not fold or may not accumulate *in vivo*). This methodology has made it possible to test the roles of individual residues in protein folding, as well as function.

Another approach to generating mutant proteins is to randomly mutagenize the gene of interest and to use a selection or screening procedure to identify protein variants with particular properties. Random mutagenesis has been used for many decades to study a variety of biological problems, and has the important advantage that it does not depend upon any prior information about which sequences are most important for the property of interest. Once mutants have been isolated by a screening procedure, the amino acid replacements that give rise to the protein phenotype can be determined by DNA sequencing. Isolating a mutant with a particular amino acid replacement provides evidence for the importance of the altered residue for the property screened, which can be studied further using physical methods and site-directed mutagenesis. Patterns observed among a large collection of mutants can provide insights about how many and what types of residues and interactions affect the properties that were originally screened for.

The major requirement for utilizing random mutagenesis is a method to identify the mutant clones of interest. Many studies have utilized simple screens for protein function *in vivo* or on petri plates after cell lysis (e.g., Alber and Wozniak, 1985; Coplen et al., 1990; Hecht et al., 1983; Shortle and Lin, 1985; Smith et al., 1980; Streisinger et al., 1961). Unfortunately, it is often not known exactly why a particular mutant is identified in a screen for function. An amino acid substitution might prevent function by altering the active site of a protein, it might prevent folding kinetically or thermo-dynamically, or it might lead to inactivation of the protein by proteolysis or aggregation. Thus, it is usually necessary to determine why particular mutants appear defective in the screen in order to interpret fully the results of a random mutagenesis experiment.

As methods for manipulating DNA sequences *in vitro* have become more powerful, the distinction between site-directed mutagenesis and random mutagenesis is becoming somewhat blurred. It is now a feasible, though formidable, project to use synthetic oligonucleotides to systematically change many, or even all, of the residues in a protein. "Degenerate" oligo-nucleotides (which are synthesized with more than one nucleotide reagent at one or more positions, so that a mixed population of sequences is generated) can be used to make multiple substitutions at each site (e.g., Reidhaar-Olson and Sauer, 1988, 1990). Alternatively, codons of interest can be changed to non-sense codons, and suppressor tRNA genes can then be used to introduce different residues at each position. There are now natural and synthetic suppressor genes to facilitate introducing as many as 13 different residues in this way (Kleina et al., 1990). These strategies provide more control over the nature of the amino acid replacements than is possible in traditional random mutagenesis experiments, while still allowing many of the sites of the protein to be examined systematically. The clones gener-ated by these procedures can be screened to identify those producing proteins with altered function or stability.

2.2 Methods for Characterizing Mutant Proteins

Often, qualitative information about the relative effects of different amino acid replacements can be inferred from simple tests of function and stability. The fact that a mutant protein has biological activity suggests that the substitution does not destabilize the protein so much that it is unable to fold to a structure roughly like that of the wild-type protein. Since different proteins have different net stabilities, however, there is probably no way to know *a priori* how much destabilization is required to prevent a given protein from folding. The ability of a mutant protein to function *in vivo* is probably also dependent upon its resistance to degradation. Correlations between thermodynamic stability and resistance to proteolysis have been observed for several proteins (e.g., Pakula et al., 1986; Parsell and Sauer, 1989; Schellman, 1986), and measuring the *in vivo* lifetime of a protein may serve as a rough estimate of its stability. Degradation is known to depend on additional factors, however, and rates of proteolysis must be interpreted with caution (Rechsteiner et al., 1987). To obtain quantitative information about the effects of substitutions on the stability and the folding mechanism, it is necessary to compare the folding thermodynamics and kinetics of the mutant protein with those of the wild-type protein.

2.2.1 Thermodynamic Measurements

As discussed in Chapter 3, the thermodynamic stability of a protein is defined by the free energy change for the unfolding of the native protein:

$$N \rightleftharpoons U. \tag{8-1}$$

Like other equilibria, the stability of a protein depends upon temperature and other solution components, and the native protein is usually greatly favored at physiological temperatures and solution conditions. In order to measure the equilibrium between the native and unfolded proteins, it is thus necessary to destabilize the native state so that measurable concentrations of the unfolded protein are present. This is most commonly done by increasing the temperature for adding chaotropic denaturants, such as urea or guanidinium chloride (GdmCl) as some physical property is used to monitor the unfolding transition. If the transition is sufficiently cooperative so that only the native and unfolded forms are present at significant concentrations, it is relatively straightforward to calculate the fraction of protein unfolded (f_u) at each denaturant concentration or temperature from the observed parameter.

Figure 8–1A illustrates hypothetical urea-induced unfolding curves such as would be obtained by following some conformational parameter, such as optical absorbance or fluorescence. Converting the primary data to a plot of f_u versus denaturant concentration (as in Figure 8–1B) requires knowing the value of the observed parameter for both the native and unfolded forms at each urea concentration. This is generally done by extrapolating the baseline regions of the curve where the native and unfolded states predominate. These extrapolations can be a significant source of uncertainty, particularly if the protein is exceptionally stable or unstable, so that one of the baseline regions is very short.

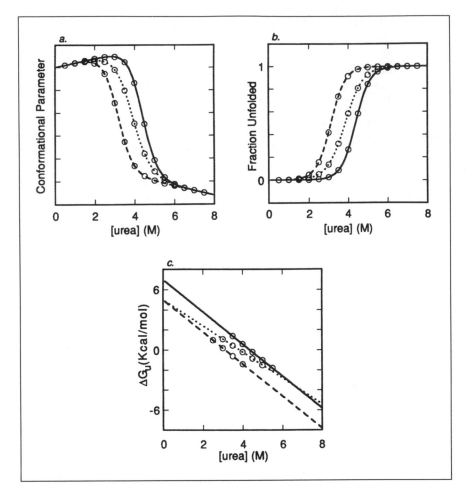

FIGURE 8–1. *Hypothetical urea-induced unfolding transitions for a wild-type protein (solid lines) and two destabilized mutants (dashed and dotted lines). (A) A plot of the primary experimental data versus urea concentration. The conformational parameter might be an optical parameter, such as UV absorbance or fluorescence, or a measure of hydrodynamic volume. (B) A plot of the fraction of protein unfolded (f_u) versus urea concentration, calculated from the observed data. This calculation assumes that only the native and unfolded protein contribute significantly to the observed parameter and requires extrapolation from the baseline regions. (C) A plot of the free energy of unfolding (ΔG_u) versus urea concentration, as calculated from Equation (8–2) of the text. Note that although the fraction unfolded can be measured at all denaturant concentrations, ΔG_u can only be determined at those urea concentrations where the concentrations of both the native and unfolded forms can be accurately measured. As a consequence, the value of ΔG_u in the absence of denaturant (ΔG_u^0) must be*

The free energy change for unfolding (ΔG_u) at each denaturant concentration or temperature can then be calculated:

$$\Delta G_u = -RT \ln K_u = -RT \ln([U]/[N]) = -RT \ln[f_u/(1-f_u)], \qquad (8\text{–}2)$$

where K_u is the equilibrium constant for unfolding and $[U]$ and $[N]$ are the equilibrium concentrations of the unfolded and native states, respectively. The change in stability arising from an amino acid replacement is the difference in the unfolding free energies of the wild-type and mutant proteins: $\Delta\Delta G_u = \Delta G_u(wt) - \Delta G_u(mutant)$. If two protein variants display overlapping unfolding transitions, then their stabilities can be compared directly under those conditions where the concentrations of both the native and unfolded states are measurable for both proteins. However, if the transitions do not overlap or if it is desirable to compare the stabilities under physiological conditions, the observed free energy changes must be extrapolated to other conditions.

Denaturant-induced unfolding transitions are most commonly interpreted assuming that ΔG_u is a linear function of denaturant concentration (C):

$$\Delta G_u(C) = \Delta G_u^o - m \times C \qquad (8\text{–}3)$$

where ΔG_u^o is the value of ΔG_u in the absence of denaturant and m describes the dependence of ΔG_u on denaturant concentration (Pace, 1975; Schellman, 1978, 1987). The two parameters defining the unfolding transition are estimated by fitting the values of ΔG_u observed in the transition region to this expression and can then be used to evaluate ΔG_u at other denaturant concentrations. Hypothetical plots of ΔG_u versus urea concentration corresponding to the unfolding curves in Figure 8–1A and B are shown in Figure 8–1C. Accurate estimates of ΔG_u can usually be obtained over only a narrow range of denaturant concentrations, and the extrapolation to zero denaturant often covers a large concentration range. As a consequence, there are often large uncertainties in the values of extrapolated stabilities.

determined by extrapolation. The solid lines in all three plots represent the results that would be expected at 25°C with a protein for which ΔG_u^o is 7,000 cal/mol and m *in Equation (8–3) is 1,600 cal/mol M. The dashed lines are the results expected for a destabilized mutant for which ΔG_u^o is 5,000 cal/mol and* m *has the same value as for the wild-type protein. The dotted lines represent the results expected for a mutant for which ΔG_u^o is also 5,000 cal/mol, but* m *is 1,300 cal/mol/M. Note that the two mutants have the same stability in the absence of denaturant, but as the urea concentration increases the difference in their stabilities increases. At the highest urea concentrations, ΔG_u is actually greater for the mutant with the altered* m *value than it is for the wild-type protein.*

Different methods for making such extrapolations have been discussed recently by Pace et al. (1990) and by Santoro and Bolen (1988).

If the value of m is the same for the wild-type and mutant proteins, then the difference in stability in the absence of denaturant will be the same as the stability difference in the transition region. For some proteins, such as barnase (Kellis et al., 1988, 1989; Serrano et al., 1990), m does not appear to be different for the wild-type and mutant proteins, allowing for straightforward comparisons of stability. On the other hand, quite large differences have been seen in m for variants of staphylococcal nuclease (Shortle and Meeker, 1986) and ribonuclease T1 (RNase T1) (Shirley et al., 1989). When m is different for two variants, the difference in stability, $\Delta\Delta G_u$, depends upon the denaturant concentration chosen for the comparison and may even change sign at different denaturant concentrations, as illustrated in Figure 8–1C.

As of early 1992, the physical origins of variations in m are uncertain. Denaturants such as urea and guanidinium ions are thought to act by favoring the solvation of nonpolar groups that become exposed when the protein unfolds, and m has been interpreted as a measure of the difference in solvent exposure of hydrophobic groups in the native and unfolded states (Creighton, 1979; Pace et al., 1990; Schellman, 1978; Shortle and Meeker, 1986; Tanford, 1968). Shortle and Meeker (1986) have suggested that the differences in m seen among staphylococcal nuclease variants may be due to differences in the distribution of conformations making up the unfolded state, with larger values of m reflecting more extended and solvent accessible conformations. Differences in m might also arise from specific binding of denaturant molecules to the native or unfolded protein. Pace et al. (1990) estimate that a single site on native RNase T1 that bound a guanidinium ion with a binding constant of 11 M^{-1} would decrease m by 170 cal mol^{-1} M^{-1}, a change of about 7%, if there were no comparable binding site on the unfolded protein. Different values of m could also arise if the degree of cooperativity changes among protein variants. If partially folded forms contribute to the measured property, then the transition cannot validly be interpreted assuming only two states, and doing so will result in lower apparent values of m.

Interpreting thermal unfolding transitions also requires a description of how ΔG_u changes, in this case with temperature (T). As discussed in Chapter 3, the variation of ΔG_u with T is determined by the enthalpy change of unfolding (ΔH_u) and the change in heat capacity (ΔC_p, the derivative of ΔH_u with respect to T) (Becktel and Schellman, 1987; Privalov and Khechinashvili, 1974; Privalov, 1989). Because ΔH_u is not constant with temperature, accurate comparisons of the stabilities of protein variants require reliable estimates of ΔC_p and ΔH_u for each protein, and the relative stabilities of variants can depend markedly on the temperature chosen for the comparison.

One of the most striking results to emerge from thermal unfolding studies is that single amino acid replacements often cause large changes in

ΔH_u accompanied by compensating changes in the entropy change for unfolding, to yield relatively small changes in ΔG_u (Connelly et al., 1991; Hawkes et al., 1984; Matthews et al., 1980; Shortle et al., 1988). The origin of this "entropy-enthalpy compensation" is not known, but may reflect differences in the solvation of the native or unfolded forms of the protein variants (Lumry and Rajender, 1970) or may arise from compensating perturbations of the folded protein (Connelly et al., 1991; Hawkes et al., 1984). Amino acid replacements have also been found to cause changes in ΔC_p, which may also reflect differences in solvation among the variants (Shortle et al., 1988).

2.2.2 Kinetic Measurements

In order to use mutant proteins to study folding intermediates and transition states, it is usually necessary to carry out kinetic experiments, as described in Chapters 5 and 7. Any change in thermodynamic stability due to a mutation must be reflected in a change in one or more rate constants for folding or unfolding. By determining which steps in a folding reaction are most affected by a substitution, the roles of the altered residue in the folding pathway can be inferred.

As discussed in Chapter 5, the kinetics of denaturant-induced unfolding/refolding reactions are usually studied by rapidly changing the denaturant concentration and following the change of a spectroscopic signal. The rate of change is measured at different denaturant concentrations, starting with both native and unfolded protein. In the simplest cases, both folding and unfolding reactions display single first-order kinetic phases. The results of a kinetic study are often summarized in a plot of relaxation time versus denaturant concentration, as illustrated in Figure 8–2 for an idealized case (Matthews, 1987b). At low denaturant concentrations, the relaxation time is dominated by the rate of refolding and increases with increasing denaturant concentration. At high denaturant concentrations, the unfolding rate constant dominates the relaxation time, which decreases with increasing denaturant. This results in an "inverted V" in which the left leg reflects the folding rate and the right leg the unfolding rate. For a simple two-state transition, the maximum in the plot occurs at the transition midpoint, where $\Delta G_u = 0$ and the folding and unfolding rate constants are equal.

If an amino acid replacement causes an increase in unfolding rate, the right leg of the inverted V is shifted to the left (Fig. 8–2A), while a decrease in folding rate is reflected as a shift of the left leg of the curve further to the left (Fig. 8–2B). In order to compare the folding and unfolding rate constants at a common denaturant concentration (usually 0 M), the observed rates must be extrapolated. Most commonly, the extrapolations are made with the assumption that the logarithm of the rate constant is a linear function of denaturant concentration (Matthews, 1987b; Matouschek et al., 1989).

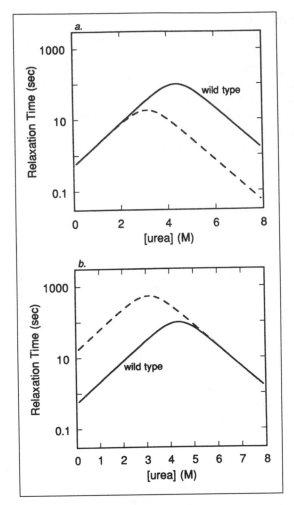

FIGURE 8–2. *Hypothetical plots of unfolding and refolding relaxation times (τ) as a function of urea concentration. In both plots, the solid lines represent the relaxation times expected for a wild-type protein with a midpoint urea concentration of 4.4 M and for which τ is 100 s at the midpoint. In (A), the dashed line represents a mutant that is 2 kcal/mol less stable than the wild-type protein and the thermodynamic destabilization is expressed kinetically as an increase in unfolding rate, so that the relaxation time decreases at the higher urea concentrations where the unfolding rate dominates the relaxation. In (B), the dashed line also represents a mutant that is destabilized by 2 kcal/mol, but the destabilization is expressed as a decrease in folding rate. The curves were calculated from transition state theory assuming that the activation energies for folding and unfolding are linear functions of urea concentration, as described by Matthews (1987b).*

Very often, the results of unfolding/refolding studies are considerably more complex than indicated in Figure 8–2 and display multiple kinetic phases, particularly during refolding. Some of these phases are due to kinetic heterogeneity in the unfolded protein, most likely arising from *cis,trans* isomerization of prolyl peptide bonds, while other phases reflect the presence of kinetic intermediates (Chapter 5; Kim and Baldwin, 1982, 1990). Fully interpreting such complex kinetics has often proven difficult, and a variety of elegant experimental approaches have been applied, as discussed in Chapter 5. The presence of multiple phases can greatly complicate the comparison of wild-type and mutant proteins, since it may not be clear which phases represent the same physical process for the different proteins.

Another method that has proven useful in studying the thermodynamics and kinetics of folding transitions is urea gradient gel electrophoresis (Creighton, 1979, 1980). In this method, protein samples are applied to the top of a slab gel containing a transverse gradient of urea concentration and subjected to electrophoresis in a direction perpendicular to the gradient. In the region of the gel containing low urea concentrations, the protein is folded, while at higher urea concentrations it unfolds and migrates more slowly because of its greater hydrodynamic volume. The position and continuity of the band between the folded and unfolded forms reflect the stability of the protein and the rates of interconversion between the native and unfolded forms. Because the method is simple and requires very small amounts of protein, it is well suited for the initial study of a collection of mutant proteins, as illustrated in Figure 8–8 for a set of T4 lysozyme variants.

The folding of proteins stabilized by disulfide bonds can often be studied using the disulfide bonds as conformational probes (Chapter 7). Many disulfide-containing proteins, including BPTI, bovine ribonuclease A (RNase A), and hen egg-white lysozyme, can be unfolded simply by reducing the disulfides. The equilibrium between the native protein and the reduced unfolded protein can be manipulated by adjusting the thiol-disulfide redox potential. As discussed in Chapter 7, disulfide-bonded intermediates can be chemically trapped, physically isolated, and characterized individually. By measuring all of the forward and reverse rate constants, a complete description of the kinetics and thermodynamics of the folding transition can be obtained. Since all of the measurements can be made at the same temperature and in the absence of denaturants, many of the complications and uncertainties associated with extrapolations are avoided.

2.3 Frames of Reference

The interpretation of thermodynamic and kinetic measurements is often limited by the fact that the measurements reflect *differences* between two equilibrium states or, in the case of rate constants, between a ground state and a transition state. Because the mutant and wild-type forms of a protein cannot be reversibly interconverted, their free energies cannot be directly compared. As a consequence, it is very difficult to determine experimentally whether a destabilizing mutation, for instance, increases the free energy of the native protein or decreases that of the unfolded protein.

One way of circumventing this ambiguity is to compare the effects of an amino acid replacement on the free energy changes associated with different steps in a folding reaction (Goldenberg et al., 1989; Klemm et al., 1991; Matouschek et al., 1989, 1990). A hypothetical comparison of this sort is illustrated in Figure 8–3A, where the reaction coordinates of a wild-type protein and a destabilized mutant are shown. For each protein, the unfolded protein is assigned a free energy of 0, recognizing that the energies of the two proteins cannot be compared directly. In this hypothetical example, the free energy change for the first step in folding (formation of Intermediate 1) is unaffected by the amino acid replacement, but Intermediate 2 and the native state are destabilized by the substitution. A structural model consistent with these hypothetical results is shown in Figure 8–3C. In this model, the first intermediate has a structure in which the altered residue is in an environment like that of the unfolded protein, while in Intermediate 2 the altered residue is in a native-like environment where it contributes to stability.

The effects of mutations on the relative stabilities of folding intermediates can be expressed quantitatively by normalizing the change in stability of each intermediate by the change in stability of the native protein.

$$\text{Relative destabilization} \equiv \frac{\Delta G_i^{\text{wild type}} - \Delta G_i^{\text{mutant}}}{\Delta G_u^{\text{wild type}} - \Delta G_u^{\text{mutant}}} \tag{8–4}$$

where $\Delta G_u^{\text{wild type}}$ and $\Delta G_u^{\text{mutant}}$ are the stabilities of the native states of the wild-type and mutant proteins and $\Delta G_i^{\text{wild type}}$ and $\Delta G_i^{\text{mutant}}$ are the stabilities of the intermediate (with all stabilities expressed relative to the respective unfolded states) for the two proteins. If the relative destabilization is 1, then the mutation destabilizes the intermediate to the same degree as it does the native protein, suggesting that the altered residue plays a similar role in the intermediate as it does in the folded structure, while a value of 0 indicates that the intermediate is not destabilized. Figure 8–3B shows a plot of relative destabilizations for the hypothetical example. This plot emphasizes the similarity between Intermediate 1 and the unfolded protein, and between Intermediate 2 and the native structure, at the site of the substitution. Analogous comparisons can be applied to measured rate constants in order to infer properties of transition states, as discussed in Section 4.

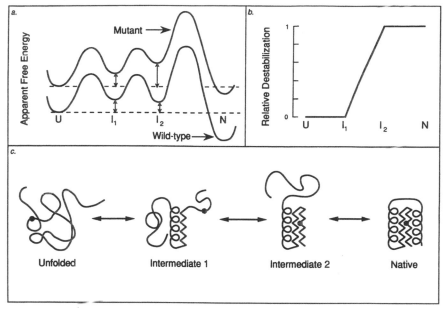

FIGURE 8–3. *A hypothetical mutational analysis of a folding pathway. In (A), the reaction coordinates for the wild-type and mutant proteins are compared. The amino acid replacement destabilizes the native protein and Intermediate 2 (I_2) with respect to the unfolded protein, but does not affect the stability of Intermediate 1 (I_1). The relative destabilizations of the intermediates, calculated according to Equation (8–4), are plotted in (B) and suggest that the site of the altered amino acid residue is in a native-like environment in Intermediate 2, but is not structured in Intermediate 1. A folding pathway that would be consistent with these results is suggested in (C). The altered residue is represented by the shaded dot.*

Studies of the structures and properties of the native and unfolded proteins can provide more direct information about the mechanisms by which amino acid substitutions alter protein stability, as discussed in Section 5.

3 MUTATIONAL ANALYSIS OF THE CONTRIBUTIONS OF INDIVIDUAL INTERACTIONS TO PROTEIN STABILITY

For some time, it has been appreciated that the folded conformations of proteins are stabilized by a variety of different types of interactions, including hydrogen bonds, ion pairs, disulfide bonds, and the hydrophobic effect. However, there is still considerable uncertainty about the relative roles of these different interactions (Alber, 1989; Baldwin, 1986; Dill, 1990a).

Until the mid 1980's, most experimental studies relied on model compounds to measure the affinities of different functional groups for one another or their tendencies to be excluded from an aqueous environment. The interpretations of model compound studies are often limited because it is not known how the environment of a protein should be compared with the bulk solution conditions where the measurements are made. The ability to make and study mutant proteins, in which individual interactions are selectively removed or altered, has opened new possibilities for the study of these interactions within the context of the protein.

A qualitative description of the number and types of residues that contribute to stability has emerged from random mutagenesis studies that ask, "Which residues can be altered to reduce stability?" Several small proteins for which stability measurements can be made—including phage T4 lysozyme (Alber et al., 1987a), staphylococcal nuclease (Shortle and Lin, 1985), the Cl (Hecht et al., 1983) and Cro (Pakula et al., 1986) repressors of phage λ, cytochrome c (Hampsey et al., 1988), and BPTI (Coplen et al., 1990)—have been subjected to studies in which randomly mutagenized clones were screened to identify those producing proteins with reduced activity or stability. The mutant genes were then sequenced to determine the amino acid substitutions responsible for the defects. In all of these studies, substitutions were identified throughout a large fraction of the polypeptide chain. While some of the mutations are likely to act by preventing function, the wide distribution of the changes in the sequences and the three-dimensional structures of the proteins suggests that many of the replacements act by destabilizing the native structures or interfering with their formation. In several cases, this has been confirmed directly by purifying the mutant proteins and demonstrating that they have reduced stabilities (e.g., Hecht et al., 1984; Shortle and Meeker, 1986).

The substitutions identified in screens for reduced stability represent virtually all of the kinds of changes that might be expected to disrupt the interactions stabilizing the native protein. This suggests that many kinds of interactions all make substantial contributions to stability. If, for instance, only the hydrophobic effect contributed significantly, one would not expect to see polar residues among the sites of destabilizing substitutions. The random mutagenesis experiments also indicate that the total number of significant interactions is large, since as many as one third of the residues in a protein have been found to be sites of destabilizing substitutions.

One can also ask the converse question, "How many residues can be altered without eliminating stability or function?" This has been done by making very large collections of clones carrying mutant genes and identifying those that produce functional proteins. From such experiments, it appears that different residues in a protein display widely varying degrees of tolerance to substitution. Some residues can be replaced by many others without preventing folding or function, while other sites must be occupied by only one or two residue types.

One of the most extensive studies of amino acid replacements is that of the *lac* operon repressor by J. Miller and his colleagues. Over 1,600 replacements at 141 of the 360 residues of this protein have been generated by suppression of nonsense mutations (Kleina and Miller, 1990). Approximately half of the substitutions do not result in an *in vivo* phenotype. Of the 141 sites tested, there are 34 where all of the replacements made result in a wild-type level of repressor activity (although some of the substitutions cause a "superrepressor" phenotype in which the repressor fails to respond to inducer). Another 76 sites can be replaced with at least one residue type to yield a wild-type level of activity. At only two sites, do all of the substitutions tested result in a complete loss of activity.

Reidhaar-Olson and Sauer (1988, 1990) observed similar patterns in the tolerance to substitution among 33 residues in two helices of the phage λ Cl repressor. These authors used targeted mutagenesis with degenerate oligonucleotides to make a large number of substitutions at each of the 33 sites, and then identified functional proteins by an *in vivo* selection. At 14 of the 33 sites examined, the wild-type residue can be replaced with eight or more other residue types. At eight of the sites, only the wild-type residue was recovered in the functional sequences.

The primary factor that determines whether or not a particular substitution will destabilize a protein appears to be the degree to which the altered residue is buried in the folded structure. Alber et al. (1987a) found that all 20 residues identified as sites of temperature-sensitive substitutions in phage T4 lysozyme are more than 40% buried in the native protein. In addition, the sites of the temperature-sensitive mutations all have less than average crystallographic thermal factors, indicating that they are relatively immobile in the native structure. Similar correlations have been observed for the destabilizing substitutions in λ repressor (Hecht et al., 1983; Reidhaar-Olson and Sauer, 1988, 1990), λ Cro (Pakula, 1986), yeast iso-1-cytochrome *c* (Hampsey et al., 1988), and BPTI (Coplen et al., 1990). This correlation indicates that buried residues generally contribute the most to stability. However, even fully buried hydrophobic residues can often be replaced without completely preventing folding, particularly if another hydrophobic residue is introduced and the total volume of the protein interior is not greatly changed (Lim and Sauer, 1989). Also, there are cases in which replacements of surface residues result in substantial changes of stability (Pakula and Sauer, 1990).

More quantitative estimates of the contributions of individual interactions have been obtained by measuring the destabilization caused by appropriate substitutions, as discussed in the following sections.

3.1 Hydrogen Bonds

Early models of the three-dimensional structures of proteins, most notably those of Pauling, emphasized the importance of hydrogen bonds (Pauling and Corey, 1951; Pauling et al., 1951). Subsequent authors, however, pointed out that the contribution from a hydrogen bond between two groups in a folded protein might not be very large, since the donor and acceptor could also form hydrogen bonds with water molecules in the unfolded protein (Kauzmann, 1959; Schellman, 1955). Amino acid replacements that alter side chains participating in hydrogen bonds have shown that hydrogen bonds can, in fact, make substantial contributions to stability. Interpreting the observed destabilizations in terms of individual hydrogen bonds is not simple, however, since a substitution may remove multiple hydrogen bonds, as well as van der Waals interactions, and the other participant in the hydrogen bond may or may not be able to form a new hydrogen bond with a solvent molecule.

An extensive analysis of the role of hydrogen bonds at a single site has been carried out by Alber et al. (1987b), who determined the crystal structures and stabilities of 13 mutant proteins with different substitutions of Thr 157 in phage T4 lysozyme. As illustrated schematically in Figure 8–4A, the side chain hydroxyl of this residue forms hydrogen bonds with the hydroxyl oxygen of Thr 155 and with the amide hydrogen of Asp 159 in the wild-type protein. Replacement of Thr 157 was found to destabilize the native protein by 0.45 to 2.9 kcal/mol. The largest destabilizations were caused by those residues, such as Ile (Fig. 8–4B) that could not form a hydrogen bond with the amide hydrogen of Asp 159. However, even those residues with side chains that could form this hydrogen bond, such as Asn (Fig. 8–4C) still resulted in destabilizations of 0.45 to 1.1 kcal/mol. Presumably these destabilizations reflect differences in the geometries of the hydrogen bonds or other perturbations of the structure. One of the most interesting mutants in this set was the protein with Gly at position 157 (Fig. 8–4D). In this protein, a water molecule takes the place of the Thr hydroxyl and forms hydrogen bonds with the side chain of Thr 155 and the amide of Asp 159, as well as a new hydrogen bond with the side chain of Asp 159. The destabilization from this substitution, 1.1 kcal/mol, might be representative of the effect of replacing two intramolecular hydrogen bonds with intermolecular hydrogen bonds with suitably positioned water molecules. In this sense, the contribution of a single intramolecular hydrogen bond would appear to be about 0.5 kcal/mol.

Serrano and Fersht (1989) have examined the role of hydrogen bonds between side chain hydroxyl oxygens and backbone amide hydrogens at the N-termini of two helices in barnase. Replacing Ser with Gly, Ala, or Val at these sites results in destabilizations of 1 to 2.3 kcal/mol. The Ser to

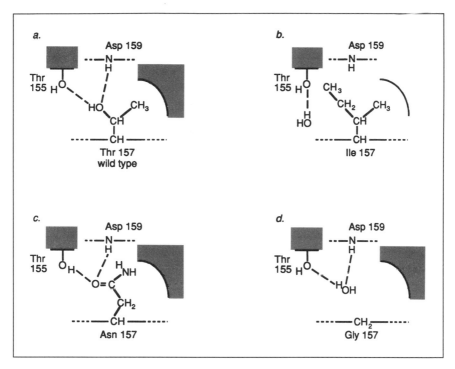

FIGURE 8–4. *Cartoon representations, based on crystal structures determined by Alber et al. (1987a), of the environment around residue 157 in the native structures of (A) wild-type phage T4 lysozyme and three mutants in which Thr 157 is replaced with (B) Ile, (C) Asn, and (D) Gly. In the wild-type protein, the side chain of Thr 157 forms hydrogen bonds with the amide hydrogen of Asp 159 and the side chain oxygen of Thr 155. Replacing Thr 157 with Ile eliminates these hydrogen bonds and destabilizes the protein by 2.9 kcal/mol. Although an Asn side chain can form hydrogen bonds with Thr 155 and Asp 159, this mutant is 0.45 kcal/mol less stable than the wild-type protein. When Gly is introduced at this site, a water molecule forms the same hydrogen bonds as does the Thr side chain in the wild-type protein. (Adapted, with permission, from Figure 1 of Alber et al., 1987a)*

Gly substitutions indicate that these hydrogen bonds contribute about 1 kcal/mol to stability. Replacements of hydrogen-bonding residues in ribonuclease T1 have been found to destabilize the native protein by as much as 2.9 kcal/mol (Shirley et al., 1992). After considering the effects of removing multiple hydrogen bonds and increasing the hydrophobicity of buried residues, these authors estimate that the average contribution of a hydrogen bond is 1.2 kcal/mol. Estimates of similar magnitudes have been obtained from mutational studies of enzyme-substrate complexes (Fersht, 1987; Lesser et al., 1990). There are, however, also examples of mutations that lead to new hydrogen bonds but do not measurably contribute to stability (Alber et al., 1988). Thus, it appears that the ability of a hydrogen bond to stabilize a protein structure can depend greatly on its environment. The reasons for the different contributions of various hydrogen bonds is not yet known, but may include differences in the orientation of the donors and acceptors, as well as entropic effects (Creighton, 1983).

3.2 *Electrostatic Interactions*

Interactions among charged groups in a protein can significantly add to or detract from stability. As with hydrogen bonds, the stabilities of electrostatic interactions are influenced by the presence of aqueous solvent and depend upon the details of the protein environment.

An important feature of the interactions between ionized groups in a protein is the thermodynamic linkage between conformational changes and the ionization equilibria (Tanford, 1961; Wyman, 1948). For instance, when a carboxyl interacts favorably with an amine:

$$-C\underset{\textstyle O^-}{\overset{\textstyle O}{\diagup}}\text{--------}^+HNH_2-- \tag{8-5}$$

the ionized form of the carboxyl and the protonated form of the amine are stabilized relative to their uncharged forms. Consequently, the pK_a of the carboxyl is lower than it would be in the absence of the interaction, as in the unfolded protein, and the pK_a of the amine is higher. Thermodynamic linkage requires that the degree of stabilization from the interaction is directly related to the magnitude of the pK_a changes. The degree of stabilization also depends upon the solution pH, since conditions that favor ionization will favor the conformational state of the protein in which the ionized state predominates.

Electrostatic interactions in proteins have been studied genetically by altering specific charged groups and measuring the effects on conformational stability and on the pK_a's of other groups in the protein. The magnitudes of the changes in stability and pK_a depend on the distances between charged groups and their local environments. In subtilisin, removing a negative charge changes the pK_a of a His residue 10 to 15 Å away by about 0.4 units (Russell and Fersht, 1987). These measurements were used to calculate an "effective dielectric constant" of about 40, indicating that the charges are very effectively shielded at the protein surface. Replacing a Glu residue in RNase T1 with Ala was found to reduce the pK_a of one His residue from 7.8 to

7.4 and that of another His from 7.9 to 7.1 (McNutt et al., 1990). The measured pK$_a$'s for the wild-type and mutant proteins were used to successfully predict the dependence of the conformational stability on pH (Fig. 8–5). At pH values below 7.5, where the histidines are protonated, the wild-type protein is more stable than the Glu → Ala mutant by as much as 1 kcal/mol. At higher pH, however, the stabilization from the electrostatic interactions in the wild-type protein are lost, and the mutant is actually more stable than wild type. In this case, it appears that the replacement has multiple effects. Electrostatic interactions with at least the two His side chains are eliminated, and other interactions may be altered to lead to the enhanced stability of the mutant at high pH where the His residues are not protonated.

In order to address this problem of multiple interactions among charged residues, Serrano et al. (1990) have used a double-mutant analysis to measure the interaction between two charged residues in barnase, Asp 12 and Arg 16. Replacing Asp 12 with Ala destabilizes the native protein by 0.76 kcal/mol at low ionic strength. When the same replacement is made in a protein in which Arg 16 has been replaced by Thr, the Asp 12 → Ala substitution results in a destabilization of 0.43 kcal/mol. The interaction energy between the two charged residues was estimated to be 0.33 kcal/mol, the difference in the destabilization from the replacement of Asp 12 in the presence and absence of Arg 16. Although this "interaction energy" might arise from effects other than electrostatics, the observation that the energy approaches 0 at high ionic strength indicates that it is due primarily to the attraction between the two charged side chains. Double-mutant analysis is discussed further in Section 3.5.

Much larger effects have been seen when more closely spaced charged groups are altered. His 31 and Asp 70 form an ion-pair in phage T4 lysozyme. Anderson et al. (1990) found that changing either of these residues to Asn decreased the stability of the folded protein by 3 to 5 kcal/mol at pH 5. As expected, the stabilization of the protein by this ion-pair is coupled to large changes in the pK$_a$'s of the interacting groups.

In addition to interactions between formally charged groups in proteins, interactions involving dipoles can be significant. In particular, the macrodipole arising from the alignment of peptide bonds in an α-helix is a significant factor in determining the stabilities of helices in small peptides and proteins (Hol, 1985; Shoemaker et al., 1987). This dipole can be thought of as placing a negative charge near the C-terminus of a helix and a positive charge near the N-terminus. His residues located at the C-termini of helices in hemoglobin (Perutz et al., 1985) and barnase (Sali et al., 1988) have anomalously high pK$_a$ values, as expected from a helix dipole effect. Replacing the His residue in barnase with Gln destabilizes the folded protein by about 1.6 kcal/mol, in approximate agreement with the pK$_a$ measurements. The stability of phage T4 lysozyme has been increased by a total of 1.6 kcal/mol

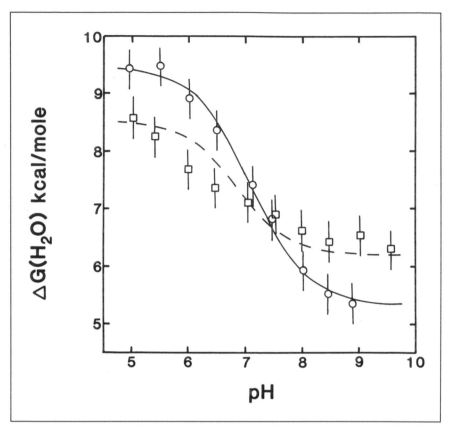

FIGURE 8–5. *The free energies of unfolding (ΔG_u) of the wild-type (solid lines and round symbols) and Glu 58 → Ala mutant (dashed lines and square symbols) forms of ribonuclease T1 as a function of pH at 25°C. The values of ΔG_u were obtained from urea unfolding curves by extrapolating to zero denaturant concentration. The curves were calculated from the measured values of the pK_a's of the three His residues in native RNase T1 and fit values for the pK_a's in the unfolded protein. The presence of Glu 58 in the wild-type protein causes increases in the pK_a's of two of the histidines in the native protein. When Glu 58 is replaced with Ala, the pK_a's in the native protein are closer to those in the unfolded state, and the effect of pH on stability is correspondingly less. At pH values where the histidine residues are protonated, they interact favorably with Glu 58 in the wild-type protein, making it more stable than the mutant. (Reproduced, with permission, from Figure 4 of McNutt et al., 1990)*

by introducing Asp residues near the N-termini of two helices (Nicholson, 1988). These studies provide important confirmation of the significance of helix dipoles.

Electrostatic interactions have been the subject of much theoretical, as well as experimental, study (Harvey, 1989; Sharp and Honig, 1990; Warshel and Russell, 1984). Some of these treatments have been very successful in predicting the pK_a's of specific residues in proteins and the effects of mutations on pK_a's (e.g., Bashford and Karplus, 1989; Gilson and Honig, 1987; Sternberg et al., 1987; Sun et al., 1989; Warshel et al., 1986). These successes indicate that the current understanding of electrostatic interactions is quite good, and illustrate how mutational studies can be used in the development of theoretical treatments.

3.3 The Hydrophobic Effect and Packing Interactions

The hydrophobic effect is one of the major factors stabilizing the native conformations of proteins (Chapter 3; Baldwin, 1986; Dill, 1990a; Kauzmann, 1959). Hydrophobicity has traditionally been measured by determining the free energies of transfer of model compounds from a nonpolar liquid to water (Tanford, 1980). Although there are some differences in the values obtained with different organic liquids, the trends seen among the amino acid side chains are consistent. Analyses of known crystal structures of proteins have shown that there is a good correlation between the hydrophobicity of a side chain as measured by transfer experiments and the likelihood that it will be buried in the native protein (Chothia, 1976; Rose et al., 1985). There is still considerable uncertainty, however, about the physical origins of the hydrophobic effect and the magnitude of its contribution to protein stability (Baldwin, 1986; Dill, 1990a, b; Murphy et al., 1990; Privalov and Gill, 1988; Privalov et al., 1990; Spolar et al., 1989). It is not yet clear how well the interior of a folded protein is modeled by nonpolar liquids, and it is difficult, both experimentally and conceptually, to distinguish between the stabilization that arises from attractive van der Waals interactions between nonpolar groups and the effect of removing nonpolar groups from water.

The contributions of the hydrophobic effect and van der Waals interactions in protein stability have been tested by using site-directed mutagenesis to alter residues that are buried in the native protein. Matsumura et al. (1988) examined the effects of replacing a buried hydrophobic residue, Ile 3, of phage T4 lysozyme with 13 other residues. Among 10 of the mutant proteins and the wild-type protein, the differences in conformational stability were similar to the differences in the hydrophobicities of the residues at position 3. Large deviations from this correlation were seen for proteins with Phe, Tyr, or Trp at residue 3, which were all less stable than expected from the residue hydrophobicities. The crystal structure of the Ile 3 → Tyr mutant suggests that the proteins with aromatic residues at this site have lower stabilities because the large side chains cannot be accommodated in the wild-type structure (Matsumura et al., 1988, 1989c).

In other cases, the changes in conformational stability have been considerably larger than expected from transfer free energy measurements. Kellis et al. (1988, 1989) measured the stabilities of barnase variants in which three buried aliphatic residues were replaced with smaller residues. Replacing Ile with Val reduces the stability of the protein by about 1 kcal/mol, while replacing Ile or Leu with Ala causes destabilizations of 3.3 to 4.3 kcal/mol. These destabilizations are approximately twice the corresponding differences in the free energies of transfer from octanol to water. In an extensive study of staphylococcal nuclease, Shortle et al. (1990) replaced each of the Leu, Ile, Val, Tyr, Met, and Phe residues with Ala and Gly. This study demonstrated that the same type of replacement can cause changes in stability that vary by more than 6 kcal/mol, depending upon the location of the altered residue. The average values for the destabilization from each type of substitution are about twice as large as expected from the free energies of transfer from octanol. Kellis et al. (1989) suggested that these large destabilizations may arise in part from the loss of favorable van der Waals interactions in the protein interior, while Shortle et al. (1990) proposed that some of the destabilization may be due to effects on residual structure in the unfolded protein.

An example of a "reverse hydrophobic effect" has been described for substitutions at position 26 of the phage λ Cro protein, where the stability of the native state is inversely correlated with residue hydrophobicity (Pakula and Sauer, 1990). In the native wild-type protein, Tyr 26 is located at a solvent-exposed turn, and the enhanced stabilities of proteins with less hydrophobic residues at this site have been explained by proposing that this site becomes less exposed to solvent when the protein unfolds. From the measured stability differences of the mutants and transfer free energies of the residues, Doig et al. (1990) have concluded that in the unfolded protein this residue has an exposure similar to that in an extended Ala-X-Ala tripeptide.

3.4 Disulfide Bonds

Shortly after the method of site-directed mutagenesis was developed, several groups attempted to increase the stabilities of proteins by introducing new Cys residues that would form disulfide bonds (Perry and Wetzel, 1984; Pantoliano et al., 1987; Sauer et al., 1986; Villafranca et al., 1983; Wells and Powers, 1986). The initial expectation was that the new disulfides would stabilize the native protein by lowering the conformational entropy of the unfolded state. Although some engineered disulfides do lead to enhanced stabilities, the effects are often less than the 3 to 5 kcal/mol predicted by considering only the unfolded state, indicating that the effects of a disulfide on the native protein must be considered as well.

Some of the effects of introducing cysteine residues and of forming a disulfide can be analyzed by considering a thermodynamic cycle:

$$
\begin{array}{ccc}
N_Y^X & \xrightarrow{\Delta G_u^{wt}} & U_Y^X \\
\Delta G_{mut}^N \downarrow & & \downarrow \Delta G_{mut}^U \\
N_{SH}^{SH} & \xrightarrow{\Delta G_u^{SH}} & U_{SH}^{SH} \\
RSSR \searrow \quad \Delta G_{S-S}^N \downarrow \quad 2RSH \nearrow & & \searrow RSSR \quad \downarrow \Delta G_{S-S}^U \quad \searrow 2RSH \\
N_S^S & \xrightarrow{\Delta G_u^{S-S}} & U_S^S
\end{array}
\qquad (8\text{–}6)
$$

where ΔG_u^{wt}, ΔG_u^{SH}, and ΔG_u^{S-S} are the free energy changes for unfolding the wild-type protein, the reduced form of the mutant protein with two Cys residues and the disulfide-bonded form, respectively. ΔG_{mut}^N and ΔG_{mut}^U are the free energy changes for hypothetical reactions changing residues X and Y to Cys in the native and unfolded proteins, while ΔG_{S-S}^N and ΔG_{S-S}^U are the free energy changes for forming the disulfide in the native and unfolded proteins by exchange with a small molecule disulfide, RSSR. Unlike most other modifications of proteins, the free energy changes for disulfide formation can be determined directly by measuring the equilibrium constants for the exchange reactions, K_{S-S}^N and K_{S-S}^U (Chapter 7). Values of these equilibrium constants are typically 0.01 M in unfolded proteins and are as large as 10^5 M in native proteins (Creighton, 1983; Creighton and Goldenberg, 1984; Goldenberg, 1985; Goto and Hamaguchi, 1982; Lin and Kim, 1989; Matsumura et al., 1989a). The low values in the unfolded protein reflect the large loss in conformational entropy due to constraining two atoms in a disordered chain.

The stabilities of the disulfide-bonded and reduced forms of the Cys-containing protein are linked to the free energies for disulfide formation according to

$$
\Delta G_u^{S-S} - \Delta G_u^{SH} = \Delta G_{S-S}^U - \Delta G_{S-S}^N = RT \ln\left(\frac{K_{S-S}^N}{K_{S-S}^U}\right),
\qquad (8\text{–}7)
$$

where R is the gas constant and T is the temperature. This linkage requires that the disulfide stabilize the native protein to the extent that forming the disulfide is more favorable in the native protein than in the unfolded protein. In comparing the disulfide bonded protein with the wild-type protein, however, the energetic differences due to introducing the Cys residues into the native and the unfolded proteins may also be significant. If, for instance, making the amino acid replacements in the native protein is less favorable than in the unfolded state, then the total stabilization from the disulfide will be decreased.

At present, the engineered disulfides for which the most thermo-dynamic data are available are those introduced into phage T4 lysozyme (Matsumura et al., 1989a). Based upon energy calculations and computer modeling, Cys residues were introduced to form four different disulfides. The stabilities of the proteins with each of the disulfides were determined by thermal unfolding, and the stabilities of the disulfides in the native proteins were measured by equilibration with reduced and oxidized forms of dithiothreitol (DTT). In each case, the midpoint temperature for un-folding (T_m) of the disulfide-bonded protein was greater than that of the mutant with the Cys thiols reduced, by 3° to 14°C. The reduced proteins, however, all had T_m's 2° to 6° lower than that of the wild-type protein, indicating that introducing the Cys residues in the native protein was destabilizing. For three of the proteins, the stabilization from the disulfide was greater than the destabilization from introducing the cysteines, so that the disulfide-bonded proteins were substantially more stable than the wild-type protein. The two most stable proteins were those in which the disulfides in the native proteins were most stable, with values of K_{S-S}^N of 20 M and 133 M, respectively.[*] These proteins were also those with the largest spacings in the sequence between the Cys residues, which are expected to have the lowest values of K_{S-S}^U, although these equilibria were not measured directly. Thus, the maximum stabilization was obtained from the disulfides for which K_{S-S}^N was the largest and K_{S-S}^U was the smallest, as expected from theory. When two or three of the most stabilizing disulfides were introduced simultaneously, the stabilization was roughly additive (Matsumura et al., 1989b), leading to a total increase in T_m of as much as 23°C.

A number of disulfides have also been engineered into subtilisin (Mitchinson and Wells, 1989; Pantoliano et al., 1987; Wells and Powers, 1986). Unfortunately, this protein does not undergo reversible unfolding,

---

[*]The values of K_{S-S}^N indicated for the engineered disulfides in T4 lysozyme and subtilisin were calculated from the published equilibrium constants for reduction of the protein disulfides with DTT. This calculation was made by dividing the equilibrium constant for formation of the disulfide of DTT with oxidized glutathione ($K = 1,100$ M) by the equilibrium constants for reducing the protein disulfides with DTT (Creighton and Goldenberg, 1984).

and thermodynamic measurements of the effects of disulfides on stability cannot be made. It is possible, however, to measure the stability of the disulfide in the native protein (i.e., K_{S-S}^N) and to measure the kinetics of inactivation, which may have considerable practical importance. Some of the disulfides introduced into subtilisin are quite stable, with values of K_{S-S}^N from 12 to 1,800 M. The stabilities of the disulfides do not, however, appear to be correlated with the rates of inactivation, indicating that the disulfides do not significantly stabilize the native protein with respect to the transition state for inactivation, even though they would be expected to stabilize it with respect to the unfolded protein.

High-resolution crystal structures have been determined for four of the disulfide-bonded variants of subtilisin (Katz and Kossiakoff, 1986, 1990) and one of T4 lysozyme (Pjura et al., 1990). The disulfides in subtilisin have dihedral angles significantly different from those of naturally occurring disulfides and were predicted to have dihedral strain energies of 2.5 to 5.4 kcal/mol. The dihedral angles of the disulfide in T4 lysozyme are similar to those found naturally in disulfides, but indicate a strain energy of 4 kcal/mol. These dihedral energies are expected to decrease the stability of the disulfide (K_{S-S}^N), and, therefore, the ability of the disulfide to stabilize the native protein. The stabilities of the disulfides in the folded proteins, however, will also depend upon entropic factors and upon strain of bond lengths and angles. Thus, it is not surprising that there is only a rough correlation between the dihedral energies and the disulfide stabilities.

The stabilization from naturally occurring disulfides has been examined by individually removing two of the three disulfides of BPTI (Hurle et al., 1990). One of the disulfides examined, 14-38, is located on the surface of the folded protein, while the other, 30-51, is buried. Although the value of K_{S-S}^N for 30-51 is approximately sevenfold greater than that for 14-38 (Creighton and Goldenberg, 1984), a mutant in which Cys 30 was replaced with Val and Cys 51 was replaced with Ala has about the same stability as one in which the 14 and 38 disulfide was removed. Proteins in which Cys 30 was replaced with Ala or Thr (and Cys 51 was replaced with Ala) were 0.5 to 2 kcal/mol less stable. Replacing the buried thiols of Cys 30 and 51 with hydrophobic residues may be particularly favorable, resulting in the relatively high stability of the Val 30/Ala 51 protein.

3.5 Synergism Among Multiple Substitutions

Although protein folding transitions are known to be highly cooperative, there is relatively little known about the ways in which the many interactions in the native protein influence one another. One way of addressing this question involves testing whether two amino acid replacements in the same protein act independently. Applications of this approach to examine interactions among residues in protein folding and function have been described by several authors (Ackers and Smith, 1985; Carter et al., 1984; Horovitz and Fersht, 1990; Hurle et al., 1986; Laskowski et al., 1989; Perry et al., 1989; Serrano et al., 1990; Wells, 1990).

The rationale of double-mutant studies can be illustrated by a set of interrelated thermodynamic cycles, as shown in Figure 8–6. This scheme describes the unfolding of four proteins; the wild-type protein, two mutants, A and B, in which single residues are replaced, and the double mutant, AB, which contains both replacements. The free energies of unfolding for these proteins are ΔG_u^{wt}, ΔG_u^A, ΔG_u^B, and ΔG_u^{AB}. The free energy changes for the hypothetical reactions to make the two amino acid replacements individually in the native protein are ΔG_{mutA}^N and ΔG_{mutB}^N, while the free energy change for making replacement A in the native protein when replacement B has already been made is $\Delta G_{mutA,B}^N$. Once replacement A has been made, the free energy change for making replacement B in the native protein is $\Delta G_{mutB,A}^N$. The free energy changes for making the replacements in the unfolded protein are represented by analogous terms with the superscript ^{U}.

If there is no interaction between the sites of the two replacements in the native protein, then the free energy change for making one of the replacements will be the same whether or not the other change has already been made. Although these free energy changes cannot be measured directly, the difference between them can be determined from the changes in stability arising from the two single substitutions and the double replacement, as discussed shortly.

The change in stability from mutation A is related to the free energy changes for making the replacement in the native and unfolded protein according to

$$\Delta\Delta G_u^{wt-A} = \Delta G_u^{wt} - \Delta G_u^A = \Delta G_{mutA}^N - \Delta G_{mutA}^U. \tag{8-8}$$

Similarly, the change in stability from mutation B is

$$\Delta\Delta G_u^{wt-B} = \Delta G_u^{wt} - \Delta G_u^B = \Delta G_{mutB}^N - \Delta G_{mutB}^U, \tag{8-9}$$

and the total change in stability from both mutations is

$$\Delta\Delta G_u^{wt-AB} = \Delta G_u^{wt} - \Delta G_u^{AB} = \Delta G_{mutB}^N + \Delta G_{mutA,B}^N - \Delta G_{mutB}^U - \Delta G_{mutA,B}^U. \tag{8-10}$$

Combining these equations gives the following expression for the difference between making substitution A in the native protein in the absence and presence of substitution B:

$$\Delta G_{mutA}^N - \Delta G_{mutA,B}^N = \Delta\Delta G_u^{wt-A} + \Delta\Delta G_u^{wt-B} - \Delta\Delta G_u^{wt-AB} + \Delta G_{mutA}^U - \Delta G_{mutA,B}^U. \tag{8-11}$$

If, in the unfolded protein, the free energy changes for making substitution A in the presence or absence of the other substitution are equal, this simplifies to

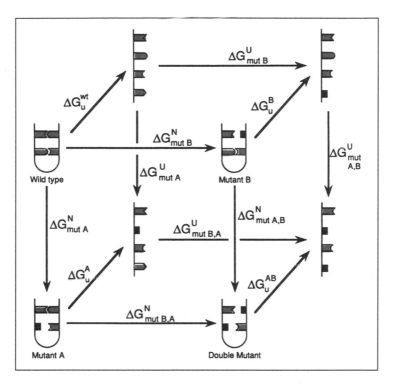

FIGURE 8–6. *A thermodynamic cycle illustrating the linkage relationships among the free energy changes for making two amino acid replacements and the free energy changes for unfolding the resulting variants. (Linkage results from the thermodynamic requirement that the sum of the free energy changes along any cyclic path be zero.) The front surface of the cube represents the hypothetical reactions for making the amino acid replacements in the native protein. The free energy changes for these reactions are: ΔG^N_{mutA} for making replacement A, ΔG^N_{mutB} for making replacement B, $\Delta G^N_{mutA,B}$ for making replacement A after replacement B has already been made, and $\Delta G^N_{mutB,A}$ for making replacement B once replacement A has been made. Reactions shown on the back face of the cube are the analogous mutational reactions in the unfolded state of the protein. The arrows pointing from the front face to the back represent the unfolding reactions for the four forms of the protein. The free energy changes for unfolding the wild-type protein, the two single mutants, and the double mutant are ΔG^{wt}_u, ΔG^A_u, ΔG^B_u, and ΔG^{AB}_u. As discussed in Section 3.5, the linkage among the reactions in the cycle can be used to determine whether or not the altered residues interact in the native protein by comparing the change in stability arising from replacing both residues with the sum of the changes due to the individual substitutions.*

$$\Delta G^N_{mutA} - \Delta G^N_{mutA,B} = (\Delta\Delta G_u^{wt-A} + \Delta\Delta G_u^{wt-B}) - \Delta\Delta G_u^{wt-AB}. \tag{8–12}$$

To a first approximation, then, the results of a double-mutant analysis can be interpreted in terms of the folded protein alone. If the free energy change for making one of the replacements in the native protein is independent of the presence of the other substitution, the left side of Equation (8–12) equals 0, and the change in stability arising from the double mutant will equal the sum of the changes in stability due to the single mutations. In this case, the effects of the two substitutions are said to be additive.

If there is an interaction between the two altered residues in the native protein, then ΔG^N_{mutB} will not equal $\Delta G^N_{mutB,A}$, and the change in stability seen in the double mutant will not equal the sum of the effects of the two single mutants. If, for instance, substitution A removes a stabilizing interaction in the native protein and, thereby, weakens surrounding interactions, then subsequently making substitution B might have a smaller effect than it would in isolation.

In many instances, the effects of multiple substitutions appear to be additive, even when the altered residues are in close proximity in the native protein. Sandberg and Terwilliger (1989), for instance, examined replacements of two hydrophobic residues in the interior of the gene V protein of phage fl. When Ile 47 was changed to Val, the protein was destabilized by 2.4 kcal/mol while changing Val 35 to Ile caused a destabilization of 0.4 kcal/mol. The double mutant is 2.9 kcal less stable than the wild-type protein, almost exactly the sum of the destabilizations from the single mutants. Additivity has also been seen among other substitutions in this protein and in others, including T4 lysozyme (Baase et al., 1986; Matsumura et al., 1989b; Nicholson et al., 1988), staphylococcal nuclease (Shortle and Meeker, 1986), λ repressor (Hecht and Sauer, 1985), kanamycin nucleotidyl transferase (Matsumura et al., 1986), RNase T1 (Shirley et al., 1989), and subtilisin (Pantoliano et al., 1989).

There are several instances, however, in which the effects of multiple substitutions are clearly nonadditive. A dramatic example, discussed earlier in Section 3.2, is that of His 31 and Asp 70 of T4 lysozyme, which form a salt bridge in the native protein. In this case, changing either residue to Asn destabilizes the protein as much as the double mutant, indicating that the predominant effect of either mutation is to eliminate the interaction with the other residue (Anderson et al., 1990). Replacements of Ala 98, Val 149, and Thr 152 in the core of T4 lysozyme also have nonadditive effects on stability (D. P. Sun et al., 1991). Crystallographic studies indicate that the energetic interdependence of these substitutions can be accounted for by local changes in the folded protein. Replacing Ala 98 with Val forces two α-helices apart, and additional changes at other sites in the interface between the helices have smaller, or even opposite, effects on stability than when they are introduced individually.

As discussed earlier, disulfide formation between Cys thiols is a protein modification for which the free energy change can be measured directly. Therefore, the effects of amino acid substitutions on disulfide formation can be determined without measuring the unfolding equilibria. One of the disulfides of native BPTI (between Cys 14 and 38) can be selectively reduced under conditions where the protein retains most of its native conformation and its activity as a trypsin inhibitor. The effects of 15 different amino acid replacements on the equilibrium constant for forming this disulfide in the native conformation have been measured (D. A. Laheru and D. P. Goldenberg, unpublished experiments). The substitutions alter seven residues with α-carbons located 5 to 15 Å away from the disulfide, and all make formation of the disulfide less favorable, by 1.3 to 5.1 kcal/ mol. In this case, one change in the protein, disulfide formation, is clearly influenced by other changes, amino acid replacements. The replacements may destabilize the disulfide by introducing strain in the folded protein, or they may increase local flexibility so that there is a greater loss of entropy upon forming the disulfide.

Together, the many mutational studies of protein stability reported in the past five years suggest that the contributions of individual inter- actions and residues to stability are very much dependent upon their context in the folded protein and, perhaps, in the unfolded state. One of the valuable uses of this growing body of data will be in testing theoretical treatments of protein energetics, as has already been the case for electro- static interactions. Double mutant studies may be particularly good sub- jects for theoretical analysis, because the experimental results can be interpreted in terms of the energetics of making the different amino acid replacements in the native protein (Equation (8–12)), and the native struc- tures can often be determined experimentally. Although the mutational reactions are entirely hypothetical, they can be analyzed by free-energy perturbation methods (Chapter 4; Dang et al., 1989; Gao et al., 1989). Similarly, measurements of disulfide-bond stabilities in closely related native proteins may be good subjects for computational treatments, since all of the relevant species have similar structures.

4 MUTATIONAL ANALYSIS OF FOLDING TRANSITION STATES AND INTERMEDIATES

In conjunction with kinetic analysis, amino acid replacements can be used as probes of the intermediates and transition states that define a protein folding mechanism. In the simplest cases, where intermediates do not accumulate significantly, the effects of substitutions on the kinetics of the unfolding/refolding reaction provide information about the major transition state separating the native and unfolded forms. As illustrated with reaction coordinate diagrams in Figure 8–7, a mutation that destabilizes the native state with respect to the unfolded protein must increase the rate of unfolding (Fig. 8–7A), decrease the rate of folding (Fig. 8–7B), or both. A decrease in folding rate indicates that the substitution destabilizes the transition state with respect to the unfolded protein and implies that the altered residue is in a different environment in the two states. An increase in unfolding rate suggests that the environment of the altered residue is different in the native protein and in the transition state. When intermediates in folding and unfolding can be detected and their relative stabilities measured, amino acid replacements can be used to infer the degrees to which the intermediates resemble the native or unfolded protein at the sites of the substitutions.

One of the first proteins for which mutations were used to characterize a folding mechanism was the α-subunit of *E.coli* tryptophan synthase. Two α-subunits and two β-subunits form the tryptophan synthase holoenzyme, which catalyzes the last two steps in tryptophan biosynthesis, but in the absence of β-subunits the α-subunit is a monomer. Early studies of the unfolding transition of the α-subunit revealed that a partially folded intermediate is detectable at equilibrium (Matthews and Crisanti, 1981; Yutani et al., 1979). Comparisons of the unfolding of the intact protein with proteolytic fragments indicated that the intermediate contains a folded amino-terminal region (corresponding to a fragment composed of residues 1 to 188) and an unfolded C-terminal region (corresponding to residues 189 to 268) (Miles et al., 1982). Amino acid substitutions at Phe 22, Tyr 175, Gly 211, and Gly 234 were all found to alter the rate at which the native protein is converted to the intermediate during unfolding (Beasty et al., 1986; Tweedy et al, 1990). Because similar kinetic effects were seen with mutations in both the N- and C-terminal regions, Beasty et al. proposed that this rate was determined by a process in which two domains of the protein dissociated and that, conversely, the last step in folding was the association of the two domains, rather than simply the folding of the second domain.

Following the initial mutational studies, the crystal structure of the tryptophan synthase tetramer was determined (Hyde et al, 1988). Surprisingly, the α-subunit was discovered to be composed of a single "α/β barrel" domain. The N-terminal proteolytic fragment, which can fold to a stable structure, corresponds to the first six β-strands and the first five α-helices of the barrel, while the unfolded C-terminal fragment corresponds to the remaining two strands and three helices of the intact protein. The sites of the mutations that alter the last step of folding are located in the β-strands that form the interfaces between the two regions of the protein. Thus, the combination of mutational and structural analyses leads to a picture in which the rate of the final step of folding is determined by the incorporation of the C-terminal two strands into the rest of the barrel, which is formed in the intermediate.

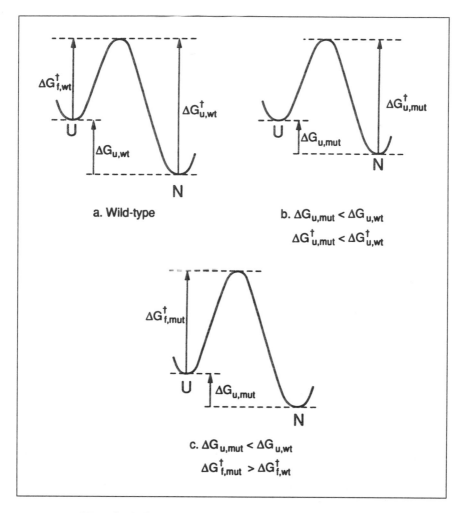

FIGURE 8–7. *Hypothetical reaction coordinates for folding transitions of a wild-type protein (A) and of thermodynamically destabilized mutants (B, C). The energies of the transition states are aligned to emphasize the differences in folding kinetics. In (B), the reduction in thermodynamic stability is expressed kinetically as an increase in unfolding rate, indicating that the native protein is destabilized with respect to the transition state. In (C), the destabilization is expressed as a decrease in folding rate, indicating that the unfolded state is stabilized with respect to the transition state. (Reproduced, with permission, from Figure 1 of Klemm et al., 1991)*

A similar approach has been used to study the folding of *E.coli* dihydrofolate reductase (DHFR). Although the equilibrium unfolding transition of this protein is well-described by a two-state model, the unfolding kinetics display two phases, which have been attributed to slowly interconverting native forms (Touchette et al., 1986). Amino acid replacements at several sites in the native protein increase or decrease the unfolding rates 2- to 10-fold (Perry et al., 1987, 1989; Garvey and Matthews, 1989). These replacements are located in three β-strands and two α-helices of the native structure, indicating that formation of the transition states during unfolding involves quite extensive perturbation of the folded protein.

The transition state for the folding and unfolding of T4 lysozyme also appears to differ from the native state throughout much of its structure. The thermodynamics and kinetics of the lysozyme unfolding transition have been studied using urea gradient gel electrophoresis (Fig. 8–8A; Klemm et al., 1991). The break between the bands of folded and unfolded wild-type protein (Fig. 8–8A) indicates that the protein folds and unfolds slowly on the time scale of the electrophoresis (20 min). Among 28 mutant forms examined, all 22 of the proteins that unfold with midpoint urea concentrations below ~5 M (versus 6.3 M for the wild-type protein) displayed continuous bands indicative of unfolding rates at least 10- to 100-fold faster than that of the wild-type protein. Since unfolding rates generally increase at higher urea concentrations, the destabilized proteins are expected to have even greater unfolding rates at 6.3 M urea, where the wild-type protein unfolds, than they do at their own midpoint urea concentrations.

The urea gradient gels indicate that amino acid replacements throughout T4 lysozyme destabilize the native protein with respect to the transition state more than does the addition of urea to yield an equivalent thermodynamic destabilization. This can be seen by comparing the pattern for the wild-type protein with that of the Ala 160 → Thr mutant (Fig. 8–8C). At the midpoint of the transition for the mutant (about 4 M urea), the native state of the wild-type protein predominates. If the urea concentration is increased to 6.3 M, $\Delta G_u = 0$ for the wild-type protein and the rate constant for unfolding is about 0.01 min^{-1}. However, if the protein is destabilized by the Ala 160 → Thr substitution, ΔG_u is again 0 (at 4 M urea), but the unfolding rate constant is at least 0.5 min^{-1}. Increases in unfolding rates upon the addition of denaturants such as urea or GuHCl have been attributed to increased solvent exposure upon formation of the major transition state (e.g., Chen et al., 1989; Kuwajima et al., 1989; Matouschek et al., 1989; Tanford, 1970). Since the mutations cause a larger relative increase in the rate of unfolding than does the addition of urea, it appears that the transition state is more similar to the folded protein with respect

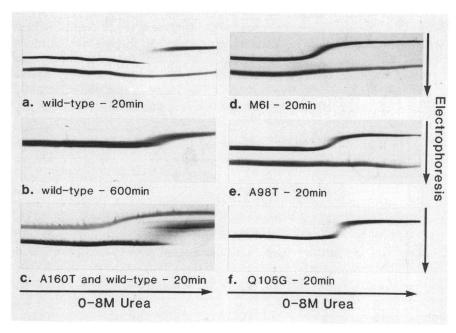

FIGURE 8–8. *Urea gradient gel electrophoresis of wild-type and mutant forms of phage T4 lysozyme. The gels contained a linear 0- to 8-M urea gradient with the urea concentration increasing from left to right. Protein samples were applied to the tops of the gels and subjected to electrophoresis toward the cathode at 22°C, pH 7. Electrophoresis was for 20 min at 250 V, except for (B), which was for 600 min at 10 V. (A) Native wild-type T4 lysozyme and native BPTI, which does not unfold under these conditions and serves as a baseline. The discontinuity between the bands of native and unfolded T4 lysozyme protein indicates that unfolding and refolding are slow on the electrophoresis time scale near the transition midpoint. (B) Wild-type T4 lysozyme subjected to electrophoresis for 600 min; unfolding and refolding are rapid on this time scale. (C) Wild-type and Ala 160 → Thr lysozyme. The wild-type protein was applied to the gel first and subjected to electrophoresis for 10 min. The mutant protein was then applied, and electrophoresis was continued for another 20 min. (D) Met 6 → Ile lysozyme and BPTI applied in a single sample. (E) Ala 98 → Thr lysozyme and BPTI applied in a single sample. (F) Gln 105 → Gly lysozyme. (Reproduced, with permission, from Figure 3 of Klemm et al., 1991)*

to overall solvent exposure than it is with respect to sensitivity to amino acid replacements. This is consistent with a picture of the transition state in which much of the folded structure is distorted so that stabilizing interactions are weakened, but the chain is relatively compact and not so exposed to solvent as in the unfolded state.

A more quantitative mutational analysis of the transition state has been carried out for barnase by Matouschek et al. (1989), who measured the kinetic and thermodynamic effects of amino acid replacements at seven sites. Destabilizing substitutions at four of these sites were found to cause large relative increases in unfolding rate, indicating that these residues are substantially perturbed during formation of the transition state from the native structure. These sites are distributed throughout the native conformation and include residues that form hydrogen bonds at the N-termini of two helices, a Tyr residue that bridges a β-loop, and a Leu residue buried in the hydrophobic core. Substitutions at the C-terminus of an α-helix and another residue in the hydrophobic core were deduced to be in more native-like environments in the transition state. Two replacements at a third buried residue, Ile 96, yielded somewhat ambiguous results, with a Val substitution causing a larger relative increase in unfolding rate than did introducing Ala at this site. Thus, formation of the transition state appears to involve disruption of several regions of native barnase, but there are also parts of the transition state that appear to have some of the interactions present in the native protein.

The same amino acid replacements in barnase have been used to characterize a kinetic intermediate that precedes the transition state during folding (Matouschek et al., 1990). Two replacements at the C-terminus of an α-helix were found to destabilize the intermediate nearly as much as they destabilize the native protein, indicating that this region contributes to stability in the intermediate. Replacements at the N-termini of two helices and a β-loop caused almost no destabilization of the intermediate, suggesting that these regions are relatively unstructured in the intermediate. Replacements of three hydrophobic residues in the core of the protein were all found to destabilize the intermediate, but to a lesser degree than the native protein. It appears that these hydrophobic residues contribute to stability early in folding, but that this contribution increases as more of the structure forms.

The intermediates and transition states in the disulfide-coupled folding pathway of BPTI have been studied using mutations that alter three residues buried in different regions of the native protein (Goldenberg et al., 1989). The major productive pathway for the folding of BPTI (Chapter 7) can be described by the following scheme:

$$R \rightleftharpoons I \rightleftharpoons II \rightleftharpoons II_N \rightleftharpoons N \qquad\qquad (8\text{--}13)$$

where R is the reduced and unfolded protein, N is the native protein with three disulfides, I and II represent populations of intermediates with one and two disulfide bonds, respectively, and II_N is a native-like intermediate with two of the disulfides of the native protein (Creighton and Goldenberg, 1984). The folding and unfolding kinetics of the mutant proteins are well described by the same scheme, and computer simulations were used to estimate the eight rate constants for each protein. Although the identities of the disulfide bonds in the intermediates have not yet been identified for the mutant proteins, the rate constants can be used to compare the energetics of the analogous steps in folding.

Reaction coordinate diagrams for wild-type BPTI and four mutant proteins are shown in Figure 8–9. The reaction coordinates in Figure 8–9A represent conditions where the intrinsic stability of a disulfide bond is such that the native wild-type protein is 7 kcal/mol more stable than the reduced unfolded protein. The free energies of the native and intermediate forms of each wild-type or mutant protein are all expressed relative to that of the reduced form of that protein. The amino acid replacements examined destabilize the native protein, by 3.5 to 6.8 kcal/mol, but the substitutions at different sites have very different effects on the folding intermediates. The replacement of Tyr 35 with Gly destabilizes the native protein by 5.4 kcal/mol, but does not affect the stabilities of any of the intermediates. Replacements at Asn 43, on the other hand, destabilize the native-like two-disulfide intermediate, II_N, as well as the native protein, but do not greatly affect the earlier intermediates. The Tyr 23 → Leu substitution appears to destabilize even the earliest intermediates.

The relative effects of each of the substitutions on the stabilities of the intermediates, calculated from Equation (8–4), are plotted in Figure 8–9C. This representation of the data indicates an "order of action" for three residues: Tyr 23 contributes to stability earliest in folding, followed by Asn 43, and, finally Tyr 35. Tyr 23 forms part of the interface between the α-helix and β-sheet of the native protein. The finding that the Tyr 23 → Leu replacement acts early in folding is very consistent with the results of model peptide experiments by Oas and Kim (1988), which indicate that this interface is formed in the major one-disulfide intermediate. The relative destabilizations of the transition states are plotted in Figure 8–9D. All of the substitutions destabilize the native protein with respect to the major transition state for the folding reaction, which involves intramolecular rearrangement of two-disulfide intermediates. Thus, it appears that all of the altered sites are disrupted in this transition state, even though it represents a relatively late stage of folding in terms of disulfide formation and compactness.

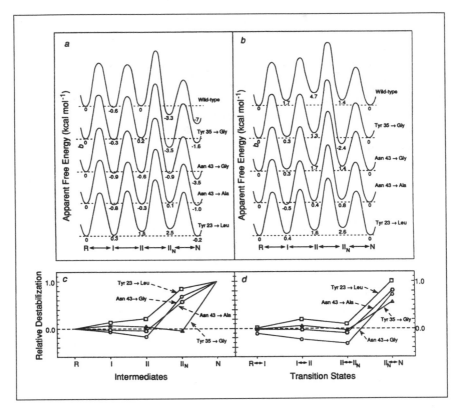

FIGURE 8–9. *Effects of four amino acid replacements on the folding energetics of BPTI. (A) Reaction coordinates for the wild-type and mutant proteins at a single thiol-disulfide redox potential. The relative apparent free-energies of the intermediates and transition states were calculated from the measured rate constants for folding and unfolding. The stabilities of the species making up the folding transition of each protein are expressed relative to the reduced unfolded state of that protein. (B) Reaction coordinates for the wild-type and mutant proteins at the midpoint redox potential for each protein, where the native and reduced states have equal apparent free energies. (C) Relative destabilizations of the folding intermediates. The effects of the amino acid replacements on the stabilities of the different intermediates are compared by dividing the destabilization of each intermediate by the destabilization of the native protein caused by the same substitution, according to Equation (8–4). (D) Relative destabilizations of the transition states, calculated as in (C). (Reproduced, with permission, from Figure 5 of Goldenberg et al, 1989)*

Because the mutations destabilize native BPTI more than they destabilize the intermediates, they reduce the cooperativity of the folding transition. This is illustrated in the reaction coordinates shown in Figure 8–9B. Unlike Figure 8–9A, these profiles represent different thiol-disulfide redox potentials set so that the native and fully reduced forms of each protein have equal apparent free energies. These conditions would correspond to the midpoint in a denaturant- or temperature-induced unfolding transition. When the native and unfolded forms of the wild-type protein have equal energies, all the intermediates are significantly less stable than the end states. Thus, the transition would be well described by a two-state model, as is seen for the denaturant- or temperature-induced transitions of most small proteins. In contrast, for the Tyr 35 $\rightarrow$ Gly mutant, II_N is 2.4 kcal/mol more stable than N or R at the midpoint. Thus, this intermediate, which is compact but inactive, would be the only species easily detected at the midpoint in an equilibrium experiment. The other mutants display similar, but less dramatic, reductions in cooperativity.

Amino acid replacements have also been found to reduce the cooperativity of the folding transitions for apomyoglobin (Hughson and Baldwin, 1989), DHFR (Perry et al., 1989), and staphylococcal nuclease (Shortle and Meeker, 1986). These decreases in cooperativity may be related to the accumulation of "molten globule" intermediates when the native states of α-lactalbumin or myoglobin are destabilized by the removal of a tightly bound ligand (Hughson et al., 1990; Kuwajima, 1989; Chapter 6 of this book). One way in which mutations may prove to be most useful in the study of protein folding is by facilitating the detection of partially folded structures that would otherwise not accumulate under the conditions required to unfold the native wild-type protein. These findings also suggest that special care may be required in analyzing equilibrium unfolding experiments with mutant proteins. Even though the unfolding of the wild-type protein may be well described by a two-state model, a mutant may deviate significantly from such behavior, and analyzing such a transition as if it were two-state would be misleading.

In summary, mutational analysis has already proven useful in characterizing protein folding mechanisms. For several proteins, it appears that the major transition state in folding and unfolding differs from the native conformation at many sites. Different amino acid replacements that destabilize the native state can have clearly distinguishable effects on the stabilities of folding intermediates. These results lend support to models in which there is a preferred order, or pathway, of folding, as opposed to models in which there are many converging pathways. Further studies of mutant proteins, in conjunction with improved physical methods for characterizing intermediates, are likely to lead to much more detailed structural descriptions of protein folding processes.

5 EFFECTS OF MUTATIONS ON THE CONFORMATIONS OF FOLDED AND UNFOLDED PROTEINS

As indicated in the previous sections, one of the current challenges in interpreting the results of mutational experiments lies in more precisely understanding the mechanisms by which amino acid replacements alter the relative stabilities of the native, unfolded, and intermediate states of a protein. This requires knowledge of how the replacements change interactions both within the protein and between the protein and solvent. A first step toward answering these questions is to determine how substitutions alter the three-dimensional structure of the native protein and the broad distribution of conformations that makes up the unfolded state.

By their very nature, the folded structures of proteins can be much more precisely defined than the conformations of partially folded and unfolded chains, and the powerful methods of X-ray crystallography and multidimensional NMR are being used to study the effects of replacements on the native state. Unfolded proteins have been examined with hydrodynamic and spectroscopic methods, but there is still much uncertainty about the nature of the unfolded state. As a consequence, the question of how amino acid substitutions affect the distribution of conformations in the unfolded state remains a subject of considerable speculation and controversy.

5.1 Effects on the Structures of Native Proteins

There are now at least three dozen examples of mutant proteins for which crystal structures have been determined at resolutions of 2 Å or better. Proteins for which mutant and wild-type structures have been compared include phage T4 lysozyme (e.g., Alber et al., 1987b, 1988; Karpusas et al., 1989; Matthews, 1987a; Matsumura et al., 1988; Nicholson et al., 1988; Pjura et al., 1990), subtilisin (Katz and Kossiakoff, 1986, 1990; Pantoliano et al., 1989), and BPTI (Eigenbrot et al., 1990). In most cases, the structural changes induced by substitutions are relatively small (atoms usually move by less than 2 Å), and often are close to the lower limits of the displacements that can be reliably measured crystallographically. Eigenbrot et al. (1990), for instance, found that the overall differences between wild-type BPTI and a mutant in which two disulfide-bonded Cys residues were replaced were smaller than the differences between two crystal forms of the wild-type protein.

The extent of structural change arising from an amino acid replacement can vary greatly depending upon the nature of the substitution and its position in the native protein. The responses of native T4 lysozyme to three different substitutions are illustrated in Figure 8–10, where the absolute values of the displacements of backbone atoms are plotted versus residue position. The largest shifts, up to 1.3 Å, are seen for the replacement of Pro 86 with Gly on the protein surface (Fig. 8–10A) (Alber et al., 1988). The largest structural change is the extension of an α-helix that is terminated by Pro 86 in the wild-type protein. Interestingly, the largest displacements are of atoms three and four residues away from the site of the substitution. Elsewhere in the protein, nearly all of the displacements are less than 0.3 Å.

More widely distributed, but smaller, changes are seen for the replacement of Ala 98, which is buried in the native protein, with Val (Fig. 8–10B; Sun et al., 1991). In this case, introducing a larger side chain in the interior of the protein causes substantial displacements of three α-helices in the C-terminal lobe of the protein. The displacements are largest closest to the site of the substitution, but can be reliably detected as far as 10 Å away. The positional shifts caused by the Thr 157 → Ile substitution are both smaller than those caused by Pro 86 → Gly and more localized than those caused by Ala 98 → Val (Alber et al., 1987b).

Shifts larger than about 1 Å are usually limited to residues close to the site of the substitution and are almost always on the surface of the folded protein. Although substitutions on the surface can perturb other surface residues several Angstroms away, they usually do not change the protein interior (Alber, 1989). On the other hand, replacements of buried residues can cause substantial changes in the interior. Often these interior displacements are distributed asymmetrically around the site of the substitution, suggesting that some local structures in a protein are more susceptible to structural perturbation than others (Eigenbrodt et al., 1990; Katz and Kossiakoff, 1990; Pjura et al., 1990; Sun et al., 1991). In subtilisin and BPTI, α-helices appear to be more easily distorted than β-sheets (Eigenbrodt et al., 1990; Katz and Kossiakoff, 1990).

There does not appear to be much correlation between the extent of structural displacement and the changes in stability arising from amino acid substitutions. The Pro 86 → Gly substitution in T4 lysozyme causes shifts greater than 1 Å, but destabilizes the native state by only about 0.5 kcal/mol (Alber et al., 1988). In contrast, the replacement of Thr 157 with Ile in the same protein causes much smaller structural changes, but destabilizes the protein by nearly 3 kcal/mol (Alber et al., 1987b).

5.2 Effects of Mutations on Unfolded Proteins

Unfolded proteins play key roles in virtually all studies of folding and stability. Measurements of thermodynamic stability use the unfolded form as a reference state, and kinetic studies of refolding begin with an unfolded polypeptide as the starting point. Given its central importance, surprisingly

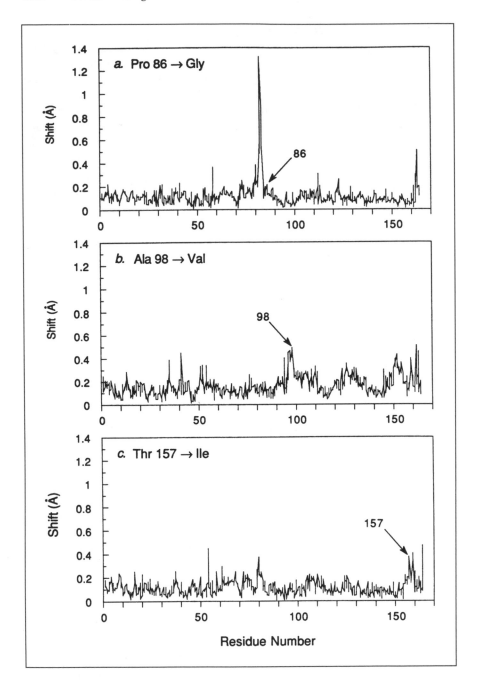

little is known about the unfolded state. A variety of experimental methods indicate that very little cooperative structure persists in denaturant-unfolded or thermally unfolded proteins, but that the distribution of conformations making up the unfolded state is not fully random (Baum et al., 1989; Broadhurst et al., 1991; Evans et al., 1989; Roder, 1989; Tanford, 1968; Tanford et al., 1967). Apparently, some conformations are energetically preferred over others, although the energetic differences may be very small, and it is not yet clear how the distribution is affected by changes in amino acid sequence.

One class of amino acid replacements that are expected to alter the unfolded state are those that replace or introduce Pro residues. Since the pyrolidine ring of a Pro residue constrains the bond between the α-carbon and amide nitrogen, changing another residue to Pro should lower the conformational entropy of the unfolded protein. If this substitution does not affect the native state, the relative stability of the folded protein should increase. In accord with this prediction, an Ala → Pro substitution was found to increase the stability of T4 lysozyme by 0.3 kcal/mol (Matthews et al., 1987). Replacing a Gly residue with any other residue may also decrease the entropy of an unfolded protein, since Gly has more backbone conformations accessible to it than other residues. Replacements of Gly have been found to stabilize λ repressor (Hecht et al., 1986) and T4 lysozyme (Matthews et al., 1987). The increase in entropy of the unfolded state is thought to be a major reason that introducing a Gly residue disfavors helix formation in peptides (Strehlow and Baldwin, 1989). The entropy of the unfolded state should also be affected by amino acid replacements that change the number of rotatable side chain bonds, and if the side chains become constrained during folding, a change in conformational stability may result.

In addition to changing its conformational entropy, replacements involving Pro residues can alter the kinetic properties of unfolded proteins. As discussed in Chapter 5, the refolding kinetics of many proteins display multiple phases that arise from slowly interconverting unfolded forms, and several lines of evidence indicate that the differences among the unfolded forms are due to *cis,trans* isomerization of the peptide bond preceding Pro residues. Site-directed mutagenesis has been used to test the roles of specific

FIGURE 8–10. (opposite) Shift plots showing the displacements of backbone atoms caused by three amino acid replacements in phage T4 lysozyme. The absolute values of positional differences were calculated after superimposing each pair of structures to minimize the differences. (A) Shifts caused by the Pro 86 → Gly substitution (Alber et al., 1988). (B) Shifts caused by the Ala 98 → Val substitution (Sun et al., 1991). (C) Shifts caused by the Thr 157 → Ile replacement (Alber et al., 1987b). (The plots were prepared from data kindly provided by Dr. T. Alber.)

Pro residues in the refolding kinetics of *E.coli* thioredoxin (Kelley and Richards, 1987), staphylococcal nuclease (Evans et al., 1987), and the two isoforms of cytochrome *c* from the yeast *Saccharomyces cerevisiae* (Ramdas and Nall, 1986; White et al., 1987; Wood et al., 1988). In some cases, replacing a Pro residue eliminates one of the slower refolding phases, indicating that isomerization of the altered Pro is responsible for that phase (Kelley and Richards, 1987; Wood et al, 1988). However, not all Pro replacements alter the refolding kinetics, suggesting that isomerization of the peptide bond of the altered Pro is not normally a limiting factor in refolding. This, and other evidence, indicates that some prolyl peptide bonds isomerize after the rate determining steps in refolding or do not isomerize in the unfolded state. (See Chapter 5.) Physical and mutational studies of *E.coli* thioredoxin (Langsetmo et al., 1989) and staphylococcal nuclease (Evans et al., 1987) indicate that some folded proteins can also accommodate different Pro isomers.

Shortle and co-workers have proposed that amino acid replacements may change the distribution of conformations and the average dimensions of unfolded staphylococcal nuclease (Shortle, 1989; Shortle and Meeker, 1986, 1989; Shortle et al., 1988). They have found that substitutions significantly change the dependence of ΔG_u on denaturant concentration (*m* in Equation [8–3]), and have suggested that these changes may arise from differences in the degree of solvent exposure of the unfolded chain. Shortle and Meeker (1989) also prepared large fragments of nuclease to serve as models of the unfolded protein and observed that many of the same replacements alter the hydrodynamic volumes and circular dichroism spectra of these fragments. The fragments have CD spectra indicative of considerable secondary structure, however, and appear to undergo cooperative unfolding transitions at low denaturant concentrations. Thus, the structure that is present in the fragments in the absence of denaturant, and is perturbed by amino acid replacements, may not be significant at the denaturant concentrations where the unfolding of the intact protein is measured.

Many proteins that are stabilized by multiple disulfide bonds can be unfolded by simply reducing the disulfides, as discussed in Chapter 7. The resulting molecules do not appear to undergo any further cooperative unfolding when denaturants are added, and spectroscopic studies indicate that they resemble random-coils at least as much as do proteins unfolded by denaturants or elevated temperature (Amir and Haas, 1988; Creighton, 1988; Kosen et al., 1980, 1981; Roder, 1989). The hydrodynamic volumes of the reduced forms of 47 different variants of BPTI have been compared by gel electrophoresis in the absence of any denaturants (D. P. Goldenberg, unpublished results). The largest differences in electrophoretic mobility among variants with the same charge are about 3%. These differences are markedly less than the changes in mobility arising from an insertion of four extra residues (a decrease of about 6%) or introduction of a cross-link between the termini of the unfolded protein (an increase of about 25%). Thus, it appears that none of these substitutions, which include a variety of different types of replacements of 22 of the 58 residues of the wild-type protein, greatly alter the hydrodynamic volume of reduced BPTI.

Another way of testing for the presence of structure in a reduced polypeptide is to measure the rate of disulfide formation. As expected for an unfolded protein, the six Cys thiols of reduced BPTI form disulfides randomly in the first step of folding, and the rate of this process is not greatly affected by the addition of 8 M urea (Creighton, 1977). This rate has also been measured for BPTI mutants with amino acid replacements at six sites, and none of the substitutions change the rate of initial disulfide formation by more than about twofold, even though they destabilize the native protein by as much as 6.8 kcal/mol (Goldenberg et al., 1989; J.-X. Zhang and D. P. Goldenberg, unpublished results). These replacements have progressively larger effects on the rates and equilibria of individual steps in folding as the protein approaches the native state. This pattern may reflect increased cooperativity among the various interactions in the protein and greater steric constraints as more of the structure forms. The energetic perturbations from amino acid replacements may generally become more significant as the native structure is approached during folding.

6 EFFECTS OF MUTATIONS ON PROTEIN FOLDING AND ASSEMBLY IN VIVO

Although most experimental studies of protein folding have focused on unfolding and refolding reactions *in vitro*, there is growing interest in the more biologically relevant folding processes that occur *in vivo*, as discussed in Chapter 10. There is now considerable evidence that cells utilize special proteins, chaperonins, to regulate folding, to direct newly synthesized polypeptides to their proper locations, and to prevent aggregation among incompletely folded proteins. Amino acid replacements may influence protein folding *in vivo* either directly, as they do *in vitro*, or by perturbing interactions between the nascent polypeptide and other components of the cell.

One class of mutations that have long been associated with defects in protein folding are temperature sensitive (ts) mutations (deWaard et al., 1965; Edgar and Lielausis, 1964; Horowitz and Leupold, 1951; Kohno and Roth, 1979; Sadler and Novick, 1965; Smith et al., 1980; Suzuki, 1970; Yura and Ishihama, 1979). These are mutations that prevent gene function at elevated temperatures but not at lower, permissive temperatures. In some cases, temperature-sensitive mutations reduce the stability of the folded protein, so that when the active protein synthesized at the permissive temperature is heated to the restrictive temperature, it is inactivated. Other temperature-sensitive mutant proteins, however, retain their activity when heated. This second class of mutants has been referred to as "temperature-sensitive synthesis" mutants to emphasize that the defective phenotype is expressed only if the protein is synthesized at the restrictive temperature (Sadler and Novick, 1965). Most mutations of this class, however, do not interfere with synthesis *per se*, but rather prevent correct folding or subunit assembly. Analysis of this class of mutations can help define the kinetic factors that influence protein folding *in vivo*.

The best studied temperature-sensitive mutations are those that block folding of the tail spike protein of phage P22. More than 50 independent temperature-sensitive mutations in the gene encoding the tail spike have been isolated by screening randomly mutagenized phage (Smith et al., 1980; Villafane and King, 1988). These mutations do not interfere with biosynthesis of the protein, but the polypeptides produced at the restrictive temperature are inactive and resemble denaturant-unfolded protein with respect to sensitivity to proteases, sensitivity to detergent, and immuno-reactivity (Smith and King, 1981; Goldenberg et al., 1983). The active proteins synthesized at the permissive temperature retain their activity at 40°C, however, and are not significantly more sensitive to thermal denaturation than is the wild-type protein, which unfolds only at temperatures greater than 80°C (Goldenberg and King, 1981; Sturtevant et al., 1989). Because these mutations interfere with formation of the native trimer, they have been called temperature-sensitive folding (tsf) mutations to distinguish them from mutations that act by preventing biosynthesis or by making the native protein thermolabile.

Pulse labeling experiments indicate that the mutations act at a stage prior to the formation of a protrimer intermediate, in which the three chains have associated but are not fully folded (Goldenberg et al., 1983). The chains that accumulate form rapidly sedimenting aggregates that resemble the "inclusion bodies" seen in bacteria producing large amounts of foreign proteins (Haase-Pettingell and King, 1988). It appears that there is a kinetic competition between productive folding and aggregation, and the temperature-sensitive mutations cause a larger fraction of molecules to follow the non-productive pathway. This might be due to a decrease in the rate of a key folding step, destabilization of a folding intermediate, or an increase in the rate of aberrant association between molecules. (See also Chapters 9 and 10.)

More than 30 sites of ts amino acid substitutions in the tail spike have been identified by DNA sequencing, and all are located in the central 350 residues of the protein (Yu and King, 1984; Villafane and King, 1988). Studies with suppressed nonsense mutations indicate that replacements in the N-terminal region of the protein are generally silent, while changes in the C-terminal region prevent correct folding at both low and high temperatures (Fane and King, 1987). The majority of the mutations alter hydrophilic residues, and 60% result in charge changes, suggesting that most of the altered residues are on the surface of the folded protein. Yu and King (1984) have suggested that the replacements are located within β-turns in the native protein and may act by preventing turn formation at crucial steps in folding. Stroup and Gierasch (1990) tested this hypothesis by measuring turn formation in peptides corresponding to the region of the protein altered by one of the temperature-sensitive substitutions. In the context of these peptides, the wild-type sequence was found to have a three- to fivefold greater tendency to form a turn than the sequence with a Gly → Arg replacement, supporting the suggestion that this substitution prevents *in vivo* folding of the intact protein by interfering with turn formation. Further details of how the temperature-sensitive mutations act may be revealed by *in vitro* folding studies with the wild-type and mutant tail spike proteins (Seckler et al., 1989).

As discussed in Chapter 10, there are now several examples of proteins that appear to facilitate protein folding and assembly *in vivo*. The two best characterized catalysts of protein folding are protein disulfide isomerase (Freedman, 1989), which enhances the rates of disulfide formation, reduction, and rearrangement, and prolyl peptide isomerase (Schmid and Fischer, 1990), which catalyzes proline peptide bond isomerization in both peptides and proteins. A number of other proteins, the chaperonins (Ellis and Hemmingsen, 1989), appear to influence the rates and efficiencies of folding, but it is not yet clear whether these proteins are true catalysts or if they act by simply binding incompletely folded chains and then releasing them. One role of chaperonins may be to bind newly synthesized polypeptides that are to be translocated across membranes and prevent their folding until they are engaged by the translocation machinery (Laminet et al., 1990; Chirico et al., 1988; Deshaies et al., 1988; Hardy and Randall, 1991). Chaperonins may also have a role in enhancing the efficiency of folding, perhaps by preventing aggregation (Gatenby and Ellis, 1990; Gloubinoff et al., 1989; Laminet et al., 1990; Ostermann et al., 1989). It seems quite possible that some mutations that lead to the accumulation of inactive polypeptide chains *in vivo* may do so by altering interactions between the nascent polypeptide and chaperonins. Analyzing mutations that alter either the chaperonins or their protein substrates may be an important means of learning how the processes of folding, subunit assembly, and translocation across membranes are regulated in the cell.

Acknowledgments

I am particularly grateful to Drs. T. Alber and C. N. Pace for sharing unpublished results and allowing me to present examples of their work in Figures 8–4, 8–5, and 8–10. Mutational studies of protein folding in my laboratory are supported by a grant from the U.S. National Institutes of Health (GM42494).

REFERENCES

Ackers, G. K., and Smith, F. R. (1985) *Ann. Rev. Biochem. 54*, 597–629.

Alber, T. (1989) *Ann. Rev. Biochem. 58*, 765–798.

Alber, T., Bell, J. A., Sun, D. P., Nicholson, H., Wozniak, J. A., Cook, S., and Matthews, B. W. (1988) *Science 239*, 631–635.

Alber, T., Sun, D. P., Nye, J. A., Muchmore, D. C., and Matthews, B. W. (1987a) *Biochemistry 26*, 3754–3758.

Alber, T., Sun, D. P., Wilson, K., Wozniak, J. A., and Matthews, B. W. (1987b) *Nature 330*, 41–46.

Alber, T., and Wozniak, J. A. (1985) *Proc. Natl. Acad. Sci. USA 82*, 747–750.

Amir, D., and Haas, E. (1988) *Biochemistry 27*, 8889–8893.

Anderson, D. E., Becktel, W. J., and Dahlquist, F. W. (1990) *Biochemistry 29*, 2403–2408.

Baase, W. A., Muchmore, D. C., and Becktel, W. J. (1986) *Biophys. J. 49*, 109a.

Baldwin, R. L. (1986) *Proc. Natl. Acad. Sci. USA 83*, 8069–8072.

Bashford, D., and Karplus, M. (1990) *Biochemistry 29*, 10219–10225.

Baum, J. L., Dobson, C. M., Evans, P. A., and Hanley, C. (1989) *Biochemistry 28*, 7–13.

Beasty, A. M., Hurle, M. R., Manz, J. T., Stackhouse, T., Onuffer, J. J., and Matthews, C. R. (1986) *Biochemistry 25*, 2965–2974.

Becktel, W. J., and Schellman, J. A. (1987) *Biopolymers 26*, 1859–1877.

Broadhurst, R. W., Dobson, C. M., Hore, P. J., Radford, S. E., and Rees, M. L. (1991) *Biochemistry 30*, 405–412.

Bycroft, M., Matouschek, A., Kellis, J. T. J., Serrano, L., and Fersht, A. (1990) *Nature 346*, 488–490.

Carter, P. J., Winter, G., Wilkinson, A. J., and Fersht, A. R. (1984) *Cell 38*, 835–840.

Chen, B. L., Baase, W. A., and Schellman, J. A. (1989) *Biochemistry 28*, 691–699.

Chirico, W. J., Waters, G., and Blobel, G. (1988) *Nature 332*, 805–810.

Chothia, C. (1976) *J. Mol. Biol. 105*, 1–14.

Connelly, P., Ghosaini, L., Hu, C.-Q., Kitamura, S., Tanaka, A., and Sturtevant, J. M. (1991) *Biochemistry 30*, 1887–1891.

Coplen, L. J., Frieden, R. W., and Goldenberg, D. P. (1990) *Proteins: Struct. Funct. Genet. 7*, 16–31.

Creighton, T. E. (1977) *J. Mol. Biol. 113*, 313–328.

Creighton, T. E. (1979) *J. Mol. Biol. 129*, 235–264.

Creighton, T. E. (1980) *J. Mol. Biol. 137*, 61–80.

Creighton, T. E. (1983) *Biopolymers 22*, 49–58.

Creighton, T. E. (1988) *Biophys. Chem. 31*, 155–162.

Creighton, T. E., and Goldenberg, D. P. (1984) *J. Mol. Biol. 179*, 497–526.

Dang, L. X., Merz, K. M. J., and Kollman, P. A. (1989) *J. Am. Chem. Soc. 111*, 8505–8508.

Deshaies, R. J., Koch, B. D., Werner-Washburne, M., Craig, E., and Scheckman, R. (1988) *Nature 332*, 800–805.

deWaard, A., Paul, A. V., and Lehman, I. R. (1965) *Proc. Natl. Acad. Sci. USA 54*, 1241–1248.

Dill, K. A. (1990a) *Biochemistry 29*, 7133–7155.

Dill, K. A. (1990b) *Science 250*, 297.

Doig, A. J., Williams, D. H., and Sauer, R. T. (1990) *Nature 348*, 397.

Edgar, R. S., and Lielausis, I. (1964) *Genetics 49*, 649–662.

Eigenbrot, C., Randal, M., and Kossiakoff, A. A. (1990) *Protein Eng. 3*, 591–598.

Ellis, R. J., and Hemmingsen, S. M. (1989) *Trends Biochem. Sci. 14*, 339–342.

Evans, P. A., Dobson, C. M., Kautz, R. A., Hatfull, G., and Fox, R. O. (1987) *Nature 329*, 266–268.

Evans, P. A., Kautz, R. A., Fox, R. O., and Dobson, C. M. (1989) *Biochemistry 28*, 362.

Fane, B., and King, J. (1987) *Genetics 117*, 157–171.

Fersht, A. R. (1987) *Trends Biochem. Sci. 12*, 301–304.

Freedman, R. B. (1989) *Cell 57*, 1069–1072.

Gao, J., Kuczera, K., Tidor, B., and Karplus, M. (1989) *Science 244*, 1069–1072.

Garvey, E. P., and Matthews, C. R. (1989) *Biochemistry 28*, 2083–2093.

Gatenby, A. A., and Ellis, R. J. (1990) *Ann. Rev. Cell. Biol. 6*, 125–149.

Gilson, M. K., and Honig, B. H. (1987) *Nature 330*, 84–86.

Gloubinoff, P., Christeller, J. T., Gatenby, A. A., and Lorimer, G. H. (1989) *Nature 342*, 884–888.

Goldenberg, D. P. (1985) *J. Cell. Biochem. 29*, 321–335.

Goldenberg, D. P. (1988) *Ann. Rev. Biophys. Biophys. Chem. 17*, 481–507.

Goldenberg, D. P., Frieden, R. W., Haack, J. A., and Morrison, T. B. (1989) *Nature 338*, 127–132.

Goldenberg, D. P., and King, J. (1981) *J. Mol. Biol. 145*, 633–651.

Goldenberg, D. P., Smith, D. H., and King, J. (1983) *Proc. Natl. Acad. Sci. USA 80*, 7060–7064.

Goto, Y., and Hamaguchi, K. (1982) *J. Mol. Biol. 156*, 911–926.

Haase-Pettingell, C. A., and King, J. (1988) *J. Biol. Chem. 263*, 4977–4983.

Hampsey, D. M., Das, G., and Sherman, F. (1988) *FEBS Lett. 231*, 275–283.

Hardy, S. J. S., and Randall, L. L. (1991) *Science 251*, 439–443.

Harvey, S. C. (1989) *Proteins: Struct. Funct. Genet. 5*, 78–92.

Hawkes, R., Grutter, M., and Schellman, J. (1984) *J. Mol. Biol. 175*, 195–212.

Hecht, M. H., Nelson, H. C. M., and Sauer, R. T. (1983) *Proc. Natl. Acad. Sci. USA 80*, 2676–2680.

Hecht, M. H., and Sauer, R. T. (1985) *J. Mol. Biol. 186*, 53–63.

Hecht, M. H., Sturtevant, J. M., and Sauer, R. T. (1984) *Proc. Natl. Acad. Sci. USA 81*, 5685–5689.

Hecht, M. H., Sturtevant, J. M., and Sauer, R. T. (1986) *Proteins: Struct. Funct. Genet. 1*, 43–46.

Hol, W. G. J. (1985) *Prog. Biophys. Mol. Biol. 45*, 149–195.

Horovitz, A., and Fersht, A. R. (1990) *J. Mol. Biol. 214*, 613–617.

Horowitz, N. H., and Leupold, U. (1951) *Cold Spring Harbor Symp. on Quant. Biol. 16*, 65–74.

Hughson, F. M., and Baldwin, R. L. (1989) *Biochemistry 28*, 4415–4422.

Hughson, F. M., Wright, P. E., and Baldwin, R. L. (1990) *Science 240*, 1544–1548.

Hurle, M. R., Marks, C. B., Kosen, P. A., Anderson, S., and Kuntz, I. D. (1990) *Biochemistry 29*, 4410–4419.

Hurle, M. R., Tweedy, N. B., and Matthews, C. R. (1986) *Biochemistry 25*, 6356–6360.

Hyde, C. C., Ahmed, S. A., Padlan, E. A., Miles, E. W., and Davies, D. R. (1988) *J. Biol. Chem. 263*, 17857–17871.

Karpusas, M., Baase, W. A., Matsumura, M., and Matthews, B. W. (1989) *Proc. Natl. Acad. Sci. USA 86*, 8237–8241.

Katz, B. A., and Kossiakoff, A. (1986) *J. Biol. Chem. 261*, 15480–15485.

Katz, B., and Kossiakoff, A. (1990) *Proteins: Struct. Funct. Genet. 7*, 343–357.

Kauzmann, W. (1959) *Adv. Protein Chem. 14*, 1–63.

Kelley, R. F., and Richards, F. M. (1987) *Biochemistry 26*, 6765–6774.

Kellis, J. T. J., Nyberg, K., and Fersht, A. R. (1989) *Biochemistry 28*, 4914–4922.

Kellis, J. T. J., Nyberg, K., Sali, D., and Fersht, A. R. (1988) *Nature 333*, 784–786.

Kim, P. S., and Baldwin, R. L. (1982) *Ann. Rev. Biochem. 51*, 459–489.

Kim, P. S., and Baldwin, R. L. (1990) *Ann. Rev. Biochem. 59*, 631–660.

Kleina, L. G., Masson, J. M., Normanly, J., Abelson, J., and Miller, J. H. (1990) *J. Mol. Biol. 213*, 705–717.

Kleina, L. G., and Miller, J. H. (1990) *J. Mol. Biol. 212*, 295–318.

Klemm, J. D., Wozniak, J. A., Alber, T., and Goldenberg, D. P. (1991) *Biochemistry 30*, 589–594.

Kohno, H., and Roth, J. (1979) *Biochemistry 18*, 1386–1392.

Kosen, P. A., Creighton, T. E., and Blout, E. R. (1980) *Biochemistry 19*, 4936–4944.

Kosen, P. A., Creighton, T. E., and Blout, E. R. (1981) *Biochemistry 20*, 5744–5754.

Kuwajima, K. (1989) *Proteins: Struct. Funct. Genet. 6*, 87–103.

Kuwajima, K., Mitani, M., and Sugai, S. (1989) *J. Mol. Biol. 206*, 547–561.

Laminet, A. A., Ziegelhoffer, T., Georgopoulos, C., and Pluckthun, A. (1990) *EMBO J. 9*, 2315–2319.

Langsetmo, K., Fuchs, J., and Woodward, C. (1989) *Biochemistry 28*, 3211–3220.

Laskowski, M. Jr., Park, S. J., Tashiro, M., and Wynn, R. (1989) In *Protein Recognition of Immobilized Ligands: UCLA Symposia on Molecular and Cellular Biology*, vol. 80 (T. W. Hutchins, ed.), A. R. Liss, New York, pp. 149–168.

Lesser, D. R., Kurpiewski, M. R., and Jen-Jacobson, L. (1990) *Science 250*, 776–786.

Lim, W. A., and Sauer, R. T. (1989) *Nature 339*, 31–36.

Lin, T. Y., and Kim, P. S. (1989) *Biochemistry 28*, 5282–5287.

Lumry, R., and Rajender, S. (1970) *Biopolymers 9*, 1125–1227.

Matouschek, A., Kellis, J. T. J., Serrano, L. Bycroft, M., and Fersht, A. (1990) *Nature 346*, 440–445.

Matouschek, A., Kellis, J. T. J., Serrano, L., and Fersht, A. R. (1989) *Nature 340*, 122–126.

Matsumura, M., Becktel, W. J., Levitt, M., and Matthews, B. W. (1989a) *Proc. Natl. Acad. Sci. USA 86*, 6562–6566.

Matsumura, M., Becktel, W. J., and Matthews, B. W. (1988) *Nature 334*, 406–410.

Matsumura, M., Signor, G., and Matthews, B. W. (1989b) *Nature 342*, 291–292.

Matsumura, M., Wozniak, J. A., Sun, D. P., and Matthews, B. W. (1989c) *J. Biol. Chem. 264*, 16059–16066.

Matsumura, M., Yasumura, S., and Aiba, S. (1986) *Nature 323*, 356–358.

Matthews, B. W. (1987a) *Biochemistry 26*, 6885–6888.

Matthews, B. W., Nicholson, H., and Becktel, W. J. (1987) *Proc. Natl. Acad. Sci. USA 84*, 6663–6667.

Matthews, C. R. (1987b) *Methods Enzymol. 154*, 498–511.

Matthews, C. R., and Crisanti, M. M. (1981) *Biochemistry 20*, 784–792.

Matthews, C. R., Crisanti, M. M., Gepner, G. L., Velicelebi, G., and Sturtevant, J. M. (1980) *Biochemistry 19*, 1290–1293.

McNutt, M., Mullins, L. S., Raushel, F. M., and Pace, C. N. (1990) *Biochemistry 29*, 7572–7576.

Miles, E. W., Yutani, K., and Ogasahara, K. (1982) *Biochemistry 21*, 2586–2592.

Mitchinson, C., and Wells, J. A. (1989) *Biochemistry 28*, 4807–4815.

Murphy, K. P., Privalov, P. L., and Gill, S. J. (1990) *Science 247*, 559–561.

Nicholson, H., Becktel, W. J., and Matthews, B. W. (1988) *Nature 336*, 651–656.

Oas, T. G., and Kim, P. S. (1988) *Nature 336*, 42–48.

Ostermann, J., Horwich, A. L., Neupert, W., and Hartl, F. U. (1989) *Nature 341*, 125–130.

Pace, C. N. (1975) *Crit. Rev. Biochem. 3*, 1–43.

Pace, C. N., Laurents, D. V., and Thomson, J. A. (1990) *Biochemistry 29*, 2564–2572.

Pakula, A. A., and Sauer, R. T. (1989) *Ann. Rev. Genet. 23*, 289–310.

Pakula, A. A., and Sauer, R. T. (1990) *Nature 344*, 363–364.

Pakula, A. A., Young, V., and Sauer, R. T. (1986) *Proc. Natl. Acad. Sci. USA 83*, 8829–8833.

Pantoliano, M. W., Ladner, R. C., Bryan, P. N., Rollence, M. L., Wood, J. F., and Poulos, T. L. (1987) *Biochemistry 26*, 2077–2082.

Pantoliano, M. W., Whitlow, M., Wood, J. F., Dodd, S. W., Hardman, K. D., Rollence, M. L., and Bryan, P. N. (1989) *Biochemistry 28*, 7205–7213.

Parsell, D. A., and Sauer, R. T. (1989) *J. Biol. Chem. 264*, 7590–7595.

Pauling, L., and Corey, R. B. (1951) *Proc. Natl. Acad. Sci. USA 37*, 729–740.

Pauling, L., Corey, R. B., and Branson, H. R. (1951) *Proc. Natl. Acad. Sci. USA 37*, 205–211.

Perry, K. M., Onuffer, J. J., Gittelman, M. S., Barmat, L., and Matthews, C. R. (1989) *Biochemistry 28*, 7961–7968.

Perry, K. M., Onuffer, J. J., Touchette, N. A., Herridon, C. S., Gittelman, M. S., Matthews, C. R., Chen, J.-T., Mayer, R. J., Taira, K., Benkovic, S. J., Howell, E. E., and Kraut, J. (1987) *Biochemistry 26*, 2674–2682.

Perry, L. J., and Wetzel, R. (1984) *Science 226*, 555–557.

Perutz, M. F., Gronenborn, A. M., Clore, G. M., Gogg, J. H., and Shih, D. T.-b (1985) *J. Mol. Biol. 183*, 491–498.

Pjura, P. E., Wozniak, J. A., and Matthews, B. W. (1990) *Biochemistry 29*, 2592–2598.

Privalov, P. L. (1989) *Ann. Rev. Biophys. Biophys. Chem. 18*, 47–69.

Privalov, P. L., and Gill, S. J. (1988) *Adv. Protein Chem. 39*, 191–234.

Privalov, P. L., Gill, S. J., and Murphy, K. P. (1990) *Science 250*, 297–298.

Privalov, P. L., and Khechinashvili, N. N. (1974) *J. Mol. Biol. 86*, 665–684.

Ramdas, L., and Nall, B. T. (1986) *Biochemistry 25*, 6959–6964.

Rechsteiner, M., Rogers, S., and Rote, K. (1987) *Trends Biochem. Sci. 12*, 390–394.

Reidhaar-Olson, J. F., and Sauer, R. T. (1988) *Science 241*, 53–57.

Reidhaar-Olson, J. F., and Sauer, R. T. (1990) *Proteins: Struct. Funct. Genet. 7*, 306–316.

Roder, H. (1989) *Methods Enzymol 176*, 446–473.

Rose, G. D., Geselowitz, A. R., Lesser, G. J., Lee, R. H., and Zehfus, M. H. (1985) *Science 229*, 834–838.

Russell, A. J., and Fersht, A. R. (1987) *Nature 328*, 496–500.

Sadler, J. R., and Novick, A. (1965) *J. Mol. Biol. 12*, 305–327.

Sali, D., Bycroft, M., and Fersht, A. R. (1988) *Nature 335*, 740–743.

Sandberg, W. S., and Terwilliger, T. C. (1989) *Science 245*, 54–57.

Santoro, M. M., and Bolen, D. W. (1988) *Biochemistry 27*, 8063–8068.

Sauer, R. T., Hehir, K., Stearman, R. S., Weiss, M. A., Jeitler-Nilsson, A., Suchanek, E. G., and Pabo, C. O. (1986) *Biochemistry 25*, 5992–5998.

Schellman, C. G. (1986) *Biophys. J. 49*, 4939a.

Schellman, J. A. (1955) *Compt. rend. lab. Carlsberg, ser. chim. 29*, 230–259.

Schellman, J. A. (1978) *Biopolymers 17*, 1305–1322.

Schellman, J. A. (1987) *Ann. Rev. Biophys. Biophys. Chem. 16*, 115–137.

Schmid, F. X., and Fischer, G. (1990) *Biochemistry 29*, 2205–2212.

Seckler, R., Fuchs, A., King, J., and Jainicke, R. (1989) *J. Biol. Chem. 264*, 11750–11753.

Serrano, L., and Fersht, A. R. (1989) *Nature 342*, 296–297.

Serrano, L., Horovitz, A., Avron, B., Bycroft, M., and Fersht, A. R. (1990) *Biochemistry 29*, 9343–9352.

Sharp, K. A., and Honig, B. (1990) *Annu. Rev. Biophys. Biophys. Chem. 19*, 301–332.

Shirley, B. A., Stanssens, P., Hahn, U., and Pace, C. N. (1992) *Biochemistry 31*, 725–732.

Shirley, B. A., Stanssens, P., Steyaert, J., and Pace, C. N. (1989) *J. Biol. Chem. 264*, 11621–11625.

Shoemaker, K. R., Kim, P. S., York, E. J., Stewart, J. M., and Baldwin, R. L. (1987) *Nature 326*, 563–567.

Shortle, D. (1989) *J. Biol. Chem. 264*, 5315–5318.

Shortle, D., and Lin, B. (1985) *Genetics 110*, 539–555.

Shortle, D., and Meeker, A. D. (1986) *Proteins: Struc. Funct. Genet. 1*, 81–89.

Shortle, D., and Meeker, A. K. (1989) *Biochemistry 28*, 936–944.

Shortle, D., Meeker, A. K., and Friere, E. (1988) *Biochemistry 27*, 4761–4768.

Shortle, D., Stites, W. E., and Meeker, A. K. (1990) *Biochemistry 29*, 8033–8041.

Smith, D. H., Berget, P. B., and King, J. (1980) *Genetics 96*, 331–352.

Smith, D. H., and King, J. (1981) *J. Mol. Biol. 145*, 653–676.

Smith, M. (1985) *Ann. Rev. Genet. 19*, 423–462.

Spolar, R. S., Ha, J.-H., and Record, M. T. (1989) *Proc. Natl. Acad. Sci. USA 86*, 8382–8385.

Sternberg, M. J. E., Hayes, F. R. F., Russell, A. J., Thomas, P. G., and Fersht, A. R. (1987) *Nature 330*, 86–88.

Strehlow, K. G., and Baldwin, R. L. (1989) *Biochemistry 28*, 2130–2133.

Streisinger, G., Mukai, F., Dreyer, W. J., Miller, B., and Horiuchi, S. (1961) *Cold Spring Harbor Symp. Quant. Biol. 26*, 25–30.

Stroup, A. N., and Gierasch, L. M. (1990) *Biochemistry 29*, 9765–9771.

Sturtevant, J. M., Yu, M. H., Haase-Pettingell, C., and King, J. (1989) *J. Biol. Chem. 264*, 10693–10698.

Sun, D. P., Alber, T., Baase, W. A., Wozniak, J. A., and Matthews, B. W. (1991) *J. Mol. Biol. 221*, 647–667.

Sun, D. P., Liao, D. I., and Remington, S. J. (1989) *Proc. Natl. Acad. Sci. USA 86*, 5361–5365.

Suzuki, D. (1970) *Science 170*, 695–706.

Tanford, C. (1961) *J. Am. Chem. Soc. 83*, 1628–1634.

Tanford, C. (1968) *Adv. Prot. Chem. 23*, 121–282.

Tanford, C. (1970) *Adv. Prot. Chem. 24*, 1–95.

Tanford, C. (1980) *The Hydrophobic Effect*, 2nd ed., Wiley, New York.

Tanford, C., Kawahara, K., and Lapange, S. (1967) *J. Am. Chem. Soc. 89*, 729–736.

Touchette, N. A., Perry, K. M., and Matthews, C. R. (1986) *Biochemistry 25*, 5445–5452.

Tweedy, N. B., Hurle, M. B., Chrunyk, B. A., and Matthews, C. R. (1990) *Biochemistry 29*, 1539–1545.

Villafane, R., and King, J. (1988) *J. Mol. Biol. 204*, 607–619.

Villafranca, J. E., Howell, E. E., Voet, D. H., Strobel, M. S., Ogden, R. C., Abelson, J. N., and Kraut, J. (1983) *Science 222*, 782–788.

Warshel, A., and Russell, S. T. (1984) *Q. Rev. Biophys. 17*, 283–422.

Warshel, A., Sussman, F., and King, G. (1986) *Biochemistry 25*, 8368–8372.

Wells, J. A. (1990) *Biochemistry 29*, 8509–8517.

Wells, J. A., and Powers, D. B. (1986) *J. Biol. Chem. 261*, 6564–6570.

White, T. B., Berget, P. B., and Nall, B. T. (1987) *Biochemistry 26*, 4358–4366.

Wood, L. C., White, T. B., Ramdas, L., and Nall, B. T. (1988) *Biochemistry 27*, 8562–8568.

Wyman, J. J. (1948) *Adv. Protein Chem. 4*, 407–453.

Yu, M. H., and King, J. (1984) *Proc. Natl. Acad. Sci. USA 81*, 6584–6588.

Yura, T., and Ishihama, A. (1979) *Ann. Rev. Genet. 13*, 59–97.

Yutani, K., Ogasahara, K., Suzuki, M., and Sugino, Y. (1979) *Biochem. (Tokyo) 85*, 915–920.

9

Folding of Large Proteins: Multidomain and Multisubunit Proteins

JEAN-RENAUD GAREL

1 INTRODUCTION

A "large" protein is defined here as any protein that is not a single-domain single-chain protein. This negative definition of a large protein uses a structural criterion, the presence of at least two compact regions linked covalently or not, and does not set a precise limit on the minimum size of a large protein. By this definition, any oligomeric protein is a large protein, independent of the size of its subunits. Single-chain proteins with more than 200 to 250 residues are often folded into domains and will thus be rated as large. Some single-chain proteins with 150 residues or even less must also be considered as "large" as soon as there is good evidence (from X-ray crystallography, sequence homologies, limited proteolysis, or any other reliable method) that they possess more than one structural domain. The size distribution of the polypeptide chains present in typical bacterial and animal cells (Kiehn and Holland, 1970) is such that the majority of existing proteins has an overall molecular weight greater than 25,000 in native conditions (in one or several polypeptide chains), and thus belongs to the class of large proteins. Consequently, many of the "interesting" proteins that are (and will be) studied because of their therapeutic, commercial, industrial, or epistemologic values are large proteins.

The effort of writing this contribution is dedicated to A. N. and R. L. Baldwin and to G. and R. Jaenicke.

1.1 Domains as Structural Units in the Folded State of Large Polypeptide Chains

A domain is a part of the chain that forms a compact globular substructure with more interactions within itself than with other parts of the chain (Janin and Wodak, 1983). Historically, the first domains were obtained by limited proteolysis of immunoglobulins, and were characterized as smaller fragments still able to bind the antigen (Porter, 1973). Subsequent results obtained with a variety of methods such as genetic complementation, X-ray crystallography, analysis of sequence homologies, and functional comparisons have confirmed the existence of domains. The consensus today is that globular polypeptides with more than 200 to 250 residues (with the exception of linear proteins like collagen, keratin, or other fibrous ones with a repetitive sequence) are folded into several domains.

The word *domain* itself is semantically ill defined, because its actual meaning depends upon the method used for its identification. A part of a large protein can be called a domain when it is:

1. *A stable unit* that can be isolated by limited proteolysis or protein engineering. This is the domain that one can "hold" and manipulate as a discrete species. This is the historical sense of a domain as encountered in immunoglobulins (Porter, 1973). Also, complementation has shown that two inactive fragments of β-galactosidase could fold by themselves because they could associate *in vivo* into a functional enzyme (Goldberg, 1969).

2. *A structural unit* visible at the atomic level on electron density maps obtained by X-ray crystallography. This is the domain that one can "see" as a compact body (Rossmann and Argos, 1981). The visual definition of a domain may not be sufficient and more "objective" computer algorithms have been proposed to identify domains in the three-dimensional structures of large proteins (Janin and Wodak, 1983).

3. *A genetic unit* that is deduced from sequence comparisons at the level of genes or polypeptides. This is the domain that one can "predict" from one-dimensional information using the consensus rule that "homologies in amino acid sequences are always associated with strong resemblances in tertiary structure" (Bajaj and Blundell, 1984; Chothia and Lesk, 1986).

4. *A functional unit* related to a given partial activity such as using NAD as a redox cofactor or ATP as a phosphate donor, or to the binding of a particular ligand (DNA, cyclic AMP, sugar, and so on) (Rossmann and Argos, 1981; Janin and Wodak, 1983). This is the domain that one can "expect" from functional studies.

The overall activity of the protein is related to its spatial organization in several domains, each with a given partial function that can be found in other proteins.

5. *An evolutionary unit* related to the complex organization into split genes that might relate protein domains and gene exons (Gilbert, 1985). This is the domain that one can "guess." The correspondence between the coding sequences and the discrete structural units could be an extension to three dimensions of the colinearity between the gene and the polypeptide.

6. *A thermodynamic unit* corresponding to the segment of a protein that can unfold and refold cooperatively in an all-or-none process. This is the domain that one can "rationalize" as having only two macroscopic states: native and denatured (Privalov, 1989).

7. *Any discrete unit* created by the independent folding of a segment of a long polypeptide (Wetlaufer, 1973). This is the domain that one can "admit," even without information on the structure, the sequence, or the function, and corresponds to the minimal definition plausible for any chain longer than 200 to 250 residues.

These different definitions partially overlap, and instead of having a precise meaning, the word *domain* corresponds to a broad concept for an intermediate level of organization of the structure, function, and evolution of proteins. In the following we will use the word *domain* in a general sense, even though several examples will be limited to stable or structural domains, i.e., the most clearly identifiable domains. Extrapolation from these examples to all the other meanings of the word *domain* outlines a general role of domains in the folding of large proteins, which will be recognized by X-ray crystallographers, biochemists, enzymologists, molecular biologists, or geneticists.

The regions of a protein corresponding to two definitions of domains have been compared in a few cases (for instance, by identifying the sites of limited proteolysis in the tridimensional structure). For several proteins such as immunoglobulins and PGK, it is found that the stable and structural units coincide. In some other cases, stable domains obtained by limited proteolysis do not apparently coincide with clearly defined structural domains: in the $\alpha_2\beta_2$ complex of tryptophan synthetase (TS),[1] the sites where proteolytic cleavages in either the α- or the β_2-subunit (αTS or βTS) produce stable domains are not located within the exposed loops connecting the structural domains seen by X-ray crystallography (Hyde et al., 1988).

---

[1]AK-HDH: aspartokinase homoserine dehydrogenase; LDH: lactate dehydrogenase; MDH: malate dehydrogenase; ODH: octopine dehydrogenase; PFK: phosphofructokinase; PGK: phosphoglycerate kinase; TS: tryptophan synthetase; αTS and βTS: α- and β_2-subunits of tryptophan synthetase.

1.2 Significance of Various Parameters as Probes for the Folding of Large Proteins

Many properties of the fully native and completely denatured states of a protein are different, and most of them can be used to follow the folding process of a large protein (Hirs and Timasheff, 1986; Jaenicke, 1987). A major difficulty encountered with large proteins, however, is that it is difficult to define an absolute probe of "nativeness." As will be described in further detail, the folding of large proteins involves multiple intermediate species, with different degrees of partial folding. At the same time, such an intermediate species may look native by one parameter, unfolded by another, and partially folded by a third one, so that many properties have only a relative value as probes of nativeness. The definition of a folded species depends upon the parameter actually used to measure folding. Folding intermediates are species that do not have *all* the properties of the native state, and only a comparison between different parameters can determine the extent of partial folding.

Changes in certain properties reveal only "local" folding (such as the burial of a given aromatic residue, the formation of an α-helix, the folding of a loop, or the formation of a specific site or epitope) and cannot monitor the formation of the native state throughout the whole protein. Other properties are related to more "global" folding (such as a cooperative binding involving interactions between distant sites) on the scale of the entire protein. Early folding stages are probably more conveniently detected by local probes, late stages by global probes.

Another distinction can be made between "physical" and "functional" probes of folding. Physical probes are related to the shape and/or geometry of the protein, such as inside-outside location of side chains, main chain conformation, and number of subunits. Functional probes are related to binding affinity, catalytic efficiency, or allosteric regulation. Such functional probes may depend crucially on small structural details that are well beyond the sensitivity of physical probes: the structural differences between the active and inactive forms of a folded protein are sometimes almost undetectable. Therefore, the more complex the protein, the less suitable are physical probes and the more adapted are functional probes!

If a single probe is used in folding studies, it should be the most stringent as an overall criterion of nativeness. Usually, it should be a functional probe such as catalytic efficiency, binding affinity, and cooperativity between sites. Measurements related to biological activity are not only a better index of minor subtle conformational adjustments than the usual physical parameters, but they are also suitable for the low protein concentrations that must be used. The appearance of a regulatory property, like cooperative binding or allosteric behavior, is a good probe of overall folding because it requires that several functional sites be formed and associated, and that they interact through a global conformational change of the entire structure.

1.3   Scope of This Chapter

Many reviews on protein folding exist, and references can be found in two 1990 articles (Creighton, 1990; Kim and Baldwin, 1990), in the other chapters of the present book, and in the several contributions on theoretical and practical aspects of protein folding gathered by Hirs and Timasheff (1986). Special mention must be given to the outstanding review by Jaenicke (1987) for its particular interest for large proteins and exhaustive coverage of the field. For results obtained before the mid-1980's, the reader will be frequently referred to that review rather than to the original references.

In the following, the word *folding* means refolding, and corresponds to the process of complete regeneration of the native state *in vitro* after partial or total disruption of the protein conformation. Such reconstitution of the native protein is also called renaturation. Reactivation refers to the reappearance of activity, and reassociation to the formation of the quaternary structure. Rather than an exhaustive survey trying to list all the cases in which renaturation of a large protein has been attempted with total or partial success, this chapter will consider only a few of the results obtained under optimized or at least controlled conditions with a limited number of proteins. These examples, listed in Tables 9–1 and 9–2, (see pp. 410 and 430), represent only a fraction of those reviewed by Jaenicke (1987). The author acknowledges this bias not only in the limited choice of examples, but also in their simplified (possibly oversimplified) presentation. By emphasizing some specific features of a few paradigmatic proteins, however, it is hoped that a more rational picture of the folding of large proteins will emerge, and that the diversity of patterns observed in the folding of individual proteins will be accommodated within a simple common framework.

2   MULTIDOMAIN SINGLE-CHAIN PROTEINS

2.1   Domains as the Folding Units of Large Proteins

The first question about the role of domains in the folding of large proteins is that of the independent folding: "Can a part of the polypeptide chain fold by itself?" Independence of folding can be established only for domains that can be isolated as fragments. At the present time, genetic engineering is making the production of isolated domains much easier. Alternatively, evidence of independent folding can be obtained by keeping a domain

**TABLE 9–1. Archetypal Single-Chain Proteins Most
Frequently Discussed in This Chapter**

Protein	Abbreviation and Chain Size	Evidence for Existence of Domains	Size and Position of Stable Fragments
Trp synthetase α chain	αTS 29 kDa	Proteolytic fragments	20 kDa N-terminal 9 kDa C-terminal
Trp synthetase β chain	βTS 45 kDa	Proteolytic fragments	29 kDa N-terminal 12 kDa C-terminal
Aspartokinase-HSer dehydrogenase	AK-HDH 89 kDa	Mutation and proteolytic fragments	45 kDa N-terminal 27 kDa N-terminal 58 kDa C-terminal
Octopine dehydrogenase	ODH 45 kDa	NAD-dependent dehydrogenase	Not determined
Phospoglycerate kinase	PGK 48 kDa	X-ray structure, mutation, and proteolytic fragments	25 kDa C-terminal
Immunoglobulin light chain	IgG 25 kDa	Proteolytic fragments and X-ray structure	2.5 kDa N-terminal 2.5 kDa C-terminal

identical and modifying the rest of the chain by genetic engineering. The large number of hybrid proteins, chimerae, or fusion constructs that fold into a functional state suggests that independent folding regions exist in large proteins (although their rates of folding and the stabilities of their final states have not been measured).

The next question is related to the influence of the rest of the chain on the folding and/or stability of a domain: "If a domain can fold by itself, is its folding the same when isolated as a fragment and when integrated into the entire chain? Are there interactions between this domain and the other parts of the chain that influence either the stability of the final folded state or the extent, mechanism, yield, or rate of folding?" Studying the role of a single domain within a larger protein requires that its folding be monitored separately from the other folding events occurring within the entire protein. This requires a probe of local folding specific for this particular domain, either a physical probe (unique tryptophan residue, bound cofactor, covalently attached reporter group, and so on) or a functional probe (binding of a specific ligand, appearance of a specific activity, formation of a specific antibody recognition site, and so on). Such a specific probe can be used to measure local folding at the same site in either an isolated stable domain or the entire original protein.

2.1.1 Spontaneous Folding of Isolated Domains

There are many examples of isolated domains that can undergo a reversible unfolding/refolding transition. This was shown for isolated domains obtained as fragments of immunoglobulin light chain, of the α and β chains of TS, of β-lactamase, of AK-HDH, of plasminogen, of PGK and of many other proteins (Jaenicke, 1987). The criteria used to measure the formation of the native structure of such isolated domains were the same as those used with small proteins, either physical probes of a folded state (UV absorption, circular dichroism, fluorescence, NMR) or functional probes of a correctly folded state (specific ligand binding, catalytic activity, antibody recognition).

2.1.2 Thermodynamic Stability of Isolated Domains

The thermodynamic stability of the folded state of a domain has been determined in a few cases where equilibrium measurements could be carried out without interference from aggregation. Several results show that the stability of an isolated domain toward heat, guanidinium ion, or urea is not markedly modified by the presence of the rest of the protein. This has been shown using calorimetry for various fragments of plasminogen (Fig. 9–1) and using fluorescence and/or circular dichroism for the fragments of αTS and βTS (Jaenicke, 1987), AK-HDH (Fig. 9–2A), of immunoglobulin light chain (Tsunenaga et al., 1987), of γIIcrystallin (Rudolph et al., 1990), and of PGK (Missiakas et al., 1990), among others. The most plausible conclusion is that the stability of a domain is due mainly to interactions within itself, and little to interactions with other domains.

This conclusion that domains in a large protein are like "solid stones held together by a weak mortar" has been reached using a set of examples strongly biased toward proteins with stable domains that denature reversibly. It is likely that the stability of domains in other less favorable cases can be strongly influenced by interactions with the rest of the chain. Calorimetry can be used to detect such cooperativity between domains (Tatunashvili et al., 1990) and to measure the energy of stabilization (Brandts et al., 1989; Privalov, 1989).

2.1.3 Rates of Folding of Isolated Domains

Even though a folded domain can be independent of the remainder of the chain, the rest of the protein can influence the rate of its folding. The folding of fragments and of whole chains have been compared for bovine serum albumin by antibody binding (Teale and Benjamin, 1977), for βTS and immunoglobulin light chain by fluorescence changes (Blond and Goldberg, 1986; Tsunenaga et al., 1987), and for AK-HDH by reactivation (Fig. 9–2B). It has been found that folding of an isolated domain takes place at the same rate or is faster than that of the same domain integrated within the intact protein. This suggests that unfavorable interactions with the rest of the

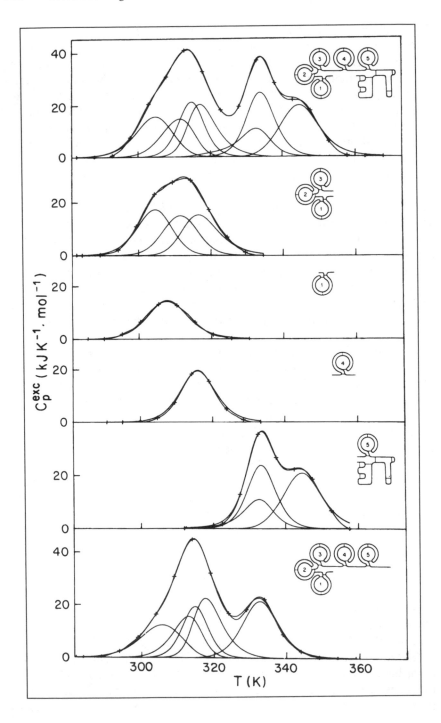

chain can occur during the folding of a domain. Results obtained with AK-HDH indicate that these "negative" interactions depend upon the solvent (unpublished results; compare also Figure 6 of Garel and Dautry-Varsat [1980a] and Table I of Dautry-Varsat and Garel [1981]). It is not known, however, if these unfavorable effects of the rest of the chain on the rate of folding of a domain involve specific interactions in defined intermediates, or result only from an unspecific polymer effect, such as that observed in the inhibition of the RNase A folding by a copolymer of tyrosine and glutamic acid (Haber and Anfinsen, 1962) and of ODH folding by polyethylene-glycol (Teschner et al., 1987).

That the folding of a part of a protein can be slowed by such a negative interaction with an "outside" segment may have a functional significance. It has been proposed that, in addition to its targeting role, the leader peptide of some exported proteins could behave as an inhibitor of folding that would give sufficient time for the partially folded precursor to bind to chaperone proteins (Park et al., 1988).

2.2 Folding Kinetics of Large Single-Chain Proteins

2.2.1 Kinetic Phases in the Folding of Large Polypeptide Chains

Complex kinetics are observed for the folding of penicillinase, αTS, βTS, AK-HDH, PGK, ODH (Fig. 9–3), and many other proteins under a variety of conditions (Jaenicke, 1987). In most cases, two kinetic phases with some common features are observed.

An initial rapid phase is apparent by large changes in several physical parameters (fluorescence, UV absorption, circular dichroism, rate of hydrogen exchange of amide protons, and so on). In the absence of denaturing agent and at room temperature, this phase occurs on the time scale of seconds or faster. This step corresponds to a gross folding of the chain, with formation of most of the secondary structure and of hydrophobic regions from which water has been expelled. The polypeptide chain reaches a state that is largely folded, but is still missing some properties of the native structure. It is more labile to proteolysis, lacks catalytic activity, and it is not recognized by all anti-native antibodies.

FIGURE 9–1. (opposite) Calorimetric measurements of the thermal unfolding of plasminogen and some of its fragments. Deconvolution of differential scanning calorimetry profiles (Privalov and Potekhin, 1986) obtained for the entire chain and several of its fragments show that the different domains of plasminogen melt independently. Interactions between domains contribute little to the stability in this protein. (Reproduced from Novokhatny et al., 1984, with permission)

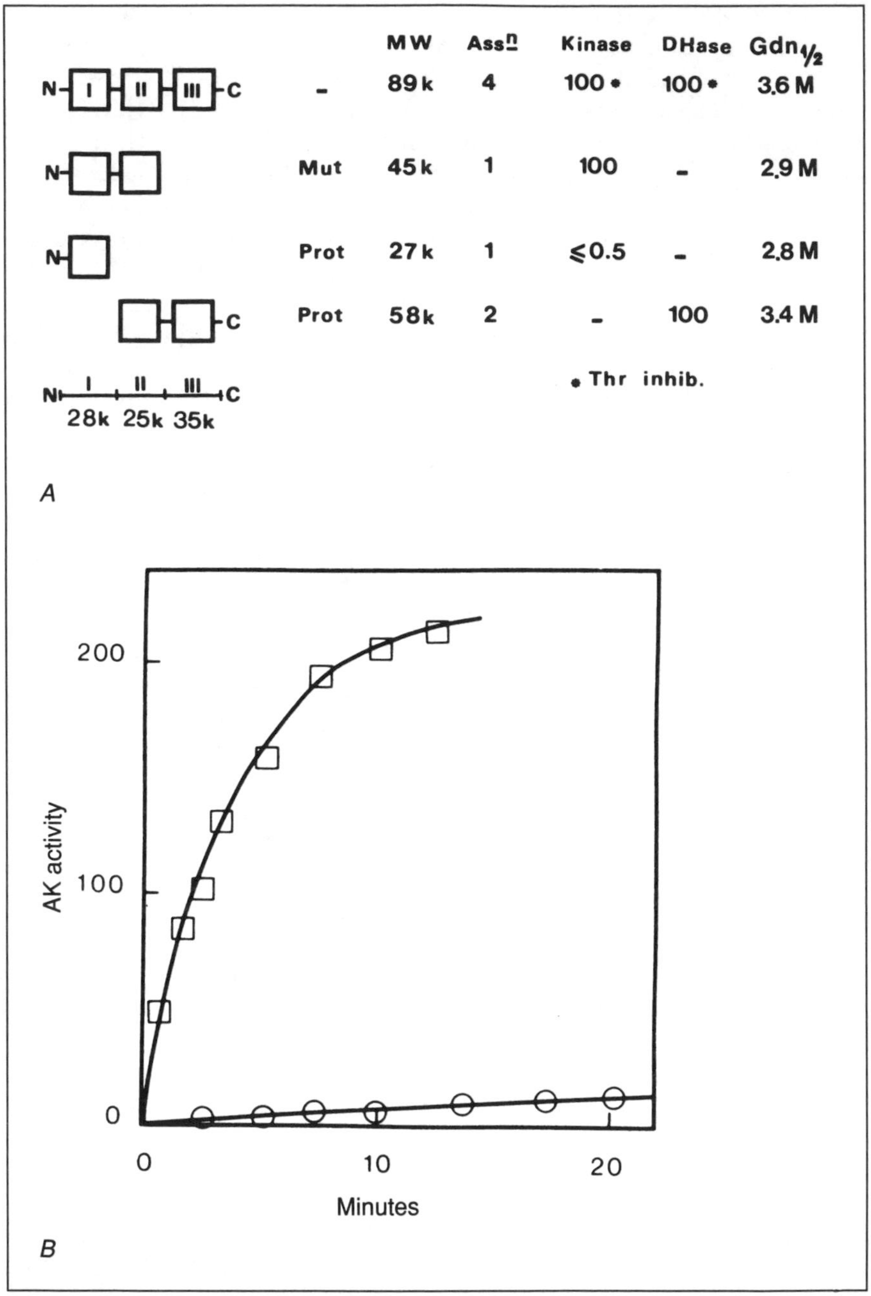

MW Assn Kinase DHase Gdn1/2
N I II III C
- 89k 4 100 • 100 • 3.6 M
Mut 45k 1 100 - 2.9 M
Prot 27k 1 ≤0.5 - 2.8 M
Prot 58k 2 - 100 3.4 M
• Thr inhib.
N I II III C
28k 25k 35k
A
AK activity
200
100
0
0
10
20
Minutes
B

A second slower phase is characterized by much smaller changes detected by the physical probes of folding, but leads to the appearance of the native epitopes, enzymatic activity, resistance to proteases, and ability to bind a specific ligand. This slower phase occurs on the minutes-to-hours time scale and is the slowest step in the complete overall folding process. This second step corresponds to the regain of native structure as a result of a conformational rearrangement within a largely folded state.

These biphasic kinetics of folding can be described by the common simple scheme:

$$\text{unfolded} \xrightarrow{\text{fast}} \text{folded intermediate} \xrightarrow{\text{slow}} \text{native.} \tag{9-1}$$

Depending upon the conditions and/or the parameters used to monitor folding, more complex kinetics have been observed with some proteins such as αTS (Matthews et al., 1983) and βTS (Blond and Goldberg, 1986). This increased kinetic complexity results from the splitting of one or the other usual main reactions into several phases.

A few proteins such as the aldolase from *Staphylococcus aureus* (R. Rudolph, personal communication quoted in Jaenicke [1987]) and the mannitol-1-phosphate dehydrogenase from *Escherichia coli* (Teschner et al., 1990) resume their native structures within seconds. It is not known if this exceptional behavior is due to the second step being faster, undetected, or absent.

2.2.2 The Slow Folding Reaction and the cis,trans Isomerization of X-Pro Peptide Bonds

This two-step folding mechanism is reminiscent of that of small proteins, where there is often a fast folding phase and a slower one that is usually

FIGURE 9–2. *(opposite) Comparison of the intact chain of AK-HDH with some of its fragments (Garel and Dautry-Varsat, 1980a; Müller and Garel, 1984). (A) Division of the polypeptide chain of AK-HDH into three segments and properties of the different fragments: localization within the chain, how produced, size in kDa, degree of association, relative enzymatic activity, as aspartokinase or dehydrogenase, whether inhibited by threonine, and midpoint of the unfolding transition induced by guanidium chloride. (Gdn$_{1/2}$; unfolding was measured by fluorescence and circular dichroism.) The interactions between the different domains are not a major contribution to stability. (B) Comparison of the rates of folding of the N-terminal moiety of AK-HDH ($\square$) and of the entire chain ($\bigcirc$), as measured by the regain of their kinase activities. The fragment corresponding to the N-terminal segments I and II has the same kinase activity as the entire chain, but refolds more rapidly than the whole chain into an active species. (Reproduced from Garel and Dautry-Varsat, 1980a, with permission)*

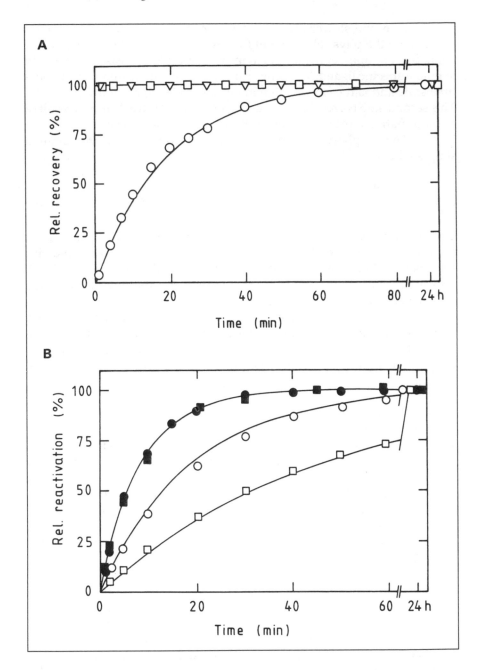

the *cis,trans* isomerization of X-Pro peptide bonds (Kim and Baldwin, 1982; Chapter 5). The rate of Pro residue–governed folding should decrease exponentially with the number of Pro residues (Creighton, 1978), and the slow folding of large polypeptide chains could simply be due to their large numbers of Pro residues. In addition, the activation energy of around 80 kJ/mole measured for the folding of some large proteins is the value associated with Pro *cis,trans* isomerization (Chapter 5). The classical test for the involvement of proline peptide bond isomerization is based on the existence of a slow reaction in unfolding (Brandts et al., 1975; Nall et al., 1978). This test, which is described in Fig. 9–4, has shown that the complete folding of the AK-HDH monomer (Vaucheret et al., 1987), of ODH (Zettlmeissl et al., 1984), and of αTS (Hurle and Matthews, 1987), under some conditions, is not rate-limited by the *cis,trans* isomerization of X-Pro peptide bonds.

It is unlikely that all Pro residues in large proteins are nonessential for folding, and thus Pro peptide bond isomerization has to occur as a slow step at some point during folding. With large proteins, however, complete folding is limited by another slower process and not by Pro peptide bond isomerization.

2.2.3 The Pairing of Domains During Folding
Several arguments suggest that the slowest step in the folding of large single-chain proteins is the pairing of already folded domains. This slowest folding step takes place within a state that is largely folded: secondary structure has formed and hydrophobic cores have been segregated from

---

FIGURE 9–3. (opposite) The kinetics of folding of ODH as measured by fluorescence, circular dichroism, enzymatic activity, and resistance to proteases. (Teschner et al., 1987) (A) Folding of ODH follows biphasic kinetics: a fast reaction is measured by the changes in "physical" probes such as fluorescence (□) and circular dichroism (▽), and a slow reaction controls the regain of a "functional" probe such as enzymatic activity (○). (B) Two slow processes take place during the slow refolding reaction of ODH. The regain of resistance to proteases (closed symbols), which is probably controlled by the cis,trans peptide bond isomerization of proline residues, occurs slightly more rapidly than does the reappearance of activity (open symbols), which is limited by the domain-pairing reaction. These two slow processes can be distinguished by their dependence on solvent viscosity. The rate of cis,trans isomerization of proline residues, measured by the reappearance of the resistance to proteolysis, is the same in the absence (circles) as in the presence (squares) of glycerol, whereas the rate of domain-pairing, measured by the reactivation, is decreased by the presence of glycerol. (Reproduced from Teschner et al., 1987, with permission)

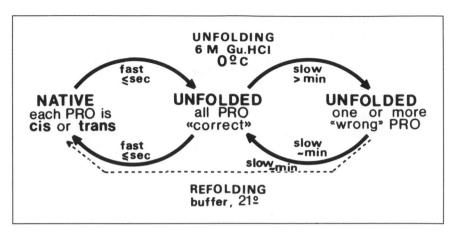

FIGURE 9–4. *The standard test for the involvement of the* cis,trans *isomerization of the peptide bonds preceding proline residues in a slow refolding reaction (Brandts et al., 1975; Nall et al., 1978). Rapid unfolding leads to an unfolded state with all the X-Pro peptide bonds in the correct isomeric state; this state can refold rapidly upon dilution of the denaturant. The formation of slow-refolding species occurs upon equilibration of the peptide bonds of proline residues between their* cis *and* trans *configurations and is slow, especially at 0°C. The double-jump procedure described here (first fast unfolding at 0°C for a given time and then refolding) detects this slow unfolding reaction due to Pro* cis,trans *isomerization.*

aqueous solvent. (See Section 2.2.1.) Extensive domain folding takes place during the initial fast phase, as indicated by the highly structured state formed rapidly by isolated domains from AK-HDH, αTS, βTS, PGK, and others (Jaenicke, 1987). Indeed, a segment of 100 to 150 amino acids, the size of a domain, can reach its native state in a fraction of second (Garel et al., 1976).

1. *Reactivation of an enzyme implies the correct arrangement of two domains.* The slowest folding step is frequently characterized by the reappearance of enzymatic activity, i.e., by the formation of a functional active site (Rossmann and Argos, 1981; Janin and Wodak, 1983). In several cases (such as NAD-dependent dehydrogenases, kinases, and other enzymes catalyzing a two-substrate reaction) this active site is located at an interface between two domains, with each domain bearing the binding site for one substrate or co-factor. Therefore, reactivation implies formation of the proper network of interactions between (at least) two domains. Kinetic results on PGK suggest indeed that assembly of folded domains is the rate-limiting step in reactivation (Vas et al., 1990).

2. *The rate of folding is inversely proportional to solvent viscosity.* The rate of the limiting step in the folding of the AK-HDH monomer, ODH (Fig. 9–5), and αTS (Chrunyk and Matthews, 1990) is sensitive to the presence of solvent additives, such as glycerol or glucose, that increase the viscosity. Interpreting this influence of solvent additives on folding rates is ambiguous because the additives also have a stabilizing effect on protein structure (Timasheff and Arakawa, 1989). In the cases of AK-HDH, ODH, and αTS, however, the changes in folding rates could not be explained entirely by stabilization effects and are attributed to the influence of viscosity. This suggests that the friction of solvent controls the relative movements of different parts of the chain during its folding. The folding rate decreases linearly with the bulk viscosity of the solvent (Fig. 9–5) as expected from the simple Stokes-Einstein model of translational and rotational diffusion of rigid bodies. This model assumes that the diffusing species are much larger than solvent molecules, which is true of folded domains, so that viscosity can be treated as a macroscopic quantity. In addition to the "double-jump" test (Fig. 9–4), this viscosity dependence is another argument against the involvement of Pro peptide bond isomerization in the rate-determining step, since the folding rate of RNase A is limited by Pro isomerization but does not depend on solvent viscosity (Baldwin, 1980).

3. *The effect on the rate of folding caused by a mutation in one domain can be compensated by a second mutation in another domain.* Single amino acid replacements in αTS modify the rate of its folding (Beasty et al., 1986; Tweedy et al., 1990). It has been observed that two single mutations, each in a different domain, decrease the folding rate by a comparable extent, but that the double mutant folds almost at the same rate as wild-type protein (Beasty et al., 1986). The simplest interpretation of this compensation between these two mutations is that the rate-limiting step involves the formation of the interface between the two domains. This could be tested experimentally by making other mutations.

All of the preceding observations are consistent with a folding mechanism in which the domains of a large polypeptide fold independently and rapidly in a first step, and then pair and adjust mutually in a second slower step:

fast domain folding   slow domain pairing

unfolded ⎯⎯⎯⎯→ folded intermediate ⎯⎯⎯⎯→ native.

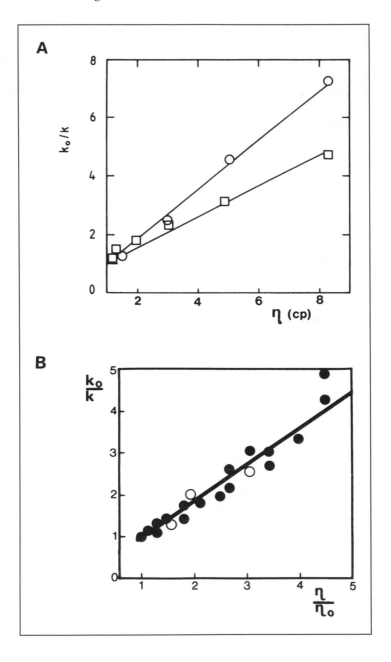

This scheme does not mean that all domains fold at the same rate in one fast step, nor that no folding occurs in the slow step. Rather, the first step involves a set of independent reactions leading to the formation of the secondary and tertiary structures in some segments of the polypeptide chain, and the second step corresponds to conformational changes (including some folding) within the entire protein in which all the interactions between folded domains are established. This mechanism is represented schematically in Figure 9–6, with an intermediate species with unpaired folded domains and unfolded segments.

The folded domains that "pair" in the last folding step are covalently attached and remain at a short distance. There is thus a high effective concentration of one domain in the vicinity of the others. It is surprising that the interaction between these two or more folded domains is at the same time so slow and viscosity-dependent. The pairing of two domains involves successively their encounter and their sticking together. (See Baldwin [1980] for a discussion of pairing two covalently linked objects.) The slowness of domain pairing could be related to the low probability of productive encounters, e.g., due to a severe angular restriction and/or to the fact that sticking occurs in a highly constrained folded state. In this case, the observed viscosity-dependence would indeed be related to the true viscosity-dependence of the relative diffusion of folded domains during their encounter. Alternatively, the changes in folding rates in the presence of solvent additives may not be related to macroscopic viscosity, because domain pairing takes place on a scale where the approximation of a continuous isotropic solvent is no longer valid. Desolvation of contact areas during domain sticking involves expulsion of individual solvent molecules from the newly formed interface, as well as conformational adjustments on the scale of solvent molecules. The presence of (even a few) solvent molecules could hinder the interactions between the pairing domains because trapping of solvent molecules between the two protein surfaces could

FIGURE 9–5. *(opposite) Increasing the solvent viscosity decreases the rate of the slow folding reaction of the single-chain proteins ODH (A) and AK-HDH (B). The folding was measured by the reappearance of activity for ODH and of the kinase activity for AK-HDH in the absence and in the presence of solvent additives that increase the viscosity; k_0 and k are the first-order rate constants of reactivation in the absence and presence of additive. (A) Reactivation of ODH. The viscosity was increased by addition of glycerol ($\square$) or glucose ($\bigcirc$). (Reproduced from Teschner et al., 1987, with permission) (B) Reactivation of AK-HDH. The viscosity was increased by addition of glycerol ($\bullet$) or sucrose ($\bigcirc$); η_0 and η are the viscosities in the absence and presence of additive. (Reproduced from Vaucheret et al., 1987, with permission)*

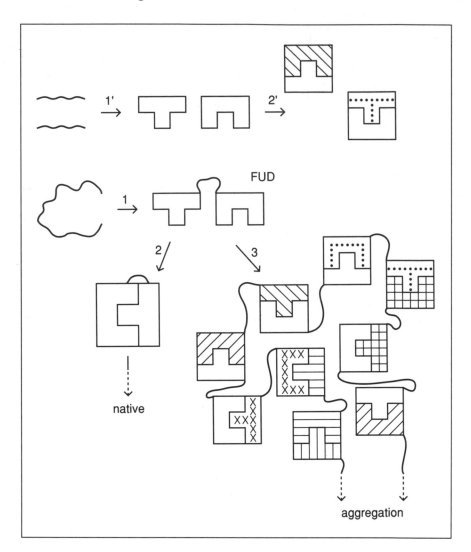

FIGURE 9–6. *A schematic illustration of the "fast-folding, slow-pairing" mechanism of folding of isolated domains of a two-domain protein (top) and of the role of the FUD intermediate in the folding of the single-chain protein (bottom). The incorrect intermolecular association and the pairing of isolated domains are represented as occurring through the same specific interacting sites as the correct intramolecular domain pairing. (Adapted from Goldberg and Zetina, 1980, with permission) (1) fast domain-folding (set of) reaction(s); (2) correct intramolecular slow domain-pairing (set of) reaction(s) leading to folded monomer; (3) incorrect intermolecular domain-pairing reactions leading to aggregates; (1') independent domain-folding reactions; (2') pairing of isolated domains leading to folded monomers.*

prevent close packing by steric hindrance and/or make some conformational adjustments very slow because of a local friction. The observed influence of solvent additives on the rate of domain pairing could then be due to both effects of the *macroscopic viscosity* on the diffusion-limited encounter and of the *"microscopic" viscosity* on the sticking process.

2.2.4 Change of Rate-limiting Step from Domain Pairing to cis,trans Isomerization of Pro Residues

During folding of ODH, the regain of resistance to proteolysis slightly precedes the domain pairing step that regenerates enzymatic activity (Fig. 9–3). Kinetic analysis suggests that it is *cis,trans* isomerization of X-Pro bonds that controls the loss of sensitivity to proteases (Teschner et al., 1987). Domain pairing has an activation energy of 112 kJ/mole (Zettlmeissl et al., 1984) and is slower in the presence of glycerol, whereas proline peptide bond isomerization has an activation energy of only 82 kJ/mole and is not influenced by glycerol (Fig. 9 3). It is thus probable that, at high temperature and in the absence of glycerol, the rate-limiting step for folding of ODH shifts from domain pairing to proline peptide bond isomerization. Such a change in mechanism also occurs in the folding of αTS, where the rate-limiting step can be either domain pairing or proline peptide bond isomerization, depending upon the concentration of the denaturing agent (Beasty et al., 1986; Hurle and Matthews, 1987).

If domains behave like small proteins (see the preceding chapters), they can form their correctly folded structures only after isomerization of the peptide bonds of most of their proline residues. In this case, the isomerization of proline residues within each domain precedes domain pairing, as in

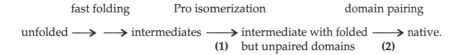

$$\text{unfolded} \xrightarrow{\text{fast folding}} \xrightarrow{\text{}} \text{intermediates} \xrightarrow{\text{Pro isomerization}} \underset{\text{but unpaired domains}}{\text{intermediate with folded}} \xrightarrow{\text{domain pairing}} \text{native.}$$
$$\text{(1)} \qquad\qquad\qquad\qquad \text{(2)}$$

The relative rates of reactions (1) and (2) determine the species that accumulate. It is proposed shortly that the intermediate with folded and unpaired domains is the major site of kinetic competition between folding and aggregation, so that the yield of renaturation would depend upon whether reaction (1) or (2) was rate-limiting. Also, the relative rates of reaction (1) and (2) could be different for different proteins, so that folding of some large proteins such as *S. aureus* aldolase (R. Rudolph, personal communication quoted in Jaenicke, 1987) or γII-crystallin (Rudolph et al., 1990) could resemble that of small proteins. Finally, the nature of the actual rate- limiting step under *in vivo* conditions is relevant to the role of peptidyl-prolyl isomerase. (See the following chapter.)

2.3 Kinetic Competition Between Folding and Aggregation

When folding is limited by the slow domain-pairing step, an intermediate accumulates in which at least some of the domains have already folded but have not yet interacted properly. The presence of this intermediate with Folded but Unpaired Domains, the FUD intermediate, leads to an important kinetic consequence, a competition between folding and aggregation.

2.3.1 Formation of Intra- or Intermolecular Interactions by the Same Contact Areas of the FUD Intermediate

In the native protein, each folded domain interacts with other parts of the chain by one or more specific sites. In the FUD intermediate, these sites are free and can give rise to two different sets of interactions (Goldberg and Zetina, 1980) (Fig. 9–6):

1. Intramolecular interactions with a legitimate partner of the same chain to produce the native state;
2. Intermolecular interactions with several illegitimate partners belonging to different chains to lead the protein into a multi-molecular network that increases in size and withdraws from the aqueous solvent.

The same contact areas are involved in both interactions, so the two reactions of the FUD intermediate are mutually exclusive and compete with each other:

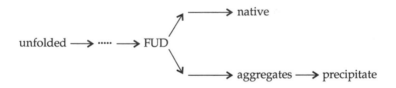

Figure 9–7A shows that there is an excellent negative correlation between folding and aggregation: almost all of the protein that has not reactivated is aggregated and vice versa.

2.3.2 Effect of Protein Concentration on the Yield of Refolding

The common experimental observation with multidomain proteins is that the yield of renaturation depends strongly upon the protein concentration (Fig. 9–7). The preceding scheme predicts that the outcome of the kinetic competition between folding and aggregation will be determined by the partition of the FUD intermediate between two pathways. This

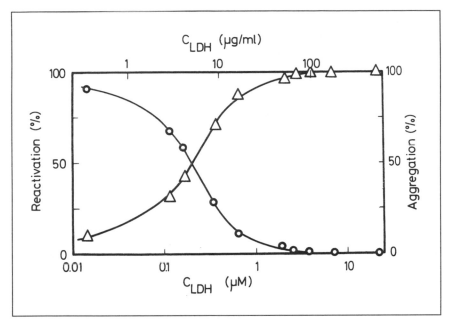

FIGURE 9–7. The influence of protein concentration on the kinetic competition between correct folding and aggregation. (A) lactate dehydrogenase (LDH): Correct folding of LDH was measured by the regain of its activity (○), and aggregation (△) was measured by light scattering (Zettlmeissl et al., 1979). The protein either is correctly folded and active or is in an aggregated state. (Reproduced from Jaenicke and Rudolph, 1989, with permission) (Continued on next page.)

partition depends upon protein concentration because the folding pathway involves only first-order reactions, and its rate does not depend upon protein concentration, whereas the aggregation pathway is a multimolecular reaction, and its rate increases as some n^{th} power of the protein concentration. For LDH, this value of n is around 2.5 (Zettlmeissl et al., 1979). As the protein concentration is increased, the FUD intermediate will shift from the productive folding pathway, which is faster at low protein concentrations, to the aggregation pathway, which is faster at high protein concentrations. Because of the [concentration]^n dependence of the relative rate, the shift in pathway is almost complete within one order of magnitude in concentration (Fig. 9–7). Such a sharp shift corresponds to a concentration threshold above which renaturation is unsuccessful under these conditions. The actual position of this threshold will depend upon the particular protein, the composition of the solvent, and the physical conditions of renaturation.

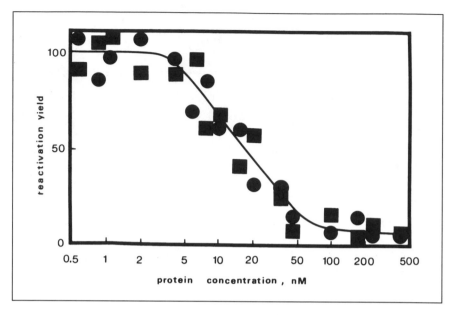

FIGURE 9–7. *(continued) (B) AK-HDH: The formation of different folded species is monitored by the reappearance of two activities. The kinase activity (●) is regained upon folding into a monomeric species, and the dehydrogenase activity (■) is regained upon association into a dimeric species (Garel and Dautry-Varsat, 1980a). Finding the same yield for both activities suggests that all the protein molecules that folded correctly as a monomer also associated correctly as a dimer. (Reproduced from Vaucheret et al., 1987, with permission)*

2.3.3 The Role of the FUD Intermediate in Aggregation

Other observations on the outcome of the kinetic competition between folding and aggregation suggest that this competition occurs at the stage of the FUD intermediate.

1. Aggregation requires *partial folding*. The extent and rate of aggregation are greater in conditions that favor a mixture of folded, partially folded, and unfolded states, than in strongly denaturing or strongly native conditions (Jaenicke, 1987; Mitraki and King, 1989; Schein, 1990).
2. The *specificity* of protein aggregation. The competition between folding and aggregation is not influenced by the presence of other proteins (London et al., 1974), even those with related sequences (Jaenicke et al., 1981). This suggests that aggregation involves specific contact areas and that it occurs only after

extensive folding has taken place (Jaenicke, 1987), such as in the FUD intermediate.

3. Aggregation occurs to a much lesser extent if there is no covalent link between the domains. The two domains of βTS can fold independently and associate with each other into a "nicked" βTS (Goldberg and Zetina, 1980), showing that the pairing of folded domains is independent of the covalent link between them. This nicked protein has a much smaller tendency to aggregate than the original intact chain (Fig. 9–6).

2.3.4 Effect of Folding Conditions on the Concentration of the FUD Intermediate

A direct consequence of the FUD intermediate being the major site of competition between folding and aggregation is that there is a negative correlation between the concentration of the FUD intermediate and the yield of renaturation. Conditions can in principle be varied to improve the renaturation yield by decreasing the concentration of the FUD intermediate, but the outcome can be difficult to predict in practice. For example, the presence of residual denaturant can increase the renaturation yield (1) by destabilizing the FUD intermediate relative to native protein and/or (2) by making the FUD intermediate more soluble and decreasing intermolecular interactions. Alternatively, residual denaturant can decrease the renaturation yield by destabilizing the native state relative to the FUD intermediate. Therefore, residual denaturant can help some proteins to renature but might favor aggregation for some others. Similarly, the influence of temperature, buffer composition, or other physical parameters on the mechanism of folding aggregation (Fig. 9–6) can be opposite for different proteins.

2.4 Formation of Disulfide Bonds During Folding of Large Single-Chain Proteins

Some large proteins have intrachain disulfide bonds, from a few, like the 4 in an immunoglobulin heavy chain, to many, such as the 17 in bovine serum albumin. It is remarkable that correct renaturation and reoxidation of serum albumin occurs (Teale and Benjamin, 1976) despite the unfavorable combinatorial statistics: the chances of forming the correct set of 17 disulfide bonds would be less than 10^{-12} if the protein did not guide the process! In several cases, the distribution of disulfide bonds is consistent with the domain structure, in the sense that disulfide bonds bridge Cys residues that belong to the same domain.

Isolated domains can fold and reoxidize correctly after reduction of their disulfide bonds (Teale and Benjamin, 1977; Goto and Hamaguchi, 1981; Trexler and Patthy, 1983; Hirose et al., 1989). The isolated constant domain of an immunoglobulin light chain forms its single disulfide bond because of the ability of this domain to fold independently (Goto and Hamaguchi, 1986). This suggests that domain folding is sufficient to drive the formation of disulfide bonds. In the renaturation of the reduced entire polypeptide chain, a rapid domain folding reaction would severely restrict the subsequent formation of disulfide bonds. In most cases, the presence of a disulfide exchange catalyst is required, which indicates that "wrong" disulfide bonds are made, but it is not known whether they are made within or between domains. The tendency of Cys residues in a random chain to interact with other Cys residues that are close in the chain suggests that a pathway similar to that observed in small proteins (and discussed in Chapter 7) tends to take place within each domain, rather than through the entire chain. Indeed, it seems that the pattern of disulfide bond formation during the folding of reduced trypsinogen (Light and Higaki, 1987) and ovotransferrin (Hirose et al., 1989) is correlated with the domain structure.

2.5 Conclusions About Single-Chain Multidomain Proteins

In the few examples selected here, domains are at the same time the folding units and the structural units, and there is a separation in time between their fast independent folding and their slow pairing into a common structure. The modular structure of large single-chain proteins thus results from their modular mechanism of folding. Most of these results have been obtained with proteins having stable isolated domains, however, so the generality of the "fast domain folding, slow domain pairing" mechanism needs to be confirmed.

2.5.1 Generalization of the Fast Domain Folding, Slow Domain Pairing Mechanism

Many multidomain proteins cannot be split into stable folded fragments, presumably because their domains have evolved toward a stronger integration within the tertiary structure and have lost part of their independence in folding. The stability of one domain may then depend upon interactions with other domains. (See Section 2.1.2.) It is still likely that each domain will begin its folding independently to generate a partially folded state. Because of their spatial proximity, this "imperfect" folding of domains can be followed by a pairing reaction, in which a larger conformational change achieves the tertiary native structure. Even in those cases where independent folding domains cannot be observed, a mechanism with "fast imperfect domain folding" and "slow domain pairing and adjusting" can be proposed. Here also, a FUD intermediate exists, with only partially folded and unpaired domains, which can either achieve its folding into the native conformation or undergo extensive intermolecular aggregation.

2.5.2 Sequential Folding of Domains

A FUD intermediate would not accumulate when one domain serves as a template for folding the other domain(s). The kinetic mechanism would then involve sequential folding of domains, with an intermediate that is half-folded and half-unfolded. This could also occur when the domains of a protein have very different stabilities and folding takes place in conditions where only one domain is stable. Some of the results obtained with αTS could correspond to this case (Matthews et al., 1983), where only *one* domain can fold.

2.5.3 Parallel Reactions in the Folding of Large Proteins

In large proteins, there is no reason why the fast folding of each independent domain must occur at the same rate, nor why folding should always begin within the same domain. The first folding reactions can then occur independently in different parts of the chain. Only the final step takes place with the same unique rate in the whole molecule. Thus, different parts of the chain might fold at different rates, so different kinetics would be observed using different "local" probes. This is indeed found for the first steps of the folding of βTS (Blond and Goldberg, 1986; Blond-Elguindi and Goldberg, 1990) and the folding of dihydrofolate reductase (Frieden, 1990). The folding of a large protein should not be considered as a strictly sequential process. Instead, a more realistic scheme would have parallel pathways at the stage of domain folding that converge toward the FUD intermediate:

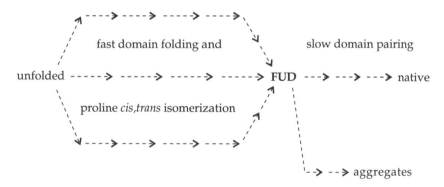

Such parallel reactions have been proposed for the early stages of βTS folding (Murry-Brelier and Goldberg, 1989). The lag period that must occur in the appearance of the native protein when there are sequential kinetic intermediates is decreased when the various steps can occur in any order. After the FUD intermediate, the protein follows a sequential pathway along which the entire structure is rearranged until the native state is formed with all its properties, in a single slow step. "Global" probes, often related to a functional ability, are more suited to monitor this late stage of folding.

3 OLIGOMERIC PROTEINS

The native structures of many proteins are composed of several polypeptide chains associated through noncovalent bonds. These chains can be identical or different, and their sizes range from 50 residues to 3,000 residues. Dimer and tetramer structures are the more frequent, but there are also trimers, hexamers, dodecamers, and even higher-order assemblies (Jaenicke, 1987). Here, we will consider only well-studied examples of oligomeric proteins (Table 9–2) where both chain folding and chain association are required for forming the native structure. Some have already been introduced as single-chain proteins because the folding steps preceding association have also been studied. These proteins allow comparison of the folding of a chain measured by the appearance of its ability to associate with other physical and/or functional probes of folding.

Protein association requires a specific site on each of the reactants, so it can therefore occur only after these sites have appeared. Each association reaction should thus occur only after separate folding of each of the reactants. Therefore, folding of an oligomeric protein from its denatured and separated chains probably begins like that of single-chain proteins and proceeds until the formation of a specific binding site that can recognize another monomer. At this critical step, *the folding pathway shifts from being intramolecular to intermolecular*, to yield a dimeric species. This homo- or heterodimer may

TABLE 9–2. *Archetypal Multichain Proteins Most Frequently Discussed in This Chapter*

Protein	Abbreviation Chain Size × Subunits	Probes of Subunit Association	Rate-Limiting Step for Complete Folding
Trp synthetase β chain	βTS 45 kDa × 2	$s_{20,w}$, ultrafiltration, fluorescence polarization	monomer → monomer'
Aspartokinase-Homoserine dehydrogenase	AK-HDH 89 kDa × 4	2nd-order reactivation, partial activities of intermediates	monomer → dimer
Phospho-fructokinase	PFK 35 kDa × 4	Partial dissociation, HPLC, 2nd-order reactivation, and fluorescence	dimer → tetramer
Lactate dehydrogenase	LDH 35 kDa × 4	Partial dissociation, fluorescence, 2nd-order reactivation, cross-linking, competition between isoenzymes	dimer → tetramer

need further folding steps to become either a native protein or an intermediate with an adequate specific site that allows a second association step to take place. The overall folding pathway of an oligomeric protein is thus a succession of monomolecular folding steps and bimolecular association steps (Jaenicke, 1987).

3.1 Specific Tools for Measuring Association During the Folding of Oligomeric Proteins

Most polypeptide chains in oligomeric proteins are sufficiently large to be composed of several domains. Therefore, the comments in Section 1.2 on the simultaneous use of physical and functional probes of folding apply, as do those on the local and global signification of these probes. In addition to all of the folding probes, specific tools can be used to study the association steps during folding of oligomeric proteins (Jaenicke and Rudolph, 1986; Jaenicke, 1987).

3.1.1 Concentration Dependence of the Rate Constants

The dependence on protein concentration of the overall rate of folding of an oligomeric protein arises from at least one of the association steps. A kinetic analysis becomes possible because the protein concentration can be experimentally controlled. Intermediates with different degrees of association will be more separated in time upon lowering the protein concentration and can therefore be characterized. All of the probes of folding do not, however, reveal this concentration-dependence of rates. Consider the simple case of two consecutive reactions with one folding and one association step:

2 unfolded monomers $\longrightarrow$ 2 folded monomers $\longrightarrow$ dimer

and consider two probes of folding: a local probe sensitive to the first folding step and a global probe specific for dimer formation. The local probe will change at the same rate, that of the monomolecular folding step, independently of the protein concentration. At sufficiently low protein concentrations, association will be slow and the global probe will demonstrate slow second-order kinetics, with the rate depending upon the square of the monomer concentration. A short lag phase corresponding to the folding step can sometimes precede the slow bimolecular step. At high protein concentrations, association becomes relatively rapid, and the global probe shows first-order kinetics, because dimer formation is limited in rate by the folding step. The kinetics observed for a global probe specific for dimer formation will shift from first- to second-order as the protein concentration is decreased. For each protein, there is often only a limited range of concentrations that can be experimentally explored and that range may or may not include this shift; in the latter case, the complete dependence on concentration of rates cannot be determined.

Most real proteins give much more complex kinetics than the preceding simple two-step mechanism, and relating the observed kinetics to an actual mechanism with accurate values for individual rate constants (rather than a "fast" or "slow" qualification) is not straightforward (Jaenicke, 1987). Thus, kinetic analysis requires the help of a computer for fitting the data, simulating progress curves, extrapolating to zero time or to infinite concentration, and designing intelligent experiments to discriminate between alternative mechanisms, so as to establish the right order of folding and association steps and to extract rate constants for individual steps from the time-courses of several parameters.

3.1.2 Chemical Cross-Linking and Rapid Chromatography

Protein association implies a change in a global property, the molecular weight. To follow directly the changes in association during the folding of an oligomeric protein, two methods are sufficiently sensitive to use with low protein concentrations, and sufficiently rapid to give an accurate picture of all the molecules implicated in association events at a given time. The size distribution of the different protein species can be determined by chemical cross-linking of associated intermediates and analyzed by electrophoresis on polyacrylamide gels in the presence of the denaturing agent sodium dodecylsulfate (Fig. 9–8). Size-exclusion chromatography using HPLC separates species with different hydrodynamic volumes (Fig. 9–9) and can also be used to monitor association during folding if it is sufficiently slow.

3.1.3 Partial Dissociation of the Native Oligomer

The quaternary structures of many oligomeric proteins are more labile than their tertiary structures, and partial or total dissociation without gross subunit unfolding can occur under some conditions, such as moderate concentrations of denaturants, chaotropic salts, or detergents, high hydrostatic pressure, low temperature, and extremes of pH (Jaenicke, 1987). At low protein concentrations, the rate of overall folding is limited by slow association steps. These steps can be identified by comparing the kinetics of refolding from separated and unfolded chains with those of reassociation of partially dissociated species. This approach has been used to show that the association of two dimers into a tetramer was the slowest step in the folding of PFK (Deville-Bonne et al., 1989), LDH (Jaenicke et al., 1981), and other proteins (Jaenicke, 1987).

In some cases, modified subunits can be used as competitors to alter the rate and/or extent of formation of the native oligomer from partially dissociated species. These modified subunits can be chemical

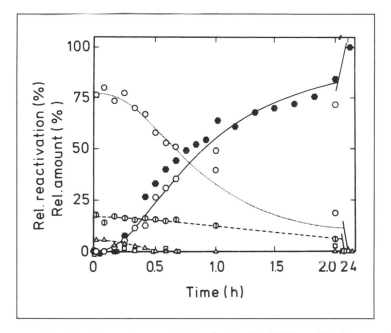

FIGURE 9–8. *Reactivation and reassociation during the folding of the hexameric enzyme UDP-glucose dehydrogenase. The relative amount of hexamers (◇), tetramers (□), trimers (△), dimers (◑), and monomers (◯) was determined by cross-linking with glutaraldehyde followed by electrophoresis in the presence of sodium dodecylsulfate. The reappearance of activity (●) occurred at the same rate as the formation of hexamers, suggesting that the monomer and dimer intermediates were inactive. The only populated species are monomers, dimers, and hexamers. The solid line was obtained by computer simulation of these results using the following mechanism:*

$$\text{6 unfolded monomers} \xrightarrow{k_1} \text{6 folded monomers} \xrightarrow{k_2} \text{3 dimers} \xrightarrow{fast} \text{native hexamer}$$

and values of $k_1 = 4.5 \times 10^{-4} \ s^{-1}$ and $k_2 = 1.6 \times 10^4 \ M^{-1} s^{-1}$. (Reproduced from Jaenicke and Rudolph, 1989, with permission)

derivatives (inactivated, labeled by a reporter group, insolubilized by coupling with a matrix, and so on), proteolytic fragments (isolated stable domains or other fragments), genetic variants (natural or engineered mutants), or natural analogs (isoenzymes or homologous proteins) (Jaenicke, 1987). The elegant studies of subunit interactions in aspartate transcarbamylase

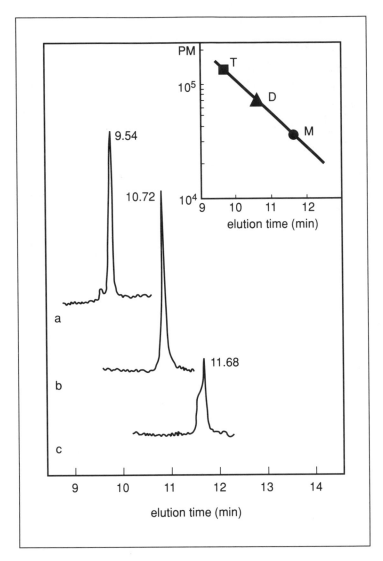

FIGURE 9–9. *Separation of intermediates with different degrees of association by rapid size-exclusion chromatography. The monomeric (trace c), dimeric (trace b), and tetrameric (trace a) states of PFK elute at different times upon chromatography on a TSK-G3000SW HPLC column. This gel filtration method will separate species with different degrees of association only when the association or dissociation reactions have half-lives longer than a few minutes. Inset: calibration of the HPLC gel filtration elution times with the molecular weights of the monomeric M, dimeric D, and tetrameric T states of PFK. (Reproduced from Le Bras et al., 1989, with permission)*

are a good example of the utilization of modified subunits and reconstituted hybrid oligomers (Eisenstein and Schachman, 1989).

3.1.4 Formation of Functional Sites and Interactions Between Distant Sites

The sites for ligand binding or enzyme activity present in several oligomeric proteins are located at interfaces between subunits and involve residues belonging to two different chains. Reappearance of the functional state of such a "sandwich" site depends upon the formation of this interface, so a specific association step can be monitored through a functional property. This applies only to oligomers of known three-dimensional structure.

Similarly, the cooperative binding of a ligand to an oligomer with identical sites (such as the binding of fructose 6-phosphate to PFK [Fig. 9–10]), the allosteric coupling between effectors and substrates (such as the inhibition of AK-HDH by threonine [Fig. 9–11]), the mutual influence of regulatory and catalytic subunits in aspartate transcarbamylase (Eisenstein and Schachman, 1989), and the catalytic enhancement of βTS upon binding αTS (Lane et al., 1984) are properties that are related to interactions between sites carried by different chains. These properties are indicative not only of association, but also of specific conformational changes responsible for the communication between sites distant within the associated state.

3.2 Kinetic Analysis of Association

3.2.1 The Number of Association Steps

In solution, encounters between more than two molecules occur so rarely that only bimolecular reactions are usually significant (Hammes, 1978). Association steps can be either a "symmetric" dimerization of identical species ($2A \rightarrow A_2$) or an "asymmetric" association of different species ($A + B \rightarrow AB$). With identical subunits, one, two, or three symmetric dimerization steps are sufficient for the assembly of dimers ($2^1 = 2$), tetramers ($2^2 = 4$), or octamers ($2^3 = 8$). In contrast, the assembly of trimers, hexamers, or dodecamers implies at least one asymmetric association step between species of different sizes, such as between a dimer and a monomer into a trimer: $A_2 + A \rightarrow A_3$.

The assembly of oligomers composed of different chains also requires at least one asymmetric association step. Except for a heterodimer, however, the nature and the number of association steps depend upon the assembly pathway. For instance, pathways with two or three association steps are possible for folding a tetramer with an A_2B_2 structure:

> $2A + 2B \rightarrow 2AB \rightarrow A_2B_2$, with an initial asymmetric step followed by a second symmetric one;

2A + 2B → A2 + 2B → A2B + B → A2B2, with an initial symmetric step followed by two successive asymmetric ones;
2A + 2B → A2 + B2 → A2B2, with two symmetric steps followed by a third asymmetric one.

Therefore the actual number of association steps cannot be determined before the pathway of assembly is known. A detailed kinetic study was required to show that the second of the three preceding mechanisms applies to the reconstitution of the $\alpha_2\beta_2$ heterotetramer of TS (Lane et al., 1984).

3.2.2 Detection of Different Kinetic Steps

In many cases, only one or two distinct kinetic phases are observed under given conditions during the folding of an oligomeric protein using a given folding probe (Fig. 9–10). The range of protein concentrations that can be explored is generally limited at the low extreme by the sensitivity of detection and by protein aggregation at the high extreme. This range is usually too restricted for the expected shift between rate-limiting folding at high protein concentrations and rate-limiting association at low concentrations (see Section 3.1.1) to be observed. Over a range of more than two orders of magnitude in protein concentration, the folding kinetics of PFK (Martel and Garel, 1984), malate dehydrogenase (MDH), LDH, and other proteins (Jaenicke, 1987) follow the same mechanism. Varying the protein

FIGURE 9–10. (opposite) Renaturation of PFK from its unfolded and separated chains measured by the regain of functional properties. (A) Reactivation measured at a saturating (◯) and a nonsaturating (▢) concentration of the cooperative substrate fructose 6-phosphate, or in the presence of both a nonsaturating concentration of fructose 6-phosphate and the allosteric effector GDP (◪). In all cases, reactivation followed biphasic kinetics: an initial lag phase corresponds to the formation of an inactive intermediate and is followed by a bimolecular reactivation reaction. These biphasic kinetics, which resemble those frequently observed for oligomeric proteins, can be described by the uni-bi mechanism:

$$\text{unfolded} \xrightarrow{\ k_1 = 10^{-2}\,s^{-1}\ } \text{inactive intermediate} \xrightarrow{\ k_2 = 10^{4}\,M^{-1}s^{-1}\ } \text{active species}$$

Note that only one second-order reaction is detected, even though the native protein is tetrameric. (B) The ratio R between the activities measured for the same nonsaturating concentration of fructose 6-phosphate in the presence and absence of allosteric effector GDP is independent of the time of renaturation.
Because of the allosteric behavior of native PFK, measuring reactivation under different conditions can be used as different probes. (i) The regain of maximum

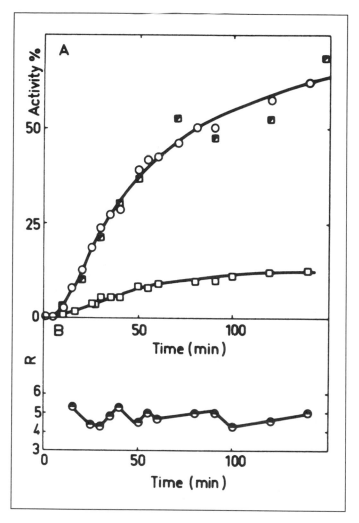

velocity at saturating substrate is a probe for the time-course of the correct functional folding of the active sites. (ii) The regain of activity at nonsaturating fructose 6-phosphate concentration is a probe not only of the affinity but also of the cooperativity during folding, i.e., is a probe of the correct interactions between active sites. (iii) The regain of activity in the presence of the allosteric activator GDP is a probe of the correct interactions between regulatory and active sites. The same folding kinetics indicate that the only active species is that with the same maximum velocity, affinity, and cooperativity for fructose 6-phosphate and the same allosteric activation by GDP as native PFK. (Reproduced from Martel and Garel, 1984, with permission)

concentration does not increase the number of observed reactions and serves only to determine which are concentration dependent.

As mentioned already for single-chain proteins, different probes can be sensitive to different folding events and can detect different steps. The different time-course observed for four different probes during the folding of AK-HDH (Fig. 9–11) show that the pathway from the unfolded to the native protein implies at least three intermediate species. Such experimental detection of several reactions along the folding pathway is sometimes helpful for subsequent kinetic analysis.

3.2.3 Association and Folding as a Rate-Limiting Step

The folding of an oligomeric protein is a succession of uni- and bimolecular reactions, and the step with the lowest rate under any given conditions will be limiting for formation of the native protein. If formation of the native oligomer (as measured by a suitable probe) is controlled by an association step, its rate will depend upon the protein concentration. If a rate does not change with protein concentration, then a unimolecular folding step is limiting.

Association-limited folding is frequently observed, but in many cases, a lag phase precedes the concentration-dependent step (Figs. 9–10 and 9–11). These biphasic kinetics correspond to a two-step mechanism (Jaenicke, 1987):

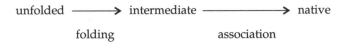

The lag phase corresponds to the time required by the folding step to form intermediate capable of associating. At early times, association is slow until the intermediate species with a specific association site has been formed in sufficient quantities. The majority of the second-order rate constants measured for the limiting association reaction are in the range of 10^3 to 10^5 $M^{-1} \cdot s^{-1}$ (Jaenicke, 1987; Rothman, 1989).

Formation of the native oligomer sometimes follows first-order kinetics, indicating that a folding step is rate limiting. Other information is needed, however, to determine if the slow folding step takes place within a monomeric or an associated state. Assembly of the catalytic trimer of aspartate transcarbamylase from its unfolded chains is limited by a folding step within a single chain (Burns and Schachman, 1982b), while formation of the native dimer of βTS is limited by a folding step within a dimeric intermediate (Blond and Goldberg, 1985).

The folding kinetics of homologous proteins are not necessarily limited in rate by the same step. The cytoplasmic and mitochondrial MDH from the same species are two closely related dimers, but they fold by different mechanisms (Rudolph et al., 1986).

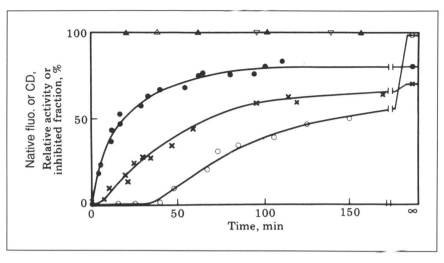

FIGURE 9–11. *Different probes to detect the successive formation of different folding intermediates. The folding kinetics of AK-HDH were measured by: (1) Fluorescence (▲) and circular dichroism (▽), which are probes for the secondary and tertiary structures. They reappear very rapidly. (2) Kinase activity (●), which is a probe of the correct folding of the monomer. This appears in a monomolecular reaction with a first-order rate constant of 6×10^{-4} s^{-1}. (3) Dehydrogenase activity (✕), which is a probe of the correct folding of the dimer. It appears in a bimolecular reaction with a second-order rate constant of 7×10^4 M^{-1} s^{-1}, after a short lag period. (4) Allosteric inhibition of the kinase activity by threonine (○), which is a probe of the correct folding of the tetramer. It appears in a bimolecular reaction with a second-order rate constant of 7×10^4 M^{-1} s^{-1}, but only after a substantial lag period. These kinetics have been interpreted with the mechanism (Garel and Dautry-Varsat, 1980b; Müller and Garel, 1984):*

$$
\overset{\text{\textit{fast}}}{\text{\textit{unfolded}} \longrightarrow} \overset{\text{\textit{1st order}}}{\underset{6 \times 10^{-4} s^{-1}}{\text{\textit{folded monomer}} \longrightarrow}} \overset{}{\underset{}{\text{\textit{active monomer}} \longrightarrow}} \overset{\text{\textit{2nd order}}}{\underset{7 \times 10^4 M^{-1} s^{-1}}{\text{\textit{dimer}} \longrightarrow}} \overset{\text{\textit{2nd order}}}{\underset{>10^{-5} M^{-1} s^{-1}}{\text{\textit{tetramer}}}}
$$

in which the rate-limiting association step is the formation of the dimer. (Adapted from Garel and Dautry-Varsat, 1980b, with permission)

3.2.4 Multiple Association Steps

Folding any oligomer more complex than a dimer involves at least two association steps, but usually only one bimolecular reaction dominates the folding kinetics. For example, the overall folding of a tetramer can be limited by the association of monomers into dimers or by that of dimers into tetramers. Identification of the rate-limiting step is achieved by characterizing the intermediate species that accumulates immediately before the slowest step. The accumulation of dimeric intermediates during the folding of LDH was established by cross-linking, which shows that tetramer formation is rate limiting (Jaenicke, 1987). During folding of PFK, a dimeric form is rapidly formed and is converted slowly to native tetramer, as monitored by molecular weight measurements using HPLC gel filtration (Le Bras et al., 1989).

An alternative procedure is to detect all the association reactions on the folding pathway, to measure their rates using a variety of probes, and to identify the rate-limiting association step as the slowest. During folding of PFK, two different association reactions can be observed, the faster using fluorescence changes and the slower using the regain of enzyme activity. The rapid association reaction corresponds to the formation of PFK dimers, which precedes the rate-limiting formation of active tetramers (Le Bras et al., 1989). The rate-limiting association step in the folding of AK-HDH is the first step, i.e., the formation of dimers, and all subsequent events take place at the same rate. The same second-order rate constant is obtained using two functional properties (dehydrogenase activity and allosteric regulation of the kinase activity), although these are not regained at the same time: the two folding reactions differ only in the length of their lag phases but not in their bimolecular phases (Fig. 9–11).

3.2.5 Computer Simulation of Folding Kinetics

The data obtained in several kinetic experiments usually lead to a minimal mechanism for the folding pathway. Quantitative computer simulations of these progress curves represent a valuable self-consistency test of this simple mechanism (Figs. 9–8 and 9–10). Also, predictions of the behavior of the protein during folding under different conditions can be checked against experiment and used to discriminate alternative mechanisms. If needed, additional complications can be introduced such as (1) other folding steps at the levels of monomer, dimer, and so on, (2) reversible steps and/or fast equilibria, (3) parallel pathways, (4) dead-end intermediates, (5) "wrong" association and reshuffling reactions, and so on (Jaenicke, 1987). At this level of kinetic complexity, each protein needs a custom-fit analysis that extends beyond the scope of this chapter, but the reader must remember that kinetic analysis cannot prove a mechanism, but can only establish whether or not it is consistent with the data.

3.3   Specificity of Association

During folding of an oligomeric protein, association steps involve the formation of interactions between recognition sites. The specificity of this recognition is important for the correct folding of oligomeric proteins.

3.3.1   Appearance of Specific Association Sites During Folding

When starting from unfolded monomers, the formation of a site on each of the two partners is presumably required for their association. In several cases, this specific association site is formed only after extensive folding has occurred. The dimerization site appears in AK-HDH at the same time as the kinase active site (Garel and Dautry-Varsat, 1980b), it appears in PFK at the same time as the ATP binding site (Martel and Garel, 1984), and it appears in βTS after some native epitopes have been formed (Blond and Goldberg, 1987). This shows that the initial steps of loose folding are not sufficient to create the association site, and that this site requires the same degree of folding as a specific functional site.

A dimeric species can also require further folding before being able to undergo the next association step and to form a trimer or a tetramer. Comparison between AK-HDH and one of its fragments suggests that an isomerization takes place within the dimer, prior to tetramer formation (Dautry-Varsat and Garel, 1981). Also, after dimeric βTS has been formed, a slow folding step must occur to form the native protein (Blond and Goldberg, 1985).

3.3.2   Competition Between Correct Association and Aggregation After the FUD Intermediate

As for single-chain proteins, the FUD (Folded but Unpaired Domains) intermediate is a major site of competition between aggregation and correct folding of an oligomeric protein (see Section 2.3). There is, however, no serious evidence that the steps subsequent to the FUD intermediate involve a competition between "good" and "wrong" association. For example, the fraction of correctly associated dimers of AK-HDH is the same as that of correctly folded monomers, i.e., of chains that have passed the FUD stage without yielding to the temptation of intermolecular aggregation (Fig. 9–7).

3.3.3   Interference of Other Proteins with the Folding of an Oligomer

Subunit recognition during folding of oligomeric proteins seems generally to be highly specific. Folding of tryptophanase into a functional tetramer was the same when the protein was by itself or in the presence of foreign proteins, such as serum albumin or all the proteins of a cellular crude extract (London et al., 1974). This shows that the association binding sites specifically recognize tryptophanase subunits. No hybrid oligomers were detected during the simultaneous *in vitro* folding of several homologous NAD-dependent dehydrogenases (Jaenicke, 1987). The α and β chains of tropomyosin have homologous sequences, and their simultaneous folding leads to association both into hybrid dimer $\alpha\beta$ and into homologous dimers $\alpha\alpha$ and $\beta\beta$. The proportion of $\alpha\beta$ hybrid depends on the initial denatured state of tropomyosin and on the procedure of renaturation, indicating that kinetic factors are involved in the choice between "symmetric" and "asymmetric" associations (Brown and Schachat, 1985; Lehrer and Qian, 1990). The existence of hybrids between the H and M isoenzymes of LDH could also be related to the particular kinetics of assembly under *in vivo* conditions (Jaenicke, 1987).

Some heat-shock or chaperone proteins can interact with folding intermediates and favor folding of both monomeric and oligomeric proteins (Rothman, 1989; Chapter 10). Because of the small number of such "helper" proteins, their interaction with the folding intermediates of a number of different proteins must not be as specific as that required for forming the proper quaternary structure.

3.4   Geometry of Association During Folding of Oligomers

Oligomeric proteins made of identical subunits have symmetric structures, and the folding pathway must take this symmetry into account. Most homotetramers have D_2 symmetry with three twofold axes, and two different dimeric intermediates can exist during their folding (Fig. 9–12). A dimeric intermediate with a defined geometry is formed during folding of PFK, in which the association sites correspond the larger areas of contact between subunits (Deville-Bonne et al., 1989; Teschner and Garel, 1989). This result suggests that the order in which the subunits associate during folding of an oligomeric protein depends upon the stability of the interfaces.

Homotrimers often have C_3 symmetry, with one threefold axis, and their folding implies two association reactions and a dimeric intermediate. This dimeric intermediate probably has no twofold symmetry, although it has two identical subunits. It has been shown in one case that this dimeric intermediate does not accumulate to substantial levels (Burns and Schachman, 1982a). The relationships between symmetry of the final structure and folding pathway are even more complex for proteins made of different subunits.

3.5   Influence of Ligand Binding on the Rate of Association Reactions

Stabilization of a protein by one of its ligands can result either from an increase in the rate of folding, or from a decrease in the rate of unfolding,

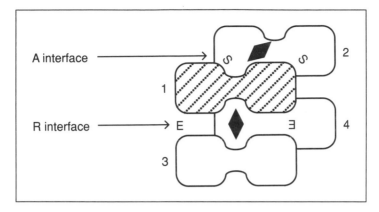

FIGURE 9–12. *A tetramer with four identical subunits arranged in D$_2$ symmetry with three twofold axes gives two different dimers upon dissociation. The "front" dimer formed with subunits 1 and 3 does not involve the same contact areas as the "top" dimer formed with subunits 1 and 2. This is a simplified version of the structure of PFK, where the different interfaces are binding sites for the substrate S (A[ctive] interfaces between subunits 1-2 and 3-4) and the allosteric effector E (R[egulatory] interfaces between subunits 1-3 and 2-4). (Reproduced from Deville-Bonne et al., 1989, with permission)*

depending upon whether the ligand is bound to the protein during the rate-limiting step. In the case of small proteins, formation of the binding site for a specific ligand generally occurs only after complete folding. Ligand binding is a probe not of "local" folding, but merely of the final state. The ligand binds only to the end product and has no influence of the rate of folding, such as the binding of 2'CMP to ribonuclease A (Garel and Baldwin, 1973).

For oligomeric proteins, folding intermediates acquire specific sites for other subunits, and they could also have acquired specific sites for other ligands. Indeed, it has been observed that the rate and/or extent of folding of several larger proteins was sensitive to the presence of a ligand (Jaenicke, 1987). The rate of folding of glyceraldehyde-3-phosphate dehydrogenase was increased by the binding of NAD or when an analog of NAD was covalently attached close to the active site (Jaenicke et al., 1980). Similarly, the folding rate of PFK was increased in the presence of ATP (Fig. 9–13). For these two proteins, the ligand binds to a monomeric intermediate and accelerates the subsequent association steps. In addition, the dimeric intermediate form of PFK can bind the allosteric inhibitor phosphoenolpyruvate, which decreases the rate of tetramer formation (Deville-Bonne et al., 1989).

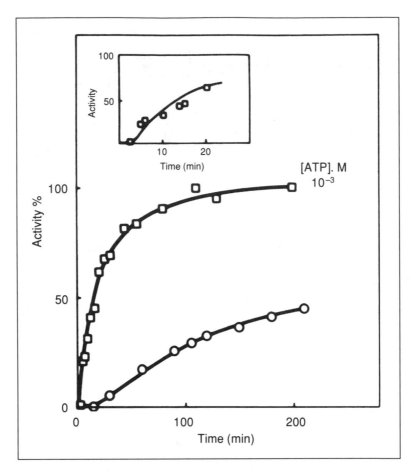

FIGURE 9–13. *The effect on the rate of folding of PFK of the presence of ligands. Refolding of PFK was faster in the presence of 1mM ATP (□) than in its absence (○) The second-order rate constant for the bimolecular reactivation was increased from 10^4 M^{-1} s^{-1} to 2×10^5 M^{-1} s^{-1}. Inset: In the presence of ATP, an initial lag phase is still present, albeit shorter than in its absence, showing that the folding of PFK takes place according to the same two-step mechanism as described in Figure 9–10. (Reproduced from Martel and Garel, 1984, with permission)*

A special situation exists when ligands are embedded in the final folded state and play a structural role in maintaining the native conformation. Such a ligand must bind to an intermediate at some point along the folding pathway. Hence Zn^{2+} ions influence in a complex way the folding of alcohol dehydrogenase (Rudolph et al., 1978), and the free amino acid valine is needed for the folding of pyruvate carboxylase (Bornmann et al., 1974).

The existence of a binding site in one intermediate does not imply that the rate of further steps is modified by binding of the ligand. None of the ligands of AK-HDH changes its rate of folding, even though some of them bind to intermediate species (Garel and Dautry-Varsat, 1980b). Finding that a ligand influences the rate of folding is, however, evidence that an intermediate or transition state not only exists but already possesses a binding site for this ligand.

3.6 Different Folding of the Polypeptide Chain in Monomers and Oligomers

A special case is that of proteins, like uteroglobin (Morizé et al., 1987) or the *trp* repressor (Zhang et al., 1987), for which X-ray crystallography shows that the conformation of the chain within the native oligomer depends crucially upon interactions between different subunits. This suggests that, even if a folded state exists for an isolated monomer, its conformation must be very different from that of the same chain within the native oligomer (Janin et al., 1988). Association between subunits must then occur together with, or be followed by, a major conformational change that folds each chain into its proper tertiary structure. If no well-folded state exists for the monomer, assembly would begin by some weak association between poorly folded chains and would proceed further by extensive folding within the dimer, each chain acting as a stabilizing ligand for the other. An equilibrium and kinetic study of the folding of *trp* aporepressor suggests that the loosely associated dimer does not accumulate to substantial levels (Gittelman and Matthews, 1990). Alternatively, each chain could exist in two well-folded states, one in the monomer and the other in the dimer. This possibility that a polypeptide chain can oscillate between different alternative structures, depending upon whether it is in an associated or a dissociated state, is attractive because of its relevance to all regulatory processes related to protein association; experimental evidence is needed, however.

4 EXTRAPOLATION TO FOLDING IN VIVO OF THE RESULTS OBTAINED IN VITRO WITH LARGE PROTEINS

In vivo, folding does not start with an unfolded chain but with a nascent chain synthesized on a ribosome (Tsou, 1988; Fischer and Schmid, 1990). In bacteria, assembly of the polypeptide chain is coupled to transcription and proceeds at the rate of about 15 amino acids per second. Domains can fold by themselves, and folding can be relatively fast (see Section 2), so it is likely that partial domain folding occurs on the ribosome. Complete chain folding, however, requires the slow domain pairing reaction, so it is likely that it can take place only after synthesis has been completed. When released from the ribosome, the polypeptide chain presumably resembles the FUD intermediate and must yet undergo the last step of monomer folding. This FUD-like intermediate would seem to have the choice between folding correctly or aggregating.

4.1 The FUD-like Intermediate and in vivo Aggregation

It seems more and more probable that folding intermediates are responsible for aggregation *in vivo* in at least some cases (Mitraki and King, 1989). The FUD-like intermediate exists in most proteins described here, is likely to accumulate during folding because it precedes a slow step, and has a structure with the appropriate sites for multiple and specific association. At 37°C, the half-time for domain pairing in AK-HDH is of the order of a minute (Vaucheret et al., 1987), i.e., comparable to the time needed for chain biosynthesis. Since chain biosynthesis is not faster than domain pairing, there will be a measurable steady-state concentration of FUD-like intermediate, which could lead to protein aggregation. A rational approach for decreasing *in vivo* aggregation would be to lower the steady-state concentration of the FUD-like intermediate.

4.2 Association Between the Subunits of an Oligomeric Protein Under Cellular Conditions

The *in vivo* folding of an oligomeric protein occurs rapidly. The first active molecules of tetrameric β-galactosidase appear three minutes after induction (Kepes and Beguin, 1966). Transcription and chain biosynthesis take around one minute (Rothman, 1989), folding of a multidomain chain may also take one minute (see Section 4.1), so association probably also occurs in about one minute. For those oligomers with association-limited folding, the second-order rate constants measured *in vitro* for subunit association are usually in the range from 10^3 to 10^5 $M^{-1} \cdot s^{-1}$ (Jaenicke, 1987; Rothman, 1989). Association would occur in one minute at protein concentrations of 10^{-5} to 10^{-7} M, i.e., close to or above the threshold for aggregation (Fig. 9–7). However, association could take place not only between free monomers but also between folded chains and nascent chains. Nascent chains would not aggregate with each other because they have only one site for domain interaction (as isolated domains [Fig. 9–6]) and are spatially separated along the polysomes. The local effective concentration of nascent chains around the polysomes could be high enough to favor rapid association at a lower concentration of free monomers. This mechanism of association would in part bypass the critical FUD-like intermediate. The intermolecular interactions already formed between the complete and the nascent chains would guide the intramolecular domain pairing occurring after completion of the nascent chain.

The volume of the sample is also involved in the competition between association and aggregation, because aggregates must reach a critical size to undergo irreversible phase separation, and the volume may be too small for this to occur. (The volume is that accessible to a macromolecule, and not the total volume.) As an example, the inside volume of an *E.coli* cell is of the order of 1 μm^3, or 10^{-15} liter, and a concentration of monomers of 10^{-7} M corresponds to fewer than 100 molecules per cell. So the smaller the volume, the more favored are those processes involving a smaller number of molecules, such as association, rather than aggregation.

5 GENERAL CONCLUSION AND PERSPECTIVES

Several specific features distinguish "large" proteins from the small ones described in the preceding parts of this book and justify a special chapter being devoted to a description of their folding. The native state of a large protein is composed of separated blocks, domains, and/or subunits. This modular structure is related to a modular assembly, and it is the independent folding of different regions of the polypeptide chain that forms the discrete spatial units. This modular assembly often involves complex kinetics and intermediate species. This is not only true for the subclass of large proteins represented by oligomers, for which the folding process involves uni- and bimolecular steps, but also for the monomeric large proteins. The rate of overall folding of a large protein can be limited by specific steps that do not exist in small proteins, such as the pairing between domains that are already folded or the association between already folded subunits of an oligomeric protein.

5.1 Practical Aspects of the Folding of Large Proteins

From a purely operational point of view, working with large proteins is frequently difficult because of irreversible side processes, ambiguous folding probes, and complex kinetics. The yield and rate of folding depends crucially upon solvent conditions, upon the procedure used for denaturation and renaturation, and upon the concentration of protein. Large proteins have a strong tendency toward precipitation and/or aggregation during unfolding/refolding studies, leading to partial or total irreversibility, so that their folding cannot be studied under all conditions. Three main limitations in the experimental approach of this folding have been found to minimize aggregation and to optimize the renaturation yield of large proteins (Jaenicke, 1987).

1. *Low concentrations of protein favor folding over aggregation.* In refolding studies of small proteins, concentrations of protein of the order of 1mg/ml, and up to more than 100 mg/ml in NMR experiments, have been used. The yield of renaturation of many large proteins decreases markedly at high protein concentrations (see Figure 9–7), and quantitative renaturation is usually obtained in the range from 0.1 to 10 μg/ml. Such low concentrations require not only that special precautions be taken to avoid adsorption (siliconization of glassware, use of a carrier protein, addition of detergent, and so on), but also that the methods used to measure folding be sensitive.

2. *Irreversible conditions minimize the presence of relatively insoluble partially folded intermediates.* At least some of the intermediate species that are present during the folding of large proteins are not very soluble and undergo aggregation (Mitraki and King, 1989; Schein, 1990). One way to minimize the accumulation of such partially folded intermediates is to carry out folding under strongly native conditions where it is essentially irreversible. In order to maximize the yield of renaturation and/or to study the rate of this renaturation, most studies have therefore been performed under conditions far from the equilibrium between the native and partially folded states.

3. *Physical and solvent conditions have to be properly chosen.* The various factors involved in the kinetic competition between correct folding and aggregation are very sensitive to the physical conditions. The yields and rates of renaturation of large proteins vary markedly with the initial denatured material (heat, extreme pH, organic solvent, denaturant, and so on), with the physical conditions of renaturation (pH, temperature, ionic strength, nature of salts, presence of other solvent additives, residual denaturant, and so on), and with the procedure of renaturation (fast dilution or slow dialysis of the denaturant, fast freezing or slow cooling, pH change, and so on). These various physical factors influence the solubilities of native, intermediate, or denatured species and the rates of interconversion between them, so the overall balance will be different, and barely predictable, for a particular protein. (See Section 2.3.4.) A slight difference in only one of these parameters can greatly enhance the efficiency of folding and make the difference between a quantitative renaturation and a milky precipitation of the same protein.

Irreversibility is not always associated with aggregation. Some proteins cannot be refolded because they have been matured after biosynthesis. Their covalent structure has been processed after folding has taken place, and their native conformation is metastable. In other cases, the impossibility of refolding a protein can arise from covalent modifications during unfolding such as oxidation, deamidation, cleavage of labile peptide bonds, removal of a prosthetic group, and so on (Volkin and Klibanov, 1989). Renaturation of ODH yields only 70% of native and 30% of an inactive and monomeric species, which corresponds to an incorrectly partially folded state (Teschner et al., 1987). Upon a second cycle of denaturation/renaturation, this inactive material gives the same ratio of 70% active to 30% inactive, which shows that it is due to the folding process and not to preexisting species. This 70 to 30 ratio is also sensitive to solvent additives. This suggests that the folding process of large proteins could lead to alternative folded states, as sometimes occurs in some small proteins. For instance, two distinct folded states of staphylococcal nuclease co-exist in solution and can interconvert without passing through the unfolded state (Evans et al., 1987).

5.2 The Folding Pathway for an "Average" Large Protein

Following the usual practice (emphasized by the heuristic value of RNase A and BPTI for the folding of small proteins), a general mechanism can be extracted from a small number of examples. Combining all the information obtained with PGK, ODH, αTS, AK-HDH, PFK, βTS, LDH, and a few others into an "average" protein leads to the scheme shown in Figure 9–14.

This succession of steps is very similar to the "consensus" pathway for protein folding (Goldberg, 1985), but all these events will not be detected during the folding of a given protein. The apparent succession of steps will be simpler and will be dominated only by the kinetically significant intermediates, i.e., those that accumulate because they precede a slow reaction and are sufficiently stable. The FUD intermediate plays an important role only when the slowness of the domain-pairing reaction causes it to accumulate. Similarly, the species preceding a rate-limiting association step will appear as a predominant intermediate.

For a given protein, observing complex kinetics either by a single parameter showing a multiphasic reaction, or by different folding kinetics measured using different probes, raises the following questions: (1) How many kinetically significant steps are there? (2) Are these steps uni- or bimolecular? (3) Are these steps reversible or irreversible under the conditions studied? (4) Are these steps arranged along a strictly sequential pathway, or do they belong to parallel pathways? (5) What is the unique set of intermediates consistent with all the different values of rates and relative

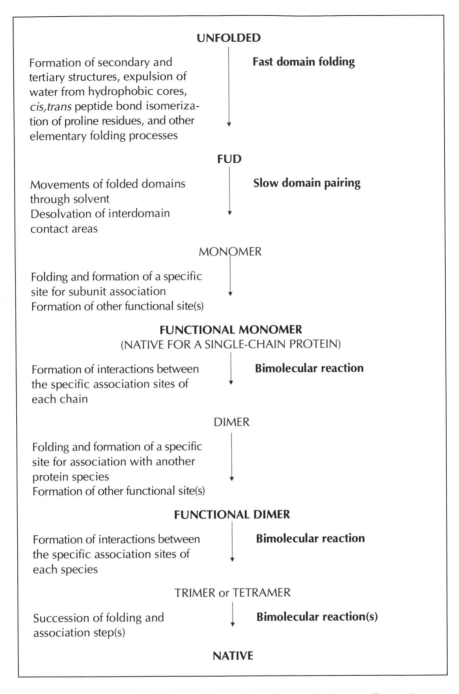

FIGURE 9-14. *The overall folding pathway of an oligomeric "average" protein. Bold letters indicate the species that have been detected in several cases because they precede slower steps.*

amplitudes obtained with different probes? (6) Can the kinetic intermediates be correlated to the structure of the folded state of the native protein? A first kinetic analysis and characterization of important intermediates of the folding of a particular protein can be achieved by comparing its behavior to that of the "average" protein.

5.3 Perspectives

The simple model of a "fast domain folding, slow domain pairing" mechanism of chain folding, followed by subunit assembly as a result of multiple "association folding" events will not apply strictly to all proteins. Exceptions exist such as sequential, nonindependent folding of domains (see Section 2.5.2), fast folding of a multidomain chain, fast association step that is not rate limiting for folding of an oligomer, or alternative structures for isolated and associated chains (see Section 3.6), but these exceptions are useful in testing the validity of the general pattern. The folding of large proteins should become much better understood in the future thanks to three lines of progress:

1. The rapid advancement of protein engineering methods and the availability of altered chains: mutants, fragments, isolated domains, hybrid proteins, and so on.
2. The growing number of structures determined by X-ray crystallography as well as the improvements in protein modeling.
3. The use of powerful techniques: rapid kinetic measurements, such as stopped-flow measurements of circular dichroism, specific probes such as monoclonal antibodies, and computerized kinetic analysis.

It is likely that a simple model derived from a biased and limited set of proteins will not be valid for all large proteins, and that many more exceptions and deviations from it will be observed. We feel however that simple models are extremely useful as references for appreciating exceptional behavior, and we hope that the simplification of the folding of an "average" large protein contributes to our understanding of the ability of one or several polypeptide chains to fold and associate into highly compact and functional structures.

REFERENCES

Bajaj, M., and Blundell, T. (1984) *Ann. Rev. Biophys. Bioeng. 13*, 453–492.

Baldwin, R. L. (1980) In *Protein Folding* (R. Jaenicke, ed.), Elsevier, Amsterdam, pp. 369–384.

Beasty, A. M., Hurle, M. R., Manz, J. T., Stackhouse, T., and Matthews, C. R. (1986) In *Protein Structure, Folding and Design* (D. L. Oxender, ed.), Alan R. Liss, New York, pp. 259–268.

Blond, S., and Goldberg, M. E. (1985) *J. Mol. Biol. 182*, 597–606.

Blond, S., and Goldberg, M. E. (1986) *Proteins: Struct. Funct. Genet. 1*, 247–255.

Blond, S., and Goldberg, M. E. (1987) *Proc. Natl. Acad. Sci. USA 84*, 1147–1151.

Blond-Elguindi, S., and Goldberg, M. E. (1990) *Biochemistry 29*, 2409–2417.

Bornmann, L., Hess, B., and Zimmermann-Telschow, H. (1974) *Proc. Natl. Acad. Sci. USA 71*, 1525–1529.

Brandts, J. F., Halvorson, H. F., and Brennan, M. (1975) *Biochemistry 14*, 4953–4963.

Brandts, J. F., Hu, C. Q., Lin, L. N., and Mas, M. T. (1989) *Biochemistry 28*, 8588–8596.

Brown, H. R., and Schachat, F. H. (1985) *Proc. Natl. Acad. Sci. USA 82*, 2359–2363.

Burns, D. L., and Schachman, H. K. (1982a) *J. Biol. Chem. 257*, 8638–8647.

Burns, D. L., and Schachman, H. K. (1982b) *J. Biol. Chem. 257*, 8648–8654.

Chothia, C., and Lesk, A. (1986) *EMBO J. 5*, 823–826.

Chrunyk, B. A., and Matthews, C. R. (1990) *Biochemistry 29*, 2149–2154.

Creighton, T. E. (1978) *J. Mol. Biol. 125*, 401–406.

Creighton, T. E. (1990) *Biochem. J. 270*, 1–16.

Dautry-Varsat, A., and Garel, J. R. (1981) *Biochemistry 20*, 1396–1401.

Deville-Bonne, D., Le Bras, G., Teschner, W., and Garel, J. R. (1989) *Biochemistry 28*, 1917–1922.

Eisenstein, E., and Schachman, H. K. (1989) In *Protein Function: A Practical Approach* (T. E. Creighton, ed.), IRL Press, Oxford, pp. 135–176.

Evans, P. A., Dobson, C. M., Kautz, R. A., Hatfull, G., and Fox, R. O. (1987) *Nature 329*, 266–268.

Fischer, G., and Schmid, F. X. (1990) *Biochemistry 29*, 2205–2212.

Frieden, C. (1990) *Proc. Natl. Acad. Sci. USA 87*, 4413–4416.

Garel, J. R., and Baldwin, R. L. (1973) *Proc. Natl. Acad. Sci. USA 70*, 3347–3351.

Garel, J. R., and Dautry-Varsat, A. (1980a) In *Protein Folding* (R. Jaenicke, ed.), Elsevier, Amsterdam, pp. 485–498.

Garel, J. R., and Dautry-Varsat, A. (1980b) *Proc. Natl. Sci. Acad. USA 77*, 3379–3383.

Garel, J. R., Nall, B. T., and Baldwin, R. L. (1976) *Proc. Natl. Acad. Sci. USA 73*, 1853–1857.

Gilbert, W. (1985) *Science 228*, 823–824.

Gittelman, M. S., and Matthews, C. R. (1990) *Biochemistry 29*, 7011–7020.

Goldberg, M. E. (1969) *J. Mol. Biol. 46*, 441–446.

Goldberg, M. E. (1985) *Trends Biochem. Sci. 10*, 388–391.

Goldberg, M. E., and Zetina, C. R. (1980) In *Protein Folding* (R. Jaenicke, ed.) Elsevier, Amsterdam, pp. 469–483.

Goto, Y., and Hamaguchi, K. (1981) *J. Mol. Biol. 146*, 321–340.

Goto, Y., and Hamaguchi, K. (1986) *Biochemistry 25*, 2821–2828.

Haber, E., and Anfinsen, C. B. (1962) *J. Biol. Chem. 237*, 1839–1844.

Hammes, G. G. (1978) *Principles of Chemical Kinetics*, Academic Press, Orlando, Fla., p. 164.

Hirose, M., Akuta, T., and Takahashi, N. (1989) *J. Biol. Chem. 264*, 16867–16872.

Hirs, C. H. W., and Timasheff, S. N. (1986) *Methods Enzymol. 131*, 3–280.

Hurle, M. R., and Matthews, C. R. (1987) *Biochim. Biophys. Acta 913*, 179–184.

Hyde, C. C., Ahmed, S. A., Padlan, E. A., Miles, E. W., and Davies, D. R. (1988) *J. Biol. Chem. 263*, 17857–17871.

Jaenicke, R. (1987) *Prog. Biophys. Mol. Biol. 49*, 117–237.

Jaenicke, R., Krebs, H., Rudolph, R., and Woenckenhaus, C. (1980) *Proc. Natl. Acad. Sci. USA 77*, 1966–1969.

Jaenicke, R., and Rudolph, R. (1986) *Methods Enzymol. 131*, 218–250.

Jaenicke, R., and Rudolph, R. (1989) In *Protein Structure: A Practical Approach* (T. E. Creighton, ed.), IRL Press, Oxford, pp. 191–223.

Jaenicke, R., Rudolph, R., and Heider, I. (1981) *Biochem. Int. 2*, 23–31.

Jaenicke, R., Vogel, W., and Rudolph, R. (1981) *Eur. J. Biochem. 114*, 525–531.

Janin, J., Miller, S., and Chothia, C. (1988) *J. Mol. Biol. 204*, 155–164.

Janin, J., Wodak, S. J. (1983) *Prog. Biophys. Mol. Biol. 42*, 21–78.

Kepes, A., and Beguin, S. (1966) *Biochim. Biophys. Acta 123*, 546–560.

Kiehn, E. D., and Holland, J. J. (1970) *Nature 226*, 544–545.

Kim, P. S., and Baldwin, R. L. (1982) *Ann. Rev. Biochem. 51*, 459–489.

Kim, P. S., and Baldwin, R. L. (1990) *Ann. Rev. Biochem. 59*, 631–660.

Lane, A. N., Paul, C. H., and Kirschner, K. (1984) *EMBO J. 3*, 279–287.

Le Bras, G., Teschner, W., Deville-Bonne, D., and Garel, J. R. (1989) *Biochemistry 28*, 6836–6841.

Lehrer, S. S., and Qian, Y. (1990) *J. Biol. Chem. 265*, 1134–1138.

Light, A., and Higaki, J. N. (1987) *Biochemistry 26*, 5556–5564.

London, J., Skrzynia, C., and Goldberg, M. E. (1974) *Eur. J. Biochem. 47*, 409–415.

Martel, A., and Garel, J. R. (1984) *J. Biol. Chem. 259*, 4917–4921.

Matthews, C. R., Crisanti, M. M., Manz, J. T., and Gepner, G. L. (1983) *Biochemistry 22*, 1445–1452.

Missiakas, D., Betton, J. M., Minard, P., and Yon, J. M. (1990) *Biochemistry 29*, 8683–8689.

Mitraki, A., and King, J. (1989) *Bio/Technology 7*, 690–697.

Morizé, I., Surcouf, E., Vaney, M. C., Epelboin, Y., Buehner, M., Fridlansky, F., Milgrom, E., and Mornon, J. P. (1987) *J. Mol. Biol. 194*, 725–739.

Müller, K., and Garel, J. R. (1984) *Biochemistry 23*, 655–660.

Murry-Brelier, A., and Goldberg, M. E. (1989) *Proteins: Struct. Funct. Genet. 6*, 395–404.

Nall, B. T., Garel, J. R., and Baldwin, R. L. (1978) *J. Mol. Biol. 118*, 317–330.

Novokhatny, V. V., Kudinov, S. A., and Privalov, P. L. (1984) *J. Mol. Biol. 179*, 215–232.

Park, S., Liu, G., Toppoing, T. B., Cover, W. H., and Randall, L. L. (1988) *Science 239*, 1033–1035.

Porter, R. R. (1973) *Science 180*, 713–716.

Privalov, P. L. (1989) *Ann. Rev. Biophys. Biophys. Chem. 18*, 47–69.

Privalov, P. L., and Potekhin, S. A. (1986) *Methods Enzymol. 131*, 4–51.

Rossmann, M. G., and Argos, P. (1981) *Ann. Rev. Biochem. 50*, 497–532.

Rothman, J. E. (1989) *Cell 59*, 591–601.

Rudolph, R., Fuchs, I., and Jaenicke, R. (1986) *Biochemistry 25*, 1662–1667.

Rudolph, R., Gerschitz, J., and Jaenicke, R. (1978) *Eur. J. Biochem. 87*, 601–606.

Rudolph, R., Siebendritt, R., Nesslaüer, G., Sharma, A. K., and Jaenicke, R. (1990) *Proc. Natl. Acad. Sci. USA 87*, 4625–4629.

Schein, C. H. (1990) *Bio/Technology 8*, 308–317.

Tatunashvili, L. V., Filimonov, V. V., Privalov, P. L., Metsis, M. L., Koteliansky, V. E., Ingham, K. C., and Medved, L. V. (1990) *J. Mol. Biol. 211*, 161–169.

Teale, J. M., and Benjamin, D. C. (1976) *J. Biol. Chem. 251*, 4603–4608.

Teale, J. M., and Benjamin, D. C. (1977) *J. Biol. Chem. 252*, 4521–4526.

Teschner, W., and Garel, J. R. (1989) *Biochemistry 28*, 1912–1916.

Teschner, W., Rudolph, R., and Garel, J. R. (1987) *Biochemistry 26*, 2791–2796.

Teschner, W., Serre, M. C., and Garel, J. R. (1990) *Biochimie 72*, 33–40.

Timasheff, S. N., and Arakawa, T. (1989) In *Protein Structure: A Practical Approach* (T. E. Creighton, ed.), IRL Press, Oxford, pp. 331–345.

Trexler, M., and Patthy, L. (1983) *Proc. Natl. Acad. Sci. USA 80*, 2457–2461.

Tsou, C. L. (1988) *Biochemistry 27*, 1809–1812.

Tsunenaga, M., Goto, Y., Kawata, Y., and Hamaguchi, K. (1987) *Biochemistry 26*, 6044–6051.

Tweedy, N. B., Hurle, M. R., Chrunyk, B. A., and Matthews, C. R. (1990) *Biochemistry 29*, 1539–1545.

Vas, M., Sinev, M. A., Kotova, N. V., and Semisotnov, G. V. (1990) *Eur. J. Biochem. 189*, 575–579.

Vaucheret, H., Signon, L., Le Bras, G., and Garel, J. R. (1987) *Biochemistry 26*, 2785–2790.

Volkin, D. B., and Klibanov, A. M. (1989) In *Protein Function: A Practical Approach* (T. E. Creighton, ed.), IRL Press, Oxford, pp. 1–24.

Wetlaufer, D. B. (1973) *Proc. Natl. Acad. Sci. USA 70*, 697–701.

Zettlmeissl, G., Rudolph, R., and Jaenicke, R. (1979) *Biochemistry 18*, 5567–5571.

Zettlmeissl, G., Teschner, W., Rudolph, R., Jaenicke, R., and Gäde, G. (1984) *Eur. J. Biochem. 143*, 401–407.

Zhang, R. G., Joachimiak, A., Lawson, C. L., Schevitz, R. W., Otwinowski, Z., and Sigler, P. B. (1987) *Nature 327*, 591–597.

10

Protein Folding in the Cell

ROBERT B. FREEDMAN

1 INTRODUCTION

1.1 The Protein Folding Problem

The "protein folding problem" has generally been understood in one of two ways: either as a problem of information or as a problem of mechanism. Classically, the problem is viewed as one of decoding information; the objective is to decipher the high-level code that would permit the inference of the native conformation from the primary structure in the same way that the amino acid sequence of a translation product can be inferred from the corresponding nucleotide sequence through knowledge of the "genetic code." The difficulty of this problem, and the slow progress made on it since its first formulation suggest that it could be illuminated by experimental studies on the process of protein folding, and hence the second, mechanistic formulation of the problem; this is to describe in kinetic, thermodynamic, and structural terms the process by which an unfolded polypeptide folds to a defined biologically active conformation. This is essentially a problem in macromolecular physical chemistry, and the majority of the chapters in this book describe such studies. Great progress has been made in identifying pathways of folding and describing the properties of intermediate species; we can begin to picture how a folding polypeptide acquires the characteristic properties of a globular protein, namely compact volume, elements of regular backbone conformation, and defined tertiary structure interactions. But in order to achieve the optimum control and definition of conditions, and to obtain the maximum information on structural changes occurring in the folding process, the vast majority of such studies have been carried out *in vitro* and have concerned the *refolding* of mature proteins after unfolding rather than the *initial folding* of newly synthesized proteins.

455

This chapter confronts a third, alternative formulation of the protein folding problem and asks how the folded states of proteins are generated *in vivo* as part of the cellular process of protein biosynthesis. Specifically, it asks how this process differs, it at all, from the well-studied model of mature protein refolding. In what ways does the context of protein folding in the cell differ from that of protein folding in the test tube? Are these differences relevant? Can we combine fragmentary evidence from direct studies of protein folding at biosynthesis with a more extensive and growing body of *in vivo* data on other co- and posttranslational events related to folding (translocation, assembly, and so on) to produce a coherent account of folding in the cell that is consistent with what we know about the process of folding when studied in isolation in the test tube?

1.2   Recent Interest in Protein Folding in the Cell

While experimental studies of protein refolding have always acknowledged that the primary interest in such work is the light that it might throw on protein folding in the cell, there has been little recognition, until recently, that cellular factors and cellular conditions might make folding in the cell significantly different from conventional refolding studies. The fact that many proteins can successfully refold under some conditions to their native states in the absence of cellular machinery implied that no cellular factors were necessary for specifying the final folded state, justifying the assumption that the cellular process could simply be inferred from what was learned from studies *in vitro*. At the same time, until the mid-1980's, there was relatively little interest from cell biologists in protein folding and assembly as a cellular process. As a result, there was for many years little direct communication between the communities studying physicochemical aspects of protein folding and the cell biological aspects of protein synthesis, translocation, and assembly. This situation changed gradually during the 1980's for two main reasons.

The belief that the functional conformation of a protein was implicit in its polypeptide sequence, and hence in its coded genetic information, inspired the objective of generating proteins of intellectual interest or commercial value by expressing coding DNA sequences in convenient microbial host cells, especially *Escherichia coli*. Such heterologous recombinant protein expression has been widely successful and is a powerful demonstration of the conservation of key biological processes and the depth of understanding now available. But in many cases, although high levels of expression of the required polypeptide are obtained, much of the resultant material does not fold to its native soluble conformation, but is found as macroscopic insoluble aggregates, or inclusion bodies. (See Section 3.2.) This outcome stimulated further work on protein refolding *in vitro*, because the recovery of biologically active proteins from this material had a major commercial incentive, but it also graphically illustrated the fact that protein folding *in vivo* does not generate the native protein in all circumstances. The fact that proteins can "direct" their own folding under appropriate conditions *in vitro* does not guarantee their ability to do so in the cell. Since around 1986, this has focused attention on the specific cellular conditions and factors necessary for the successful folding of proteins *in vivo*.

Parallel with this development has been a striking upsurge in cell biological interest in the processes intervening between the biosynthesis of a translation product and its ultimate functional expression, possibly as part of a multicomponent complex and possibly in a remote intracellular or extracellular location. The analysis of protein traffic and traffic control in the cell began with the classic description of the secretory pathway (see Palade [1975] for review) and received a major impetus from the development of *in vitro* systems in which translation of mRNA's for secretory proteins was coupled to the translocation of nascent chains into microsomal vesicles and their segregation in the internal (luminal) compartment (Blobel and Dobberstein, 1975). From this beginning, powerful model systems have been developed for analyzing co- and posttranslational events in the lives of all classes of proteins. During the period 1975 to 1985, these studies were primarily descriptive, but since then there has been greater attention on issues of mechanism, and this has focused attention on the conformation of the protein as it undergoes the complex sequence of events leading to its eventual appearance in a functional state at the appropriate compartment. At the same time, powerful genetic systems have been developed for analysis of secretion and other aspect of protein traffic, especially in lower eukaryotes. The combination of genetic and subcellular studies has revealed the existence of several major families of proteins that apparently function in cellular protein folding processes.

The desire to express functional recombinant proteins in heterologous systems, and the wish to understand protein folding in the context of the other processes undergone by a newly synthesized protein, have emphasized that *in vitro* refolding studies on isolated unfolded proteins provide an incomplete model for describing protein folding in the cell. But such model studies do provide essential background against which cellular events can be understood.

1.3 Scope of the Chapter

This chapter continues with a summary of the cell biology of protein synthesis, and an analysis of how protein folding in this context may differ from that in convenient experimental model systems. It then asks what can be learned from such model studies about the problems that a protein faces in folding to a defined globular conformation, and describes evidence indicating that significant protein misfolding may occur in cells, both under normal conditions and in various perturbed conditions. The chapter then describes enzymes capable of catalyzing specific rate-determining steps in protein folding, and reviews the evidence concerning their role in the cell. Next we consider the cell biological and genetic evidence that has indicated that temporal and spatial control of protein folding is crucial to the complete network of protein traffic in cells and particularly to translocation across membranes, assembly of oligomers and complexes, and passage from compartment to compartment through the secretory pathway. This work has identified several families of conserved and abundant proteins that have been implicated in ensuring productive protein folding. The chapter concludes with a review of these factors and of current information on their roles and modes of action.

The relation of *in vitro* refolding studies to protein folding in the cell was first reviewed by Epstein et al. (1963) and relevant early studies were considered by Wetlaufer and Ristow (1973). In addition to more specific reviews cited through the chapters, the following excellent recent reviews should be noted, which relate traditional protein folding studies to the accumulating information on cellular factors participating in protein folding (Fischer and Schmid, 1990; Gething and Sambrook, 1992; Jaenicke, 1991; Nilsson and Anderson, 1991).

2   PROTEIN FOLDING IN THE CONTEXT OF PROTEIN BIOSYNTHESIS

In the cell, protein folding occurs not as an isolated event, but as one of a set of linked and overlapping processes including translation, translocation, proteolytic processing, posttranslational modification, assembly, and association with ligand or co-factor. The temporal, spatial, and causal relationships between these processes are complex and, in some cases, difficult to resolve. In what way does folding in this cellular context differ from that observed in refolding studies?

2.1   Translation

In relation to translation, is it important to the folding process that proteins are constructed stepwise by the sequential linkage of amino acids from the N-terminus, with the growing end linked to the elongation machinery of the ribosome? Rates of elongation are of the order of several residues per second, so that the time for synthesis of a complete polypeptide falls in the range of 10 to 1,000 secs. Individual processes in protein folding can occur with half-times in the millisecond range or lower, and some small proteins can refold from the fully unfolded state in times within the preceding range. So it is reasonable to ask whether the product of co-translational protein folding differs significantly from that observed *in vitro*. Does the fact that N-terminal portions of the polypeptide chain are constructed earlier than C-terminal portions influence the folding conformation generated? The evidence is now clear that the vectorial nature of protein synthesis does not determine the outcome of protein folding. If folding of N-terminal regions during biosynthesis were to constrain the outcome of folding of more C-terminal regions of the polypeptide, one would not expect unfolding and refolding of the intact mature protein to yield the native conformation. Since such refolding can occur for most proteins from a wide variety of conditions, it is clear that the outcome of folding in the cell is not directly controlled by the vectorial nature of translation.

This point has been made more dramatically by a number of studies in which proteins have been permuted so that connectivity has been altered, although the linear sequence of amino acid residues has been conserved. Thus, Goldenberg and Creighton (1983) linked the N- and C-termini of bovine pancreatic trypsin inhibitor (BPTI) by chemical condensation to form a peptide bond, so generating a cyclic polypeptide; they subsequently cleaved this molecule enzymatically between residues 15 and 16 to generate a new linear peptide comprising BPTI with the N-terminal 15 residues transposed to the C-terminus. Both of these modified forms were capable of reversible reduction and refolding to generate the native set of disulfide bonds.

In the case of a larger protein, phosphoribosyl anthranilate isomerase (PRAI), circularly permuted forms have been generated at the genetic level. Yeast PRAI is a monomeric protein with the characteristic eightfold "TIM" barrel conformation of alternating regions of α-helix and extended conformation giving rise to a barrel made up of eight strands of parallel β-structure surrounded by eight α-helices. Luger et al. (1989) generated constructs coding for proteins in which the natural N- and C-termini of yeast PRAI were linked by a short connecting peptide and in which new N- and C-termini were generated in loops elsewhere in the protein (Fig. 10–1). When these constructs were expressed to high levels in *E.coli*, they generated significant PRAI activity, indicating that these circularly permuted forms folded *in vivo* to active conformations. Residues from several of the β-strands are juxtaposed to form the active site of this enzyme, so the generation of activity is a sensitive test for correct folding of the polypeptide as a whole. Furthermore, significant quantities of the recombinant protein were found in insoluble inclusions that could be recovered, solubilized by denaturation, and renatured to form active enzyme. The recovered circularly permuted forms could be reversibly unfolded in guanidinium chloride (GdmCl) and were identical to the natural protein by several spectroscopic criteria. These results amply demonstrate that the

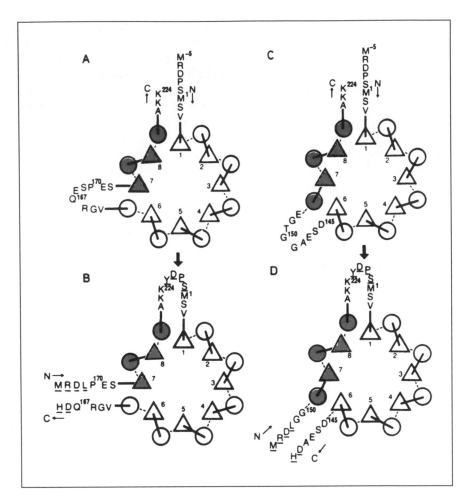

FIGURE 10–1. *Circular permutation of the sequence of a* βα *barrel protein. The figures show schematic structures of an "extended" form of yeast phosphoribosyl anthranilate isomerase (xPRAI) and of alternative circularly permuted forms of the protein. In all cases, triangles represent* β *strands, numbered in sequential order from the amino-terminus (N), perpendicular to the plane of the page, and running downward. Circles are perpendicular* α *helices coming upward, numbered as the preceding* β *strands. Connections from* α *to* β *are shown as bold lines and from* β *to* α *as dashed lines. C, carboxyl terminus. Secondary structure regions that were transferred to yield the circularly permuted proteins are shaded. (A) xPRAI showing partial sequences at the amino and carboxyl terminus (note, residue 1 is the N-terminus of the wild-type enzyme), and in the loop between helix 6 and strand 7. (B) cPRAI-1 (circularly permuted mutant 1) showing the new residues (underlined) linking the wild-type N- and C- termini and the new*

folded conformation is specified by the polypeptide as a whole and is not constrained by the order in which its elements are synthesized.

It remains possible, however, that the route or path by which a protein folds could differ between initial folding at biosynthesis, where N-terminal regions are already present while synthesis of C-terminal regions continues, and refolding *in vitro* after denaturation, when all parts of the protein begin to fold at the same time (Chantrenne, 1961; Wetlaufer and Ristow, 1973). Thinking on this question has been intimately associated with the analysis of the experimental question of whether the whole of a polypeptide chain is required for folding or whether fragments are capable of autonomous folding. For many years (Wetlaufer, 1981; Tsou, 1988) the answer to this question appeared controversial because in analyzing experimental data, the distinction was not clearly made between studies on single-domain proteins and those on proteins comprising several structural domains. The high-resolution structural analysis of most proteins comprising more than 200 residues indicates that they are composed of distinct structural elements or domains, in each of which the component amino acids interact more closely with each other than with other parts of the polypeptide. (See Chapters 2 and 9.)

Individual domains from multidomain proteins can be isolated, following selective proteolysis, or expressed, following mutation or recombinant DNA manipulations. It is quite clear that these distinct structural units within proteins also function as distinct folding units, so that isolated domains are capable of reversible unfolding and refolding to the native conformation, and the process of refolding of an unfolded multidomain protein can be represented as a sequence of domain folding events and interdomain associations. (See Chapter 9.) For this reason, it is entirely reasonable to assume that in the biosynthesis of a multidomain polypeptide, native folding of domains can occur as soon as each domain is synthesized and available for folding, and need not await the completion of the entire polypeptide. This is consistent with the limited evidence available from studies on folding in intact cells. (See p. 468.)

To this extent then, folding is a co-translational event. Is this also the case for smaller single-domain proteins? Tsou (1988) has proposed that significant folding generally occurs in the growing polypeptide chain with minor adjustments in the folding of N-terminal regions as the C-terminal

N- and C-terminal residues introduced into the previously highlighted loop. (C) xPRAI, as (A) but showing the partial sequence of the loop between strand 6 and helix 6. (D) cPRAI-2 (circularly permuted mutant 2), comparable to (B) but with the new N- and C-terminal residues now introduced into the loop between strand 6 and helix 6. (From Luger et al., 1989; copyright 1989 by the AAAS)

residues are added. In his argument, however, he does not distinguish between evidence from single- and multi-domain proteins. Data relevant to this case come from studies on the conformations and stabilities of peptide fragments, and from the folding properties of C-terminally truncated short proteins. There is plentiful evidence that even peptide fragments of 10 to 20 residues can have defined conformations in aqueous solution at low temperatures, and CD and NMR studies show that peptides corresponding to folded regions of small proteins often fold in isolation to a conformation that the corresponding region shows in the intact protein (Kim and Baldwin, 1984; Oas and Kim, 1988). Might such regions fold essentially to their final conformation co-translationally and in advance of the completion of the polypeptide? Two arguments suggest not.

First, the cooperativity of protein folding is demonstrated by the fact that native folding can be perturbed by some quite small truncations of the polypeptide. Thus, the removal of six residues from the C-terminus of bovine pancreatic ribonuclease to generate peptide 1 to 118 completely abolishes the ability of this polypeptide to refold and reform the native four disulphide bonds after reductive denaturation (Taniuchi, 1970; Andria and Taniuchi, 1978). Likewise, staphylococcal nuclease, a small stable protein, is significantly perturbed by removal of a few C-terminal residues (Shortle and Meeker, 1989). CD and NMR studies show that the resultant protein is compact in native conditions but lacks stable secondary and tertiary structure, except in the presence of substrates (Flanagan et al., 1992). Such trunctated species may be good models for incomplete domains during the course of elongation.

Second, studies on the kinetics of folding of proteins indicate that the rate-determining step is late in the folding pathway and that earlier intermediates, even those with relatively compact dimensions and regions of native-like secondary structure, are in rapid equilibrium with fully unfolded species (Chapters 5 to 7). Hence, the two lines of evidence suggest that, before the completion of the full sequence of amino acids specifying a distinct folded domain, a nascent chain does not acquire native conformation; it may comprise a complex mixture of species, some of which may include elements of folded structure, some of which may be native-like, but the whole ensemble will be in rapid equilibrium and, hence, will be equilibrating with fully unfolded species. The major activation energy barrier separating these forms from the native state will only be crossed after completion of the polypeptide segment specifying the entire domain. So the folding pathway would not be significantly affected by the finite time taken to elongate the chain; productive folding of a small protein or domain does not begin before the complete corresponding polypeptide chain has been formed and, in this respect, refolding *in vitro* is a satisfactory model of initial biosynthetic folding.

2.2 Translocation

Apart from the small minority of proteins encoded in organellar genomes and translated within mitochondria and chloroplasts, all proteins are translated by ribosomes located within the cytoplasmic compartment. But these translation products may be destined for a wide variety of intracellular locations or for secretion from the cell; all transfers to locations outside the cytoplasmic compartment will require at least one step in which the protein is translocated across a membrane. In the case of proteins transferred across the endoplasmic reticulum into the secretory pathway, it appears that translocation is essentially co-translational (consistent with the abundance of membrane-bound ribosomes in cells synthesizing secretory proteins), even though posttranslational translocation across isolated microsomal membranes can be observed in some circumstances. (See Section 6.3.) In the case of import into mitochondria and chloroplasts, and of secretion from bacteria, posttranslational translocation may be the norm. But there is now powerful evidence indicating that translocation is dependent on the translocated protein being unfolded or incompletely folded and that significant machinery exists to ensure that this is the case. This is discussed in more detail in Sections 5 and 6; its significance here is that proteins undergo translocation either during or following translation, but that in either case folding to the native state follows translocation. Provided that studies on protein folding *in vitro* use polypeptides chemically identical to the translocated form, such *in vitro* studies can be informative models of folding within various cellular compartments. The situation is made more complex, however, by the changes in covalent structure that can accompany translocation.

2.3 Proteolytic Processing

Translocation across the endoplasmic reticulum (ER) membrane and across mitochondrial and chloroplast membranes is usually followed by specific proteolysis to remove an N-terminal sequence. The functions of such sequences are still being actively explored (see Section 6), but they include retardation of folding, targeting to the appropriate membrane, and facilitating translocation. Removal of these sequences (processing) occurs during co-translational translocation (in the case of ER) or immediately following translocation (in the case of mitochondria); in either case, removal of the sequence precedes folding to the native state. In such circumstances, *in vitro* studies on refolding using the unfolded processed protein are the appropriate model for the ultimate folding of the processed translation product to the native state.

In many cases, however, proteins undergo further proteolytic processing subsequent to folding. In such cases, folding studies using the mature protein cannot model the actual folding process undergone during biosynthesis. The classic example is the reoxidation of reduced A and B chains of insulin, which under physiological conditions does not yield significant quantities of correctly folded insulin with the three native disulfide bonds (see references in Givol et al., 1965), although modest yields have been reported under other conditions (Tang et al., 1988; Tang and Tsou, 1990). Insulin is a classic secretory protein coded and synthesized as a single polypeptide chain that folds after removal of the N-terminal signal sequence, to generate proinsulin with three intrachain disulfide bonds. Prior to secretion, it is further processed at paired basic residues to remove an internal sequence, the C peptide, so that the mature, biologically active, secreted form comprises distinct A and B chains linked by specific disulfide bonds. The inability of mature insulin to undergo reversible reduction and unfolding indicates that the protein is in a metastable state; its conformation is retained by kinetic barriers, rather than by its intrinsic stability.

The same may be said of subtilisin. This bacterial protease has not been successfully refolded from the unfolded state. The protein is initially biosynthesized with a 77-residue N-terminal extension. When this full-length prosubtilisin is expressed in *E.coli*, it is capable of folding to the native state after unfolding in 6-M GdmCl and can then process itself to the mature form (Ikemura and Inouye, 1988; Zhu et al., 1989). With other zymogens of serine and aspartic proteases, similar results have been obtained. All of these findings indicate that the folded conformation is specified by the complete amino acid sequence of the species that actually folds in the cell. If this is subsequently processed to remove significant regions of sequence, the resultant protein may well be metastable and incapable of reversible unfolding and refolding.

Conversely, the observation that a protein is incapable of reversible unfolding and refolding may be used (with care) to suggest that it is a metastable protein derived from a precursor. Thus, from studies on the effects of enzyme-catalyzed disulfide interchange on chymotrypsinogen, chymotrypsin, insulin, ribonuclease A, and various proteolytic derivatives of ribonuclease A, Givol et al. (1965) proposed that disulfide interchange provided a thermodynamic probe for testing the stability of protein conformations, and inferred (in advance of the direct demonstration *in vivo*) that insulin "is originally synthesized as a single chain protein later converted to the mature form by a zymogen-like conversion." A similar argument was made in the case of collagen, a triple-stranded fibrous protein with a characteristic triple-helical conformation in the mature form. The difficulty in generating this conformation from unfolded collagen polypeptides led to the proposal that it was generated in a larger precursor, and this was subsequently demonstrated directly.

The conclusion is that proteolytic processing events may precede and/or follow folding in the course of generation of a mature protein; in order to study the folding process *in vitro*, the appropriately processed form of the protein must be selected (that is proinsulin, rather than preproinsulin or insulin).

2.4 Posttranslational Modification

The proteolytic processing events just described are specific examples of posttranslational modification; the proteins that result differ at the level of chemical structure and covalent bonds from the initial translation product encoded genetically. Very many other posttranslational modifications are known, and, in the context of this discussion, it is important to know whether they occur *in vivo* before, during, or following protein folding, in order to appreciate their significance for folding *in vivo* and for model studies *in vitro*.

The catalogue of posttranslational modifications is vast and growing, so not all can be considered here. The majority clearly occur subsequent to folding; even where this has not been demonstrated directly, the fact that specificities of modification are rarely defined by simple sequence characteristics implies that specificity is not expressed toward unfolded peptides, but is determined by conformational properties, and, hence, that folding must precede modification. Modifications of proteins in the secretory pathway offer many insights because of their number and variety and because some modifications clearly precede folding. Some, such as O-glycosylation, sulfation and the processing and modification of N-linked carbohydrates, occur at intermediate and late stages in the secretory pathway, at a point after folding is known to be complete. But a number occur within the lumen of the ER and can be shown to occur on nascent chains; these modifications—N-glycosylation, disulfide formation, γ-carboxylation, and Pro- and Lys-hydroxylation—therefore are contemporary and co-localized with folding, and the interaction between modification and folding can be complex (Freedman, 1989a).

2.4.1 N-glycosylation
The amino acid sequence -Asn-Xaa-Ser/Thr- is necessary, but not sufficient, to specify N-glycosylation of proteins within the lumen of the ER. N-glycosylation can also occur on small peptides, so it is probable that this modification precedes folding. Because a substantial number of these sequences in secretory proteins are not glycosylated, however, and because some are glycosylated to substoichiometric levels, it is likely that folding and glycosylation occur in parallel, with some sites becoming unavailable for modification for steric reasons as folding proceeds. Thus, the extent of modification at some sites would be a function of the rate of folding, and

there is clear evidence for this in studies on folding and disulfide formation following translation *in vitro*, where the rate of folding can be controlled by the concentrations of thiol and disulfide species present (Bulleid et al., personal communication).

If N-glycosylation occurs predominantly before folding of the nascent chain, it is clearly possible for this modification to affect the pathway, kinetics, yield, and product of folding. Several studies have indicated that there is no dominant influence of N-glycosylation on the nature of the folded product or on the kinetics of productive folding. This is borne out by the many successful refoldings, now reported, of mammalian proteins expressed in *E.coli* and hence free of glycosylation (Sarmientos et al., 1989; Kohno et al., 1990; Rudolph, 1990). Through its effect on the solubility of the unfolded protein and of folding intermediates, however, glycosylation may influence the competition between folding and aggregation processes, favoring formation of the native protein. Thus, in studies on mammalian ribonucleases with varying carbohydrate contents, it was shown that the kinetics of folding were entirely independent of the presence of N-glycosyl groups (Krebs et al., 1983; Grafl et al., 1987).

In a very thorough study, Schülke and Schmid (1988a,b) compared the unfolding and folding properties of the cytoplasmic (unmodified) and secreted (glycosylated) forms of yeast invertase. The protein is large, and the secreted form contains approximately 50% by weight of carbohydrate, so this is a challenging example. The two forms of the protein showed identical stabilities to inactivation as a function of temperature and similar susceptibilities to unfolding by GdmCl. While the glycosylated protein showed well-defined spectroscopic transitions in thermal or GdmCl-induced unfolding, however, these could not be observed for the nonglycosylated protein because of its tendency to aggregate in the thermally unfolded state and at intermediate denaturant concentrations. When refolding was studied from the GdmCl-denatured state, the yields were identical for the two proteins at pH values well above their isoelectric points and at concentrations below 0.003 mg/ml. At higher concentrations and lower pH values, glycosylated invertase refolded in moderate yield, while little renaturation was observed with the nonglycosylated form (Fig. 10–2). These results show that glycosylation is of minor significance for the molecular mechanism of folding of native invertase; the presence of carbohydrate slightly decelerates late events in renaturation that are probably subsequent to the formation of folded domains. Through its effect on the solubility of unfolded and partly folded protein in native conditions, however, glycosylation suppresses the tendency to aggregate and thus favors productive folding.

These *in vitro* studies account for many of the observed effects *in vivo* of glycosylation and inhibition of glycosylation. Thus, secretion of invertase is markedly reduced by treatment of yeast growing at 37 °C with

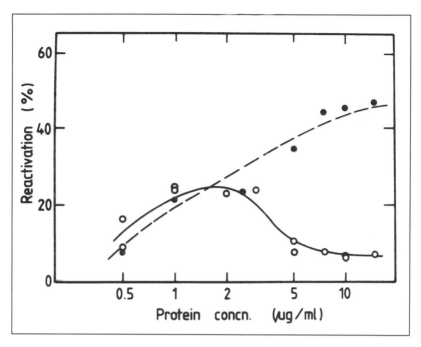

FIGURE 10–2. *Effect of glycosylation on the yield of protein refolding. The yield of reactivation of glycosylated (●) and nonglycosylated (○) invertase is plotted as a function of the protein concentration during refolding. Proteins were denatured by incubation in 4 M GdmCl pH 6.5 at 20°C for 1 hr at a concentration of 1.2 mg/ml; aliquots were then diluted with 4 M GdmCl to various protein concentrations and renatured by 80-fold dilution with pH 6.5 buffer to the final protein concentrations shown. Reactivation was determined after 48 hrs. (From Schülke and Schmid, 1988b)*

tunicamycin to block glycosylation; but glycosylation *per se* is not required for secretion, because secretion of native invertase occurs in the presence of tunicamycin when yeast are grown at 26°C (Ferro-Novick et al., 1984). Both glycosylated and nonglycosylated invertase are stable at 37°C, so it must be concluded that glycosylation is required for folding at 37°C; in the absence of glycosylation, folding is temperature-dependent because the nonglycosylated protein aggregates at the nonpermissive temperature. This result parallels many findings in animal cells (see Section 3.3), where perturbation of glycosylation prevents progress through the secretory pathway (Rose and Doms, 1988; Hurtley and Helenius, 1989; Freedman, 1989a); early experiments of this kind were frequently misinterpreted as showing that the carbohydrate groups functioned directly as targeting signals for secretion.

2.4.2   Disulfide Formation

There are cases where disulfide bond formation occurs just prior to secretion or at the cell surface (see Freedman and Hillson, 1980), but, for the most part, the formation of native disulfide bonds occurs within the lumen of the ER and cannot be distinguished in time and space from the process of folding. Intrachain disulfide bonds are, in one sense, only a particularly stable instance of the sterically specific interactions between remote protein side chains that define the tertiary structure of a protein; hence, the formation of these bonds must be viewed as an intrinsic part of the folding process. It is also an experimentally useful part of the folding process, because the stability of the disulfide bonds can be controlled by the imposed redox conditions, as exemplified in the work of Creighton. (See Chapter 7.)

In complex, multichain or multidomain proteins, formation of disulfides within individual domains will be part of the folding of those domains. Thus, in the biosynthesis of immunoglobulin light chains, the formation of the intrachain disulfide in the N-terminal domain occurs before the complete translation of the C-terminal domain (Bergman and Kuehl, 1979b). In procollagens, by contrast, folding of the central collagenous domain into the characteristic triple-helical conformation cannot be initiated until the three polypeptide chains are assembled and brought into correct register by the folding of the C-terminal domains and the associated formation of interchain disulfides (Bächinger et al., 1981).

Much important information about mechanisms of protein folding has been gained from studies on the refolding *in vitro* of mature disulfide-bonded proteins, where the disulfide bonds remain intact throughout the process of unfolding and refolding. Thus, key experiments on the kinetics of folding and the significance of prolyl-*cis,trans* isomerization were carried out on the refolding of ribonuclease A, and the definition of the "molten globule" state and of its role in protein folding equilibria and kinetics was dependent on studies on α-lactalbumin (Ikeguchi et al., 1986; Kuwajima, 1989). These model studies clearly do not claim to represent or mimic the folding process accompanying the initial folding of these proteins at biosynthesis; the unfolded forms used retain the specific set of disulfide bonds characteristic of the native state and, hence, constrain the subsequent folding events. Nevertheless, such experiments have been highly significant in the development of general ideas about the process of protein folding.

2.4.3   γ-Carboxylation

Proteins of the blood coagulation system, and other secreted proteins with high affinity for Ca^{2+}, are co-translationally modified by γ-carboxylation of Glu residues, in a reaction that requires a reduced form of vitamin K as co-factor. The reaction can also occur with short model peptides, but no obvious local sequence features define the specificity. It has been shown recently that the presence of an 18-residue prosequence N-terminal to the modified sequence, and cleaved from the mature protein, is essential for γ-carboxylation. The structural details of this specificity are not yet clear, and there is no direct information on the interaction between this modification and protein folding. When γ-carboxylation is inhibited *in vivo* by vitamin K antagonists, such as warfarin, however, nonglycosylated forms of the proteins, such as prothrombin, are found in the serum, indicating that the modification is not essential for secretion and presumably has a minimal effect on folding (Vermeer, 1990).

2.4.4 Prolyl-4-Hydroxylation

The collagens are a complex family of proteins characterized by regions of extended triple helical conformation involving three associated polypeptides with repeating sequence (-Xaa-Yaa-Gly-)$_n$, where Pro is abundant at both the Xaa and Yaa positions. In such collagenous sequences, there is a significant level of 4-hydroxylation of the Pro residue at the Yaa position. The modification is clearly co-translational and is catalyzed by an O_2-, Fe^{2+}-, and α-ketoglutarate–dependent enzyme that is a soluble resident of the ER lumen. The enzyme is prone to inactivation *in vivo,* and the reversal or prevention of this by ascorbic acid is one of the major functions of this vitamin. When the prolyl-4-hydroxylase is inactivated *in vivo* by iron chelators, the procollagen polypeptides produced are deficient in 4-hydroxylation. Although these "protocollagen" chains assemble into trimers through the formation of globular disulfide-bonded domains involving their C-terminal regions, the central domain does not fold to the triple helix at physiological temperatures. Thus, prolyl-4-hydroxylation is essential for the stability of the triple-helical conformation, and folding to this conformation requires both disulfide formation, to assemble the three chains, and hydroxylation to stabilize the conformation (Bächinger et al., 1981).

Different classes of procollagen assemble at different rates and are hydroxylated to different extents. Assembly does not depend on hydroxylation, so the consensus is that the rate of assembly determines the extent of hydroxylation, rather than vice versa (Kao et al., 1983). More slowly assembled procollagens are available within the ER in the unfolded state for longer and hence are more extensively hydroxylated; triple-helical procollagens are not substrates for hydroxylation.

2.4.5 Summary

Consideration of these co-translational covalent modifications indicates that the interactions between modification and folding are complex. This does not of itself, however, undermine the usefulness of refolding studies *in vitro* for understanding folding *in vivo*. Provided that the appropriate covalent form of the protein is used in these *in vitro* studies, they can be valuable models of the process *in vivo*.

2.5 Assembly

So many proteins function as oligomers or as components of complexes that the assembly of these entities from individual polypeptide chains is a major aspect of protein folding in the cell. Refolding and assembly of oligomers following unfolding and dissociation of mature oligomers have been extensively studied *in vitro* (Jaenicke, 1987). The results indicate, in general, that significant folding of individual chains occurs first, in order to provide the specific structural features recognized in the assembly step; however, significant further conformational change may follow the assembly step (as in the case of procollagens described previously). The refolding and reassembly studies model the *in vivo* process effectively, except in the case of those hetero-oligomers where the components are initially synthesized in different compartments or where perturbations occur in the relative timing or level of synthesis of the components. Studies *in vivo* indicate the existence of significant cellular mechanisms to facilitate successful assembly in these circumstances. (See Sections 6.5 and 6.6.)

2.6 Binding of Co-factors

Where proteins function in association with a tightly bound ligand, the binding of the ligand might influence the folding process, through an effect on either the yield or the rate. The limited studies on co-factor binding *in vivo* do not yield easy generalizations. In the case of most of the components of the mitochondrial inner membrane electron transfer chain, it is not known at what stage of import attachment of prosthetic groups occurs (Hartl et al., 1989). Translocation and processing of polypeptides occur in absence of the co-factors and prosthetic groups, but in some cases assembly of subunits into functional complexes requires prior association with co-factors, implying that interaction with co-factors is involved in folding to the native conformation. In the case of cytochrome *c*, where the heme group is covalently attached to the polypeptide, this step appears to occur early in the import of the protein into mitochondria and to be an essential part of the folding process; apocytochrome *c* lacks detectable secondary structure under physiological conditions.

Some of the complexities of the interaction between ligands and folding polypeptides are illustrated by the work of Jaenicke and colleagues on the interaction of Zn^{2+} ions with liver alcohol dehydrogenase (Gerschitz et al., 1978; Rudolph et al., 1978; Jaenicke, 1987). Refolding of dissociated and unfolded ADH polypeptides in absence of Zn^{2+} leads to the formation of partly folded, inactive, monomeric species that tend to aggregate. Addition of low concentrations of Zn^{2+} initiates an assembly process leading to the functional tetramer, through stabilization of a dimeric folding intermediate.

3 PROTEIN MISFOLDING: INSIGHT INTO PROTEIN FOLDING IN THE CELL

3.1 *Pathology of Protein Refolding* in vitro

Although most proteins can be refolded *in vitro* under appropriate conditions, and even large oligomers and multicomponent complexes can be reassembled from their subunits, most refolding experiments regenerate the native protein in yields significantly below 100%; in many experiments the yield of native material is negligible. That isolated proteins are capable of folding and self-assembly is central to the theoretical understanding of protein folding, but it is not sufficient to guide the selection of productive and efficient refolding protocols. Nor does it throw sufficient attention on the real questions that must be answered about the process of protein folding in the cell. These center on how rapid and efficient folding can be achieved in the defined solution conditions and high protein concentrations found in cellular compartments. Experimentalists aiming to optimize the rate and yield of protein refolding can manipulate temperature, pH, the concentration of refolding protein, thiol/disulfide redox components, ionic strength, and the concentrations of added denaturants, surfactants, cosolvents, solutes, and so on. They can change these conditions abruptly or gradually, and in the case of hetero-oligomers they can manipulate the order and timing of addition of the various components. Analysis of the effects of these variables on the rate and yield of refolding, coupled with study of what happens to the fraction of protein that is not refolded correctly, can usefully illuminate the more obscure and experimentally intractable questions about protein folding in the cell.

The key result from studies of this kind is that productive refolding from the unfolded state competes with aggregation. Unfolded proteins, and proteins at intermediate states of folding, are usually extremely insoluble under the conditions in which native proteins are stable and soluble. Hence, the return of unfolded proteins to conditions that generate the native state can lead to rapid aggregation. Even dialysis to effect a slow adjustment of the conditions can lead to aggregation if an incompletely folded species accumulates at intermediate denaturant concentrations and is insufficiently soluble in these conditons. (For early experimental results, see Jaenicke, 1974; and London et al., 1974. For reviews see Goldberg, 1985; Mitraki and King, 1989; Jaenicke and Rudolph, 1989; Chapter 9.) How can this aggregation be avoided in refolding *in vitro*?

Obviously, reduction of the concentration of refolding protein to very low levels can be effective. Concentrations in the range 10^{-7} to 10^{-5} g/l are generally required, however, and this may cause complications in the case of oligomeric proteins, because dilution will reduce the rate of slow second- and higher-order association steps and will leave partly folded monomers, which are frequently insoluble, unassociated for long periods. Aggregation processes are usually of higher order still, however, so the yield, if not the rate, of native protein recovery increases with dilution. Several detailed studies have been carried out of the direct kinetic competition between productive folding and aggregation. In a classic experiment with porcine lactate dehydrogenase, Zettlmeissl et al. (1979) showed that the distribution of product of refolding between reactivated oligomers and inactive aggregates varied simply with the concentration at which refolding occurred; below 1 µg/ml, reactivation was close to 100%, while above 100 µg/ml, aggregation was total. In most cases where the nature of aggregated protein has been studied, or the conditions affecting the balance between folding and aggregation have been analyzed, it appears that aggregates are derived from partly folded intermediates, so the intermolecular interactions involve contacts between partly folded elements of chains that have not yet docked correctly with their intramolecular partners (e.g., hydrophobic β-sheets that have not yet been shielded by surface helices) or domains that have not yet interacted intramolecularly (Goldberg, 1985; Mitraki et al., 1987; Horowitz and Criscimagna, 1986; Havel et al., 1986; Brems, 1988; Chapter 9). Conditions for resolubilization of such aggregates indicate that they are stabilized by hydrophobic interactions and can only be resolubilized in strongly denaturing conditions.

A variety of approaches are available for facilitating productive refolding at moderate protein concentrations and, hence, avoiding the inconvenience (and expense) of recovering refolded protein from high dilution. Denaturants such as urea and GdmCl are effective solubilizers of refolding intermediates. Hence, it is often useful to ensure the continued presence of low concentrations of these agents; refolding is initiated by diluting the unfolded protein into denaturant concentrations where the native form is stable, but intermediate forms are soluble. In other cases, the presence of specific surfactants to solubilize intermediates has proved valuable; Horowitz and colleagues have carried out a careful study of the optimal refolding of rhodanese in the presence of various surfactants (Horowitz and Simon, 1986; Tandon and Horowitz, 1986). A recent case (Samuelsson et al., 1991) showed that the presence of a highly soluble fusion partner can facilitate refolding. Insulin-like growth factor I fused to IgG-binding domains from protein A, could be refolded at higher concentrations and in significantly higher yield than the wild-type protein. Incorrectly disulfide-linked multimers formed in both cases, but those formed from the fusion protein were soluble and could readily be recycled to correctly folded species.

Alternatively, low final concentrations of folded protein can be avoided by taking account of the fact that it is only essential to minimize concentrations of unfolded or partly folded protein. Hence, repeated additions of small aliquots of unfolded protein to a large volume of refolding solution, allowing each aliquot to refold before the next is added, can give minimal aggregation with acceptable final concentrations of folded protein, as can comparable protocols based on continuous slow addition of a stream of unfolded protein (Rudolph and Fischer, 1989). Finally, the whole problem of aggregation can be sidestepped by immobilizing the unfolded protein on a matrix and altering the solution conditions to bring about refolding of the immobilized protein in circumstances where, by definition, it cannot aggregate (Sinha and Light, 1975; Mozhaev and Martinek, 1981). The problem here is to ensure that the protein interacts with the immobile matrix through groups that are exposed in the refolded protein and are not required to interact with other parts of the molecule during the folding process. Creighton (1986) has demonstrated that this can be achieved by binding unfolded proteins reversibly to ion-exchange resins.

All of these methods are aimed at ensuring either that nonnative interactions, principally hydrophobic interactions, do not occur during refolding, or that they occur only under conditions where they are readily reversible and can be replaced by native interactions later in the folding process. Appreciation of the basis for these *in vitro* strategies is invaluable for understanding the role of some of the cellular factors that facilitate folding.

The significance of nonnative interactions that cannot be reversed is also illustrated by considering the particular case of the formation of native disulfide bonds during the refolding and reoxidation of unfolded and reduced proteins. If reducing conditions are retained when such proteins are transferred out of denaturing conditions, the protein can refold, in principle. However, the native conformation must be stabilized by the correct disulfides and the equilibrium in their absence generally favors the unfolded conformation, which is, in most cases, highly insoluble. If an attempt is made to refold such proteins in strongly oxidizing conditions (e.g., excess of low molecular mass disulfide compounds) so as to stabilize protein disulfides, the yields of correctly folded protein are also usually low. This apparently surprising result is an example of kinetic rather than thermodynamic control; it reflects the fact that under such conditions any protein disulfides formed will be stabilized, whether or not they occur in the native protein. Because the number of intra-and intermolecular disulfide permutations is always large (105 alternative intramolecular pairings for a molecule with 8 Cys residues, 10,395 alternatives for a molecule with 12 Cys) the probability of forming "wrong" disulfides far exceeds the probability of forming "right" ones in the early stages of folding, when other structural constraints are not yet expressing themselves. Hence, the probability of folding to the native, correctly disulfide bonded conformation is not enhanced by operating in conditions that strongly stabilize disulfides.

As first demonstrated by Wetlaufer, the preferred conditions are those in which there are significant concentrations of both simple disulfides and simple thiols (Saxena and Wetlauger, 1970); this favors native disulfide bond formation thermodynamically, because it exploits the difference in stability between "correct" disulfides (which are cooperatively stabilized by the "correct" protein conformation) and "incorrect" ones, and also favors it kinetically. A mixture of simple disulfide oxidant and simple thiol reductant facilitates both protein disulfide formation and the reduction of incorrect protein disulfides and, hence, catalyzes the process of formation of the correct set of disulfides. This lesson again can readily be transferred to the cell.

Understanding of the sources of aggregation and other pathologies of refolding *in vitro* not only provides guidance on the constraints that operate on protein folding *in vivo*, but also permits the production-scale generation of native biologically active proteins from unfolded proteins. The commercial importance of this in the context of recovery of active recombinant proteins following expression in heterologous hosts has already been noted. While early refolding protocols in this field often appeared to be the result of trial and error, the underlying principles are now widely understood and have been applied systematically to devise successful refolding protocols for complex multichain or multisubunit proteins such as albumin, tissue plasminogen activator (Rudolph and Fischer, 1987; Sarmientos et al., 1989; Rudolph, 1990), or immunoglobulins (Buchner and Rudolph, 1991).

3.2 Pathology of Protein Folding in the Cell

The stringent conditions required for achieving high yields of correctly folded protein in refolding studies *in vitro*, and the variation in the required conditions for different proteins, focuses attention on the success of initial protein folding *in vivo*, where many different proteins fold within the same compartment and where there is not the freedom to vary at will the protein concentration, the pH, and so on, that is enjoyed by the *in vitro* refolder. But it should be not be supposed that mystical and vital forces are at work that ensure that protein folding is successful when proteins are synthesized in a cellular context. Just as analysis of failures of refolding *in vitro* illustrates the nature of the requirement for successful folding, so it is valuable to consider situations in which proteins fail to fold correctly when expressed *in vivo*.

The most dramatic and widely known case of failure of protein folding *in vivo* concerns the production of recombinant heterologous proteins in *E.coli*. Because of their pharmaceutical potential, the majority of the initial targets for over-expression in *E.coli* were mammalian secretory proteins, including hormones, cytokines, proteases, protease inhibitors, and immunoglobulins. With some exceptions, these proteins gave low yields of active material following high-level expression in *E.coli*, but significant amounts of the recombinant protein were found in large aggregates or inclusion bodies, which were clearly visible as refractile granules in the over-expressing cells. (For review, see Marston, 1986; Kane and Hartley, 1988; Mitraki and King, 1989; Schein, 1989.) What was initially surprising was the difficulty in producing satisfactory generalizations about the classes of protein that tended to form inclusions (Marston, 1986). Although basic proteins were highly likely to form inclusions, there was no clear dependence on the size or other gross properties of the protein. Nor was the phenomenon restricted to heterologous proteins; it was therefore, not a response to some intrinsic "foreignness." (For recent, and contrasting, insights into the difficulty of predicting inclusion body formation, see Wilkinson and Harrison, 1991, and Wetzel et al., 1991.) Mutant and covalently damaged *E.coli* proteins are frequently found to form inclusion bodies, but even wild-type *E.coli* proteins can do so within the normal host, under some expression conditions (Gribskob and Burgess, 1983; Botterman and Zabeau, 1985). Although inclusion bodies are usually encountered in cases of high-level expression, there is again no simple generalization; some proteins form active soluble products at expression levels well above those at which others form aggregates. Growth temperature appears to be a major factor in determining the fate of proteins; in many cases, recombinant proteins can be successfully expressed as active soluble products at 30 °C or lower, while they form inclusion bodies at 37 °C (Schein, 1989).

The significance of this finding can be understood by reference to the illuminating study carried out by King and colleagues on the folding and assembly of the thermostable tail-spike protein of the bacteriophage P22 (Goldenberg and King, 1981; Yu and King, 1984; King, 1986; Villafane and King, 1988; Haase-Pettingell and King, 1988; Mitraki and King, 1989). This work has shown that significant fractions of the expressed wild-type protein fail to fold and assemble into mature trimers at 40 °C and that a large class of "temperature-sensitive synthesis mutants" exists that produce biologically active folded trimers when expressed at low temperatures but not at 37 °C. Crucially, the trimeric mutant proteins generated at the permissive temperature are highly stable at the restrictive temperature; the defect only occurs if synthesis occurs at high temperature. From a detailed analysis of this case, King has inferred that an initial intermediate in the folding pathway is highly thermolabile; in the case of the wild-type protein, the majority of this intermediate passes on to a more stable monomer intermediate that is competent to assemble into trimers and to complete folding to the native product, but even marginally higher temperatures destabilize the initial intermediate converting it to a form that readily aggregates. The mutations that generate the temperature-sensitive for folding phenotype are then those that have no effect on the stability of the final assembled trimer but specifically destabilize the initial intermediate. Consequently, the pathway leading to aggregation competes more favorably with productive folding, except at low temperatures (Fig. 10–3).

This system demonstrates clearly that intermediates in protein folding pathways may be only marginally soluble under normal conditions, and that mutations and many other perturbations may deflect them towards precipitation and away from productive folding. This interpretation, if applied to inclusion body formation, implies that inclusions are generated from protein folding intermediates, rather than from native or fully unfolded proteins, and this is consistent with direct analyses of inclusion bodies.

What then are the specific perturbations of the normal folding pathway that cause mammalian secretory proteins, in particular, to form aggregates when expressed within the cytoplasm of *E.coli?* (1) In some cases, the nucleotide sequence coding for the protein's natural secretory signal sequence is still present in the expression construct, or other amino acids deriving from the subcloning strategy are present as N-terminal extension; the presence of such sequences might obstruct or retard folding so that the product will accumulate in an incompletely folded form. (2) The more reducing environment of the cytoplasm, where reduced glutathione predominates over oxidized, will diminish the stabilizing contributions of the native protein disulfide bonds. The native conformation of most extracellular proteins is stabilized by disulfide bonds (Chapter 7) so the product may accumulate in an unfolded, reduced form. (3) The absence of an appropriate glycosylation machinery in the *E.coli* cytoplasm rules out this modification of the newly synthesized protein and, hence, may minimize the solubility of unfolded and partly folded forms. Therefore, in a cell engineered to overexpress such proteins within the cytoplasm, the translation products will accumulate to high concentrations in misfolded forms of low solubility, and the generation of insoluble aggregates is not surprising. In most cases, the details have not been studied as thoroughly as for the P22 tail-spike protein model system, so it is not so clear how a lowered growth temperature suppresses aggregation; a change in temperature could alter the steady-state concentration of folding intermediates by differential effects on the rate of elongation and folding, or it could directly affect partitioning of folding species between productive and aggregative pathways.

While the formation of inclusions is a striking and commercially important instance of protein misfolding *in vivo*, other examples can be cited. The case of the P22 tail-spike protein emphasizes that significant levels of misfolding can occur with prokaryotic wild-type proteins expressed in their natural host, and there are examples of this for eukaryotic proteins also. (See Hurtley and Helenius, 1989, for review.) Thus, up to 10% of influenza virus hemaglutinin (HA) expressed in cultured mammalian cells is retained and degraded in the ER, and the fraction of protein so

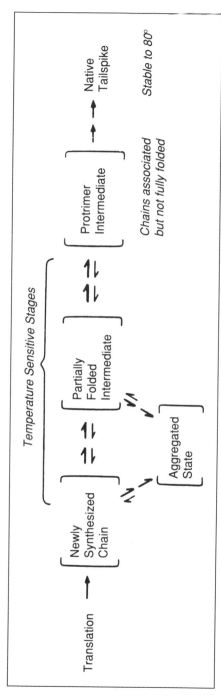

FIGURE 10–3. Folding and assembly pathway for the phage P22 tailspike protein. Newly synthesized chains form partially folded intermediates that convert to species sufficiently structured for intersubunit recognition. These interact to form protrimers, in which the chains are associated but not fully folded; protrimers then fold further to yield mature thermostable tailspikes. The intermediates preceding the protrimer are thermolabile; the mature tailspike alone is resistant to SDS-denaturation. (From Haase-Pettingell and King, 1988)

retained increases with growth temperature. The degree of aggregation and retention of this and other proteins can also be manipulated by varying their glycosylation; inhibition of glycosylation or the removal of N-glycosylation sites by mutation can lead to retention and aggregation in the ER, especially at elevated growth temperatures. (See Rose and Doms, 1988, for review.) The fact that the influence of glycosylation is not direct, but is expressed through a more general influence on folding and the solubility of folding species, is suggested by the fact that glycosylation at either one or the other of the two normal glycosylation sites in the VSV G protein is sufficient to ensure normal maturation and transport from the ER and by the fact that, in mutants lacking these two glycosylation sites, the introduction of such sites elsewhere in the protein restores secretion in some cases (Machamer et al., 1985; Machamer and Rose, 1988).

All of these data on protein expression *in vivo* emphasize that protein folding in the cell faces similar physicochemical problems to protein folding in the test tube. Given the unavoidable constraints of the cellular milieu, it is not surprising that protein folding in the cell is not always 100% efficient, or that numerous perturbations, either of the protein or of the cell in which it is expressed, can lead to dramatic failures in folding at biosynthesis. The question to be answered now is the nature of the mechanisms that act in the cell to maximize the rate and yield of initial protein folding.

4 CELLULAR CATALYSIS OF PROTEIN FOLDING

4.1 Rate-Limiting Chemical Reactions in Protein Refolding in vitro

The detailed analysis of the kinetics of protein refolding *in vitro* has shown that protein refolding can be very rapid, provided that specific classes of slow process are absent. (See Chapters 5, 7, and 9.) In considering the folding of individual domains or single-domain proteins, two distinct chemical processes have been identified as limiting refolding rates. (1) In the refolding from the reduced, denatured state of proteins that contain disulfide bonds in the native state, it is clear that disulfide bond formation and isomerization are rate determining. Rates of refolding are influenced by the presence of thiol and disulfide compounds and are favored by high pH through ionization of thiol (-SH) to thiolate (-S⁻), which promotes thiol disulfide interchange (Chapter 7; for review of earlier data, see Creighton, 1978; Freedman and Hillson, 1980). In the best cases, of small proteins with relatively few disulfides, optimal *in vitro* refolding of such proteins often occurs with half-times in the range of 10^3 to 10^5 secs and is slower at physiological pH; for larger proteins, *in vitro* refolding may take several hours. (2) The second, more recently recognized, slow step in protein refolding is the *cis,trans* isomerization of Xaa-Pro peptide bonds. In unfolded proteins, peptide bonds involving the imino groups of Pro residues exist in equilibrium between the *cis* and *trans* conformers with the *trans* form in excess; in folded proteins, by contrast, each bond has a specific conformation and, although *trans* conformers predominate, there are numerous well-defined *cis* peptide bonds to Pro residues (amounting to 5% to 8% of all peptide bonds to proline in the current Brookhaven database [Stewart et al., 1990]). Hence, in any protein containing Pro residues, refolding from the fully equilibrated unfolded state will involve isomerization of Xaa-Pro peptide bonds. (See Chapter 5.) This process is intrinsically slow at neutral pH in model peptides, although the rate within a protein can be strongly influenced by folding of neighboring parts of the polypeptide. It is clear, however, that the overall folding rates of proteins vary approximately with the number of Pro residues, and that in some cases two populations of unfolded protein molecules can be distinguished, namely, rapidly refolding forms that require no isomerization of Pro peptide bonds, and slowly refolding forms where such isomerization is required. For a significant number of proteins, it is now established that Pro peptide isomerization is rate determining in refolding *in vitro,* and the rates of refolding can have half-times of many minutes (Lang et al., 1987; Lin et al., 1988).

How can the existence of these well-defined slow chemical steps in protein refolding *in vitro* be reconciled with the rate of appearance of functional activity during protein synthesis *in vivo*? Time scales for the latter do not greatly exceed the time taken to biosynthesize the polypeptide chain. Even for complex multimeric proteins, folding and assembly *in vivo* is very rapid; for some viral surface membrane oligomers (VSV G protein and influenza HA) the times quoted by Helenius and colleagues (Copeland et al., 1988; De Silva et al., 1990) are: synthesis and co-translational modification, 1 to 2 min; folding of monomers, 3 to 4 min; trimerization, 7 to 9 mins; transport to the Golgi compartment, up to 20 min. The general discrepancy between the *in vivo* folding rate and *in vitro* refolding rates was noted 30 years ago and led to the proposal that specific catalysis of protein folding processes might occur in cells. In view of the preceding discussion, we might expect catalysis of the two identified classes of rate-limiting reactions, namely disulfide interchange and Pro peptide bond *cis,trans* isomerization. A catalyst of thiol-disulfide interchange with the potential to act on protein folding in the cell was first detected in 1963 (Goldberger et al.,1963; Venetianer and Straub, 1963) and its role is now firmly established. (See section 4.2.) Enzymic catalysis of Pro peptide bond *cis,trans* isomerization was demonstrated more recently, but there is now very active study of such enzymes (see Section 4.3); their role in initial protein folding remains controversial.

Disulfide bond isomerization and prolyl-peptide isomerization are well-defined chemical processes for which specific assays can be established and catalysts identified. In refolding of large multidomain or multisubunit proteins, the association steps between domains or subunits are dependent upon some prior folding of the individual domains, but the appearance of the final native state is generally slow under most accessible conditions, and it is clear that the rate-limiting steps are late assembly steps (Jaenicke, 1987; Chapter 9). Domain interactions and subunit assembly have also been identified as slow steps in protein refolding *in vivo*, but these are a physical interaction that is less easy to use as the basis for a well-controlled assay system; some cellular factors have been claimed to facilitate protein association and assembly *in vivo*, but their precise roles and mechanism remain less well defined.

4.2 Protein Disulfide Isomerase (PDI, EC 5.3.4.1)

PDI is an enzyme that catalyzes thiol-disulfide interchange processes in protein substrates and is an abundant component of the luminal volume of the ER in secretory cells. It was discovered in the early 1960's (Goldberger et al., 1963; Venetianer and Straub, 1963), first purified soon after (De Lorenzo et al., 1966), and its catalytic properties have been characterized over many years. (See Freedman and Hawkins, 1977; Freedman and Hillson, 1980; Freedman et al., 1984.) PDI shows a very broad substrate specificity and acts to catalyze native disulfide bond formation in many proteins, including multidomain and multisubunit proteins, starting from either the reduced or the "incorrectly disulfide-bonded" states. Studies on its cellular distribution and on its developmental properties in a range of systems indicated a close correlation between the presence of PDI in cells and their activity in the biosynthesis and secretion of disulfide-bonded extracellular proteins. Furthermore, both biochemical and morphological studies indicated that PDI is located uniquely within the lumen of the ER, the site at which disulfide formation occurs in newly synthesized proteins. In view of these findings (reviewed by Freedman, 1984, 1987; Freedman et al., 1984) there has been general acceptance of the proposal, first made by Anfinsen and colleagues (Epstein et al., 1963), that PDI catalyses the rate-determining folding steps associated with native disulfide formation in protein biosynthesis. Evidence for this role has been strengthened by the findings that PDI can be cross-linked *in vivo* to nascent polypeptide chains (Roth and Pierce, 1987), that the presence of PDI is essential for efficient co-translational disulfide bond formation in a reconstituted *in vitro* translation and translocation system (Bulleid and Freedman, 1988) and that PDI is encoded by a single-copy gene in *Saccharomyces cerevisiae* that is essential for viability (Farquhar et al., 1991; Scherens et al., 1991). PDI is thus unique at present in being, on the one hand, a well-characterized enzyme at the level of sequence, catalytic activity, and mechanism in purified systems, and at the same time identified with a specific action in protein folding phenomena in the cell.

4.2.1 Catalytic Activity

PDI exhibits a broad substrate specificity, but the details of its action have not been studied in detail. Those results that bear directly on its mode of action (summarized in Freedman, 1991) indicate: (1) that it is a true catalyst, capable of turning over many molecules of substrate per molecule of enzyme; (2) that it catalyses all thiol-disulfide interchange steps that also involve conformational change in the protein substrate; (3) that it shows no local sequence specificity, so that the product disulfide bonds generated are those permitting the most stable folded conformation of the system as a whole at the applied thiol/disulfide redox potential; and (4) that, in cases where the system has been optimized, half-times of PDI-catalyzed folding reactions can approach the values of the corresponding reactions in the cell.

When analyzed in terms of catalytic center activity toward standard *in vitro* substrates, however, catalysis by PDI is modest. Catalytic center activity in the generation of native RNase A from the reduced protein is approximately 0.43 min^{-1} (Lyles and Gilbert, 1991a), which is comparable to those calculated by Hawkins et al., (1991a) for the catalysis of folding of BPTI (3 min^{-1}), for reduction of insulin (12 min^{-1}), and for the regeneration of native RNase A from the incorrectly oxidized "scrambled" protein (less than 0.1 min^{-1}). The value of k_{cat}/K_m for the reoxidation of RNase A by PDI is of the order of $5 \times 10^4 \, M^{-1} \, min^{-1}$, representing an acceleration of 100-fold over the second-order rate constant for typical thiol-disulfide interchange processes between simple thiols and disulfides (Lyles and Gilbert, 1991a).

4.2.2 Structure and Mechanism: Homology to Thioredoxin

Sequences of cDNA's coding for PDI from a number of vertebrate species are now known (Parkkonen et al., 1988; Bassuk and Berg, 1989; Freedman et al., 1989), and the corresponding yeast sequence has now been determined (Farquhar et al., 1991; Scherens et al., 1991). As first pointed out by Edman et al.(1985), the key sequence feature for understanding the thiol-disulfide interchange activity of the enzyme is the presence of two homologous regions (*a* and *a'* domains, 47% identical) that are strongly homologous to the small dithiol-disulfide oxidoreductase, thioredoxin. Thioredoxin acts in a variety of oxidoreduction processes, and its active center is a short disulfide loop involving the sequence -Cys-Gly-Pro-Cys-. (See Holmgren, 1985, for review.) In the high-resolution structure of *E.coli* thioredoxin in the disulfide state (Katti et al., 1990), this tetrapeptide sequence forms the first turn of a long α-helix (Fig. 10–4), and the S atom

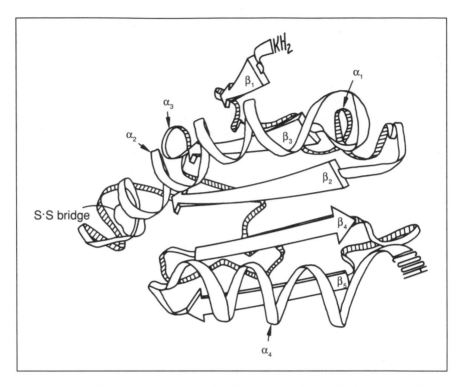

FIGURE 10–4. *Schematic three-dimensional structure of* E.coli *thioredoxin. The model is based on the low-resolution structure derived from X-ray crystallographic studies on the disulfide form of the protein (Holmgren, 1985) and is consistent with a more recent study at higher resolution (Katti et al., 1990). Arrows represent β-strands. (Reproduced, with permission, from the* Annual Review of Biochemistry, *vol. 54 ©1985 by Annual Reviews Inc.)*

of the more N-terminal Cys residue is located at the positive pole of the helix dipole, which accounts for the unusually low pK and high reactivity of this Cys residue in the dithiol state of the protein.

The high level of homology between the PDI *a* and *a'* domains and thioredoxin, plus preliminary modeling studies (Freedman et al., 1988), leave no doubt that the conformations of these domains are similar to that of thioredoxin. Chemical modification studies identify the sequences -Cys-Gly-His-Cys-, which occur in the *a* and *a'* domains at positions homologous to the active site of thioredoxin, as the active sites of PDI, and indicate that the essential Cys residues in the reduced state have a pK of 6.7, identical to that of reduced thioredoxin (Hawkins and Freedman, 1991). Mutation of these residues also generates inactive enzyme (Vuori et al., 1990). Moreover, the close family relationship is confirmed by the fact that both enzymes are active in protein:thiol–disulfide interchange (Pigiet and Schuster, 1986; Lundström and Holmgren, 1990; Hawkins et al., 1991a) and that the PDI active sites are substrates for reduction by NADH catalyzed by thioredoxin reductase (Lundström and Holmgren, 1990).

Clear differences between PDI and thioredoxin are evident, however, in their specific activities in protein disulfide interchange activities (Hawkins et al., 1991a) and in the redox properties of the active site disulfides. In a direct comparison under identical conditions, the catalytic center activity of PDI was found to be more than 50-fold greater than that of thioredoxin (Hawkins et al., 1991a). By studying the reaction between thioredoxin and PDI, Lundstrom and Holmgren (1990) concluded that the standard redox potential of the PDI disulfides was significantly less negative than that of thioredoxin, certainly above -230 mV compared to -260 mV for thioredoxin. By taking alternative approaches to monitoring the equilibrium betwen PDI and a redox mixture of oxidized and reduced glutathione (GSH + GSSG), Lyles and Gilbert (1991a) and Hawkins et al. (1991b) deduced equilibrium constants of 0.06 mM and 0.05 mM, respectively. These values convert to standard redox potentials of approximately -110 mV, indicating that the active site disulfide-dithiol couples of PDI are unusually strongly oxidizing compared to standard protein disulfides (Gilbert, 1990).

Recent site-directed mutagenesis studies indicate that the local sequence within the active site region plays a significant part in this difference in properties between PDI and thioredoxin. Krause et al. (1991) mutated Pro34 of *E.coli* thioredoxin to His, to give the active-site sequence -Trp-Cys-Gly-His-Cys-Lys-, as in the *a* and *a'* domains of PDI. The mutant thioredoxin had a more oxidizing standard redox potential than the wild-type, and its catalytic properties were intermediate between those of wild-type thioredoxin and those of PDI.

4.2.3 PDI as a Component of the Reticuloplasm

Conventional biochemical analyses indicate that PDI is a soluble enzyme located within the luminal volume of the ER (Lambert and Freedman, 1985). This location within the reticuloplasm (Koch, 1987) has been confirmed by immunocytochemical techniques at both the electron microscope and fluorescence microscope level (Macer and Koch, 1988; Akagi et al., 1988). PDI and other soluble luminal content proteins can be readily released from isolated microsomal vesicles (Paver et al., 1989). A small number of proteins dominate the resident protein content of the ER lumen, and PDI is one of these (Mills et al., 1983; Kaderbhai and Austen, 1984; Pelham, 1989). The C-terminal sequence of mammalian PDI is -Lys-Asp-Glu-Leu, or -KDEL in the one letter abbreviation, and the subsequent discovery of this sequence at the C-terminus of other luminal content proteins led to the proposal that this sequence functions as a signal for ER retention (or recycling) of soluble resident proteins, blocking their movement to distal compartments of the secretory pathway. (See Pelham, 1989, for review.) The nature of the system for retention or recycling of these proteins is under intense investigation. That the system can be saturated is demonstrated by the observation that PDI may be found at the cell surface or secreted into the medium in the case of cells that generate very high levels of the protein (Yoshimori et al., 1990; Dorner et al., 1990).

The question arises as to whether the catalytic action of PDI is modulated by other components present in the ER lumen. By immunological criteria, PDI comprises 0.35% to 0.4% of the protein content of a mammalian liver homogenate, and its specific activity in such a homogenate, after sonication, is approximately 0.4% of that of the homogeneous enzyme, implying that the other proteins present have no influence on the activity of PDI. Likewise, the specific activity of a total reticuloplasm extract in catalyzing refolding and reoxidation of reduced BPTI was 11% of that of homogeneous PDI, and PDI was shown to comprise 11% of the extract, again suggesting no synergistic effects due to the presence of other luminal components (Zapun et al., 1992). These results involve simple model assay systems, and low protein concentrations compared to those found within the ER lumen in the cell, but they suggest that PDI acts alone in catalyzing protein folding in the ER lumen. On the other hand, there is evidence that PDI participates as a component of other enzyme activities within the ER.

4.2.4 Multiple Roles?

Cloning approaches to identifying proteins involved in a number of ER functions have indicated that the PDI polypeptide may participate in a range of other ER processes concerned with the modification and assembly of functional proteins. (For review, see Freedman, 1989b; Bulleid and Freedman, 1990; Freedman and Tuite, 1992.) The first and best-established case is prolyl-4-hydroxylase, an $\alpha_2\beta_2$ tetrameric enzyme of which the β subunits are identical to PDI (Pihlajaniemi et al., 1987; Koivu et al., 1987). More recently good protein structural evidence has appeared to establish that PDI is a component of the heterodimeric microsomal triglyceride transfer protein complex, which facilitates incorportion of triglyceride into nascent very low density lipoprotein (VLDL) particles within the ER (Wetterau et al., 1990,1991). The possible role of PDI in these activities is considered shortly. Persuasive photoaffinity labeling data at one time identified a luminal protein as an essential component of the N-glycosyltransferase system of the ER responsible for the specific interaction with -Asn-X-Ser/Thr- sequences in nascent proteins (Geetha-Habib et al., 1988); this glycosylation-site binding protein (GSBP) appears to be closely related to PDI, but its role in N-glycosylation has since been disproven (Bulleid and Freedman, 1990; Noiva et al., 1991a,1991b).

4.2.5 Biochemical and Genetic Approaches to Establishing the Role(s) of PDI

While there seems little doubt that PDI acts in the cell to catalyze disulfide interchange reactions and hence to facilitate the initial folding of disulfide-containing extracellular proteins, its role in the other activities identified in Section 4.2.4 is less easy to establish. In none of these cases has resolution and reconstitution of the oligomeric enzymes been successful, so it is not possible to test the details of the requirement for the PDI polypeptide in the overall activity. The approach of resolution and reconstitution of PDI in microsomal vesicles established that the presence of PDI was necessary for efficient co-translational disulfide bond formation (Bulleid and Freedman, 1988), and this approach could be adapted to study the role of PDI in other activities. Thus, PDI-depleted microsomes show no defect in the co-translational glycosylation of a number of nascent proteins (Bulleid and Freedman, 1990). In all such studies, however, it is difficult to ensure that PDI is quantitatively removed and that a specific subpopulation, such as might be associated in a specific complex, is not retained.

A more elegant answer to such questions may come from a genetic approach. Although PDI has been mainly studied in vertebrates, it has been detected in a wide range of organisms and has been purified from wheat (Bulleid et al., 1991), the alga *Chlamydomonas reinhardii* (Kaska et al., 1990), and the yeast *Saccharomyces cerevisiae* (Mizunaga et al., 1990). The power of genetic analysis in *S. cerevisiae* to identify and analyze the roles of components of the secretory apparatus has been amply demonstrated by the work of Schekman and other (Novick et al., 1980; Schekman, 1988). On the basis of the detection of PDI as a latent activity in yeast microsomes (Murant, 1989) and an assumed homology between yeast and mammalian PDI, we cloned a yeast genomic DNA sequence that (1) shows a high level of homology in sequence and overall organization to mammalian PDI's, (2) indicates an amino acid composition in close agreement with that obtained directly from purified yeast PDI (Mizunaga et al., 1990), and (3) led to a 10-fold overexpression of PDI activity when transformed into wild-type yeast on a multicopy plasmid (Farquhar et al., 1991). A gene disruption has been constructed, and analysis of the diploid and haploid yeast bearing the disrupted gene indicates that the gene is essential for viability. Southern blotting analysis reveals that this gene exists in a single copy in the yeast genome; the sequence has recently been mapped to the left arm of chromosome III (Scherens et al., 1991). There is now clearly the potential to develop conditional lethal mutations in the yeast PDI gene in order to analyze more closely its role in the cell.

4.2.6   Unanswered Questions

Although it may reasonably be claimed that PDI is the best understood of the protein factors thought to act as facilitators of folding in the cell, this does not indicate a high level of understanding. There are a number of obvious questions that need to be answered before we can provide a crude picture of the action of PDI as a catalyst of protein folding in the cell.

4.2.6.1   Nature of substrates and source of oxidizing equivalents.   While it is clear that the lumen of the ER (and correspondingly the periplasmic compartment in bacteria) are likely to be more oxidizing than the cytoplasm, we have no real information on what are the major redox components within the ER lumen, how the differential in redox potential with the cytoplasm is maintained, or what is the immediate source of oxidizing equivalents for protein disulfide formation. We can imagine that an exogenous oxidant could act in the way that oxidized DTT does in the detailed model studies of the action of PDI on reduced BPTI (Creighton et al., 1980), but this does not reveal whether in the cell PDI acts mainly as a redox carrier in a net oxidation, with the oxidant acting on reduced PDI active sites, or whether it acts essentially in isomerizations, with the oxidant acting directly on the nascent protein as substrate. This may well vary between proteins, but our ignorance of redox processes and components in the luminal compartment makes speculation hazardous (Lyles and Gilbert, 1991b; Hawkins et al., 1991b; Freedman, 1990). Vermeer (1990) has made the interesting proposal that the vitamin K cycle, which is usually studied in connection with γ-carboxylation processes, may also operate in the ER membrane uncoupled from γ-carboxylation to act as a source of oxidizing equivalents for protein disulfide formation. Model studies (Soute et al., 1992) indicate that this proposal merits further study.

4.2.6.2   Is PDI more than a simple catalyst of thiol-disulfide interchange?   Much of the work on the mechanism of action of PDI (Section 4.2.2) stresses the similarities between PDI and thioredoxin. But there is a major difference; PDI is a large polypeptide of approximately 500 residues and exists in solution as a homodimer. Hence, an intact molecule of PDI is approximately 10-fold larger than a molecule of thioredoxin and contains four thioredoxin-like domains, accounting for 40% of the total molecule. What is the role in PDI function of the remainder of the molecule, in particular of the homologous *b* and *b'* domains and of the *e* domain, which has some homology to a region of the estrogen receptor? Does the enzyme interact with its protein substrates only through the dithiol/disulfide active sites within the thioredoxin-like domains, or is conformational change in the substrate proteins facilitated through interactions with the remainder of the enzyme? Does the coexistence of four active site domains within a single molecule facilitate complex thiol-disulfide interchanges in substrates with many disulfide bonds? Is there interdomain flexibility in the PDI molecule that allows it to bring its four active sites to bear on a single substrate molecule? Do the differences in catalytic activity and redox potential between PDI and thioredoxin derive only from local sequence differences (Krause et al., 1991) or also from the fact that multiple sites coexist in a PDI molecule? As yet we have little sound kinetic evidence by which to judge whether the active sites in the PDI molecule are independent or cooperative. None of the recent detailed studies indicates a clear difference in properties between the sites in the *a* and *a'* domains (Lundström and Holmgren, 1990; Hawkins and Freedman, 1991a,b; Lyles and Gilbert, 1991a), but study of the equilibrium between PDI active sites and a mixture of GSH and GSSG indicated a deviation from the hyperbolic curve expected if the PDI sites are equivalent and independent; Hill analysis indicated positive cooperativity with a Hill coefficient of 1.5 (Hawkins et al., 1991b). Behind these questions, of course, lies the question of the tertiary and quaternary structures of PDI; other than the modeling of the *a* and *a'* domains, there is no information on this topic apart from the observation that the two active sites within a PDI polypeptide can be cross-linked by bifunctional reagents with a span of at least 1.6 nm (Hawkins et al., 1991b).

4.2.6.3 What is the role of the PDI polypeptide in proteins with other functions? As noted previously, we have no real clues as to the role that the PDI polypeptide plays in the other proteins in which it has been implicated. These are all posttranslational modification or maturation processes, however, and it is possible that PDI plays a common role in them. Geetha-Habib et al. (1988) proposed that the role of the glycosylation site binding protein (GSBP) is to maintain an open or unfolded conformation in substrate proteins, permitting glycosylation of appropriate sites, and it is certainly possible that the PDI subunits within prolyl-4-hydroxylase and triglyceride transfer protein complex play such a role.

4.2.6.4 Does PDI interact significantly with other major components of the reticuloplasm? The major proteins within the ER lumen, PDI, BiP (see Section 6.6), endoplasmin, and reticulin exist at high concentrations, show affinity for Ca^{2+} ions, and are secreted from fibroblasts when Ca^{2+} levels are perturbed by addition of Ca^{2+}–ionophores (Booth and Koch, 1989). Does this imply some general interaction between the proteins to form a Ca^{2+}–mediated matrix? This has been suggested on several occasions, but there is no direct evidence to support it, and it has been challenged by a microinjection study showing that newly synthesized BiP can diffuse across an oocyte at a rate comparable to that of a secretory protein (Ceriotti and Colman, 1988). Likewise, as noted in Section 4.2.3, there is no evidence to date for functional synergy between PDI and these other major luminal components. Nevertheless, it remains the case that we have only limited knowledge of conditions and interactions within the ER lumen and, hence, a limited picture of the environment in which PDI functions.

4.2.6.5 Is there a family of PDI-like proteins?   The recent literature has reported several sequences of proteins that show marked sequence similarity to PDI, in particular the presence of the active site -Trp-Cys-Gly-His-Cys-Lys- sequence within a region strongly homologous to thioredoxin and to the PDI *a* and *a'* domains. These proteins include the already-mentioned GSBP (Geetha-Habib et al., 1988) and another protein that is clearly located within the ER lumen, namely *Erp*72 (Mazzarella et al., 1990). This protein was first identified as one of the set of major proteins synthesized as the ER proliferated in mouse plasmacytoma cells; the others were subsequently recognized as identical to proteins studied in other contexts (PDI, BiP, endoplasmin), but (as of early 1992) *Erp*72 has only been identified in this context. Cloning and sequencing of cDNA for *Erp*72 indicated that the protein contained three domains with clear homology to thioredoxin and to the PDI *a* and *a'* domains; in outline, the protein resembles a molecule of PDI with an additional *a*-type domain at the N-terminus. The protein has not been isolated, and it is not known whether it has PDI activity, but overexpression of this cDNA in CHO cells does not lead to overexpression of PDI activity (Dorner et al., 1990).

Two further examples are intriguing. A rat protein identified as a PI-specific phospholipase C was cloned and sequenced and shown to resemble PDI; it has two thioredoxin-like domains at positions corresponding to the *a* and *a'* domains of PDI, but its N- and C-terminal sequences imply targeting to cellular locations other than the ER (Bennett et al., 1988). An analysis of genes differentially expressed between bloodstream and culture forms of *Trypanosoma brucei* identified one (BS2) that sequencing revealed to resemble PDI; two thioredoxin-like domains are present at corresponding positions to those in PDI (although one has the sequence -Gly-Cys-Gly-Tyr-Cys-Gln-) while the other has the characteristic -Trp-Cys-Gly-His-Cys-Lys-, and there is some homology throughout the protein. The function and subcellular location of this protein are not known. Overall the data suggest that there exists a family of PDI-like proteins that resemble the prototype both in local sequence and in overall architecture. Until the other members of this family are better characterized in terms of function, however, the significance of these conserved structural features will not be apparent.

4.3 *Prolyl-Peptidyl* cis,trans *Isomerase (PPI)*

The clear indication that the *cis, trans* isomerization of Xaa-Pro imide bonds was rate limiting in the refolding of some proteins (or, more precisely, that the unfolded forms of some proteins contained significant populations whose refolding was rate-limited by slow prolyl peptide isomerization; Chapter 5) led to a search for potential enzymic catalysts of this process. Discovery of such PPI's depended upon the development of simple model assays, which was first achieved by Fischer et al. (1984) using peptides of the class -Ala-Pro-Phe-p-nitroanilide. These molecules are chromophoric substrates for chymotrypsin, which cleaves the anilide linkage with predictable specificity, but only if all preceding peptide bonds are in the *trans* conformation; hence, the small fraction of substrate molecules that have the Ala-Pro bond *cis* are not chymotrypsin-sensitive. Catalysis of the isomerization of these molecules, which renders them chymotrypsin-sensitive, was the basis for an assay for PPI that led to its purification from mammalian tissues (Fischer et al., 1989b; Takahashi et al., 1989).

4.3.1 Catalytic Activity in Protein Folding

Although PPI's from other cellular sites have been inferred from DNA sequencing (see Section 4.3.3), and although other families of proteins with PPI activity, but different substrate and inhibitor specificities, have been characterized (see Section 4.3.4), most study of enzymic catalysis of Pro peptide bond isomerization in protein folding has involved the abundant cytosolic mammalian PPI first identified by Fischer et al. (1984) and isolated by Fischer et al. (1989b) and by Takahashi et al. (1989). The discussion here will focus on this enzyme. Initial studies of its action in catalyzing protein folding were not promising; there was little catalysis of the refolding of RNase A (Fischer and Bang, 1985), the classic case in which the analysis of refolding kinetics had implicated prolyl peptide bond isomerism as a key step. (See Chapter 5 and Kim and Baldwin, 1982, for reviews.) Subsequent studies on a range of proteins, however, gave a wider picture (Bächinger et al., 1987; Lang et al., 1987; Lin et al., 1988). PPI catalyzes the refolding of some proteins whose folding is limited by prolyl isomerization, but is not effective toward others. Thus the refolding of RNase T_1 and of Ig light chains is substantially catalyzed by PPI, there is moderate catalysis of the refolding of procollagen, cytochrome c, pepsinogen, bovine RNase S, and porcine RNase A, and no effect on thioredoxin and bovine RNase A (Schmid et al., 1991). PPI also had no catalytic effect on the refolding to the native oxidized state of reduced RNase A (Lang and Schmid, 1988; Lin et al, 1988). This latter reaction is dependent upon thiol-disulfide interchange and is catalyzed by PDI. Lang and Schmid (1988) demonstrated formally that PPI and PDI were distinct in their mode of action; PDI did not catalyze prolyl-isomerization-dependent folding reactions, and PPI did not catalyze thiol-disulfide interchange-dependent processes. Furthermore, there was no synergistic effect on either reaction when the two enzymes were present together.

The catalytic effect of PPI on protein folding reactions is not dramatic; in the best case a 30 to 100-fold increase in rate constant is observed for the major slow phase of folding of RNase T_1, at enzyme molar concentrations approaching those of the substrate. It is also clear that many prolyl isomerizations in protein folding are not catalyzed by the enzyme. Lang et al. (1987) proposed that this was due to the inaccessibility of many Pro peptide bonds; isomerization of Pro peptide bonds is often a late step occurring within an already significantly folded intermediate, rather than being required in order to initiate folding. (See Chapter 5 and Schmid and Blaschek, 1981.) In such circumstances, the catalytic potential of PPI in protein folding reactions will be limited to the isomerization of Pro peptide bonds that are accessible in these intermediates.

The alternative explanation, that catalysis by PPI of protein folding is not related to its ability to isomerize prolyl peptide bonds, has been eliminated by a careful study of the refolding of RNase T_1 (Kiefhaber et al., 1990a,b,c). In the native folded protein there are two *cis* peptide bonds preceding Pro39 and Pro55. The overall kinetics of refolding from the unfolded state are complex and have been interpreted in terms of alternative folding pathways (Kiefhaber et al., 1990b,c). The fast phase of folding, which is on the time scale of milliseconds, is not accelerated by PPI, but the major slow phase is; there is a small catalytic effect on the minor slow phase of refolding. Mutation of the sequence Ser54-Pro55 to Gly-Asn (the sequence found in the corresponding position in a homologous microbial ribonuclease) led to a simpler pattern of folding kinetics, lacking the major slow phase; folding of this mutant was not significantly accelerated by PPI. This indicates the source of the major slow phase of refolding is isomerization of the Ser54-Pro55 peptide bond and confirms that the catalytic role of the enzyme in protein folding is due to its effect on such isomerizations (Kiefhaber et al., 1990a).

In a recent study, Schönbrunner et al. (1991) have shown that homologous PPI's from man, *Neurospora crassa, S.cerevisiae,* and *E.coli* all show effective and comparable catalysis of the major slow phase of refolding of RNase T_1 and indeed show that the *E.coli* enzyme accelerates the minor very slow phase, which is believed to involve isomerization of the Tyr38-Pro39 peptide bond.

4.3.2   Structure and Mechanism

The mammalian cytoplasmic PPI's are abundant monomeric enzymes of 17 kDa molecular weight. Enzymes that are homologous appear to be ubiquitous in all species. The enzymes act catalytically on peptide substrates by noncovalent stabilization of a transition state representing a species in which partial rotation has occurred so that the peptide bond unit is no longer planar. A nucleophilic mechanism with a Cys residue acting analogously to the attacking nucleophile in cysteine proteinases was proposed (Fischer et al., 1989a), but this now appears to be ruled out by detailed studies of enzyme kinetics and isotope effects (Harrison and Stein, 1990a). Also, site-directed mutagenesis in which each of the Cys residues of human

PPI was mutated to Ala found all the mutant species to have the same catalytic activity as the wild-type enzyme in the standard peptide assay (Liu et al., 1990). The values of k_{cat}/K_m lie in the range of 1.2 to 1.7 x 10^7 $M^{-1}s^{-1}$ at 10 °C compared to the wild-type value of 1.4 x 10^7 $M^{-1}s^{-1}$, indicating that catalysis with these substrates is close to the diffusional limit.

The structure of the cytosolic human PPI (CyPA) has been determined by X-ray diffraction (Ke et al., 1991). The overall structure is of an eight-stranded antiparallel β-barrel, with a hydrophobic core, and there is no resemblance to any other known protein in topology or overall form. A shallow hydrophobic pocket on one face of the barrel is thought to be the binding site for cyclosporin.

4.3.3 Distribution and Subcellular Distribution

As noted previously, enzymes homologous to the mammalian PPI have been detected in a number of microbial sources. There is now evidence that the family contains members in cellular locations other than the cytosol. Thus in *N.crassa*, PPI was found in both cytosolic and mitochondrial compartments, but only one coding gene was detected (Tropschug et al., 1990). Two forms of PPI have been detected in *E.coli*, and the enzymes have been resolved, characterized, and sequenced at the N-terminus (Hayano et al., 1991). One form, PPIa, was found to be periplasmic in its subcellular location, and the corresponding gene sequence indicated the presence of a 24-residue N-terminal signal sequence that is removed proteolytically. PPIb lacked such a signal sequence and was found in the cytoplasmic fraction. Two sequences homologous to mammalian PPI were detected in *S.cerevisiae*, of which one closely resembled the mammalian prototype (Haendler et al., 1989), while the other differed in having an N-terminal extension with the characteristic properties of a secretory signal sequence (Koser et al., 1990). Subsequently, a human PPI homologue sequence with an apparent secretory signal sequence has been identified (Price et al., 1991). A *Drosophila* gene cloned on the basis of its role in the production of functional rhodopsin was sequenced and found to have a major domain that is homologous to the PPI family, but with extensions at both the N- and C- termini (Shieh et al., 1989). This protein, the *nina A* gene product, has since been shown to be an integral membrane protein with a cell-type specific pattern of expression, and to be required for functional expression of certain rhodopsins in *Drosphila* (Stammes et al., 1991). These results suggest that PPI-related proteins have wide-ranging cellular roles in a variety of locations, including, presumably, the secretory pathway.

4.3.4 Inhibition by Immunosuppressants

Its unusual catalytic activity and potential role in protein folding would not, in itself, have been sufficient to generate the rapid expansion of work on PPI since 1989. This expansion was, rather, a response to the surprising finding that PPI was identical to an already known protein, cyclophilin or CyP (Fischer et al., 1989b; Takahashi et al., 1989), the major cellular binding protein for the drug cyclosporin A, which is a cyclic undecapeptide of fungal origin that is widely used as an immunosuppressant in transplant therapy (Fig. 10–5). The high affinity of CyP for cyclosporin, and its abundance, make it the major cellular binding site for the drug and an obvious candidate for the drug's target. Hence, it was natural to test for an effect of cyclosporin on the PPI activity of CyP/PPI, and the drug was found to be an effective inhibitor. Earlier work had already indicated,

FIGURE 10–5. *Structures of immunosuppressants that inhibit PPI activity. The compounds are not closely related structurally, although both contain imide bonds. They bind to distinct classes of immunophilin and inhibit the PPI activity of these proteins. (From Rosen et al., 1990; ©1990 by the AAAS)*

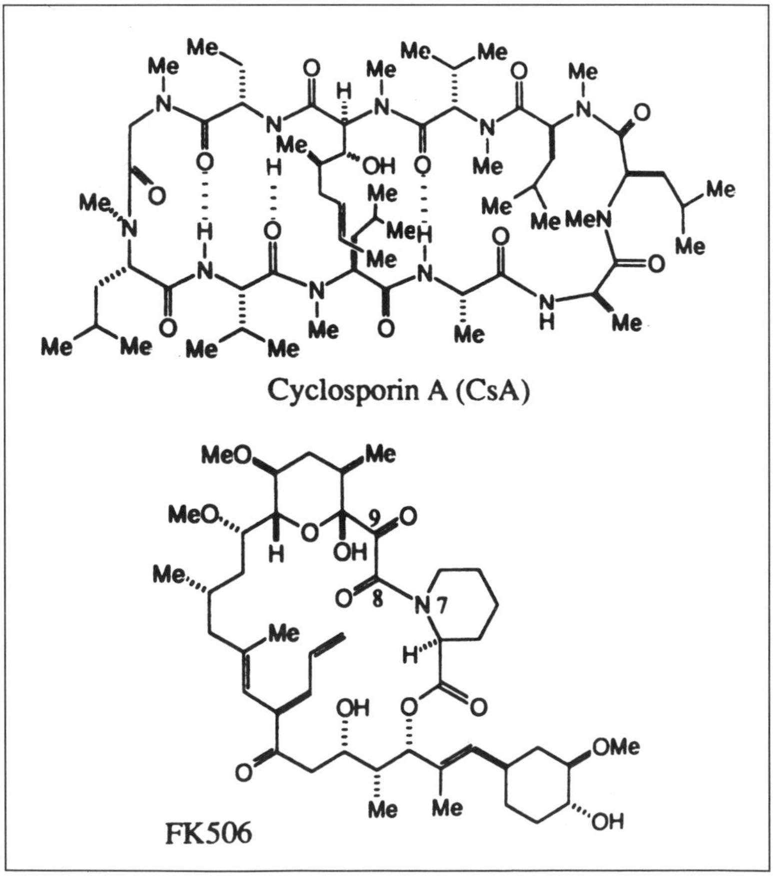

Cyclosporin A (CsA)

FK506

however, that the major pharmacological action of cyclosporin was to block the transcription of a subset of T-cell genes crucial to the expression of the immune response, and it is not immediately obvious how the inhibition of a general PPI activity could have so specific an effect.

A connection between PPI activity and immunosuppression was made firmer, however, by the discovery that a second class of immunosuppressants, of which the prototype is FK506 (Fig. 10–5), also binds to an abundant cytosolic protein, and that this protein is distinct from CyP but also has PPI activity (Siekierka et al., 1989; Harding et al., 1989). It is now clear that two major classes of proteins with PPI activity exist, with different substrate and inhibitor specificities; these proteins have been generically termed "immuno-philins," and their properties have been reviewed (Schreiber, 1991). The original class, the cyclophilin PPI's, are more active toward the conventional peptide substrate suc-Ala-Ala-Pro-Phe-pNA, but they do not discriminate sharply between substrates of this class and are inhibited by cyclosporin A. The second class, the FK506 binding proteins (FKBP's), show 10^3-fold differences in kinetic properties toward various peptide substrates (with a preference for suc-Ala-Leu-Pro-Phe-pNA) and are inhibited by FK506 (Harrison and Stein, 1990b). Neither class is affected by the diagnostic inhibitor of the other class. The structures of FKBP and of the complex between FKBP and its ligand, FK506, have been determined by NMR and X-ray diffraction, respectively (Michnik et al., 1991; Van Duyne et al., 1991). The structure of the complex confirms the proposal (Rosen et al., 1990) that the bound ligand differs in conformation from its structure when free: a *cis* amide bond in the free ligand is found to have the *trans* conformation in the complex. Similarly, NMR data on bound cyclosporin show that it undergoes a comparable isomerization on binding to its cognate immunophilin CyPA (Fesik et al., 1991), although CyPA and FKBP have quite different tertiary structures.

4.3.5 Physiological Role of PPI's

Isomerization of Pro peptide bonds is rate-determining in the refolding of some proteins, and PPI catalyzes this process in a subset of such proteins. The protein synthesis apparatus is assumed to generate all peptide bonds in the *trans* conformation initially, so it is clearly attractive to propose that the role of PPI is to catalyze isomerization to generate *cis* bonds in initial protein folding at biosynthesis. There is no strong concensus in favor of this view, however, and as of early 1992, all proposals on the cellular role of PPI remain speculative. Hayano et al. (1991) infer from the existence of multiple forms of PPI with different cellular locations that the enzyme accelerates protein folding in different compartments. Circumstantial evidence for a role for PPI in folding is provided by the observation that treatment of chick and human fibroblasts with cyclosporin A led to a delay in the folding of newly synthesized procollagen to the triple-helical protease-resistant conformation and to its consequent overhydroxylation (Steinman et al., 1991). The result is clearly significant, although the inhibitory effect was small, and there is no direct evidence yet for the presence of PPI in the lumen of the ER where procollagen folding occurs; the amino acid sequence evidence only indicates that members of the CyP/PPI family are found within the secretory pathway.

Analysis of the biological action of cyclosporin A should provide clues to the role of CyP/PPI. In *N.crassa*, genetic evidence shows that CyP/PPI is the target for the cytotoxic effect of cyclosporin A (Tropschug et al.,1989). It is generally assumed that the major CyP/PPI is the target for the action of cyclosporin A in higher organisms, although it is possible that minor, as yet uncharacterized members of the family are specially sensitive targets in specific cell types (Price et al., 1991). The major immunosuppressive action of cyclosporin is a reduction in transcription of lymphokines in T cells, and this appears to derive from an effect on the binding of specific transcription factors (Emmel et al., 1989; Randak et al, 1990). Hence, the drug disrupts signal transduction between cell surface receptors and nuclear events. It seems unlikely that this is an action on initial protein folding at biosynthesis, and Schmid et al. (1991) speculate that the role of PPI is to regulate the action of proteins by catalyzing *cis,trans* isomerizations of peptide bonds that interconvert proteins between states of different activity. There is evidence from drug specificities and from the relative concentrations required, however, that inhibition of the PPI activity of these proteins by immunosuppressants is insufficient to induce their biological effects as inhibitors of T-cell stimulation (Bierer et al., 1990; Schreiber, 1991).

5 PROTEIN FOLDING AND PROTEIN TRAFFIC IN THE CELL: GENERAL CONSIDERATIONS

5.1 The Cell Biological Discovery of Protein Folding

Analysis of the process of protein movement within cells, especially of movement of newly synthesized proteins from the cytoplasm to the secretory pathway or to organellar destinations, became an active topic in the 1970's following the development of effective *in vitro* systems in which the process could be reconstituted and analyzed. In this development, the major initial emphasis was on the timing of events, particularly the timing of translocation with respect to translation. Was the movement of proteins across membranes coupled obligatorily to the process of translation; was it coupled for some membrane systems or for some proteins, but not for others? The question of the conformation of the protein before or during translocation did not immediately emerge as a distinct question, but when techniques were developed that could probe, however indirectly, the conformations of proteins undergoing translocation across membranes, discussion of the relationship between protein folding and protein movement became a central theme in cell biological research.

Although translocation across membranes is a major aspect of protein traffic in cells, the relationship of folding to other important events has also come under consideration. Thus, when a protein is composed of more than one gene product, the expression of the ultimate functional protein depends not only on the generation of all the translation products, but also upon their assembly; if they are synthesized at different times, in different amounts, or in different locations, assembly will not necessarily be a simple process. Recent work has emphasized the importance of the integration of folding of individual polypeptides with the assembly of the complex.

Another aspect of protein traffic where a relationship has emerged between protein conformation and protein movement is the passage of proteins between compartments of the secretory pathway. Although progress through this pathway does not require movement across membranes (molecules move from one recognizable compartment to the next by a directed traffic of vesicles), it is clear that the traffic machinery is selective and that molecules are accepted or excluded from movement by criteria related to their state of folding and assembly. Major control is exerted at the point of exit from the ER, and there has been considerable progress in analyzing "quality control" at this point in the secretory pathway and how it reflects the state of protein folding and assembly.

Work in these fields has advanced rapidly since the mid-1980's. Progress has been reviewed frequently, many cellular factors implicated in these events have been identified, and hypotheses have been advanced as to their precise roles and mechanisms of action (Rothman and Kornberg, 1986; Zimmermann and Meyer, 1986; Randall et al., 1987; Eilers and Schatz, 1988; Meyer, 1988; Wickner, 1988,1989; Rothman, 1989; Bernstein et al., 1989; Ellis and Hemmingsen, 1989; Ellis, 1990; Wiech et al., 1990). In these analyses, specific proposals have been made, or implied, as to the effects of these factors on protein conformation; they have been proposed to be "foldases," "unfoldases," "antifoldases," or "chaperones." The ability to define these terms clearly and to substantiate them in particular cases has inevitably been limited. Analysts of protein refolding can apply very powerful physical techniques and can select protein substrates, their concentrations, and other folding conditions in order to gain increasingly precise information about the conformations of intermediates in protein refolding processes or about peptide analogues of such intermediates. (See Chapters 5 to 9.) Studies *in vivo*, or in systems where cellular events are reconstituted *in vitro*, provide far fewer opportunities for detailed conformational analysis. As a result, it is rarely possible to offer more than a

two-state characterization of proteins in such systems; the data are interpreted as indicating that the protein is either folded or unfolded, with little
room for subtle distinctions between manifestations of folding such as
compactness, secondary structure, overall tertiary fold, or the generation
of biologically functional sites. Nevertheless, the techniques available,
when supported by adequate controls, do provide considerable insight.

5.2 Methods for Analysis of Protein Folding in the Cell

While methods of analysis of protein conformation *in vivo* or in reconstituted studies have varied with the individual protein, the amount of
material available and the cellular process under study, a brief catalogue
can include most of the methods in widespread use. The methods are
generally dictated by the fact that the protein under study is synthesized in
the system in a radiolabeled form, and the quantity available is sufficient
only to detect the protein following specific extraction or precipitation, or
as a labeled band on a gel, and insufficient for direct physical characterization. Suitable methods are as follows:

5.2.1 Susceptibility to Proteolysis
Most native proteins in their fully folded state show some resistance to
proteases, while being very susceptible to proteolysis in the unfolded state.
If an appropriate protease and appropriate conditions of digestion can be
defined, then SDS-PAGE and autoradiography or fluorography after protease
treatment will define whether the polypeptide remains at its intact M_r
(hence, resistant to proteolysis, folded) or appears as small fragments (sensitive
to proteolysis, unfolded). For proteins in the presence of organelles or
membrane vesicle preparations, susceptibility to proteases in the absence or
presence of surfactants provides an assay for translocation, since segregation of
the protein behind a membrane barrier will confer resistance to proteolysis.

5.2.2 Use of Specific Antibodies
Some antibodies raised against linear peptide sequence epitopes do not
bind to that sequence in a folded protein, dependent upon its accessibility.
Antibodies raised against a folded protein include some that recognize
conformational epitopes that are dependent on the juxtaposition of side
chains in the folded protein. The former class of antibodies will be specific
for unfolded forms of a protein, while the latter will be specific for folded
forms. In a study of protein folding in the cell, the ability of a labeled protein
to be immunoprecipitated by such antibodies or to bind to such an antibody
on an immunoaffinity column can readily discriminate the overall conformational state of the protein.

5.2.3 Solubility and State of Association

A protein detected in a system in the form of large nonspecific aggregates will be presumed to have failed to fold on a reasonable time-scale to its native soluble conformation. Likewise for individual proteins, solubility and, hence, extractability in specific pH conditions, or with a specific cosolvent, may be used diagnostically.

5.2.4 Conformationally Sensitive Electrophoretic Methods

A protein with intact disulfide bonds has a smaller hydrodynamic volume than the same protein in a reduced state, even when both are unfolded. Thus, SDS-PAGE under nonreducing conditions can discriminate between reduced and disulfide-bonded forms of a protein, since the disulfide-bonded form shows characteristically greater mobility. Other forms of electrophoresis, including urea gradient gel electrophoresis, can also distinguish conformational states of a protein on the basis of their hydro-dynamic volume (Goldenberg and Creighton, 1984).

5.2.5 Biological Activity

Enzymic assay methods are rarely sufficiently sensitive to determine whether a protein synthesized in a cell-free system is folded to its native conformation. The specific affinity of the native state of a protein for a specific ligand may be exploited, however, to develop affinity absorption or precipitation methods to monitor the conformational state of a protein made in such a system.

5.2.6 Specific Assembly

If control work establishes that the association of a given protein with a specific partner in a complex depends upon the inital native folding of the first protein, then the folding of the protein may be determined by immuno-precipitation with antibodies directed against the partner.

5.2.7 Effects of Stabilization or Destabilization of the Folded State

The hypothesis that a protein has to be in a specific state (folded or unfolded) in order to undergo a subsequent step in protein traffic can be tested by modifying the protein in various ways and testing for an effect on the relevant step. Thus, if a process is blocked by the addition of specific ligands that bind to and stabilize the folded state, it is inferred that the process requires the unfolded state; similar conclusions can be drawn from experiments in which disulfide or other cross-links are introduced into the protein. If prior treatment of a protein with denaturants or the introduction of destabilizing mutations stimulate a process, it is also inferred that the process requires the unfolded state of the protein.

In Section 6, studies on the folding of proteins by all these methods are reviewed in relation to the cellular events of translocation, assembly, and secretion.

5.3 Protein Families Implicated in Protein Folding in the Cell

Many of the studies to be reviewed in Section 6 suggest not only that the timing of protein folding is important in relation to steps in intracellular protein traffic, but that the folding of a protein is modulated or facilitated by the action of other proteins. Protein folding factors have been implicated in three different ways. They may be found associated with proteins at various stages of biosynthesis, translocation, and/or assembly. They may be identified by a classical biochemical approach, namely that a cell extract is shown to have a folding modulating activity and is fractionated and resolved until a specific protein responsible for the activity is purified. Thirdly, they may be identified as the product of a gene that is recognized as having a role in modulating folding on the basis of the phenotype of cells in which this gene is deleted, disrupted, or mutated.

A key result since the mid-1980's is that in widely disparate systems, where different aspects of protein traffic have been studied by varied approaches, the proteins implicated as modulators of folding have been observed to fall into a relatively small number of identifiable families, suggesting a strong conservation of important and ubiquitous functions. In eukaryotic cells, two major classes of proteins have been implicated, namely members of the heat-shock protein 70 *(hsp70)* family that are found in many cellular compartments, and homologues of the bacterial protein groEL, which are termed chaperonins, *cpn60's,* or *hsp60's*, and are found within mitochondria and chloroplasts (Fig. 10–6). We will use here the abbreviations *hsp70* and *cpn60*. The actions of these and other proteins implicated as modulators of folding are considered in Section 7.

5.3.1 The hsp70 Family

The *hsp70* family comprises a highly conserved and ubiquitous set of genes and proteins originally defined by the fact that some members of the family are among the most abundant proteins synthesized when organisms of many kinds (e.g., *Drosophila*, yeast) are exposed to 'heat-shock,' i.e., temperatures significantly above their normal growth temperatures (Pelham, 1986; Lindquist and Craig, 1988; Gething and Sambrook, 1990; Craig and Gross, 1991; Schlesinger, 1991). The proteins are of molecular weight 70 to 80 kDa and are hence described as heat shock protein 70's *(hsp70's)*. In bacteria only one member of the family is known, the *dnaK* protein, while in higher organisms several members of the family co-exist. Not all members

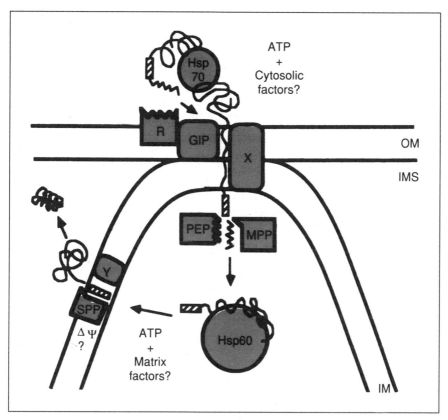

FIGURE 10–6. *Protein folding factors in mitochondrial import. The scheme emphasizes the role in mitochondrial biogenesis of members of the major protein families implicated in protein folding in the cell; it outlines the pathway of import and sorting of cytochrome* b_2 *to the intermembrane space (IMS) after passage across outer (OM) and inner (IM) mitochondrial membranes. The conformation of the protein throughout the process is modulated by members of the* hsp70 *and* cpn60 *families. (See also Section 6.1 and 6.5.) R, GIP, and X represent, respectively, an import receptor, insertion protein, and putative translocation protein. MPP is the mitochondrial processing peptidase, PEP is associated processing-enhancing protein, SPP is the inner membrane signal peptidase, and Y is a putative translocation protein. (From Hartl and Neupert, 1990; ©1990 by the AAAS)*

of the family are, in fact, heat-shock inducible, some being constitutively synthesized, while others are induced by stresses other than heat. In yeast, where the genetics are well characterized, there are several members of the family expressed in the cytosol and also present in the nucleus, one expressed in the mitochondrion and one in the ER; viability is dependent on at least one *hsp70* gene being expressed in each of the cytosol, mitochondria, and ER. In most higher eukaryotic cells there is good evidence for at least one major constitutive and one heat-inducible *hsp70* protein in the cytoplasm, in addition to those in mitochondria and the ER.

The proteins exist as free dimers, although they have a tendency to form higher aggregates. They bind ATP and demonstrate weak ATPase activity that is thought to play an important role in their action. Both sequencing data and sensitivity to proteolysis indicate a domain structure in native *hsp70's*, with a protease-sensitive region separating a highly conserved N-terminal region of approximately 44 kDa from a less well conserved and smaller C-terminal region (Chappell et al., 1987). The N-terminal region has ATPase activity, and the structure of this domain of a representative *hsp70* has been determined by X-ray diffraction (Flaherty et al., 1990). Remarkably, the structure of this domain is homologous to that of actin and also has some similarity in topology to hexokinase (Kabsch et al., 1990; Flaherty et al., 1991). The structure of the more variable C-terminal domain is not established, but a study based on sequence alignment, secondary structure prediction, and molecular modeling has suggested that it may be structurally homologous to the peptide-binding domain of HLA class I molecules (Rippmann et al., 1991).

Before the accumulation of data (see Section 6) implicating *hsp70's* in protein folding and assembly processes, their functions were guessed from observations on the behavior of *hsp70's* in heat-stressed cells. In such cells, cytosolic *hsp70's* migrate to the nucleus, specifically to the nucleolus, and bind to protein aggregates that are rich in the components of partially assembled ribosomes (Pelham, 1984). Association of *hsp70's* with these aggregates is stable to high salt, but reversed by ATP. A further influential discovery was the observation that the protein responsible for the ATP-dependent removal of clathrin from clathrin-coated vesicles (clathrin-uncoating ATPase) is a constitutive *hsp70* referred to as heat-shock cognate protein or h*sc70* (Schlossmann et al., 1984; Ungewickell, 1985). It has been suggested that the constitutive cytoplasmic h*sp70's* interact with most, if not all, nascent proteins in a transient, ATP-sensitive interaction (Beckmann et al., 1990). The role of such interactions in mediating translocation into organelles will be considered in Sections 6.1 and 6.3. The bacterial *hsp70*, *dnaK*, was originally defined as a protein essential for the assembly of λ phage, but it also has other functions, not yet fully defined, in the uninfected cell.

5.3.2 The cpn60 Family

The groEL protein of *E.coli* is the prototype of this family, but homologues from mitochondria and chloroplasts show closely related sequences with more than 45% identity (Georgopoulos and Ang, 1990; Hallberg, 1990;

Hemmingsen,1990; Ellis and van der Vies, 1991). The *groE* system was originally defined genetically; the gene was required for head and/or tail assembly in the replication of various bacteriophage. In fact, the system involves two genes in a single operon, producing two proteins: groEL and groES (to be discussed shortly). GroEL and its homologues have a polypeptide Mw of approximately 60 kDa and occur as an unusual oligomer, a tetra-decamer comprising two stacked rings of seven subunits each. The chloroplast homologue, originally identified as a protein involved in the biosynthesis of the enzyme ribulose bisphosphate carboxylase (Rubisco) is similar. There appear, however, to be two homologous *cpn60* proteins in chloroplasts, and the oligomers appear to contain equal numbers of both types of chain; the functional significance of this is not understood. The *cpn60's* bind ATP and have weak ATPase activity; their quaternary structure appears to be ATP sensitive in that high levels of Mg^{2+} and ATP (MgATP) cause dissoci-ation to dimers. Both groEL and its mitochondrial homologue are heat-inducible and are found associated with ribosomes. Currently *cpn60's* are known and have been characterized from bacteria and from those eukaryotic organelles believed to derive from prokaryotic endosymbionts. It remains controversial whether there exist cytosolic members of the family in eu-karyotes. Interestingly, the genes for the organellar *cpn60's* are located in the nuclear genome.

The groES protein binds directly to the groEL oligomer in the presence of MgATP, but not in its absence, and inhibits the ATPase activity of groEL. The groES protein has a subunit Mw of 10 kDa and exists as a heptamer. Mitochondria contain a protein that is functionally homologous to groES (Lubben et al., 1990), and it is assumed that chloroplasts do also, although in neither case is the protein well defined; the putative family of related proteins is referred to as *cpn10's*.

Bacterial *cpn60's* appear to be major targets of the immune response to bacterial infection (Young, 1990). This response may be particularly important in leprosy and tuberculosis, and the close similarity between bacterial *cpn60's* and their eukaryotic homologues suggests that the response to chaperonins may be important in autoimmunity and the development of autoimmune disease.

6 PROTEIN FOLDING AND PROTEIN TRAFFIC IN THE CELL: SPECIFIC SYSTEMS

6.1 Protein Import into Mitochondria

Characterization of the processes and components involved in the import of precursors of mitochondrial proteins from the cytoplasm is now very well advanced; the combination of *in vivo*, genetic, and biochemical studies in systems reconstituted *in vitro* has provided a detailed stepwise picture of the overall process and of the detail of targeting of proteins to the matrix, the inner membrane, and the intermembrane space (Attardi and Schatz, 1988; Hartl et al., 1989; Pfanner and Neupert, 1990; Hartl and Neupert, 1990; Glick and Schatz, 1991).

Initial recognition of proteins to be imported is mediated by an N-terminal import sequence that interacts with specific import machinery and is cleaved, after import is initiated, by a metalloproteinase located within the mitochondrial matrix. Considerable information has been obtained through the use of fusion proteins in which N-terminal import sequences of authentic mitochondrial proteins are fused to "passenger" proteins that would not otherwise be targeted to the mitochondrion. Manipulation of the passenger protein can then throw light on the requirements for a passenger to be translocated efficiently. There are clear indications that stably folded passengers that are not able to unfold cannot be translocated in *in vitro* posttranslational import assays.

Eilers and Schatz (1986) constructed a fusion between a mitochondrial import signal and mouse dihydrofolate reductase (DHFR). The fusion protein was expressed in *E.coli* and purified; it was enzymatically active, suggesting that the N-terminal import signal did not perturb its native conformation and that the signal and passenger formed independent domains. The purified fusion (or the fusion generated in an *in vitro* translation system) was imported posttranslationally into mitochondria with reasonable efficiency, and the import signal was cleaved normally. Addition of methotrexate, a specific inhibitor of DHFR with a dissociation constant for the wild-type and fusion proteins of less than 10 nM, blocked import with the same apparent affinity. The conformation and dynamics of the fusion protein in the absence and presence of the inhibitor were not determined directly, but it was inferred that inhibitor binding stabilized the folded conformation, because it conferred resistance to thermolysin. This conclusion is strengthened by the finding that when the isolated precursor was unfolded in 8-M urea, and then added directly to the import assay system, its import was more rapid and more efficient, occurred even at low temperatures, and was not blocked by methotrexate (Eilers et al., 1988). Dilution of the unfolded precursor into import buffer, before addition of mitochondria to initiate import, led to a time-dependent loss of import competence in the presence of methotrexate with $t_{1/2}$ of less than 1 min, and the loss of import competence correlated with the appearance of resistance to proteolysis of the DHFR domain.

Comparable evidence has been obtained in other systems; a yeast Cu^{2+}-binding metallothionein can be imported into mitochondria as a passenger in a fusion protein, but this is blocked by addition of Cu^{2+} (Chen and Douglas, 1987), and passenger proteins destabilized by mutation or truncation are more efficiently imported than are wild-type constructs (Ness and Weiss, 1987; Vestweber and Schatz, 1988a). Conversely, a small stable disulfide-bonded passenger protein, such as BPTI, could not be imported, and it blocked the translocation machinery (Vestweber and Schatz, 1988b).

An alternative line of evidence that indicates that ability to unfold is required for a protein to be imported into mitochondria was the earlier finding of Schleyer and Neupert (1985), with authentic mitochondrial precursors, that it was possible at low temperatures, to trap intermediates that extended across both the inner and outer mitochondrial membranes; the N-terminus of each of these proteins had penetrated into the matrix and had been processed by the metalloproteinase located there, while the C-terminus was still exposed at the mitochondrial surface and accessible to exogenous proteases. Given the dimensions of the membranes and of the folded forms of the proteins undergoing transport, substantial unfolding of parts of these proteins would be necessary to span the two membranes (Hartl et al., 1986; Pfanner et al., 1987; Schwaiger et al., 1987).

In subsequent work, the unfolding step has been resolved from the translocation process itself, and their separate requirements have been identified. It is clear that steps are present that are dependent upon both ATP and membrane potential (see Hartl et al., 1989, Section 7 for discussion) and attention is now concentrated on the cytosolic and membrane-associated proteins mediating these steps. For several years it has been clear that cytosolic preparations can stimulate the posttranslational *in vitro* import of mitochondrial precursor proteins. Biochemical and genetic approaches indicate that cytosolic members of the *hsp70* family are important in this action of cytosolic preparations and that the effect of *hsp70's* is ATP-dependent. (See Sheffield et al., 1990, and papers cited therein.) The genetic evidence derives from studies on yeast mutants in which three members of the cytosolic *hsp70* family were disrupted; these mutants are essentially nonviable, but viability is restored by transformation with a plasmid that conditionally expresses one of the disrupted genes, under galactose-inducible control (Deshaies et al., 1988). Cells grown on galactose, where the gene is expressed, import and process mitochondrial precursors normally, but when the cells are transferred to a glucose medium the gene is repressed, and the cells now accumulate unprocessed forms of mitochondrial precursors (and unprocessed precursors of secretory proteins, see Section 6.4). A range of controls demonstrate that the defect in the glucose-grown cells is a general block on translocation of precursors across both the mitochondrial and the ER membranes. In biochemical studies, *hsp70* plus ATP was found to be necessary, but not sufficient, to reproduce the effect of whole cytosol in stimulating mitochondrial import (Murakami et al., 1988; Sheffield et al., 1990). The data suggest that a higher-Mw factor that is NEM-sensitive is also required. Possible mechanisms accounting for the role of *hsp70* and its homologues are considered in Section 7; that import of denatured precursors does not show a dependence on *hsp70* and ATP suggests that the latter are involved in unfolding the precursor.

If proteins for import into mitochondria are unfolded or prevented from folding in the cytoplasm, and cross the mitochondrial membranes in an unfolded condition, then they must subsequently fold in the matrix prior to their incorporation into biologically active complexes within the matrix or inner membrane. This step is considered in Section 6.5.

6.2 *Protein Export from* E.coli

Translocation across the cytoplasmic membrane of bacterial proteins destined for the outer membrane or the periplasmic space was initially a difficult area of study, since cell-free systems were not readily developed. (See Lee and Beckwith, 1986.) In the last few years, however, there has been considerable progress in the definition of the translocation-competent state of bacterial proteins destined for export, and in the identification of cytoplasmic factors that facilitate protein export.

Evidence that translocation across the bacterial cytoplasmic membrane required a translocation-competent protein conformation that was unfolded or incompletely folded was presented simultaneously with the demonstration of this requirement for mitochondrial import. The key data (Randall and Hardy, 1986) were derived from a study in which export to the periplasm of the maltose binding protein (MBP) precursor was blocked by addition of an uncoupler, and the conformation of intracellular, unprocessed precursor was monitored by its sensitivity to proteolysis. Initially all the precursor was protease-sensitive, but protease sensitivity was slowly lost as the wild-type precursor folded; subsequent work established that this folding was to a state closely resembling the native state of the mature MBP (Park et al., 1988). A mutant MBP precursor with a Pro→Ala mutation at position -14 in the leader sequence was exported inefficiently, and study of this mutant precursor in similar experiments showed that it folds much more rapidly to a protease-resistant conformation. In this case protease-sensitivity and translocation-competence appear to be directly correlated, and it was proposed that the loss of these properties with time reflected premature folding of the precursor to a stable conformation within the cytoplasm.

The finding that a translocation-defective precursor with a mutation in the leader sequence folded to a translocation-incompetent state more rapidly than did wild-type precursor implies that, in addition to mediating translocation itself, one function of the leader sequence may be to delay folding of the precursor to allow the newly synthesized precursor to dock with export machinery, while yet in a competent conformation. This proposal was supported by direct studies on the kinetics of folding of mature and precursor forms of MBP (using fluorescence spectroscopy) and ribose-binding protein (RBP) (Park et al.,1988) and by showing that a mutation in the mature MBP that restores export efficiency to a mutant with an export-inefficient leader sequence also has the effect of reducing the rate of folding of the precursor (Liu et al., 1988).

Genetic studies on secretion from *E.coli* implicated the products of three genes as being of major importance for secretion; SecB and SecA are now known to be cytoplasmic proteins, while Sec Y is a membrane protein (Kumamoto and Beckwith, 1983). Further analysis of the folding and export of MBP has strongly implied that this process is modulated by SecB *in vivo*. A functional *SecB* gene is required for export of MBP, and the nature of this requirement has now been analyzed, both *in vitro* and *in vivo*. Weiss et al. (1988) showed that preMBP synthesized in an *in vitro* system could be translocated into inverted cytoplasmic membrane vesicles in the presence of cytoplasmic extracts, but only when the cytoplasm was derived from $SecB^+$ cells; addition of purified SecB to the cytoplasmic extract of $SecB^-$ cells made this extract capable of stimulating translocation *in vitro*. Furthermore, addition of purified SecB protein, or of cytosol from proteins overexpressing this protein, retarded the folding of newly synthesized preMBP to a protease-resistant conformation, and this retardation of folding was correlated with a prolongation of translocation-competence. These results suggest that both the presence of an appropriate leader peptide and interaction with SecB function to slow precursor folding and to deliver precursor to the export machinery in a translocation-competent conformation. In fact, SecB will bind in vitro to a form of MBP without a leader sequence, provided that it carries a mutation that retards folding (Liu et al., 1989). This implies both that SecB does not interact directly with the leader sequence and that one of the roles of the wild-type leader sequence is to retard folding in order to ensure interaction between SecB and features of the mature protein that are lost or buried on folding (Randall et al., 1990). The components of this system are now available in purified form, and it is one of the cases where the structural basis for modulation of protein folding can be studied in most detail. (See Section 7.)

A surprising aspect of the bacterial export system is that a variety of soluble proteins seem to play comparable roles, displaying distinct but overlapping specificities toward exported proteins. Thus, while SecB clearly plays a role in MBP export, genetic evidence implies that it has no role in export of RBP, and no effect of SecB protein on RBP export was noted in any of the studies just described. There is similarly clear evidence that SecB is not involved in the *in vivo* transport of pre-β-lactamase and, correspondingly, that there is no effect of SecB on the folding *in vitro* of isolated pre-β-lactamase (Laminet et al., 1991). On the other hand, different approaches have implicated other cellular factors in bacterial export. Bochkareva et al. (1988) used cross-linking to identify cytosolic components that interact with newly translated presecretory proteins and found that pre-β-lactamase binds (in an apparently unfolded state) to the protein groEL, which forms a rapidly sedimenting (20S) tetradecameric complex in *E.coli.* This protein is the prototype of the *cpn60* family and appears to have wide-ranging roles *in vivo*, since overexpression of the groE operon (which codes for both groEL and groES) suppresses a wide range of mutant phenotypes, including heat-sensitive and other missense mutations. Analysis of the diversity of the effects and of the target proteins (Van Dyk et al., 1989) led to the conclusion that the single common process in which the groE gene products could be involved that would allow them to interact with so many diverse proteins was in their folding and/or assembly. In the case of pre-β-lactamase, there is now clear evidence that the groE proteins modulate its folding, and this system is proving useful for mechanistic studies (Laminet et al., 1990).

Direct attempts to purify, from bacterial cytosolic preparations, a factor capable of maintaining translocation-competence in precursor proteins led to the identification of yet another cytosolic protein, termed "trigger factor," which is distinct from SecB and groEL (Crooke et al., 1988; for reviews see Wickner, 1989, and Wiech et al., 1990). The substrate protein used in the isolation of trigger factor was the precursor for the bacterial outer membrane protein, OmpA. Lecker et al. (1989) reviewed the evidence that the modes of action of SecB, groEL, and trigger factor were comparable (although the studies had been made with different precursor proteins as translocation substrates) and showed that all three could form soluble complexes with proOmpA and could stabilize translocation-competent forms of proOmpA. Only SecB and groEL showed such interactions with another outer-membrane protein, prePhoE, and none of the three interacted with mature, globular soluble proteins. The basis of the specificity of interaction here remains to be elucidated thoroughly.

Overall the requirements for export of bacterial proteins can be summarized as being a leader sequence, ATP, a membrane potential, functional *SecA* and *SecY* gene products, plus, in most cases, one of SecB, groEL, or trigger factor. The popular interpretation of the data presented here is that the leader sequence retards folding of the precursor, which permits its interaction with one of the three proteins, which retains the protein in a translocation-competent form that is distinct from the fully folded form. Hardy and Randall (1991) have presented evidence suggesting that there is direct kinetic competition between folding to the native state and interaction with SecB.

6.3 Protein Translocation Across the ER Membrane

Translocation of nascent secretory proteins across the ER membrane has perhaps been studied more intensively than any other protein translocation process, and it was certainly the first system in which certain key questions could be posed experimentally. The demonstrations that secretory proteins were synthesized on ribosomes firmly associated with the ER membrane (Palade, 1975) and that nascent proteins made on isolated rough microsomal vesicles could be discharged into the inner luminal compartment when synthesis was terminated by puromycin (Redman and Sabatini, 1966) suggested that synthesis of secreted protein occurred at the membrane and that nascent proteins had direct access to the luminal space. The subsequent establishment by Blobel and Dobberstein (1975) of an *in vitro* translation system in which secretory proteins could be co-translocated opened the way to the full characterization of this system and stimulated developments in the study of mitochondrial import and bacterial export reviewed in the previous sections.

Work in whole cells and on the classical *in vitro* translocation systems (wheat germ extract or rabbit reticulocyte lysate, plus target rough ER membrane vesicles derived from dog pancreas) indicates that cleavage of the N-terminal signal sequence and core N-glycosylation can both occur on ribosome-associated nascent polypeptides, and the known topology of signal cleavage and glycosylation indicate that the nascent protein must extend across the membrane. Thus, it was initially envisaged that the nascent polypeptide chain emerging from the ribosome passed directly into some translocation apparatus in the membrane, and questions about the conformation of the chain were not meaningful.

It became apparent, however, that the initial targeting to the ER membrane of ribosomes that translate secretory or cell-surface proteins is not directly effected by the emerging signal sequence alone, but by a large cytoplasmic particle, the signal recognition particle, SRP (Walter et al., 1984; Walter and Lingappa, 1986). This particle, comprising six distinct polypeptides and a 300-nucleotide RNA molecule, acts catalytically to transfer a ribosome with an emerging secretory signal sequence to the ER membrane and to dock it with integral membrane components. The SRP interacts directly with a "docking protein" or SRP-receptor, but this is a transient interaction; continuing engagement with the membrane of the translating, translocating ribosome is effected by stoichiometric interaction with a signal sequence receptor and a ribosome receptor, both integral membrane proteins (Rapoport, 1990). In the majority of studies, this whole system has been shown to function in translocation only co-translationally, and it is generally assumed that co-translational translocation is the norm *in vivo*. But a number of experimental systems have been established in which posttranslational translocation across the ER is observed, and these demonstrate both that translocation across the ER has much in common with translocation across other membranes and that the conformation of the substrate protein is crucial in this case also.

The situations in which posttranslational translocation have been demonstrated include (1) small higher eukaryotic peptides (reviewed by Zimmermann et al., 1990); (2) proteins that remain associated with the ribosome after completion of elongation (through deletion of the appropriate termination signals) (Perara et al., 1986); (3) completed and discharged proteins in strongly reducing conditions (Maher and Singer, 1986); and (4) some yeast proteins in a homologous translocation system with yeast microsomal membranes (Rothblatt and Meyer, 1986). The interpretation of these findings appears to be that there are alternative modes of translocation across the ER, one of which is dependent on the whole machinery of ribosomes, SRP, docking protein, and so on, and one of which is independent of these factors (Zimmermann et al., 1990; Schlenstedt et al., 1990). The SRP-dependent pathway ensures that the substrate protein is in an unfolded conformation by effecting interaction between the signal sequence and SRP at an early stage; the nascent chain cannot fold because the N-terminus and the growing C-terminus are attached to SRP and to the ribosome, respectively. *In vitro* unfolding and refolding studies on complete precursor proteins, analogous to those carried out on mitochondrial precursors, have shown that SRP can interact with unfolded secretory precursors to retard folding and to maintain a translocation-competent conformation (Sanz and Meyer, 1988).

By contrast to the SRP-dependent pathway, the SRP-independent pathway requires ATP and specifically ATP, not nonhydrolyzable analogues. This requirement is comparable to that for import into mitochondria, and the cytosolic components mediating this requirement appear to be the same. Thus, genetic deletion of functional *hsp70's* in the yeast cytosol causes an accumulation not only of mitochondrial precursor proteins, but of secretory precursors also (Deshaies et al., 1988). Parallel biochemical studies have demonstrated that *hsp70's* are required for posttranslational translocation of precursors across yeast microsomal membranes (Chirico et al., 1988).

Thus, despite the unique nature of the SRP-dependent translocation system, the requirement for efficient translocation across the ER membrane appears to be similar to that for translocation across bacterial and mitochondrial membranes; the targeting systems may be distinct (although bacterial and ER signal sequences are frequently interchangeable), but there is an underlying requirement that the passenger protein is not folded and this requirement is met by interactions with cytoplasmic components, which ensure that folding is prevented.

Some results have suggested that subsequent interactions with factors within the ER lumen may be necessary for continuous translocation; this possibility is considered in Section 6.6.

6.4 What Is the Translocation-Competent State?

The preceding discussion has introduced the concept of "translocation-competence," which has been widely used without being fully defined in physicochemical terms. The data indicate that polypeptides are translocation-competent when newly synthesized or when newly diluted from denaturing conditions, that polypeptides stabilized by disulfide bonds or by interactions with tight-binding ligands are not translocation-competent, that translocation-competence can be retained for long periods through interaction with some cytosolic factors but is otherwise lost with $t_{1/2}$ of the order of 1 min, and that translocation-competence in some cases is dependent upon cellular factors plus ATP.

Few studies have defined the conformation and dynamics of a translocation-competent protein. In the light of our general knowledge of protein folding, it is reasonable to assume that translocation-competence is not equivalent to a fully unfolded state, because proteins on dilution from denaturant rapidly collapse to a compact globular conformation with some defined elements of secondary structure. (See Chapter 6 and Kuwajima, 1989). Thus, Bychkova et al. (1988) have proposed that the translocation-competent state is equivalent to the molten globule state, a state energetically close to the unfolded state, highly dynamic, and yet compact. Such direct evidence as exists is consistent with this interpretation. Thus, Lecker et al. (1990) studied the conformation of pro-OmpA immediately after dilution from denaturant and found that its far-UV CD spectrum indicated a content of secondary structure elements equivalent to that of the native state. On the other hand, the extensive studies by Hardy, Randall, and colleagues (Section 6.2) on the correlation between translocation *in vivo* and folding *in vitro* of MBP precursors have shown that loss of translocation-competence correlates with "folding" as monitored by the fluorescence of tryptophan residues. This folding process monitored by tryptophan side chains was well defined in kinetic and thermodynamic terms and appeared to represent a reversible two-state transition. A protein that is translocation-competent and unfolded by these criteria, and yet has some secondary structure and a relaxation time to the folded state of about 1 min, is likely to be in the molten globule state.

6.5 *Protein Assembly Within Chloroplasts and Mitochondria*

The energy-transducing and metabolic functions of mitochondria and chloroplasts depend on numerous large protein complexes, both soluble and membrane-bound. The assembly of several of these multicomponent species is further complicated by the fact that both mitochondria and chloroplasts encode a small number of polypeptides that are synthesized within the organelle, whereas others are synthesized in the nucleus. So generation of functional complexes can involve import processes (Section 6.1) and assembly steps in which components with different subcellular origins interact. Study of the assembly of Rubisco has been particularly significant (Ellis and Hemmingsen, 1989; Ellis, 1990; Gatenby and Ellis, 1990).

Rubisco is the major protein component of chloroplasts and is possibly the most abundant protein in the world; its function is the key step of fixation of CO_2 in the Calvin cycle of photosynthesis. Chloroplast Rubisco has an Mw of over 500 kDa and comprises eight large subunits and eight small subunits. The Rubisco from some photosynthetic prokaryotes is similar, while other prokaryotes have a simpler dimeric enzyme whose subunits resemble the large subunits of the chloroplast enzyme. In most plants, the large (L) subunits of Rubisco are encoded and synthesized within the chloroplast while the small (S) subunits are nuclear-encoded and synthesized in the cytoplasm. In studies on the biosynthesis of L subunit by isolated chloroplasts, it was found that the newly synthesized subunits were associated with a distinct abundant oligomeric protein within chloroplasts before being incorporated into Rubisco oligomers. It subsequently was shown that the binding protein also bound newly synthesized S subunits after their import. Rubisco from the cyanobacterium *Anacystis nidulans* has a similar composition and quaternary structure to that from higher plants. The cyanobacterial enzyme can be dissociated and reassociated from its subunits *in vitro* and is generated in active assembled form when expressed in *E.coli* (Andrews and Lorimer, 1987; Bradley et al., 1986), whereas higher plant Rubisco does not assemble correctly in either of these situations, i.e., in the absence of its subunit binding protein.

The Rubisco subunit binding protein was characterized (Musgrove et al., 1987), and its role in the assembly of Rubisco confirmed by the finding that antibodies toward the binding protein blocked incorporation of newly synthesized Rubisco L subunits into Rubisco holoenzyme (Roy, 1989). Ellis (1987) proposed that the subunit binding protein was essential for correct assembly of functional Rubisco oligomers and proposed that proteins with such functions be termed "Molecular chaperones"; the Rubisco subunit binding protein is now know as *cpn60*. Subsequent cloning and sequencing of Rubisco binding proteins from higher plants demonstrated that their polypeptide sequences are strongly homologous (46% identical over 550 residues) with those of the protein groEL from bacteria (Hemmingsen et al., 1988). GroEl and groES had been discovered as gene products essential for assembly of some bacteriophages, and these systems had been long quoted as the exceptions to the rule that complex oligomeric proteins can self-assemble. It was, therefore, striking that proteins with comparable functions of facilitating assembly in chloroplasts and bacteria should be homologous in sequence.

The importance of Rubisco in the global carbon cycle, as the major enzyme of photosynthetic CO_2 fixation, has led to many attemps to engineer it, which have focused on expression of Rubisco in *E.coli* or other convenient organisms. The fact that cyanobacterial Rubisco is readily assembled when expressed in *E.coli* prompted the question whether groEL is implicated in this process. Goloubinoff et al. (1989a) demonstrated that in *E.coli* expressing cyanobacterial Rubisco the yield of fully assembled and active Rubisco could be increased 10-fold by the simultaneous overexpression of the groE genes; the key step appeared to be the assembly of the core octamer of L subunits. Conversely, mutation or deletion of the groEL and groES genes blocked assembly of the cyanobacterial Rubisco. Hence, it appears that the failure of chloroplast Rubisco to assemble correctly when expressed in *E.coli* arises from a difference in specificity, whereby the bacterial chaperonin functions toward cyanobacterial Rubisco but not toward the higher plant homologue.

The role of the chloroplast *cpn60* has not been widely explored apart from its function in Rubisco assembly, but it appears that it interacts with a wide range of polypeptides imported into chloroplasts, but not with all of them (Lubben et al., 1990; Hemmingsen, 1990; Ellis and van de Vies, 1991). In contrast to chloroplasts, there has been extensive study of the import system in mitochondria (see Section 6.1), indicating that proteins are translocated in an unfolded conformation. So in both organelles there is a requirement for folding and assembly of imported polypeptides. In the case of mitochondria, current evidence suggests that this folding process in the matrix is mediated by matrix-located members of both the *hsp70* family and the *cpn60* family. The role of *cpn60* has been inferred from several lines of evidence (reviewed in Ostermann et al., 1989; Hartl and Neupert, 1990). *Cpn60* is a matrix-located protein in mitochondria from a wide range of sources and is constitutively expressed. It is encoded by a nuclear gene, and current data suggest that preexisting functional *cpn60* complexes are required in the mitochondrial matrix for effective import and assembly of newly synthesized *cpn60* subunits (Cheng et al., 1990). Thus, spontaneous self-assembly is not observed *in vivo*, although the homologous groEL protein is capable of self-assembly *in vitro* (Lissin et al., 1990). Mutants defective in *cpn60* function can import and process mitochondrial precursors, but do not incorporate them into functional complexes. Ostermann et al. (1989)

used a model system with DHFR as a passenger protein to study refolding of processed protein after import; folding is blocked at low temperatures and after ATP-depletion or after NEM-treatment. In such cases, the unfolded protein accumulates in a protease-sensitive state (that is not recognized by antibodies specific for native DHFR) as aggregates or complexes, the major component of which is *cpn60*.

Hartl and Neupert (1990) propose that interaction with *cpn60* is the crucial stage in folding and assembly of proteins within the matrix and that initial binding of precursors to *cpn60* could occur co-translocationally and facilitate import. The role of ATP, however, and the specific requirements for folding and release from *cpn60* remain uncertain. Glick and Schatz (1991) propose that the initial interaction of newly imported precursors is with a matrix-located member of the *hsp70* family, the product of the *SSCI* gene, because a temperature-sensitive mutation in this gene causes partly imported precursors to be blocked in the import machinery, and authentic precursors in transit can be cross-linked to, or co-immunoprecipitated with, the matrix *hsp70* (Kang et al., 1990; Scherer et al, 1990). As of early 1992, the details of the requirement for mitochondrial *cpn60* and *hsp70* for import, folding, and assembly remain unclear, and the sequence in which they act on imported precursors is controversial. Both of these proteins are clearly implicated, however, in the complex process by which an unfolded precursor matures to a folded and assembly-competent form in the mitochondrial matrix.

6.6 *Protein Assembly and Quality Control in the Endoplasmic Reticulum*

The lumen of the ER is the primary site of folding for proteins destined for secretion, the cell surface, or lysosomes, and for proteins that are to be permanent residents of the various compartments of the secretory pathway. Study of protein folding in this pathway has been intimately involved with a key feature of the secretory system, namely its selectivity. The half-time taken for a newly synthesized protein to be secreted or to reach the cell-surface varies widely from protein to protein, and the major part of this variation is in the time taken for proteins to exit from the ER. (See e.g., Lodish et al., 1983.) So there is clearly some selectivity in the process by which proteins pass from the ER to the *cis* Golgi compartment and, hence, on through the secretory pathway. There was a long debate on whether selection for secretion was mainly driven by positive signals or whether secretion was the default outcome, with signals being required for retention or deflection from the secretory pathway; this has now clearly been resolved in favor of secretion as the default outcome. (See Pfeffer and Rothman, 1987; Rothman, 1987; Rose and Doms, 1988.) Proteins for branching destinations (such as lysosomes) and for retention within the pathway (e.g., for retention within the lumen of the ER itself) require positive signals that are now being defined. (See Pelham, 1989.) But resolution of this controversy does not immediately clarify the variation in residence time within the ER of proteins that will eventually leave it and pass down the pathway. Understanding here has come from studies on oligomeric proteins that are assembled within the ER, both soluble secreted proteins, such immunoglobulins, and membrane proteins, such as the proteins of enveloped viruses, especially VSV G protein and the influenza virus HA. These proteins are inserted into the ER membrane at biosynthesis and pass along the secretory pathway as transmembrane proteins oriented with their exocellular domains within the lumen of the secretory compartments.

Study of the biosynthesis of immunoglobulins produced two early results that have been of continuing importance in this context. In a pioneering study of the timing of disulfide bond formation *in cellulo*, Bergman and Kuehl (1979a,b) established, first, that the disulfide bond in the N-terminal (variable) domain of immunoglobulin light chains formed co-translationally, essentially as soon as the nascent chain was long enough for the whole N-terminal domain to pass into the ER lumen, and, second, that the assembly of immunoglobulins could begin co-translationally through intermolecular disulfide formation between nascent heavy chains and completed light chains within the ER. This established the ER lumen as the site of the major protein folding and assembly processes, and demonstrated that folding could be co-translational in the case of multi-domain proteins, where a whole domain can be available within the ER lumen while the remainder of the protein is still being synthesized.

The other key early discovery was that, in cells that synthesized immunoglobulin heavy chains but not the light chains for them to assemble with, heavy chains were retained within the ER, rather than being secreted, and were found in association with an intracellular protein, known as heavy-chain binding protein or BiP (Morrison and Scharff, 1975; Haas and Wabl, 1983). It was subsequently shown that BiP also interacts with Ig heavy chains in cells synthesizing both light and heavy chains and secreting intact immunoglobulins. In this case immunoprecipitation with antibodies against either BiP or Ig heavy chains showed that BiP is associated with assembly intermediates (Ig heavy chains, heavy chain dimers, and so on) but not with intact immunoglobulins (Bole et al., 1986). At the same time, it was shown that inhibition of N-glycosylation decreased assembly of heavy and light chains and decreased Ig secretion, but prolonged the interaction between heavy chains and BiP. Interestingly, Ig heavy chain mutants lacking the C_H1 domain do not interact with BiP, and these mutant heavy chains are secreted unassembled when expressed in cells that do not synthesize Ig light chains (Hendershot et al., 1987). These findings implied that BiP might function either to facilitate immunoglobulin assembly, or to retain unassembled components to ensure assembly. They also suggested that at least part of the variation in time for secretory products to leave the ER might lie in their variable interaction with such factors within the ER lumen.

Studies on biosynthesis and transport of membrane proteins have exploited a wide range of tools to analyze folding and oligomerization within the secretory pathway. The methods include chemical cross-linking, sedimentation, or other hydrodynamic methods, loss or acquisition of specific antigenic epitopes, immunoprecipitation, acquisition of ligand binding functions, and sensitivity to proteolysis (Hurtley and Helenius, 1989). Using a combination of these techniques, two groups characterized the assembly and transport of the trimeric influenza virus surface protein HA and demonstrated, in particular, that mutants in the protein that fail to move to the cell surface fail to fold and trimerize in the ER (Gething et al., 1986; Copeland et al., 1986). They demonstrated with the wild-type protein that trimers are folded and appear with a half-time of less than 10 min, and that trimers are formed before the protein leaves the ER. Furthermore, they showed that a small minority of wild-type molecules fail to fold and assemble, and remain for a long period within the ER in this state. A panel of mutants, with modifications distributed throughout the HA molecule, included many that did not move to the cell surface; these mutants, by a number of criteria, fail to trimerize and/or fail to fold, and they do not move from the ER. In subsequent more detailed work, Copeland et al., (1988) showed directly that folding, trimerization, and transport were sequential events in HA expression; some unassembled monomers are folded by the criterion that they contain native disulfide bonds and display epitopes characteristic of the native conformation. Two alternative disulfide-bonded intermediates can be distinguished that chase through to the native disulfide-bonded monomer with the same kinetics (Braakman et al., 1992a,b). By varying the steady-state level of translated chains within the ER lumen, it was shown that the rate of monomer folding was not influenced by the expression level, but that trimerization was accelerated at higher expression levels, as expected for a higher-order reaction (Braakman et al., 1992b).

The finding that complex folding and oligomerization events occur within the ER, and are necessary for movement of the protein from the ER, appears to be general (Hurtley and Helenius, 1989). Thus, the conclusion holds not just for viral surface proteins but also for a range of endogenous cell surface receptors, where the acquisition of ligand binding activity occurs within the ER but is clearly subsequent to a complex chain of modification, folding, and assembly processes (Slieker et al., 1986; Olson and Lane, 1987,1989). Different receptor proteins display different activation and assembly kinetics, but in each case movement to the Golgi occurs soon after oligomerization (Fig. 10–7; Olson and Lane, 1989); conversely, when

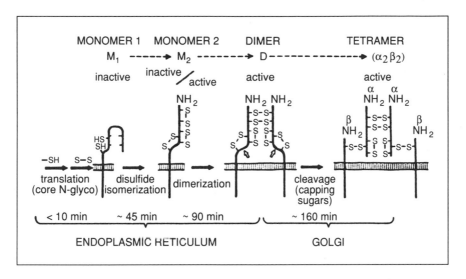

FIGURE 10–7. Folding and oligomerization in the secretory pathway. Schematic pathway of maturation of the insulin proreceptor indicating posttranslational modification, folding, and assembly steps. Core N-glyco: co-translational addition of N-linked oligosaccharide chains; NH_2: protein amino terminus; α and β: the α- and β-subunits of the mature receptor. -SH and $-S_{2-}$ (and -S-S-) represent thiol groups and disulfide bonds, respectively, but do not indicate the number or positions of such groups precisely. (From Olson and Lane, 1989)

the same protein (HA) was synthesized in a variety of cells or using a variety of expression systems, the kinetics of folding were widely variable, implying that both intrinsic properties of the protein and endogenous cellular factors influence the rate of folding (Braakman et al., 1992b).

The possibility of generalizing about mechanisms involved in "quality control" and the retention of misfolded or unassembled proteins in the ER came from the observation that many proteins that are retained in the ER, either through a block in glycosylation or through a mutation that prevents native folding, are found associated with BiP, which therefore is not specifically associated with immunoglobulin assembly (Gething et al., 1986; Dorner et al., 1987; Machamer et al., 1990; Hurtley et al., 1989; Ng et al., 1989). Cloning of a member of the *hsp70* family located in the ER indicated that this *hsp70* was identical to BiP (Munro and Pelham, 1986). BiP is not strongly heat-inducible, but it is induced by any of a number of stress stimuli (glucose starvation, inhibition of glycosylation, treatment with amino acid analogues, or with calcium ionophores), which have in common that they cause the accumulation of misfolded proteins within the

ER lumen (Kozutsumi et al., 1988). Indeed, BiP is identical to a previously described protein, *grp78*, whose synthesis is stimulated by glucose starvation (Munro and Pelham, 1986; Hendershot et al., 1988), and which is one of the major resident proteins within the ER of normal secretory cells (Macer and Koch, 1988). Using the conventional *in vitro* translation and processing system based on dog pancreas microsomes, Kassenbrock et al. (1988) demonstrated that association of BiP with newly synthesized proteins was specific for chains that were aberrant in some way, either through a failure of glycosylation or failure to fold and form native disulfides.

From all of the preceding data, it is apparent that a wide range of aberrant polypeptides that do not fold or assemble correctly are retained within the ER and that interaction with BiP, a luminal *hsp70*, is a key feature of this retention system. But the data do not clearly establish what is the role of BiP, and a number of proposals have been put forward. Proposals range from essentially passive functions (filtering misfolded proteins and thus blocking their secretion, possibly with the additional role of identifying such aberrant polypeptides for later proteolysis) to a more active role either in normal folding and assembly or in rescuing and refolding proteins that have deviated from the productive folding pathway (Hurtley and Helenius, 1989; Gething and Sambrook, 1990).

This remains an area of controversy, and detailed characterization of the properties of BiP *in vitro* with defined protein substrates will be necessary to establish whether BiP can function as a catalyst or facilitator either of initial folding or of refolding. (See Section 7.) The relevant findings on whole cell or subcellular systems include the following:

1. Transient association of BiP with normal folding intermediates is observed in some cases but not in others.
2. Association with BiP is preferentially observed in cases where ATP has been depleted, and addition of ATP can lead to dissociation of complexes between newly synthesized proteins and BiP.
3. The proportion of proteins associated with BiP increases with growth temperature in several systems.
4. The reduction of the level of expression of BiP in a cell overexpressing a recombinant secretory protein increased the level of secretion of that protein (Dorner et al., 1988).
5. In many cases association with BiP is clearly reversible, and molecules apparently in long-term association with BiP can be dissociated and can then move normally through the secretory pathway.

This last point can be illustrated by the dissociation of Ig heavy chain/BiP complexes and assembly of intact immunoglobulins in cells where light chain synthesis is transiently initiated (Hendershot, 1990) or by studies on a temperature-sensitive folding mutant of VSV G protein (de Silva et al., 1990). In the latter case the mutant protein folded, assembled, and was transported efficiently to the cell surface in cells grown at 32°C, but not in cells maintained at 39°C; at the restrictive temperature, the protein failed to form native disulfide bonds, associated with BiP, and was retained in the ER, but these effects could all be reversed by a shift down to 32°C. Interestingly, when a shift up from 32° to 39°C was applied, protein that had already left the ER was already thermostable and competent for further transport, whereas protein within the ER (even protein that had already folded correctly) misfolded at the higher temperature.

In view of the findings (Sections 6.1, 6.3, and 6.5) of the role of *hsp70's* in protein translocation from the cytoplasm and in assembly within the mitochondrial matrix, a comparable role for BiP within the ER lumen is not surprising. It is interesting, however, that genetic evidence suggests that the role of BiP is not confined to proteins that are already located within the ER lumen. The gene coding for BiP in *S. cerevisiae* has been cloned (Normington et al., 1989; Rose et al., 1989) and has been shown to be essential for viability. In cells in which BiP expression is temperature-sensitive, transfer to the restrictive temperature to reduce the level of functional BiP not only reduces protein secretion but appears also to block translocation of precursors into the ER lumen (Vogel et al., 1990; Nguyen et al., 1991); this suggests that BiP may be actively involved in translocation, as well as acting at later posttranslational and posttranslocational stages in folding and assembly.

7 THE MODE OF ACTION OF PROTEIN FOLDING FACTORS

7.1 Actions of Factors on Unfolded Proteins in vitro

From 1989 onward, the discussion of the mode of action of potential factors in facilitating protein folding has begun to approach a more detailed and sophisticated level through the appearance of experimental studies with purified factors in well-defined *in vitro* systems. The picture is by no means yet complete, but the activities can now begin to be described in molecular terms.

7.1.1 Studies with SecB

As described in Section 6.2, SecB is the product of a gene essential for secretion in *E.coli*. The protein was purified (Weiss et al., 1988) and shown to be an oligomer of subunits of 17 kDa Mw with a very acidic isoelectric point. Many studies with the purified SecB protein have documented its effect in retarding or blocking the folding of bacterial secretory protein precursors (Hardy and Randall, 1991, and references therein). SecB forms soluble complexes with unfolded forms of presecretory proteins (Lecker et al., 1989), and studies on the refolding of proOmpA showed that formation of a complex with SecB blocked the generation of insoluble aggregates (Lecker et al., 1990).

Hardy and Randall (1991) have defined in detail the equilibrium between SecB and unfolded MBP precursor, as manifested by the SecB-dependent blockage of refolding of urea-denatured MBP (precursor or mature protein), which is monitored by the intrinsic fluorescence of MBP. This blockage of refolding is not affected by ATP and is not reversed by ATP. Complete blockage of the slow refolding process required a significant molar excess of SecB over MBP. Various unfolded proteins competed for binding to SecB and, hence, diminished the ability of SecB to block the refolding of MBP; this was used to determine the affinity of SecB for various unfolded proteins. The interaction with reduced BPTI occurred with a K_d of approximately 5 nM, and those for other unfolded proteins were determined to lie in the range of 1 to 50 nM. Dissociation constants of SecB for native proteins were higher by several orders of magnitude and could not be determined. Interestingly, however, SecB interacts only weakly with the unfolded form of RBP, and does not detectably retard folding of pre-β-lactamase (Laminet et al., 1991), implying some specificity in the interaction even with unfolded polypeptides. The nature of such specificity for very general features of precursor protein structure has been discussed by Landry and Gierasch (1991).

Complete blockage of the slow refolding of MBP precursor and its variants requires a significant molar excess of SecB over refolding protein, the precise quantity varying with temperature and the folding species. Hardy and Randall (1991) propose that the crucial factor is the rate of folding of the protein relative to its rate of association with SecB. In the case of a mutant refolding very slowly at 5°C, complete blockage of refolding required a fivefold molar excess of SecB. Kinetic partitioning between continued folding, aggregation of unfolded or partially folded species, and stable association of the unfolded form with SecB would account for the observations on SecB *in vitro* and clarify its role *in vivo*. Although no studies have been done on the dissociation of proteins from their complexes with SecB, it is assumed that unfolded proteins are delivered as complexes with SecB to subsequent components of the export machinery. Hence, the role of SecB is not to catalyze folding, nor to bring about unfolding, but to stabilize and solubilize the unfolded protein; hence it retards spontaneous folding sufficiently for the unfolded species to engage with other factors and consequently to undergo translocation. The term *unfolded* here is deceptive, of course; the intrinsic fluorescence methodology used to follow the refolding process allows one to say only whether or not tertiary structure interactions in the vicinity of the fluorescent tryptophan side chains are formed. Much remains to be learned about the structural features that SecB recognizes and about the conformations of proteins bound in complex with SecB.

7.1.2 Studies with GroE Proteins

The mode of action of the *groE* proteins is understood in greater detail than any other protein folding factor, mainly through the availability of over-expressing strains from which the proteins can be purified (Goloubinoff et al., 1989a; Fayet et al., 1989). Detailed studies on the action of the *groE* proteins have been carried out with an increasing number of proteins as they refold and assemble *in vitro* from the unfolded state.

Lorimer and colleagues have studied in detail the refolding and reassembly of the Rubisco from the cyanobacterium *Rhodospirillum rubrum* (Goloubinoff et al., 1989b; Viitanen et al., 1990); this enzyme is a dimer of subunits that are homologous to the large subunits of higher plant Rubisco. Assembly of recombinant *R.rubrum* Rubisco in *E.coli* is dependent on the presence of endogenous *groE* proteins (Goloubinoff et al., 1989a), a finding that indicated that this might be a suitable system for study *in vitro*. Denatured *R.rubrum* Rubisco could refold in good yield after dilution and incubation for 24 hr at temperatures below 20 °C (Fig. 10–8; Viitanen et al., 1990). By contrast, no conditions could be found that permitted spontaneous

FIGURE 10–8. *Effect of temperature on the yield of Rubisco refolding. The protein was unfolded in 4.8 M GdmCl, 80 mM Tris-HCl (pH 7.7), 0.8 mM EDTA, and 0.1 M DTT; portions were diluted 200-fold into solutions of 0.1 M Tris-HCl (pH 7.8), 10 mM MgCl₂, 10 mM KCl, and 5 μM bovine serum albumin held at the indicated temperatures. Rubisco activity was determined after 44 hrs of refolding. (From Viitanen et al., 1990)*

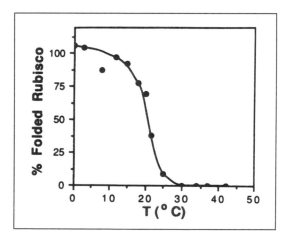

reassembly and reactivation of acid- or urea-denatured Rubisco at temperatures above 25 °C, but significant reassembly and reactivation occurred when denatured Rubisco was diluted into solutions containing groEL, groES, and MgATP (Goloubinoff et al., 1989b). These factors also increased the rate and yield of reactivation at lower temperatures. The rate of reactivation was independent of the method of initial denaturation, even though acid-denatured Rubisco undoubtedly retained some secondary structure that was absent from urea- or GdmCl denatured enzyme, implying that the formation of a common refolding intermediate precedes the *groE*-mediated step. The reactivation showed a lag period consistent with a requirement for the association of folded monomers to generate active dimeric enzyme. Both groEL and groES were essential; optimal reactivation required equimolar groEL and groES, and a three- to fourfold molar excess of *groE* oligomers over Rubisco subunits.

As with SecB, it appears that the productive interaction of refolding subunits with *groE* proteins competes with a reaction that diverts folding intermediates into aggregates (Fig. 10–9). Experiments with individual components indicate that the initial interaction of Rubisco subunits is with groEL and that a stable binary complex of groEL tetradecamer with Rubisco subunit is formed. Discharge of this subunit and subsequent folding requires groES, MgATP (which cannot be replaced by nonhydrolyzable analogues), and K^+ ions. Apart from the ATP- and groES-dependent discharge, the findings resemble those with SecB. There was no evidence in this work that the *groE* system could unfold native Rubisco or solubilize and refold aggregated Rubisco. It appears that the role of groES is to couple an endogenous K^+-dependent ATPase activity of groEL to the release of bound proteins. As of early 1992, it is uncertain whether the released Rubisco is in the same conformation as that which initially binds to groEL or whether the interaction with the *groE* system induces any conformational change in the bound protein, but this question has been addressed with other proteins (to be discussed shortly).

Study of the reactivation of denatured mammalian mitochondrial citrate synthase by *E.coli groE* proteins (Buchner et al., 1991) confirms and extends the findings with Rubisco. The spontaneous refolding of this dimeric enzyme is very inefficient and concentration-dependent; at concentrations above 0.3 μM the yield of spontaneous reactivation is negligible. The presence of a several-fold molar excess of groEL plus groES increased the yield of refolded and reactivated citrate synthase at all citrate synthase concentrations up to 1 μM (Fig. 10–10), but without affecting the rate of reactivation (Buchner et al., 1991). In the spontaneous refolding, aggregation of citrate synthase competes with refolding and leads to the formation of aggregates that can be detected by light scattering; this aggregation is inhibited by the presence of groEL/groES/MgATP but cannot be reversed.

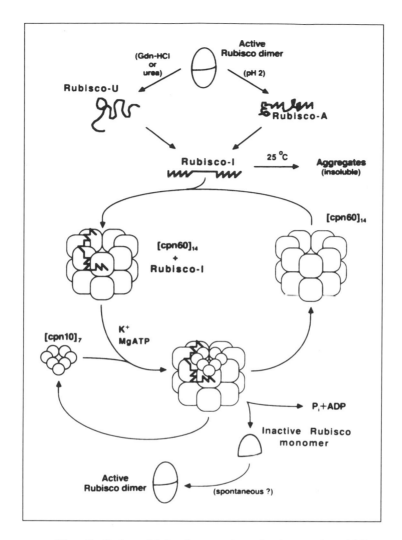

FIGURE 10–9. *Hypothetical model for chaperonin action in protein refolding. Model for chaperonin-dependent reconstitution of dimeric Rubisco from unfolded polypeptides. The active Rubisco dimer (top) can be denatured with 8 M urea or 6 M guanidinium chloride (Gdn-HCl) to give an unfolded polypeptide with little secondary structure (Rubisco-U). The dimer can also be acid-denatured to give a polypeptide that still retains elements of secondary structure (Rubisco-A). It is suspected that a common intermediate forms from Rubisco-U or Rubisco-A on removal of the denaturant; this intermediate is labeled Rubisco-I. In the absence of chaperonins, dilution of denatured Rubisco into buffer at 25°C results in precipitation. If cpn60 is present during dilution, a stable binary complex is formed between it and Rubisco-I. When cpn10 and MgATP are added to this complex in the presence of K^+ ions, active Rubisco dimers are formed. (From Gatenby et al., 1990)*

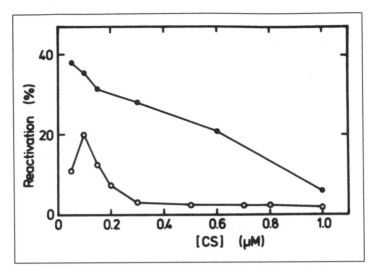

FIGURE 10–10. Suppression by groE complex of aggregation during refolding. Concentration-dependence of the yield of reactivation of citrate synthase (CS) in the presence (●) and absence of (○) of the GroE complex at 25°C. CS was completely denatured in a buffer containing 6.0 M GdmCl, 0.1 M Tris/HCl, pH 8.0, and 20 mM DTE at various protein concentrations. Refolding was initiated by 100-fold dilution of the unfolded protein into a buffer of 0.1 M Tris/HCl, 10 mM MgCl$_2$, 10 mM KCl, and 2 mM ATP, pH 8.0, to the indicated CS concentration. For refolding in the presence of GroE, a sixfold molar excess of GroE complex over CS was added to the refolding solution. The extent of reactivation was measured after 2 hrs of refolding. (From Buchner et al. 1991)

The time course of aggregation is considerably faster than that of activation, implying that aggregation competes with the formation of a critical folding intermediate that is formed before the rate-determining step in the folding pathway. GroEL alone suppresses aggregation, but the complete system is required for reactivation. The action of the *groE* system in this case therefore appears to be primarily as a suppressor of aggregation through binding of folding intermediates that is modulated by ATP hydrolysis.

The general conclusions from the work on the interaction of groEL/groES with Rubisco and citrate synthase have been confirmed and extended by studies on a wider variety of proteins. Whereas Rubisco and citrate synthase are oligomers, the action of the *groE* system has now been studied on simple monomeric enzymes, in particular DHFR and rhodanese (Martin et al., 1991; Mendoza et al., 1991; Viitanen et al., 1991). The data indicate that groEL alone blocks the spontaneous refolding of DHFR by binding and stabilizing an intermediate; likewise, with rhodanese, for which the yield of spontaneous refolding is very low, groEL binds an early folding intermediate and, hence, blocks its aggregation. Interaction with groES, MgATP, and K^+ then leads to discharge of the bound intermediates and refolding. It appears that the precise requirements may differ from protein to protein (e.g., with DHFR, the presence of groES is not essential, but it potentiates the effect of MgATP), but the overall picture is consistent.

The key questions concern the conformation of the folding intermediate bound to groEL and whether the substrate protein is released from groEL still in this state, or whether active folding occurs while the intermediate remains bound to groEL. By use of intrinsic protein fluorescence, protease sensitivity, and dye-binding properties, Martin et al., (1991) monitored the conformation of groEL-bound DHFR and rhodanese and concluded that the bound protein folding intermediates were essentially in the molten globule state. A similar conclusion has been reached with a variety of other refolding proteins that are bound to groEL and, thus, inhibited from completing refolding (Semisotnov et al., 1991). By monitoring the MgATP-dependent and groES-stimulated refolding of groEL-bound DHFR and rhodanese, Martin et al. (1991) concluded that this refolding occurs in close association with the groEL, possibly by the binding of a single protein molecule by many sites on the groEL oligomer, so that release from individual binding sites may occur while the protein as a whole remains bound to the groEL oligomer. A structural model to account for this has been proposed (Creighton, 1991).

In the case of pre-β-lactamase, the overall picture is similar, but some distinctive aspects should be highlighted. This is a monomeric enzyme that can spontaneously refold to the active state in good yield. The effect of addition of groEL is to inhibit this spontaneous refolding by the formation of a stable stoichiometric soluble complex (Laminet et al., 1990). Subsequent addition of groES plus MgATP leads to refolding that gives a higher yield than the spontaneous process. The slowest step in refolding followed release of bound precursor. Surprisingly, there appeared to be a direct effect of groEL (or groEL/groES) on refolded precursor, which led to a slow unfolding; this could be reversed by the addition of ATP. This is the first indication that the *groE* system can act to unfold a folded protein, but because the effect is on a precursor rather than on a mature protein, it is not an effect on a truly native protein. In the study on pre-β-lactamase, the conformation of the protein could be studied both in terms of both enzyme activity and sensitivity to proteases; both approaches confirm that the *groE* system binds the protein in an unfolded conformation.

7.1.3 *Studies with* hsp70 *Proteins*
Studies on the detailed mechanism of the *hsp70* family of factors have employed not only the prokaryotic prototype, the *dnaK* gene product, but

also the mammalian cytosolic constitutive homologue (the heat-shock cognate protein or *hsc70*) and the mammalian homologue BiP located in the ER. As of early 1992, there is every reason to believe that all members of the family share a common mechanism of action, so that studies on any one are relevant to understanding the whole family. Thus, following the discovery from cDNA sequencing that BiP was a member of the *hsp70* family, its ATPase activity was characterized (Kassenbrock and Kelly, 1989). This work showed that BiP had an endogenous ATPase activity with an acidic pH optimum, a K_m for ATP of less than 1 µM, and a slow turnover, k_{cat} being less than 1 min^{-1}. These properties are close to those previously observed for the mammalian *hsc70* (uncoating ATPase) and for the *dnaK* protein.

The nature of the BiP ATPase activity has been further explored by Flynn et al. (1989,1991), who have demonstrated that it is stimulated by a wide variety of short peptides, which they regard as models for structural elements in unfolded or partly folded proteins. In the absence of ATP, BiP showed affinity for a 15-mer hydrophilic oligopeptide, but addition of ATP led to peptide release; this was accompanied by a peptide-dependent ATPase activity. The efficiency of synthetic peptides in stimulating this activity varied widely. (K_m values were in the range of 10^{-2} to 10^{-5} M.) The peptide-stimulated ATPase activity had a k_{cat} of 0.2 min^{-1}, characteristic of the low turnover number of *hsp70* ATPases. Binding peptides have a minimum length of 7 amino acids and a predominance of hydrophobic residues.

The peptides used in this work were viewed as models for elements exposed on nascent or heat-denatured proteins, or on native proteins whose conformation and state of assembly is modulated by *hsp70's*. Subsequent studies with *hsc70* (clathrin-uncoating ATPase) have indicated which parts of the clathrin light chains interact with *hsc70* and stimulate its ATPase activity (DeLuca-Flaherty et al., 1990).

Using another system, the enzyme DHFR coupled to a mitochondrial targeting sequence as a model of a mitochondrial import precursor, Sheffield et al. (1990) studied the effects of the constitutive cytoplasmic *hsc70* on folding in order to interpret the results (Section 6.1) that indicate a requirement for this factor to ensure import-competence into mitochondria. Their data indicate that the *hsc70* inhibits refolding of the protein on dilution from the urea-unfolded state and, in particular, that it inhibits aggregation during refolding *in vitro*; this inhibition of aggregation is ATP-dependent. Interestingly, however, these effects are not sufficient to retain the protein in a translocation-competent state; for that, an additional cytosolic component is required, namely a protein of 200 kDa Mw that is sensitive to NEM.

A key question in models of the action of "folding factors" is whether they can dissociate and reactivate proteins from aggregates of unfolded protein. Two studies with the *dnaK* protein from *E.coli* suggest that this may be possible. A temperature-sensitive mutant of the λcI repressor is inactivated at 42 °C, but reactivates when cells are transferred to 32 °C. Gaitanaris et al. (1990) have shown that this reactivation requires the *dnaK* protein *in vivo* and in whole cell extracts *in vitro*; no reactivation was observed in controls or upon treatment with extracts of cells bearing mutations in the *dnaK* gene. The physical state of the heat-denatured repressor was not established in this work, and the use of whole cell extracts limits mechanistic interpretation. In the work of Skowyra et al. (1990), however, similar results were obtained in a fully defined system. Active and mutant *dnaK* proteins were purified and studied for their effect on heat denaturation and renaturation of RNA polymerase. Active *dnaK* protected the polymerase from inactivation on heating at 45 °C, and this effect did not require ATP. More strikingly, the *dnaK* protein reactivated heat-denatured RNA polymerase, and this effect was entirely dependent on ATP. The heat-denatured enzyme was in the form of an aggregate with $Mw > 5 \times 10^6$, but incubation with *dnaK* protein plus ATP led to disaggregation, generating active dimeric species of $Mw = 1 \times 10^6$. This work is currently unique in documenting an *hsp70*- and ATP-dependent disaggregation and reactivation of a heat-denatured protein aggregate.

Understanding of the role of *hsp70's* will inevitably require some understanding of their structure. As mentioned in Section 5.3, they appear to contain a variable C-terminal protein-substrate binding domain and an N-terminal well-conserved ATP-binding domain. The solution of the structure of this latter domain by X-ray diffraction (Flaherty et al., 1990) has indicated that it has a complex two-lobe structure with a significant similarity in topology to hexokinase, and more strikingly, to actin (Kabsch et al., 1990; Flaherty et al., 1991). Understanding of substrate-induced conformational changes in hexokinase and of the mechanism of force generation in actin make it tempting to speculate on mechanisms by which hydrolysis of ATP by the N-terminal domain of *hsp70's* could transmit a conformation change to peptides bound to the C-terminal domain. In the absence of any conclusive structural information on this domain, however, or evidence on conformational changes in peptides bound to it, such proposals remain speculative.

7.1.4 Summary

The actions of proteins implicated as facilitators of folding in the cell have been clarified by studies on their action, in the pure state, on the process of protein refolding *in vitro*. In most cases the refolding processes under study are inefficient in the absence of the added factors, leading to aggregation; productive refolding is favored by low temperatures and very low protein concentrations. The addition of the putative folding factors initially blocks refolding by the formation of complexes with unfolded or partially folded proteins. In the case of the *cpn60's* and *hsp70's*, release of bound protein from such complexes is coupled to ATP hydrolysis and can lead to continued folding and assembly of the released protein. In all cases, facilitated refolding requires a considerable molar excess of folding factor over refolding substate protein. Although the factors increase the yield of folded protein, in one case only is the overall factor-dependent refolding *faster* than that observed in the absence of factor (Viitanen et al., 1990). In one case, there is evidence that the aggregation that competes with productive folding can be *reversed* in the presence of the folding factor (Skowyra et al., 1991). In one case, there is evidence that the protein undergoes a *conformational change* while bound to the factor, rather than being released in the same state as that which bound initially (Ostermann et al., 1989). The majority of cases as of early 1992 suggest that the folding factors simply bind poorly soluble unfolded and partially folded states of proteins, release them in an ATP-hydrolysis–dependent step to continue their refolding, and cannot recover proteins that have already deviated from the productive folding route and aggregated.

7.2 Models of the Action of Protein Folding Factors

Since the appearance of the cellular evidence that protein factors facilitate protein folding and assembly, various models for their action have been described, differing in emphasis and detail.

The reviews of Ellis (Ellis and Hemmingsen, 1989; Ellis, 1990; Ellis and van der Vies, 1991) emphasize a minimal definition of such factors as "molecular chaperones"; such factors facilitate folding and assembly events but are not components of the final folded and/or assembled product. Ellis emphasizes that "interactive surfaces"—hydrophobic or polar—are exposed naturally in the course of protein biosynthesis and translocation and in other physiological processes, and argues that the action of chaperones is to recognize and combine with such surface features in order to prevent the formation of nonnative, irreversible interactions. No requirement is imposed that the chaperone should turn over many molecules of the substrate on which it acts. Indeed, Ellis includes in his tabulations of chaperones the "pro" sequences of some proteins that are required for their folding to the native conformation (Section 2.3); such sequences act with a stoichiometry of 1:1 and play a role in determining the nature of the final product, rather than simply facilitating the folding process.

The model proposed by Pelham for the action of the *hsp70* family (Pelham, 1986) is clearly influenced by the behavior of the heat-inducible members of the family during heat-shock, and by the action of *hsc70* in uncoating clathrin-coated vesicles. In this model, the *hsp70's* act as enzymes catalyzing ATP-driven unfolding or disaggregation reactions. The *hsp70's* recognize and bind to characteristic exposed regions on unfolded or aggregated proteins (presumed to be hydrophobic regions). In the case of a protein aggregate, it is then proposed that binding and hydrolysis of ATP by the *hsp70* induces a conformational change in the bound protein, which detaches it from the aggregate and buries the previously exposed hydrophobic region, allowing dissociation of the *hsp70*.

The model of Rothman (Rothman and Kornberg, 1986; Rothman, 1989) draws on the same concepts but suggests more mechanistic detail. He suggests that, in some cases, simple reversible binding of an unfolded protein may be sufficient to block aggregation and permit the spontaneous correct folding of the released polypeptide. In such a case, the folding factor would effectively be an antifolding factor, and the hydrolysis of ATP would function to regulate interaction of factor with its unfolded substrate. He also suggests, however, that in other cases folding may occur at the surface of the folding factor through the formation of multiple interactions between substrate and factor, which are then released in a specific sequence that permits productive folding through a pathway that is less favorable when the whole unfolded protein is available for folding at the same time. This model emphasizes the peptide-binding and peptide-dependent ATPase activities of folding factors, and its formulation in greater detail will require further characterization of the structure and peptide-binding properties of such factors. For example, do *hsp70* factors have multiple binding sites that could interact with different exposed regions of a polypeptide? If refolding is facilitated by an *hsp70* oligomer, is a bound substrate protein unfolded to the extent that it can interact with the peptide binding sites on several *hsp70* protomers simultaneously? Do the differences in affinity of *hsp70's* for short peptides reflect differences in dissociation rates?

The models of Pelham and Rothman imply that folding factors can function in repeated cycles of binding and release and, hence, can turn over many molecules of substrate, but they do not require that folding and assembly in their presence necessarily occur more rapidly than in their absence.

7.3 Definitions, Models, and Observations

In concluding an account of the mode of action of protein folding factors, it is important to emphasize that their actions *in vivo* are relatively newly described and that by early 1991 few studies had been published describing their action in well-defined systems *in vitro*. So we should not be especially disturbed by our inability to describe a detailed model that is well supported by experimental observations. On the other hand, it is important to be aware of the limitations of what can be supported by the data and to avoid loose descriptions of the actions of these factors.

It may be useful to recall the basic features of an enzyme as a catalyst: (1) it is not a component of the reaction product; (2) as a result of its release at the end of the catalytic cycle, it is free to participate in another cycle; hence, each enzyme molecule can turn over many molecules of substrate; and (3) it increases the rate of reaction over that seen in its absence under comparable conditions. By these definitive criteria, none of the factors described here has been shown to be a catalyst of folding, assembly, or disaggregation. In the *in vitro* studies, SecB has been observed only to bind and prevent folding, and the studies on *cpn60's* and *hsp70's* that demonstrate an increased yield of folded product have all involved a considerable excess of factor over substrate, implying that no active turnover of substrate protein is occurring. So it is inappropriate, or premature, to describe these proteins as catalysts of folding and/or assembly. It would certainly be interesting to check on the state and activity of the *cpn60's* or *hsp70's* at the end of an experiment in which they have facilitated refolding. Are they free and in the same conformation and state of association as at the start of the experiment? Are they still active? Could they facilitate the refolding of a new batch of unfolded substrate with a similar yield? If it could be shown that refolding experiments such as those described could be run repeatedly with the same batch of factor so that, overall, a molar excess of product over factor could be generated, this would be a major step in establishing their mode of action. The excess of factor that is now required for refolding experiments *in vitro* could then be compared to the molar excess over unfolded newly synthesized protein substrates that exists in the cell at any one moment.

In defining the possible actions of folding factors, it is worth emphasizing the growing consensus on the general pathway of spontaneous protein refolding *in vitro*. This suggests that a fully unfolded extended polypeptide upon transfer to "native" conditions, rapidly collapses to an ensemble of relatively compact states in which some elements of secondary structure are formed; in subsequent slower steps the protein becomes more compact through extrusion of interior water and formation of optimal tertiary structure interactions, with the major transition state that defines the slowest step in the pathway lying close to the native state. This initial collapsed, or "molten globule," state is characterized by considerable internal mobility (Chapter 6), so that, for example, spontaneous thiol-disulfide interchange is as rapid in this state as in the fully unfolded state (Ewbank and Creighton, 1991). Hence, it is probable that there are no significant activation barriers to the breakage of individual nonnative hydrophobic or polar interactions in this state. In that case there will be no cellular requirement for factors that can catalyze disruption of such interactions within a single folding domain, and the cellular problem can be defined as the prevention of nonnative interactions between domains and subunits.

It may also be helpful to look back at the difficulties encountered in achieving high yields of refolding *in vitro*, the processes that compete with refolding, and the experimental tactics that can be used to overcome these. These will suggest the kinds of capability that might have evolved under pressure to optimize the efficiency of initial protein folding and assembly *in vivo*. Reduction in temperature or in the concentration of refolding protein are not options open to the cell. Other strategies for favoring productive refolding rather than aggregation are basically of two kinds: (1) addition of detergents or other solutes that render the unfolded and partly folded states more soluble, and (2) reversible binding of the unfolded protein to a matrix so that folding can occur in an essentially isolated state. At this stage it appears that the cellular folding factors operate in similar ways; by binding unfolded proteins reversibly, they reduce the effective free concentration while maintaining the bound intermediates in a folding-competent state. If these intermediates are then simply released, the action is effectively one of improving the solubility of the protein substrate and reducing the likelihood of aggregation. On the other hand, if more general evidence can be produced that protein substrates actually undergo conformational transitions to a more folded state before release, as is suggested by recent work on groEL (Section 7.1.2), then the action of the factors is comparable to a matrix that isolates individual refolding molecules and permits folding in the immobilized state.

Overall, it is clear that protein folding in the cell can no longer be thought of as simply identical to the experimental refolding of a mature protein in isolation. But the folding process, in whatever cellular circumstances, is constrained by the physicochemical properties of polypeptide chains in an aqueous environment. Work on refolding *in vitro* has been of extraordinary value in defining pathways of folding, properties of folding intermediates, rate-determining steps, and processes that compete with productive refolding. Cell biological studies have identified proteins that clearly have a role in ensuring the efficiency of folding to the native state in the cell. There is now the clear opportunity to characterize in more detail the properties of these proteins and their actions. Such work will undoubtedly deepen our understanding of the process of protein folding and assembly in cells, a process that is integral to the central paradoxes of change, order, discontinuity, and organization that characterize the activities of cells.

Acknowledgments

I wish to thank Mary-Ann Preston for her valued assistance in the preparation of this chapter.

I am grateful to many colleagues for sending me reprints and to the following for sending me advance copies of papers in press: Drs. Ineke Braakman, John Ellis, Mary-Jane Gething, Sasha Girshovich, Franz-Ulrich Hartl, Arne Holmgren, George Lorimer, Anna Mitraki, Taina Pihlajaniemi, Andreas Pluckthun, Oleg Ptitsyn, Linda Randall, Tom

Rapoport, Gottfried Schatz, Franz Xaver Schmid, Stuart Schreiber, Ross Stein, and Richard Zimmermann.

I acknowledge continuing research grant support from the Science and Engineering Research Council (GR/F/41839 and GR/G/09030), the Agriculture and Food Research Council (LRG 174), and the Medical Research Council (G8911277CB).

REFERENCES

Akagi, S., Yamamoto, A., Yoshimori, T., Masaki, R., Ogawa, R., and Tashiro, Y. (1988) *J. Histochem. Cytochem. 36,* 1069–1074.

Andrews, T. J., and Lorimer, G. H. (1987) *J. Biol. Chem. 260,* 4632–4636.

Andria, G., and Taniuchi, H. (1978) *J. Biol. Chem. 253,* 2262–2270.

Attardi, G., and Schatz, G. (1986) *Ann. Rev. Cell Biol. 4,* 289–333.

Bächinger, H. P. (1987) *J. Biol. Chem. 262,* 17144–17148.

Bächinger, H. P., Fessler, L. I., Timpl, R., and Fessler, J. H. (1981) *J. Biol. Chem. 256,* 13193–13199.

Bassuk, J. A., and Berg, R. A. (1989) *Matrix 9,* 244–258.

Beckman, R. P., Mizzen, L. A., and Welch, W. J. (1990) *Science 248,* 850–853.

Bennett, C. F., Balcarek, J. M., Varrichio, A., and Crooke, S. T. (1988) *Nature 334,* 268–270.

Bergman, L. W., and Kuehl, W. M. (1979a) *J. Biol. Chem. 254,* 5690–5694.

Bergman, L. W., and Kuehl, W. M. (1979b) *J. Biol. Chem. 254,* 8869–8876.

Bernstein, H. D., Rapoport, T. A., and Walter, P. (1989) *Cell 58,* 1017–1019.

Bierer, B. E., Somers, P. K., Wandless, T. J., Burakoff, S. J., and Schreiber, S. L. (1990) *Science 250,* 556–559.

Blobel, G., and Dobberstein, B. (1975) *J. Cell Biol. 67,* 852–862.

Bochkareva, E. S., Lissin, N. M., and Girshovich, A. S. (1988) *Nature 336,* 254–257.

Bole, D. G., Hendershot, L. M., and Kearney, J. F. (1986) *J. Cell Biol. 102,* 1558–1566.

Booth, C., and Koch, G. L. E. (1989) *Cell 59,* 729–737.

Botterman, J., and Zabeau, M. (1985) *Gene 37,* 229–234.

Braakman, I., Helenius, J., and Helenius, A. (1992a) *Nature 536,* 260–262.

Braakman, I., Helenius, J., and Helenius, A. (1992b) *EMBO J.* (in press).

Bradley, D., van der Vies, S. M., and Gatenby, A. A. (1986) *Phil. Trans. R. Soc. Lond. B313,* 447–458.

Brems, D. N. (1988) *Biochemistry 27,* 4541–4546.

Buchner, J., and Rudolph, R. (1991) *Bio/Technology 9,* 157–162.

Buchner, J., Schmidt, M., Fuchs, M., Jaenicke, R., Rudolph, R., Schmid, F. X., and Kiefhaber, T. (1991) *Biochemistry 30,* 1586–1591.

Bulleid, N. J., Bessel-Duby, R. S., Freedman, R. B., Sanbrook, J. F., and Gething, M.-J. (1992) *Biochem. J.* (in press).

Bulleid, N. J., and Freedman, R. B. (1988) *Nature 335,* 649–651.

Bulleid, N. J., and Freedman, R. B. (1990) *EMBO J. 9,* 3527–3532.

Bulleid, N. J., Shewry, P. R., and Freedman, R. B. (1991) In *Plant Protein Engineering* (P. R. Shewry and S. Gutteridge, eds.), Edward Arnold, in press.

Bychkova, V. E., Pain, R. H., and Ptitsyn, O. B. (1988) *FEBS Letters 238,* 231–234.

Ceriotti, A., and Colman, A. (1988) *EMBO J 7*, 633–638.

Chantrenne, H. (1961) *The Biosynthesis of Proteins*, Pergamon, New York, p. 122.

Chappell, T. G., Conforti, B. B., Schmid, S. L., and Rothman, J. E. (1987) *J. Biol. Chem. 262*, 746–751.

Chen, W. J., and Douglas, M. G. (1987) *J. Biol. Chem. 262*, 15605–15609.

Cheng, M. Y., Hartl, F.-U., and Horwich, A. L. (1990) *Nature 348*, 455–458.

Chirico, W. J., Waters, M. G., and Blobel, G. (1988) *Nature 332*, 805–810.

Copeland, C. S., Doms, R. W., Bolzau, E. M., Webster, R. G., and Helenius, A. (1986) *J. Cell Biol. 103*, 1179–1191.

Copeland, C. S., Zimmer, K.-P., Wagner, K. R., Healey, G. A., Mellman, I., and Helenius, A. (1988) *Cell 53*, 197–209.

Craig, E. A., and Gross, C. A. (1991) *Trends Biochem. Sci. 16*, 135–140.

Creighton, T. E. (1978) *Prog. Biophys. Mol. Biol. 33*, 231–297.

Creighton, T. E. (1986) A process for the production of a protein. International patent no. WO86/05809.

Creighton, T. E. (1991) *Nature 352*, 17–18.

Creighton, T. E., Hillson, D. A., and Freedman, R. B. (1980) *J. Mol. Biol. 142*, 43–62.

Crooke, E., Guthrie, B., Lecker, S., Lill, R., and Wickner, W. (1988) *Cell 54*, 1003–1011.

DeLorenzo, F., Goldberger, R. F., Steers Jr., E., Givol, D., and Anfinsen, C. B. (1966) *J. Biol. Chem 241*, 1562–1567.

DeLuca-Flaherty, C., McKay, D. B., Parham, P., and Hill, B. L. (1990) *Cell 62*, 875–887.

Deshaies, R. J., Koch, B. D., Werner-Washburne, M., Craig, E. A. and Schekman, R. (1988) *Nature 332*, 800–805.

De Silva, A. M., Balch, W. E., and Helenius, A. (1990) *J. Cell Biol. 111*, 857–866.

Dorner, A. J., Bole, D. G., and Kaufman, R. J. (1987) *J. Cell Biol. 105*, 2665–2674.

Dorner, A. J., Krane, M., and Kaufman, R. J. (1988) *Mol. Cell Biol. 8*, 4063–4070.

Dorner, A. J., Wasley, L. C., Raney, P., Haugejorden, S., Green, M., and Kaufman, R. J. (1990) *J. Biol. Chem. 265*, 22029–22034.

Edman, J. C., Ellis, L., Blacher, R. W., Roth, R. A., and Rutter W. J. (1985) *Nature 317*, 267–270.

Eilers, M., Hwang, S., and Schatz, G. (1988) *EMBO J. 7*, 1139–1145.

Eilers, M., and Schatz, G. (1986) *Nature 322*, 228–232.

Eilers, M., and Schatz, G. (1988) *Cell 52*, 481–483.

Ellis, R. J. (1987) *Nature 328*, 378–379.

Ellis, R. J. (1990) *Seminars Cell Biol. 1*, 1–10.

Ellis, R. J., and Hemmingsen, S. M. (1989) *Trends Biochem. Sci. 14*, 339–343.

Ellis, R. J., and van der Vies, S. M. (1991) *Ann. Rev. Biochem. 60*, 321–347.

Emmel, E. A., Verweij, C. L., Durand, D. B., Higgins, K. A., Lacy, E., and Crabtree, G. R. (1989) *Science 246*, 1617–1620.

Epstein, C. J., Goldberger, R. F., and Anfinsen, C. B. (1963) *Cold Spring Harbor Symp. Quant. Biol. 28*, 439–449.

Ewbank, J. J., and Creighton, T. E. (1991) *Nature 350*, 518–520.

Farquhar, R., Honey, N., Murant, S. J., Bossier, P., Schultz, L., Montgomery, D., Ellis, R. W., Freedman, R. B., and Tuite, M. F., (1991) *Gene 108*, 81–89.

Fayet, O., Ziegelhoffer, T., and Georgopoulos, C. (1989) *J. Bacteriol. 171*, 1379–1385.

Ferro-Novick, S., Hansen, W., Schauer, I., and Schekman, R. (1984) *J. Cell Biol. 98*: 44–53.

Fesik, S. W., Gampe Jr., R. T., Holzman, T. E., Egan, D. A., Edalji, R., Luly, J. R., Simmer, R., Helfrich. R., Kishore, V., and Rich, D. H. (1990) *Science 250*, 1406–1408.

Fischer, G., and Bang, H. (1985) *Biochim. Biophys. Acta 828*, 39–42.

Fischer, G., Bang, H., and Mech, C., (1984) *Biomed. Biochim, Acta 43*, 1101–1111.

Fischer, G., Berger, E., and Bang, H. (1989a) *FEBS Letters 250*, 267–270.

Fischer, G., and Schmid, F. X. (1990) *Biochemistry 29*, 2205–2212.

Fischer, G., Wittmann-Liebold, B., Lang, K., Kiefhaber, T., and Schmid, F. X. (1989b) *Nature 337*, 476–478.

Flaherty, K. M., DeLuca-Flaherty, C., and McKay, D.B. (1990) *Nature 346*, 623–628.

Flaherty, K. M., McKay, D. B., Kabsch, W., and Holmes, K. C. (1991) *Proc. Natl. Acad. Sci. USA 88*, 5041–5045.

Flanagan, J. M. Kataoka, M., Shortle, O., and Engelman (1992) *Proc. Natl. Acad. Sci. USA 89*, 748–752.

Flynn, G. C., Chappell, T. G., Rothman, J. E. (1989) *Science 245*, 385–390.

Flynn, G. C., Pohl, J., Flocco, M. T., and Rothman, J. E. (1991) *Nature 353*, 726–730.

Freedman, R. B. (1984) *Trends Biochem. Sci. 9*, 438–441.

Freedman, R. B. (1987) *Nature 329*, 294–295.

Freedman, R. B. (1989a) *Biochem. Soc. Trans. 17*, 331–335.

Freedman, R. B. (1989b) *Cell 57*, 1069–1072.

Freedman, R. B. (1990) In *Glutathione: Metabolism and Physiological Functions* (J. Vina, ed.), CRC Press, Boca Raton, Fla.9, pp. 125–134.

Freedman, R. B. (1991) In *Conformation and Forces in Protein Folding* (B. T. Nall and K. A. Dill, eds.), AAAS, Washington, D.C., pp. 204–214.

Freedman, R. B., Brockway, B. E., and Lambert, N. (1984) *Biochem. Soc. Trans. 12*, 929–932.

Freedman, R. B., Bulleid, N. J., Hawkins, H. C., and Paver, J. L. (1989) *Biochem. Soc. Symp. 55*, 167–192.

Freedman, R. B., and Hawkins, H. C. (1977) *Biochem. Soc. Trans. 5*, 348–357.

Freedman, R. B., Hawkins, H. C., Murant, S. J., and Reid, L. (1988) *Biochem. Soc. Trans. 16*, 96–99.

Freedman, R. B., and Hillson, D. A. (1980) In *The Enzymology of Post-Translational Modification of Proteins*, vol. 1 (R. B. Freedman and H. C. Hawkins, eds.), Academic Press, London, pp. 157–212.

Freedman, R. B., and Tuite, M. F. (1992) In *Guidebook to the Secretory Pathway* (P. Novich, J. A. Rothblatt, and T. Stevens, eds.), Sambrooke and Tooze Scientific Publications, in press.

Gaitanaris, G. A., Papavassiliou, A. G., Rubock, P., Silverstein, S. J., and Gottesman, M. E. (1990) *Cell 61*, 1013–1020.

Gatenby, A. A., and Ellis, R. J. (1990) *Ann. Rev. Cell Biol. 6*, 125–149.

Gatenby, A. A., Viitanen, P. V., and Lorimer, G. H. (1990) *Trends Biotechnol. 8*, 354–358.

Geetha-Habib, M., Noiva, R., Kaplan, H. A., and Lennarz, W. J. (1988) *Cell 54*, 1053–1060.

Georgopoulos, C., and Ang, D. (1990) *Seminars Cell Biol. 1*, 19–25.

Gerschitz, J., Rudolph, R., and Jaenicke, R. (1978) *Eur. J. Biochem. 87*, 591–599.

Gething, M.-J., McCammon, K., and Sambrook, J. (1986) *Cell 46*, 939–950.

Gething, M.-J., and Sambrook, J. (1990) *Seminars Cell Biol. 1*, 65–72.

Gething, M.-J., and Sambrook, J. (1991) *Nature 355*, 33–45.

Gilbert, H. C. (1990) *Adv. Enzymol. 63*, 69–172.

Givol, D., de Lorenzo, F., Goldberger, R. F., and Anfinsen, C. B. (1965) *Proc. Natl. Acad. Sci. USA 53*, 676–684.

Glick, B., and Schatz, G. (1991) *Ann. Rev. Genetics, 21*, 25–44.

Goldberg, M. E. (1985) *Trends Biochem. Sci. 10*, 388–391.

Goldberger, R. F., Epstein, C. J., and Anfinsen, C. B. (1963) *J. Biol. Chem. 238*, 628–635.

Goldenberg, D. P., and Creighton, T. E. (1983) *J. Mol. Biol. 165*, 407–413.

Goldenberg, D. P., and Creighton, T. E. (1984) *Anal. Biochem. 138*, 1–18.

Goldenberg, D. P., and King, J. (1981) *J. Mol. Biol. 145*, 633–651.

Goloubinoff, P., Gatenby, A. A., and Lorimer, G. H. (1989a) *Nature 337*, 44–47.

Goloubinoff, P., Christeller, J. T., Gatenby, A. A., and Lorimer, G. H. (1989b) *Nature 342*, 884–889.

Grafl, R., Lang, K., Vogl, H., and Schmid, F. X. (1987) *J. Biol. Chem. 262*, 10624–10629.

Gribskob, M., and Burgess, R. R. (1983) *Gene 26*, 109–118.

Haas, I. G., and Wabl, M. (1983) *Nature 306*, 387–389.

Haase-Pettingell, C. A., and King, J. A. (1988) *J. Biol. Chem. 263*, 4977–4988.

Haendler, B., Keller, R., Hiestand, P. C., Kocher, H. P., Wegmann, S., and Rao Movva, N. (1989) *Gene 83*, 39–46.

Hallberg, R. L. (1990) *Seminars Cell Biol. 1*, 37–45.

Harding, M. W., Galat, A., Uehling, D. E., and Schreiber, S. L. (1989) *Nature 341*, 758–760.

Hardy, S. J. S., and Randall, L. L. (1991) *Science 251*, 439–443.

Harrison, R. K., and Stein, R. L. (1990a) *Biochemistry 29*, 1684–1689.

Harrison, R. K., and Stein, R. L. (1990b) *Biochemistry 29*, 3813–3816.

Hartl, F.-U., and Neupert, W. (1990) *Science 247*, 930–938.

Hartl, F.-U., Pfanner, N., Nicholson, D. W., and Neupert, W. (1989) *Biochim. Biophys. Acta 988*, 1–45.

Hartl, F.-U., Schmidt, B., Wachter, E., Weiss, H., and Neupert, W. (1986) *Cell 47*, 939–951.

Havel, H. A., Kauffman, E. W., Plaisted, S. M., and Brems, D. N. (1986) *Biochemistry 25*, 6533–6538.

Hawkins, H. C., Blackburn, E. C., and Freedman, R. B. (1991a) *Biochem. J. 275*, 349–353.

Hawkins, H. C., de Nardi, M., and Freedman, R. B. (1991b) *Biochem. J. 275*, 341–348.

Hawkins, H. C., and Freedman, R. B. (1991) *Biochem. J. 275*, 335–339.

Hayano, T., Takahashi, N., Kato, S., Maki, N., and Suzuki, M. (1991) *Biochemistry 30*, 3041–3048.

Hemmingsen, S. M. (1990) *Seminars Cell Biol. 1*, 47–54.

Hemmingsen, S. M., Woolford, C., van der Vies, S. M., Tilly, K., Dennis, D. T., Georgopoulos, C. P., Hendrix, R. W., and Ellis, R. J. (1988) *Nature 333*, 330–334.

Hendershot, L. (1990) *J. Cell Biol. 111*, 829–837.

Hendershot, L., Bole, D., Köhler, G., and Kearney, J. F. (1987) *J. Cell Biol. 104*, 761–767.

Hendershot, L. M., Ting, J., and Lee, A. S. (1988) *Mol. Cell Biol. 8*, 4250–4256.

Holmgren, A. (1985) *Ann. Rev. Biochem. 54*, 237–271.

Horowitz, P., and Criscimagna, N. L. (1986) *J. Biol. Chem, 261*, 15652–15658.

Horowitz, P., and Simon, D. (1986) *J. Biol. Chem. 261*, 13887–13891.

Hsu, M. P., Muhich, M. L. , and Boothroyd, J. C. (1989) *Biochemistry 28*, 6440–6446.

Hurtley, S. M., Bole, D. G., Hoover-Litty, H., Helenius, A., and Copeland, C. S. (1989) *J. Cell. Biol. 108*, 2117–2125.

Hurtley, S. M., and Helenius, A. (1989) *Ann. Rev. Cell Biol. 5*, 277–307.

Ikeguchi, M., Kuwajima, K., Mitani, M., and Sugai, S. (1986) *Biochemistry 25*, 6965–6972.

Ikemura, H., and Inouye, M. (1988) *J. Biol. Chem. 263*, 12959–12963.

Jaenicke, R. (1974) *Eur. J. Biochem. 46*, 149–155.

Jaenicke, R. (1987) *Prog. Biophys. Mol. Biol. 49*, 117–237.

Jaenicke, R. (1991) *Biochemistry 30*, 3147–3161.

Jaenicke, R., and Rudolph, R. (1989) In *Protein Structure: A Practical Approach* (T. E. Creighton, ed.), IRL Press, Oxford, pp. 191–223.

Kabsch, W., Mannherz, H. G., Suck, D., Pai, E. F., and Holmes, K. C. (1990) *Nature 347*, 37–49.

Kaderbhai, M. A., and Austen, B. M. (1984) *Biochem. J. 217*, 145–157.

Kane, J. F., and Hartley, D. L. (1988) *Trends Biotech 6*, 95–101.

Kang, P. J., Ostermann, J., Shilling, J., Neupert, W., Craig, E. A., and Pfanner, N. (1990) *Nature 348*, 137–143.

Kao, W. W.-Y., Mai, S. H., Chou, K.-L., and Ebert, J. (1983) *J. Biol. Chem. 256*, 7779–7787.

Kaska, D. D., Kivirikko, K. I., and Myllylä, R. (1990) *Biochem J. 268*, 63–68.

Kassenbrock, C. K., Garcia, P. D., Walter, P., and Kelly, R. B. (1988) *Nature 333*, 90–93.

Kassenbrock, C. K., and Kelly, R. B. (1989) *EMBO J. 8*, 1461–1467.

Katti, S. K., LeMaster, D. M., and Eklund, H. (1990) *J. Mol. Biol. 212*, 167–184.

Ke, H., Zydowski, L. D., Liu, J., and Walsh, C. T. (1991) *Proc. Natl. Acad. Sci. USA 88*, 9483–9487.

Kiefhaber, T., Grunert, H.-P., Hahn, V., and Schmid, F. X. (1990a) *Biochemistry 29*, 6475–6480.

Kiefhaber, T., Quaas, R., Hahn, U., and Schmid, F. X. (1990b) *Biochemistry 29*, 3053–3061.

Kiefhaber, T., Quaas, R., Hahn, U., and Schmid, F. X. (1990c) *Biochemistry 29*, 3061–3070.

Kim, P. S., and Baldwin, R. L. (1982) *Ann. Rev. Biochem. 51*, 459–489.

Kim, P. S., and Baldwin, R. L. (1984) *Nature 307*, 329–334.

King, J. (1986) *Bio/Technology 4*, 297–303.

Koch, G. L. E. (1987) *J. Cell. Sci. 87*, 491–492.

Kohno, T., Carmichael, D. R., Sommer, A., and Thompson, R. C. (1990) *Methods Enzymol. 185*, 187–195.

Koivu, J., Myllylä, R., Helaakoski, T., Pihlajaniemi, T., Tasanen, K., and Kivirikko, K. I. (1987) *J. Biol. Chem. 262*, 6447–6449.

Koser, P. L., Sylvester, D., Livi, G. P., and Bergsman, D. J. (1990) *Nucleic Acid Res. 18*, 1643.

Kozutsumi, Y., Segal, M., Normington, K., Gething, M.-J., and Sambrook, J. (1988) *Nature 332*, 462–464.

Krause, G., Lundström, J., Barea, J. L., De la Cuesta, C. P., and Holmgren, A. (1991) *J. Biol. Chem. 266*, 9494–9500.

Krebs, H., Schmid, F. X., and Jaenicke, R. (1983) *J. Mol. Biol. 159*, 619–635.

Kumamoto, L. A., and Beckwith, J. (1983) *J. Bact. 154*, 253–260.

Kuwajima, K. (1989) *Proteins: Struct. Funct. Genet. 6*, 87–103.

Lambert, N., and Freedman, R. B. (1985) *Biochem. J. 228*, 635–645.

Laminet, A. A., Kumamoto, C. A., and Plückthun, A. (1991) *Mol. Microbiol. 5*, 117–122.

Laminet, A. A., Ziegelhoffer, T., Georgopoulos, C., and Plückthun, A. (1990) *EMBO J. 9*, 2315–2319.

Landry, S. J., and Gierasch, L. M. (1991) *Trends Biochem. Sci. 16*, 159–163.

Lang, K., and Schmid, F. X. (1988) *Nature 331*, 453–455.

Lang, K., Schmid, F. X., and Fischer, G. (1987) *Nature 329*, 268–270.

Lecker, S. H., Driessen, A. J. M., and Wickner, W. (1990) *EMBO J. 9*, 2309–2314.

Lecker, S. H., Lill, R., Ziegelhoffer, T., Georgopoulos, C., Bassford Jr., P. J., Kumamoto, C. A., and Wickner, W. (1989) *EMBO J. 8*, 2703–2709.

Lee, C., and Beckwith, J. (1986) *Ann. Rev. Cell Biol. 2*, 315–336.

Lin, L. N., Hasumi, H., and Brandts, J. F. (1988) *Biochim. Biophys. Acta 956*, 256–266.

Lindquist, S., and Craig, E. A. (1988) *Ann. Rev. Genet. 22*, 631–677.

Lissin, N. M., Venyaminov, S. Y., and Girshovich, A. S. (1990) *Nature 348*, 339–342.

Liu, J., Albers, M. W., Chen, C.-N., Schreiber, S. L., and Walsh, C. T. (1990) *Proc. Natl. Acad. Sci. USA 87*, 2304–2308.

Liu, G., Topping, T. B., Cover, W. H., and Randall, L. L. (1988) *J. Biol. Chem. 263*, 14790–14793.

Liu, G., Topping, T. B., and Randall, L. L. (1989) *Proc. Natl. Acad. Sci. USA 86*, 9213–9217.

Lodish, H. F., Kong, N., Snider, M., and Strous, G. I. A. M. (1983) *Nature 304*, 80–83.

London, J., Skrzynia, C., and Goldberg, M. E. (1974) *Eur. J. Biochem. 47*, 409–415.

Lubben, T. H., Donaldson, G. K., Viitanen, P. V., and Gatenby, A. A. (1989) *Plant Cell 1*, 1223–1230.

Lubben, T. H., Gatenby, A. A., Donaldson, G. K., Lorimer, G. H., and Viitanen, P. V. (1990) *Proc. Natl. Acad. Sci. USA 87*, 7683–7687.

Luger, K., Hommel, U., Herold, M., Hofsteenge, J., and Kirschner, K. (1989) *Science 243*, 206–210.

Lundström, J., and Holmgren, A. (1990) *J. Biol. Chem. 265*, 9114–9120.

Lyles, M. M., and Gilbert, H. F. (1991a) *Biochemistry 30*, 613–619.

Lyles, M. M., and Gilbert, H. F. (1991b) *Biochemistry 30*, 619–625.

Macer, D. R. J., and Koch, G. L. E. (1988) *J. Cell Sci. 91*, 61–70.

Machamer, C. E., Doms, R. W., Bole, D. G., Helenius, A., and Rose, J. K. (1990) *J. Biol. Chem. 265*, 6879–6883.

Machamer, C. E., Florkiewicz, R. Z., and Rose, J. K. (1985) *Mol. Cell. Biol. 5*, 3074–3083.

Machamer, C. E., and Rose, J. K. (1988) *J. Biol. Chem. 263*, 5948–5954.

Maher, P. A., and Singer, S. J. (1986) *Proc. Natl. Acad. Sci. USA 83*, 9001–9005.

Marston, F. A. O. (1986) *Biochem. J. 240*, 1–12.

Martin, I., Langer, T., Boteva, R., Schramel, A., Horwich, A. L., and Hartl, F. U. (1991) *Nature 352*, 36–42.

Mazzarella, R. A., Srinivasan, M., Haugejorden, S. M., and Green, M. (1990) *J. Biol. Chem. 265*, 1094–1101.

Mendoza, J. A., Rogers, E., Lorimer, G. H., and Horowitz, P. M. (1991) *J. Biol. Chem. 266*, 13044–13049.

Meyer, D. I. (1988) *Trends Biochem. Sci. 13*, 471–474.

Michnik, S. W., Rosen, M. K., Wandless, T. J., Karplus, M., and Schreiber, S. L. (1991) *Science 252*, 836–842.

Mills, E. N. C., Lambert, N., and Freedman, R. B. (1983) *Biochem. J. 213*, 245–248.

Mitraki, A., and King, J. (1989) *Bio/Technology 7*, 690–697.

Mitraki, A., Betton, J. M., Desmadril, M., and Yon, J. (1987) *Eur. J. Biochem. 163*, 29–34.

Mizunaga, T., Katakura, Y., Miura, T., and Marutama, Y. (1990) *J. Biochem. 108*, 846–851.

Morrison, S. L., and Scharff, M. D. (1975) *Immunol. 114*, 655–659.

Mozhaev, V. V., and Martinek, K. (1981) *Eur. J. Biochem. 115*, 143–147.

Munro, S., and Pelham, H. R. B. (1986) *Cell 46*, 291–300.

Murakami, H., Pain, D., and Blobel, G. (1988) *J. Cell Biol. 107*, 2051–2057.

Murant, S. J. (1989) "A Molecular Biological Study of Protein Disulphide-Isomerase." Ph.D. thesis, University of Kent at Canterbury, England.

Musgrove, J. E., Johnson, R. A., and Ellis, R. J. (1987) *Eur. J. Biochem. 163*, 529–534.

Ness, S. A., and Weiss, R. L. (1987) *Proc. Natl. Acad. Sci. USA 84*, 6692–6696.

Ng, D. T. W., Randall, R. E., and Lamb, R. A. (1989) *J. Cell Biol. 109*, 3273–3289.

Nguyen, T. H., Law, D. T. S., and Williams, D. B. (1991) *Proc. Natl. Acad. Sci. USA 88*, 1565–1569.

Nilsson, B., and Anderson, S. (1991) *Ann. Rev. Microbiol. 45*, 607–635.

Noiva, R., Kaplan, H. A., and Lennarz, W. J. (1991) *Proc. Natl. Acad. Sci. USA 88*, 1986–1990.

Noiva, R., Kinura, H., Roos, J., and Lennarz, W. J. (1991b) *J. Biol. Chem. 266*, 19645–19649.

Normington, K., Kohno, K., Kozutsumi, Y., Gething, M. -J., and Sambrook, J. (1989) *Cell 57*, 1223–1236.

Novick, P., Field, C., and Schekman, R. (1980) *Cell 21*, 205–215.

Oas, T. G., and Kim, P. S. (1988) *Nature 336*, 42–48.

Olson, T. S., and Lane, M. D. (1987) *J. Biol. Chem. 262*, 6816–6822.

Olson, T. S., and Lane, M. D. (1989) *FASEB J. 3*, 1618–1624.

Ostermann, J., Horwich, A. L., Neupert, W., and Hartl, F.-U. (1989) *Nature 341*, 125–130.

Palade, G. (1975) *Science 189*, 347–358.

Park, S., Liu, G., Topping, T. B., Cover, W. H., and Randall, L. L. (1988) *Science 239*, 1033–1035.

Parkkonen, T., Kivirikko, K. I., and Pihlajaniemi, T. (1988) *Biochem. J. 256*, 1005–1011.

Paver, J. L., Hawkins, H. C., and Freedman, R. B. (1989) *Biochem. J. 257*, 657–663.

Pelham, H. R. B. (1984) *EMBO J. 3*, 3095–3100.

Pelham, H. R. B. (1986) *Cell 46*, 959–961.

Pelham, H. R. B. (1989) *Ann. Rev. Cell Biol. 5*, 1–23.

Perara, E., Rothman, R. E., and Lingappa, V. R. (1986) *Science 232*, 348–352.

Pfanner, N., Hartl, F.-U., Guiard, B., and Neupert, W. (1987) *Eur. J. Biochem. 169*, 289–293.

Pfanner, N., and Neupert, W. (1990) *Ann. Rev. Biochem. 59*, 331–353.

Pfeffer, S. R., and Rothman, J. E. (1987) *Ann. Rev. Biochem. 56*, 829–852.

Pigiet, V. P., and Schuster, B. J. (1986) *Proc. Natl. Acad. Sci. USA 83*, 7643–7647.

Pihlajaniemi, T., Helaakoski, T., Tasanen, K., Myllylä, R., Huhtala, M.-L., Koivu, J., and Kivirikko, K. I. (1987) *EMBO J. 6*, 643–649.

Price, E. R., Zydowsky, L. D., Jin, M., Baker, C. H., McKeon, F. D., and Walsh, C. T. (1991) *Proc. Natl. Acad. Sci. USA 88*, 1903–1907.

Randak, C., Brabletz, T., Hergenrither, I., and Serfling, E. (1990) *EMBO J. 9*, 2529–2536.

Randall, L. L., and Hardy, S. J. S. (1986) *Cell 46*, 921–928.

Randall, L. L., Hardy, S. J. S., and Thom, J. R. (1987) *Ann. Rev. Microbiol. 41*, 507–541.

Randall, L. L. Topping, T. B., and Hardy, S. J. S. (1990) *Science 248*, 860–863.

Rapoport, T. A. (1990) *Trends Biochem. Sci. 15*, 355–358.

Redman, C. M., and Sabatini, (1966) *Proc. Natl. Acad. Sci. USA 56*, 608–615.

Rippmann, F., Taylor, W. R., Rothbard, J. B., and Green, N. M. (1991) *EMBO J. 10*, 1053–1059.

Rose, J. K., and Doms, R. W. (1988) *Ann. Rev. Cell Biol. 4*, 257–288.

Rose, M. D., Misra, L. M., and Vogel, J. P. (1989) *Cell 57*, 1211–1221.

Rosen, M. K., Standaert, R. F., Galat, A., Nakatsuka, M., and Schreiber, S. L. (1990) *Science 248*, 863–865.

Roth, R. A., and Pierce, S. B. (1987) *Biochemistry 26*, 4179–4182.

Rothblatt, J. A., and Meyer, D. I. (1986) *EMBO J. 5*, 1031–1036.

Rothman, J. E. (1987) *Cell 50*, 521–522.

Rothman, J. E. (1989) *Cell 59*, 591–601.

Rothman, J. E., and Kornberg, R. D. (1986) *Nature 322*, 209–210.

Roy, H. (1989) *Plant Cell 1*, 1035–1042.

Rudolph, R. (1990) In *Modern Methods in Protein and Nucleic Acid Research* (H. Tschesche, ed.), Walter de Gruyter, Berlin, pp. 149–171.

Rudolph, R., and Fischer, S. (1987) Verfahren zur Aktivierung von gentechnologisch hergestellten, heterologen Disulfidbrücken aufweisenden eukaryontischer Proteinen nach Expression in Prokaryonter. European patent application 0 219 874.

Rudolph, R., and Fischer, S. (1989) Process for obtaining renatured proteins, US patent 4,933,434.

Rudolph, R., Gerschitz, J., and Jaenicke, R. (1978) *Eur. J. Biochem 87*, 601–606.

Samuelsson, E., Wadensten, H., Hartmanis, M., Moks, T., and Uhlen, M. (1991) *Bio/Technology 9*, 363–366.

Sanz, P., and Meyer, D. I. (1988) *EMBO J. 7*, 3553–3557.

Sarmientos, P., Duchesne, M., Denèfle, P., Boiziau, J., Fromage, N., Delporte, M., Parker, F., Lelièvre, Y., Mayaux, J.-F., and Cartwright, T. (1989) *Bio/Technology 7*, 495–501.

Saxena, V. P., and Wetlaufer, D. B. (1970) *Biochemistry 9*, 5015–5022.

Schein, C. H. (1989) *Bio/Technology 7*, 1141–1148.

Schekman, R., (1988) *Ann. Rev. Cell Biol. 1*, 115–143.

Scherens, B., Dubois, E., and Messenguy, F. (1991) *Yeast 7*, 185–193.

Scherer, R. E., Krieg, U. C., Hwang, S. T., Vestweber, D., and Schatz, G. (1990) *EMBO J. 9*, 4315–4322.

Schlenstedt, G., Hudmundsson, G. H., Boman, H. G., and Zimmermann, R. (1990) *J. Biol. Chem. 265*, 13960–13968.

Schlesinger, M. J. (1990) *J. Biol. Chem. 265*, 12111–12114.

Schleyer, M., and Neupert, W. (1985) *Cell 43*, 339–350.

Schlossman, D. M., Schmid, S. L., Braell W. A., and Rothman, J. E. (1984) *J. Cell Biol. 99*, 723–733.

Schmid, F. X., and Blaschek, H. (1981) *Eur. J. Biochem. 114*, 110–117.

Schmid, F. X., Lang, K., Kiefhaber, T., Mayer, S., and Schönbrunner, E. R. (1991) in *Conformations and Forces in Protein Folding* (B. T. Nall and K. A. Dill, eds.), AAAS, Washington, D.C., pp 198–203.

Schönbrunner, E. R., Mayer, S., Tropschug, M., Fischer, G., Takahashi, N., and Schmid, F. X. (1991) *J. Biol. Chem. 266*, 3630–3637.

Schreiber, S. L. (1991) *Science 251*, 283–287.

Schülke, N., and Schmid, F. X. (1988a) *J. Biol. Chem. 263*, 8827–8831.

Schülke, N., and Schmid, F. X. (1988b) *J. Biol. Chem. 263*, 8832–8837.

Schwaiger, M., Herzog, V., and Neupert, W. (1987) *J. Cell. Biol. 105*, 235–246.

Semisotnov, G. V., Sokolovsky, I. V., Bochkareva, E. S., and Girshovich, A. S. (1991) Unpublished observations.

Sheffield, W. P., Shore, G. C., and Randall, S. K. (1990) *J. Biol. Chem. 265*, 11069–11076.

Shieh, B.-H., Stamnes, M. A., Seavello, S., Harris, G. L., and Zuker, C. S. (1989) *Nature 338*, 67–70.

Shortle, D., and Meeker, A. K. (1989) *Biochemistry 28*, 936–944.

Siekierka, J. J., Staruch, M. J., Hung, S. H. Y., and Sigal, N. H. (1989) *Nature 341*, 755–757.

Sinha, N. K., and Light, A. (1975) *J. Biol. Chem. 250*, 8624–8629.

Skowyra, D., Georgopoulos, C., and Zylicz, M. (1990) *Cell 62*, 939–944.

Slieker, L. I., Mortensen, T. M., and Lane, M. D. (1986) *J. Biol. Chem. 261*, 15233–15241.

Soute, B. A. M., Groenen-van Dooven, M. M. C., Holmgren, A., Lundström, J., and Vermeer, C. (1992) *Biochem. J. 281*, 255–259.

Stamnes, M. A., Shieh, B.-H., Chuman, L., Harris, G. L., and Zuker, C. S. (1991) *Cell 65*, 219–227.

Steinmann, B., Bruckner, P., and Superti-Furga, A. (1991) *J. Biol. Chem. 266*, 1299–1313.

Stewart, D. E., Sarkar, A., and Wampler, J. E. (1990) *J. Mol. Biol. 214*, 253–260.

Takahashi, N., Hayano, T., and Suzuki, M. (1989) *Nature 337*, 473–475.

Tandon, S., and Horowitz, P. (1981) *J. Biol. Chem. 261*, 15675–15681.

Tang, J.-G., and Tsou, C.-L. (1990) *Biochem J. 268*, 429–435.

Tang, J.-G., Wang, C.-C., and Tsou, C.-L. (1988) *Biochem. J. 255*, 451–455.

Taniuchi, H. (1970) *J. Biol. Chem. 245*, 5459–5468.

Tropschug, M., Bartholmess, I. B. , and Neupert, W. (1989) *Nature 342*, 953–955.

Tropschug, M., Wachter, E., Mayer, S., Schönbrunner, E. R., and Schmid, F. X. (1990) *Nature 346*, 674–677.

Tsou, C. L. (1988) *Biochemistry 27*, 1809–1812.

Ungewickell, E. (1985) *EMBO J. 4*, 3385–3391.

Van Duyne, G. D., Standaert, R. F., Karplus, P. A., Schreiber, S. L., and Clardy, J. (1991) *Science 252*, 839–842.

Van Dyk, T. K., Gatenby, A. A., and LaRossa, R. A. (1989) *Nature 342*, 451–453.

Venetianer, P., and Straub, F. B. (1963) *Biochim. Biophys. Acta 67*, 166–168.

Vermeer, C. (1990) *Biochem. J. 266*, 625–636.

Vestweber, D., and Schatz, G. (1988a) *EMBO J. 7*, 1147–1151.

Vestweber, D., and Schatz, G. (1988b) *J. Cell Biol. 107*, 2037–2043.

Viitanen, P. V., Lubben, T. H., Reed, J., Goloubinoff, P., O'Keefe, D. P., and Lorimer, G. H. (1990) *Biochemistry 29*, 5665–5671.

Viitanen, P. V., Donaldson, G. K., Lorimer, G. H., Lubben, T. H., and Gatenby, A. A. (1991) *Biochemistry* (in press).

Villafane, R., and King, J. (1988) *J. Mol. Biol. 204*, 607–619.

Vogel, J. P., Misra, L. M., and Rose, M. D. (1990) *J. Cell. Biol. 110*, 1885–1895.

Vuori, K., Myllylä, R., Pihlajaniemi, T., and Kivirikko, K. (1990) *J. Cell Biol. 111*, 105a, Abstract 582.

Walter, P., Gilmore, R., and Blobel, G. (1984) *Cell 38*, 5–8.

Walter, P. and Lingappa, V. R. (1986) *Ann. Rev. Cell Biol. 2*, 499–516.

Weiss, J. B., Ray, P. H., and Bassford Jr., P. J. (1988) *Proc. Natl. Acad. Sci. USA 85*, 8978–8982.

Wetlaufer, D. B. (1981) *Adv. Protein Chem. 34*, 61–92.

Wetlaufer, D. B., and Ristow, S. (1973) *Ann. Rev. Biochem. 42*, 135–158.

Wetterau, J. R., Combs, K. A., Spinner, S. N., and Joiner, B. J. (1990) *J. Biol. Chem. 265*, 9800–9807.

Wetterau, J. R., Aggerbeck, L. P., Laplaud, P. M., and McLean, L. R. (1991) *Biochemistry 30*, 4406–4412.

Wetzel, R., Perry, L. J., and Veilleux, C. (1991) *Bio/Technology 9*, 731–737.

Wickner, W. (1988) *Biochemistry 27*, 1081–1086.

Wickner, W. (1989) *Trends Biochem. Sci. 14*, 280–283.

Wiech, H., Stuart, R., and Zimmermann, R. (1990) *Seminars Cell Biol. 1*, 55–63.

Wilkinson, D. L., and Harrison, R. G. (1991) *Bio/Technology 9*, 443–448.

Yoshimori, T., Semba, T., Takemoto, H., Akagi, S., Yamamoto, A., and Tashiro, Y. (1990) *J. Biol. Chem. 265*, 15984–15990.

Young, D. B. (1990) *Seminars Cell Biol. 1*, 27–35.

Yu, M.-H., and King, J. (1984) *Proc. Natl. Acad. Sci. USA 81*, 6584–6588.

Zapun, A., Creighton, T. E., Rowling, P. J. E., and Freedman, R. B. (1991) *Proteins: Struct., Funct., Genet.*, in press.

Zettlmeissl, G., Rudolph, R., and Jaenicke, R. (1979) *Biochemistry 18*, 5567–5571.

Zhu, X., Ohta, Y., Jordan, F., and Inouye, M. (1989) *Nature 339*, 483–485.

Zimmermann, R., and Meyer, D. I. (1986) *Trends Biochem. Sci. 11*, 512–515.

Zimmermann, R., Zimmermann, M., Wiech, H., Schlenstedt, G., Muller, G., Morel, F., Klappa, P., Jung, C., and Cobet, W. W. E. (1990) *J. Bioen. Biomem. 22*, 711–723.

Index